Advanced Renewable Energy Sources

Advanced Renewable Energy Sources

G. N. Tiwari and R. K. Mishra
Centre for Energy Studies, Indian Institute of Technology Delhi, New Delhi, India

RSC Publishing

ISBN: 978-1-84973-380-9

A catalogue record for this book is available from the British Library

Published by The Royal Society of Chemistry,
Thomas Graham House, Science Park, Milton Road,
Cambridge CB4 0WF, UK

Registered Charity Number 207890

For further information see our web site at www.rsc.org

Printed in Great Britain by CPI Group (UK) Ltd, Croydon, CR0 4YY

Preface

ROLE OF THE SUN

The Sun provides solar energy that is clean and environmentally friendly. It can be directly used to produce thermal as well as electrical power. However Sun is also directly the main source of all renewable energy sources discussed in this book. Furthermore, the Sun can also be the indirect source of other nonrenewable energy sources. Solar energy is in the form of electromagnetic wave as well as photons. Electromagnetic waves produce thermal energy and wind energy. The photons produce electrical energy through semiconductor devices. Photons were also used for photosynthesis of all plants on the Earth. Thermal energy available on the Earth is able to defreeze frozen water on the Earth for the growth of bacteria. Thus, living organism exists on the Earth. One can say that there exists conservation of living organism of all species in the Universe, like conservation of energy and mass.

The motion of the Earth and the moon around the Sun is responsible for creating waves and tides in oceans in the form of kinetic energy, which is also a renewable source of energy.

IMPORTANCE OF RENEWABLE-ENERGY SOURCES

Due to global industrialisation, particularly after the Second World War (WWII) in developed countries, there was a sharp increase of CO_2 concentration from 270 ppm to 450 ppm. This was due to using polluting fossil fuels in the form of solid, liquid and gases available inside the Earth for power production. This has challenged the existence of human beings and living plants. If it is not controlled, then survival of human beings and other living organism will face a dangerous situation in the coming years. In order to balance the ecological system for clean air, water and food, less-polluting renewable energy sources should be used to meet the energy demand of human beings across developed, developing and underdeveloped countries.

ABOUT THIS BOOK

On the basis of the demands of such a book and our teaching experience in this subject over many years, we decided to upgrade our existing book of **Fundamental of Renewable Energy Sources** published by Narosa Publishing House, New Delhi, India by incorporating many subjects such as photovoltaic thermal (PVT) systems, energy and exergy analysis, CO_2 credit, energy matrices, life cycle analysis with and without CO_2 credit. The status of energy data has also been upgraded.

Advanced Renewable Energy Sources
G. N. Tiwari and R. K. Mishra

Published by the Royal Society of Chemistry, www.rsc.org

Some materials have been taken from various sources and due acknowledgement for the same has been given at the appropriate places, particularly from **Fundamentals of Photovoltaic Modules and their Applications**, Royal society of Chemistry, UK. If such acknowledgement for referred research work is omitted, it will be without any intention and thus will be rectified in the next edition. We also invite readers to note errors occurring in the text where possible, to rectify the revised edition.

OBJECTIVE

The main purpose of writing this book is to provide a suitable text for teaching the subject to engineering and science students, as well as a reference book for scientists and professionals doing self-study about renewable energy sources. Some solved examples and problems are also given at appropriate places. The objective question for better understanding of the subject is also given at the end of each chapter.

The book contains twelve chapters. Chapter 1 deals with the definition and classification of renewable and nonrenewable energy and their status at global level. This chapter also contains the basic heat-transfer mechanisms and laws of thermodynamics. Chapter 2 deals with the availability of solar radiation at different latitudes and the energy and exergy analysis of flat-plate collectors, solar air collectors, solar concentrators, evacuated tube collectors, solar water heating systems, solar distillation kits and solar cookers. Chapter 3 discusses the basics of semiconductors, their characteristics, working, characteristics of solar cells in the dark and daylight situations, fundamentals of characteristic curves of semiconductors, fundamentals of PV modules and arrays and some PVT systems. The detail discussions on biomass, biofuels and biogas and their applications is discussed in Chapter 4. The power produced by biomass, biofuels and biogas, namely biopower is discussed in Chapter 5. However, other renewable energy sources like hydropower, wind and geothermal are discussed in Chapters 6–8. Chapter 9 deals with the working principle, basic theory and capability to produce power from ocean thermal, tidal, wave and animal energy-conversion system. Net CO_2 mitigation, carbon credits, climate change and environmental impacts of all renewable energy resources are covered in Chapter 10. The technoeconomic feasibility of any energy sources is the backbone of its success and hence energy and economic analysis are carried out in chapters 11 and 12, respectively.

Most of the chapters deal with the overall exergy of renewable energy sources by using the thermal and mechanical power and electrical energy as output wherever possible.

SI units will be used throughout the book in solving various exercises in each chapter. Conversion units of various physical and chemical parameters of metals and nonmetals will also be given in the appendices.

It is our immense pleasure to express our heartfelt gratitude to Padmashree Prof. M.S. Sodha, India; Prof. Brian Norton, Ireland; Prof. Ibrahim Dincer, Canada; Prof. T. Muneer, UK; Prof. Yogi Goswami, USA; Prof. T.T. Chow, Hong Kong and Prof. Christophe Ménézo, France and our other colleagues in India and abroad.

We duly acknowledge with thanks the financial support by the Curriculum Development Cell (CD Cell), IIT Delhi for preparation of the book.

Full credit is due to our publishers, RSC Publishing, Cambridge, UK, for producing a nice print of the book.

Last but not least, we express our deep gratitude to the late Smt. Bhagirathi Tiwari; the late Shree Bashisht Tiwari; Smt. Sharda Mishra and Shree Shashi Bhusan Mishra for their blessing during writing of the book. Further, we also thank Smt. Kamalawati Tiwari; Smt Premlata Mishra and Shree Sri Vats Tiwari for keeping our morale high during this work.

Gopal N. Tiwari
Rajeev K. Mishra

Contents

Advanced Renewable Energy Sources
G. N. Tiwari and R. K. Mishra

Published by the Royal Society of Chemistry, www.rsc.org

Dedication

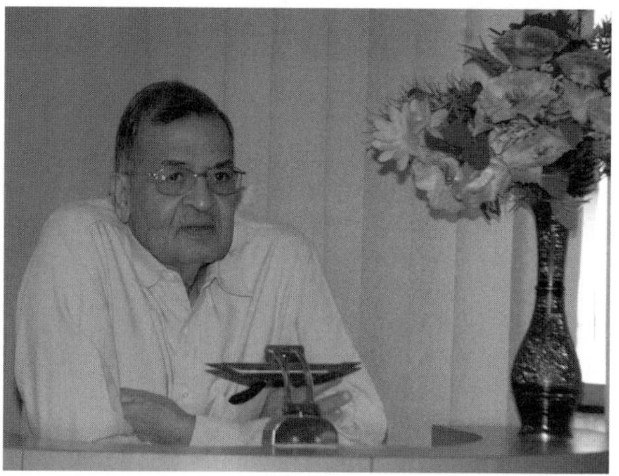

Our respected teacher and guru ji
Padmashri Professor M.S. Sodha, FNA
For his 80th birthday (February 08, 2012)

Professor M.S. Sodha was born in Ajmer on February 08, 1932, Rajashthan, pursued his higher education at Allahabad University, Allahabad (UP), India. He obtained his D.Phil. in 1955. During 1956–1964, he worked in Canada and the USA after a brief service at the Defence Science Laboratory in New Delhi. He joined as a Professor in the Physics Department at IIT Delhi in 1964. He has made outstanding/significant contributions in the area of fusion plasma, solid-state physics, optics, combustion, ballistics and renewable energy sources. He has published more than 500 research papers, supervised 75 Ph.D. students, written 14 books and 12 reviews. During his stay at IIT Delhi, Professor Sodha served in many administrative capacities including Head of Department of Physics, Energy and ITMEC; Dean of Science, Students and PG studies and Deputy Director.

He has also been Vice Chancellor of Devi Ahilya Vishwavidyalaya, Indore; Lucknow University, Lucknow and Bakatullah University, Bhopal during 1988–2000. He has been instrumental in higher education through setting up of the Centre for Energy Studies, IIT Delhi and the Faculty of Engineering Science in DAVV, Indore in India. Professor Sodha was awarded the S.S. Bhatnager award (1974), Hari Om ashram Prerit S.S. Bhatnager award (1978) and Padmashree in 2003. He is a Ramanna Fellow and visiting Professor at Lucknow University.

Authors' Profiles

Professor G. N. Tiwari
Centre for Energy Studies,
Indian Institute of Technology Delhi
Ph. 011-26591258
E-mail: gntiwari@ces.iitd.ac.in

Professor Gopal Nath Tiwari was born on July 01, 1951 at Adarsh Nagar, Sagerpali, Ballia (UP), India. He received postgraduate and doctoral degrees in 1972 and 1976, respectively, from Banaras Hindu University (B.H.U.). Over several years since 1977, he has been actively involved in the teaching programme at the Centre for Energy Studies, IIT Delhi. His research interest in the field of solar-energy applications are solar distillation, water/air heating system, greenhouse technology for agriculture as well as for aquaculture, Earth to air heat exchanger, passive building design and hybrid photovoltaic thermal (HPVT) systems, climate change, energy security, *etc*. He has guided about 65 Ph.D. students and published over 450 research papers in journals of repute. He has authored twenty books associated with reputed publishers namely Pergamon Press UK, CRC Press USA, Narosa Publishing House, Ahsan (UK), Alpha-science (UK), Royal Society of Chemistry (UK), *etc*. He is a corecipient of the "Hariom Ashram Prerit S.S. Bhatnagar" Award in 1982. Professor Tiwari has been recognised both at national and international levels. He visited the University of Papua, New Guinea in 1987–1989 as an Energy and Environment Expert. He was also a recipient of a European Fellowship in 1997 and visited the University of Ulster (UK) in 1993. In addition, he had been nominated for an IDEA award in the past. He is responsible for the development of the "Solar Energy Park" at IIT Delhi and Energy Laboratory at the University of Papua, New Guinea, Port Moresby. Dr. Tiwari had visited many countries namely Italy, Canada, USA, UK, Australia, Sweden, Germany, Greece, France, Thailand, Singapore, PNG, Hong Kong and Taiwan, *etc*., for invited talks, chairing international conferences, acting as an expert in renewable energy, presenting research papers, *etc*. He has successfully co-coordinated various research projects on Solar distillation, water heating system, Greenhouse technology, hybrid photovoltaic thermal (HPVT) *etc*. funded by the Government of India in the recent past.

Dr. Tiwari was Editor of the International Journal of Agricultural Engineering for three years (2006–2008). He is an Associate Editor for the Solar Energy Journal (SEJ) in the area of Solar Distillation since 2007. He is also Editor of the International Journal of Energy Research.

Professor Tiwari has been conferred "Vigyan Ratna" by the Government of U.P., India on March 26, 2008 and Valued Associated Editor award by the J. of Solar Energy. He organised SOLARIS 2007, the third international conference on "Solar Radiation and Day lighting" held at IIT Delhi, New Delhi, India from February 7 to 9, 2007. He is also president of the Bag Energy Research Society (BERS:www.bers.in) which is responsible for energy education in rural India.

Mr. Rajeev Kumar Mishra
Research Scholar
Centre for Energy Studies,
Indian Institute of Technology (IIT) Delhi
Ph. No. 09717720464
E-mail: bhu.rajeev@gmail.com

Mr. Rajeev Kumar Mishra was born on July 3, 1982 at Ballia (U.P.), India. He received a Bachelor of Science degree (B.Sc) with Mathematics, Physics and Chemistry in 2003 and Master of Science degree (M.Sc.) in Physics from the University of Allahabad, Allahabad, UP, India in 2005. He received a Master of Technology degree (M. Tech.) in Material Science and Technology from the Institute of Technology, Banaras Hindu University (BHU), Varanasi, India in 2008. His M. Tech dissertation title was "Synthesis and characterisation of CNT reinforced PC-PMMA blends for optical storage applications". During his master's program, he published one research paper in the area of material science and technology published in Chemical Physics Letters. Recently, he has presented a research paper in the area of solar photovoltaic thermal applications in an international conference "World Renewable Energy Congress (WREC-2011)" in Linkoping University, Linkoping, Sweden. He has published a review article on photovoltaic thermal system in the International Journal of Applied Energy.

Presently, he is pursuing his Ph.D. under the supervision of Prof. G. N. Tiwari. His areas of research interest are solar thermal, photovoltaics, heat and mass transfer, exergy, CO_2 mitigation, climate change and carbon trading.

Approximate Values of Some Constants in Renewable-Energy Sources

	Constants	*Actual value*	*Approximate value*
1	Diameter of the Sun ($2R_S$)	1.39×10^9 m	1.5×10^9 m
2	Distance of the Sun from the Earth	1.5×10^{11} m	150×10^9 m
3	Black body temperature of the Sun	5777 K	6000 K
4	Central core ($0 - 0.23R$) temperature	$8 - 40 \times 10^6$ K	$9 - 30 \times 10^9$ K
5	Energy generated in centre core	90%	90%
6	Diameter of the Earth ($2R_E$)	1300 km	1.5×10^6 m
7	Solar constant	1367 W/m^2	1500 W/m^2
8	Short-wavelength radiation	0.23 – 2.6 μm	0.3 – 3.0 μm
9	Average temperature of the Earth ($\approx 25\,°C$)	298 K	300 K
10	Stefan–Boltzmann constant	5.67×10^{-8} W/m^2 K^4	60×10^{-9} W/m^2 K^4
11	Wein's displacement law	$\lambda T = 2897.6$ μm K	3000 μm K
12	Long-wavelength radiation from Earth	10 μm	9 μm
13	Wavelength radiation from the Sun Mean Sun–Earth angles	0–30 μm	0–30 μm
14	(three w.r.t. centre of Earth and three w.r.t. observer on Earth)	6	6
15	Sunshine hours at equator	12 h	12 h
16	Sunshine hours at north pole	24 h	24 h
17	Type of radiation (beam, diffuse and reflector)	3	3
18	Optimum till angle for maximum solar radiation (°C)	$\Phi \pm 15$	$\Phi \pm 15$
19	Basic heat transfer (Conduction, convection and radiation)	3	3
20	Convective heat-transfer coefficient for air	2.8 + 3V	3 + 3V
21	Sky temperature (°C)	$(T_a - 12)$	$(T_a - 12)$
22	Long-wavelength radiation exchange (ΔR) between ambient and sky	60 W/m^2	60 W/m^2
23	Order of radiation heat-transfer coefficient		6 W/m^2 K
24	Convective and radiative heat-transfer coefficient for air	$(5.7 + 3.8V)$ W/m^2 K	(6 + 3V) W/m^2 K

Advanced Renewable Energy Sources
G. N. Tiwari and R. K. Mishra

Published by the Royal Society of Chemistry, www.rsc.org

	Constants	*Actual value*	*Approximate value*
25	Order of convective heat-transfer coefficient between hot plate and water		$(90 - 300)\mathrm{W/m^2\,K}$
26	Overall heat-transfer coefficient for glazed FPC single		$6\ \mathrm{W/m^2\,K}$
27	FPC efficiency factor (F')		0.9
28	Insulation thickness	0.10 m	0.09 m
29	Fin efficiency		0.9
30	Flow rate factor (F_R)		< 0.9
31	Transmissivity of window glass		0.9
32	Threshold intensity		
	(a) Winter		$> 300\ \mathrm{W/m^2}$
	(b) Summer		$> 300\ \mathrm{W/m^2}$
33	FPC connected in series		≤ 3
34	Rate of evaporation from free water surface ($\dot{q}_{ew}$)	$0.016 \times h_{cw} \times (P_w - rP_a)$	$0.015 \times h_{cw} \times (P_w - rP_a)$
35	Thermal conductivity of insulating material (K)	0.03 – 0.04 W/m K	0.03 W/m K
36	Maximum temperature in concentrating collector		3000 °C
37	Ideal efficiency of solar still	60%	60%
38	Optimum depth of basin water	0.02 – 0.03 m	0.03 m
39	Effect of climatic parameters on yield		9 – 12%
40	Broad classification of thermal comfort (Physical, Physiological, Intermediate) parameters	3	3
41	Physical comfort parameters	9	9
42	Intermediate comfort parameters	6	6
43	Physiological comfort parameters	6	6
44	Heating concepts (Direct, Indirect and Isolated)	3	3
45	The rate of ventilation/infiltration	$0.33\ NV\ (T_r - T_a)$	$0.33\ NV\ (T_r - T_a)$
46	Emissivity of surface	0.9	0.9
47	Optimum water depth in collection-cum storage water heater	0.10 m	0.09 m
48	Optimum temperature for fermentation of slurry for biogas production	35 – 37 °C	36 °C
49	Cooking time by solar cooker	2 – 3 h	3 h
50	Latent heat of vaporisation	2.3×10^6 J/kg	3×10^6 J/kg
51	Bandgap for silicon	1.16 eV	1.2 eV
52	Boltzmann constant	1.38×10^{-23} J/K	12×10^{-24} J/K
53	V group impurity concentration	$10^{15}\ \mathrm{cm^3}$	$10^{15}\ \mathrm{cm^3}$
54	Effective density of states in conduction bands	$2.82 \times 10^{19}\ \mathrm{cm^3}$	$27 \times 10^{18}\ \mathrm{cm^3}$
55	Saturation current in reverse bias (I_o)	$10^{-8}\ \mathrm{A/m^2}$	$0.1 \times 10^{-9}\ \mathrm{A/m^2}$
56	Thickness of n-type semiconductor in silicon solar cell	0.2 μm	0.3 μm
57	Thickness of p-type semiconductor in silicon solar cell	0.50 mm	0.60 mm
58	Diffusion path length in Si	50 – 100 μm	60 – 90 μm
59	Junction near n-type semiconductor in Si	0.15 μm	0.15 μm
60	Solar intensity in terrestrial region		$900\ \mathrm{W/m^2}$
61	Number of solar cell in standard PV module		36
62	Efficiency of solar cells in standard conditions	15%	15%

	Constants	*Actual value*	*Approximate value*
63	Efficiency of PV module with Si-solar cell	12%	12%
64	Standard test condition	1000 W/m^2 and 25 °C	900 W/m^2 and 24 °C
65	Climatic zone in India	6	6
66	Specific heat of water	4190 J/kg °C	4200 J/kg °C
67	Specific heat of air	1 kJ/kg K	1 kJ/kg K
68	Density of air	1.2 kg/m^3	1.2 kg/m^3
69	Absorptivity of base surface		0.3
70	Absorptivity of blackened surface		>0.9
	(a) Heating value of coal	29 000 kJ/kg	30 000 kJ/kg
	(b) Biogas	20 000 kJ/kg	21 000 kJ/kg
71	(c) Wood/Straw	15 000 kJ/kg	15 000 kJ/kg
	(d) Gasolene/Kerosene	42 000 kJ/kg	42 000 kJ/kg
	(c) Methane	50 000 kJ/kg	51 000 kJ/kg
72	The energy contained in infrared region	51.02% (697.4 W/m^2)	51%
73	The energy contained in visible region	36.76% (502.6 W/m^2)	36%
74	The energy contained in ultraviolet (UV) region	12.22% (167 W/m^2)	12%
75	The rate of heat generated by healthy person during sleeping	60 W	60 W
76	The rate of heat generated by healthy person during hard work	600 W/m^2	600 W/m^2
77	The body temperature of human being	37 °C	36 °C
78	Geothermal energy from Earth	300×10^{12} W	300×10^{12} W
79	Tidal power from planetary motion	3×10^{12} W	3×10^{12} W
80	Thermal solar energy absorbed by Earth	120×10^{15} W	120×10^{15} W
81	Solar energy for photosynthesis	30×10^{12} W	30×10^{12} W
82	Solar energy for wind and wave conversion	300×10^{12} W	300×10^{12} W
83	Solar energy for hydropower	40×10^{15} W	30×10^{15} W
84	Solar energy for sensible heating	80×10^{15} W	90×10^{15} W
85	Altitude of O_3 present in stratosphere	12–25 km	12–24 km
86	Mature tree consumes	12 kg of CO_2	12 kg of CO_2
87	Vehicle using gasoline produces	2.5 kg of CO_2/litre	2.4 kg of CO_2/litre
88	Propane in LPG	90%	90%
89	Dry biomass in biosphere	250×10^9 ton/year	240×10^9 ton/year
90	Energy produced in one fusion reaction inside Sun	26.7 MeV	24 MeV
91	Methane presence in biogas	60%	60%
92	Gas-turbine operates	600–1200 °C	600–1200 °C
93	For maximum hydropower	$u_t = u_j/3$	$u_t = u_j/3$
94	Hydropower system efficiency	60%	60%
95	Optimum velocity for wind power	10 m/s	9 m/s
96	Maximum wind power extraction	$a = 1/3$	$a = 1/3$
97	Maximum power coefficient (WECS)	59%	60%
98	Average heat flow from centre of Earth	0.06 W/m^2	0.06 W/m^2
99	Low-temperature geothermal well	≤ 150 °C	≤ 150 °C
100	High-temperature geothermal well	≥ 150 °C	≥ 150 °C
101	For OTEC power generation, ΔT	≥ 15 °C	≥ 15 °C

To Sustain Ecosystem and Climate, Every Human Being on Planet Earth should Use Renewable Energy Sources to Meet their Basic Energy Need.

CHAPTER 1

General Introduction

1.1 ENERGY: ITS DEFINITION AND BASIC CONCEPT

Energy is one of the major building blocks of society and it is needed to create goods from natural resources. Global economics development and improved standards of energy are complex processes that share a common denominator *i.e.* the availability of an adequate and reliable supply of clean energy. With an oil embargo in 1973, continuing with the Iranian revolution of 1979 and the Persian Gulf War of 1991, political events had made many people aware of how crucial clean energy is for everyday functioning of our society. The energy crises of the 1970s were almost forgotten by the 1980s, that period brought an increased awareness of other environmental issues. The global warming, acid rain and radioactive waste are still very much with us today, and each of these topics is related to our energy security.

In the present scenario all sectors of society, *e.g.* labour, environment, economics and international relations, *etc.* in addition to our own personal livings, *i.e.* housing, food, transportation, recreation and communication, *etc.*, strongly depends on energy. The use of energy resources has relieved us from much drudgery and made our efforts more productive. Human beings once had to depend on their own muscle energy to provide the energy necessary to do the daily work. Today, muscle energy supplies less than 1% of the work done in the industrialised world.

Energy is a globally conserved quantity, *i.e.* the total amount energy in the universe is constant. Energy can neither be created nor destroyed. It can only be transformed from one state to another. Two billiard balls colliding, for example, may come to rest, with the resulting energy becoming sound and perhaps a bit of heat at the point of collision.

Energy, environment and economic development are closely related. The proper use of energy requires consideration of social impact as well as technological ones. Indeed, sustained economic growth of a country in this century along with improvements in the quality of everyone's lives may be possible only by the well planned and efficient use of fossil fuel and other resources and the development of new renewable energy technologies.

1.1.1 Basis Concept of Energy

In physics, energy is defined as "the capacity of a physical system to perform work". The word is used by each of us with many different connotations, but in physics, it has a very definite meaning:

$$\textbf{Work} = \textbf{Force} \times \textbf{Displacement along the direction of force}$$

Advanced Renewable Energy Sources
G. N. Tiwari and R. K. Mishra

Published by the Royal Society of Chemistry, www.rsc.org

or, **"Work is the product of force and displacement through which the force acts".** Mathematically, work can be expressed as;

$$W = F \times d(\mathrm{J})$$

where F is the acting force in "Newtons" and d is the displacement along the direction of force in "meters".

A force of one Newton (N) acting through a distance of one meter in the same direction performs an amount of work equivalent to one joule (J). It is also important to note that work, the capacity for doing work and energy has the same units. A system may possess energy even when no work is being done. Since energy is measured by the total amount of work that the body can do, hence energy is expressed in the same unit of work as mentioned above.

1.2 DIFFERENT FORMS OF ENERGY

There is an important law known as the **"Law of conservation of energy"** that states that the total amount of energy in a closed system remains constant. Energy may change from one form to another, but the total amount in any closed system does not change. This law is extremely important in order to understand a variety of phenomenon. We will begin by identifying various forms of energy and transformations of energy from one form to another as mentioned below.

(a) Kinetic Energy (KE)

The kinetic energy of an object is the extra energy that it possesses due to its motion. It is defined as the work needed to accelerate a body of a given mass from rest to its current velocity. Mathematically, kinetic energy can be expressed an

$$\mathrm{KE} = \frac{mv^2}{2}\ (\mathrm{J})$$

where m is the mass of the object in kg and v is its velocity in m/s.

(b) Potential Energy (PE)

This is the energy of an object due to its elevation in a gravitational field. Mathematically, potential energy can be expressed an

$$PE = mgh\,(\mathrm{J})$$

where m is the mass of object in kg, g is acceleration due to gravity in m/s^2 and h is its height in m.

(c) Chemical Energy

Chemical energy is the energy stored in the chemical bonds of molecules. It arises out of the capacity of atoms to evolve heat as they combine or separate. It is the chemical energy in coal, natural gas, wood, oil that heats our homes, powers cars and is used to generate electricity.

(d) Electrical Energy

This is defined as the capacity of moving electrons to evolve heat, electromagnetic radiation and magnetic fields.

(e) Heat Energy

This is the energy of a material due to the random motion of its particles. It is also called thermal energy. The word "heat" is used when energy is transferred from one substance to another.

(f) Radiant Energy

Radiant energy is the energy emitted by electrons as they change orbit and by atomic nuclei during fission and fusion; on striking matter, such energy appears ultimately as heat.

(g) Nuclear (Mass) Energy

This is the energy stored in the nucleus of an atom. According to Einstein when the mass of some system is reduced by an amount Δm, as in nuclear reaction, then the amount of energy released is

$$E = \Delta mc^2$$

where c is the velocity of light (3×10^8 m/s). The above simple equation is the basis of the energy derived when a ^{235}U nucleus fissions, as in a nuclear reactor or when a deuteron and a trition (^{2}H and ^{3}H) fuse in a thermonuclear reaction.

EXAMPLE 1.1

Calculate the energy liberated per fission in the fission of ^{235}U. The fission reaction is as follows:

$$^{235}_{92}\text{U} + ^{1}_{0}\text{n} \rightarrow ^{141}_{56}\text{Ba} + ^{92}_{36}\text{Kr} + 3^{1}_{0}\text{n} + Q$$

Given,

Mass of $^{235}_{92}$U = 235.045733u
Mass of $^{1}_{0}$n = 1.008665u
Mass of $^{141}_{56}$Ba = 140.9177u
Mass of $^{92}_{36}$Kr = 91.885u
Mass of 3 neutrons = 3.025995u

1 u $\times c^2$ = 931.5 MeV.

Solution

Total initial mass = (235.045733 + 1.008665)u = 236.054398u
Total final mass = (140.9177 + 91885 + 3 × 1.008665) u = 235.829095u
Mass decrease in the fission reaction Δm = (236.054398–235.829095) u = 0.2253u
This decrease in mass is converted into energy in accordance with Einstein's equation.
Energy released = 0.2253u $\times c^2$ = 0.2253 × 931.5 MeV.
= 209.8 MeV.
Thus, in the process of fission of one nucleus of uranium, about 200 MeV energy is released.

The various forms, sources and end users of energy are summarised in Table 1.1. Further, Table 1.2 shows a number of devices to illustrate conversions of energy from one form to another. For example, a solar cell illustrates the conversion of light energy to electrical energy; a battery

Table 1.1 Various forms of energy.

Forms of energy
Mechanical (kinetic and potential)
Chemical
Electrical
Heat
Radiant
Nuclear

Primary Sources		*End users*
Coal	Chemical	Chemical processes
Oil	Chemical	Motion
Natural gas	Chemical	Electricity
Uranium-nuclear		Heat
Sun radiant/solar		Light

Table 1.2 Energy conversions.

	To Chemical	*To Electrical*	*To Heat*	*To Light*	*To Mechanical*
From Light	plant (photosynthesis) Camera film	solar cell	Heat lamp radiant solar	laser	photoelectric, door opener
From Chemical	food plants	battery fuel cell	fire food	candle phosphorescence	rocket animal muscle
From Electrical	battery electrolysis electroplating	transistor transformer	toaster heat lamp spark plug	fluorescent lamp light-emitting diode	electric motor relay
From Heat	gasification Vapourisation	thermocouple	heat pump heat exchanger	fire	turbine gas engine Steam engine
From Mechanical	heat cell (crystallisation)	generator alternator	friction brake	flint spark	flywheel pendulum Water wheel

converts chemical energy into electrical energy. The kinetic part of mechanical energy of a car converts into heat when the brakes are applied.

1.3 RENEWABLE/NONRENEWABLE ENERGY SOURCES

1.3.1 Definitions

Energy that is used by the human beings on planet Earth can be classified as: (i) renewable energy sources and (ii) nonrenewable energy sources.

(i) **Renewable energy sources:** These are the energy sources that are derived from natural sources that replenish themselves over short periods of time. These resources include the Sun, wind, moving water, organic plant and waste material (biomass), and the Earth's heat (geothermal). These resources are also called nonconventional sources of energy. This renewable energy sources can be used to generate electricity as well as for other applications. For example, biomass may be used as a boiler fuel to generate steam heat; solar energy may be used to heat water or for passive space heating; and landfill methane gas can be used for heating or cooking.

(ii) **Nonrenewable energy sources:** These are the energy sources that are derived from finite and static stocks of energy. It cannot be produced, grown, generated or used on a scale that can sustain its consumption rate. These resources often exist in a fixed amount and are consumed much faster than nature can create them. Examples of these types of resources are fossil fuels such as coal, petroleum, and natural gas and nuclear power (uranium). Due to its exhaustibility in nature, these types of energy resources are sometimes also called conventional sources of energy.

The basic different between renewable and nonrenewable energy are presented in Figure 1.1. Table 1.3 provides the features of comparison between renewable and nonrenewable sources of energy.

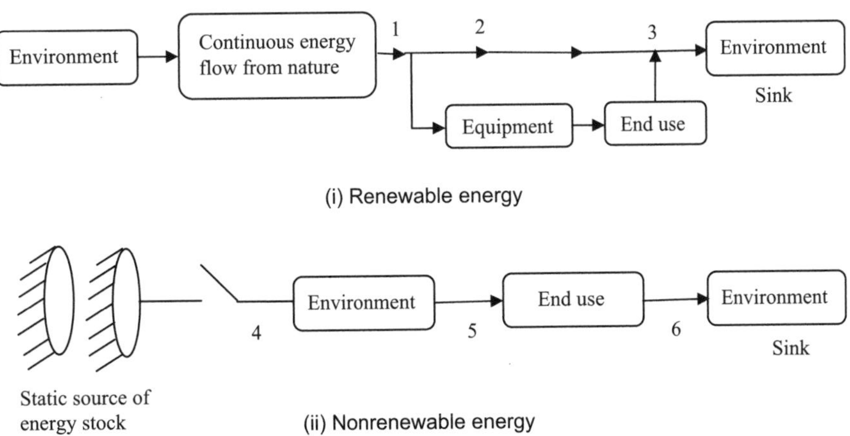

Figure 1.1 Basic differences between renewable and nonrenewable energy sources. Environmental energy flow 1→2→3. Used energy flow 4→5→6.

Table 1.3 Comparison of renewable and nonrenewable energy sources.

Important features	*Renewable energy*	*Nonrenewable energy*
Source	Natural local environment	Static stock
Supply time	Infinite	Finite
Normal state	Continuous energy flow	Finite source of energy
Location	Site and society specific	General and commercial use
Cost effectiveness	Free	Increasingly expensive
Scale potential	Small scale	Large scale
Skill requirement	Interdisciplinary and varied wide range of skill	Strong link with electric and mechanical engineering with specific range of skills
Dependence	Self-sufficient system encouraged	Systems dependent on Outside inputs
Area specific	Rural and decentralised industry	Urban centralised industry
Effects on environment	Little environmental harm	Environmental pollution Particularly for air and water
Safety	Less hazards	Most dangerous when faulty
Examples	Solar, wind, biomass, minihydro, tidal, *etc.*	Coal, oil, natural gas, *etc.*

1.3.2 Sources of Energy

There are six sources of useful energy utilised by human beings on planet Earth. These sources are given below:

(i) the Sun (thermal and electric);
(ii) geothermal energy from cooling, chemical reactions and radioactive decay in the Earth (thermal and electric);
(iii) the gravitational potential and planetary motion among Sun, Moon and Earth;
(iv) chemical energy from reactions among mineral sources;
(v) fossil fuels such as coal, petroleum products and natural gases (thermal and electric); and
(vi) nuclear energy from nuclear reactions on the Earth.

Renewable energy is obtained from sources (i), (ii) and (iii), whereas nonrenewable energy is derived from sources (iv), (v) and (vi).

1.3.3 Environmental Energy

The continuous flow of natural energy as renewable energy on Earth is shown in Figure 1.2. The total solar flux incident on Earth at sea level is about 1.2×10^{17} W and the solar flux per person (for around 6×10^{9} persons) is approximately 20 MW. This is equivalent to the power generating

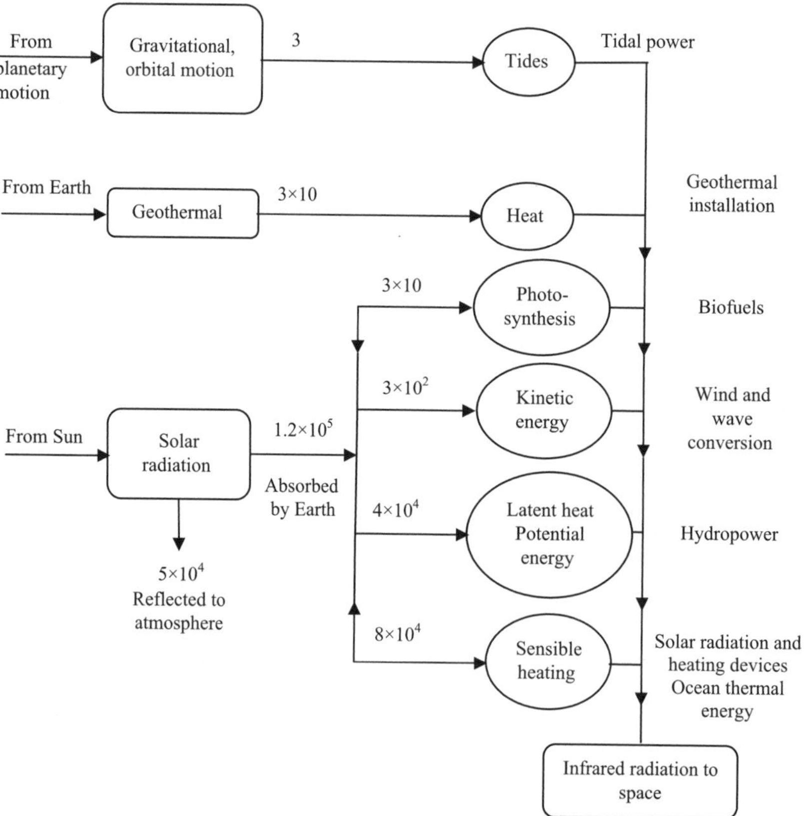

Figure 1.2 Continuous flow of natural energy as nonrenewable energy on Earth. Units, terawatts (10^{12} W).

capacity of seven very large sized diesel power plants. The maximum solar flux density perpendicular to the solar beam on Earth is around 1 kW/m^2.

Figure 1.2 provides a rough idea of the availability of natural energy on Earth. They have little value for practical purposes as renewable energy, which is site specific. Each region has a different environment that determines the amount of energy to be harnessed from the renewable energy sources.

1.4 ENERGY AND ENVIRONMENT

The conversion of energy from one form to another generally affects the environment. Hence, without considering the impact of energy on the environment, the study of energy is not complete. Fossil fuels have been powering the industrial growth and the amenities of modern life that we are enjoying now since the 1700s. But this has not been without any undesirable side effects. From the soil we farm, the water we drink and the air we breathe, the environment has been paying a heavy price for it.

During the combustion of fossil fuels the emitted pollutants are strongly responsible for smog, acid rain, global warming and climate change. The environmental pollution has reached such a high level that it becomes a serious threat for vegetables growth, wild life and human health. Air pollution can cause health problems and it can also damage the environment and property. It has caused thinning of the protective ozone layer of the atmosphere, which is leading to climate change.

Hundreds of elements and compounds such as benzene and formaldehyde are known to be emitted during the combustion of coal, oil, natural gas, engine of vehicles, furnaces and even fireplaces. Air pollution results from a variety of causes, not all of which are within human control. Dust storms in desert areas and smoke from forest fires and grass fires contribute to chemical and particulate pollution of the air. The source of pollution may be in one country but the impact of pollution may be felt elsewhere. Major air pollutants and their sources are listed below:

Carbon monoxide (CO): This is a colourless, odourless gas that is produced by the incomplete burning of carbon-based fuels including petrol, diesel and wood. It is also produced from the combustion of natural and synthetic products such as cigarettes. It lowers the amount of oxygen that enters our blood. It can slow our reflexes and make us confused and sleepy.

Carbon dioxide (CO_2): This is the principle greenhouse gas emitted as a result of human activities such as the burning of coal, oil, and natural gases.

Chlorofluorocarbons (CFC): These are gases that are released mainly from air conditioning systems and refrigeration. When released into the air, CFCs rise to the stratosphere, where they come in contact with other gases, which lead to a reduction of the ozone layer that protects the Earth from the harmful ultraviolet rays of the Sun.

Lead: This is present in petrol, diesel, lead batteries, paints, hair dye products, *etc.* Lead affects children in particular. It can cause nervous system damage and digestive problems and, in some cases, cause cancer.

Ozone (O_3): This occurs naturally in the upper layers of the atmosphere. This important gas shields the Earth from the harmful ultraviolet rays of the Sun. However, at the ground level, it is a pollutant with highly toxic effects. Vehicles and industries are the major source of ground level ozone emissions. Ozone makes our eyes itch, burn, and water. It lowers our resistance to colds and pneumonia.

Nitrogen oxide (NO_x): This causes smog and acid rain. It is produced from burning fuels including petrol, diesel, and coal. Nitrogen oxides can make children susceptible to respiratory diseases in winters.

Suspended particulate matter (SPM): This consists of solids in the air in the form of smoke, dust, and vapour that can remain suspended for extended periods and is also the main source of haze, which reduces visibility. The finer of these particles, when breathed in can lodge in our lungs and cause lung damage and respiratory problems.

Sulfur dioxide (SO_2): This is a gas produced from burning coal, mainly in thermal power plants. Some industrial processes, such as production of paper and smelting of metals, produce sulfur dioxide. It is a major contributor to smog and acid rain. Sulfur dioxide can lead to lung diseases.

The major areas of environmental problems may be classified as follows:

- water pollution;
- ambient air quality;
- hazardous air pollutants;
- maritime pollution;
- solid waste disposal;
- land use and siting impact;
- acid rain;
- stratospheric ozone depletion;
- global climate change (greenhouse effect).

Among these environmental issues, the internationally most vital problems are the acid precipitation, the stratospheric ozone depletion and the global climate change.

1.4.1 Acid Rain

Acid rain is a widespread term used to describe all forms of acid precipitation (rain, snow, hail, fog, *etc.*) Atmospheric pollutants, particularly oxides of sulfur and nitrogen, can cause precipitation to become more acidic when converted to sulfuric and nitric acids, hence the term acid rain. Motor vehicles also contribute to SO_2 emissions since petrol and diesel fuel also contains small amounts of sulfur.

The sulfur oxides (SO_2) and nitric oxides (NO) react with water vapour (H_2O) and other chemicals in the atmosphere in the presence of sunlight to form sulfuric acid (H_2SO_4) and nitric acid (HNO_3) as follows:

(a) Sulfur dioxide reacts with water to form sulfurous acid (H_2SO_3):

$$SO_2(g) + H_2O(l) \rightleftharpoons H_2SO_3\,(aq)$$

Sulfur dioxide (SO_2) can be oxidised gradually to sulfur trioxide (SO_3):

$$2SO_2(g) + O_2(g) \dashrightarrow 2SO_3\,(g)$$

Sulfur trioxide (SO_3) reacts with water to form sulfuric acid (H_2SO_4):

$$SO_3(g) + H_2O(l) \dashrightarrow H_2SO_4\,(aq)$$

(b) Carbon dioxide reacts with water to form carbonic acid:

$$CO_2(g) + H_2O(l) \rightleftharpoons H_2CO_3\,(aq)$$

Since carbonic acid is a weak acid it partially dissociates:

$$CO_2(g) + H_2O(l) \rightleftharpoons H^+(aq) + HCO_3^-\,(aq)$$

Nitrogen dioxide reacts with water to form a mixture of nitrous acid and nitric acid:

$$2NO_2(g) + H_2O(l) \rightleftharpoons HNO_2(aq) + HNO_3(aq)$$

These are shown in Figure 1.3. The acids formed usually dissolve in the suspended water droplets in clouds or fogs. These acid-laden droplets are washed from the air to the soil by rain or snow onto

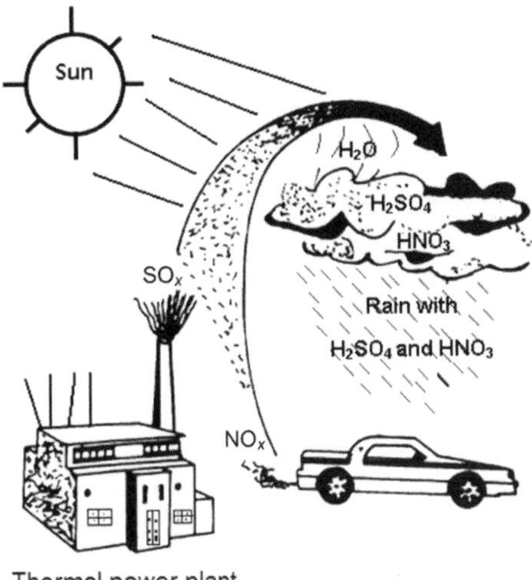

Figure 1.3 Formation of sulfuric acid and nitric acid when sulfur oxides and nitric oxides react with water vapour and other chemicals in atmosphere.

the Earth. This is known as acid rain, which is as acidic as lemon juice. The soil is capable of neutralising a certain amount of acid. However, the power plant, which uses high-sulfur coal, pollutes many lakes and rivers in industrial areas that have become too acidic for fish to grow. Forests in different regions of the Earth also experience a slow death due to absorption of acids from acid rain through the leaves, needles and roots of the trees.

1.4.2 Depletion of Ozone Layer

It is well known that the natural build up of oxygen in the atmosphere gradually led to the formation of the ozone layer. This layer is found between 19 and 30 kilometres (km) above the ground. The ozone layer filters out incoming radiation from the Sun that is harmful to life on Earth. The development of the ozone layer allowed more advanced lifeforms to evolve. Most ozone is produced naturally in the stratosphere, a layer of atmosphere between 10 and 50 km above the Earth's surface, but it can be found throughout the whole of the atmosphere. The ozone layer plays a natural and equilibrium maintaining role for the Earth through the absorption of ultraviolet (UV) radiation (240–320 nm) and absorption of infrared radiation. A global environmental problem is the distortion and regional depletion of the stratospheric ozone layer. This effect is shown in Figure 1.4 due to the emissions of NO_x and CFCs, *etc.* Ozone depletion in the stratosphere can lead to increased levels of damaging ultraviolet radiation reaching the ground. This increases rates of skin cancer, eye damage and other harm to many biological species. Chlorofluorocarbons (CFCs) and NO_x emissions are produced by fossil fuel and biomass combustion processes and play the most significant role in ozone depletion. Hence, the major pollutant, NO_x emissions, needs to be minimised to prevent stratospheric ozone depletion.

1.4.3 Global Warming and Climate Change (Greenhouse Effect)

1.4.3.1 Greenhouse Effect

The greenhouse effect is a process by which radiative energy leaving a planetary surface is absorbed by some atmospheric gases, called greenhouse gases. They transfer this energy to other components of the atmosphere, and it is reradiated in all directions, including back down towards the surface. This transfers energy to the surface and lower atmosphere, so the temperature there is higher than it would be if direct heating by solar radiation were the only warming mechanism.

The greenhouse effect is also experienced on a larger scale on Earth. This warms up as a result of the absorption of solar energy (short wavelength) during the day, cools down at night by radiating part of its energy into deep space as infrared radiation (long wavelength). Carbon dioxide (CO_2), water vapour and trace amounts of some other gases such as methane (CH_4) and nitrogen oxides act like a blanket and keep the Earth warm at night by blocking the heat radiation from the Earth, as shown in the Figure 1.5. Therefore, they are called "greenhouse effect" gases. In this case, the CO_2 is the primary component.

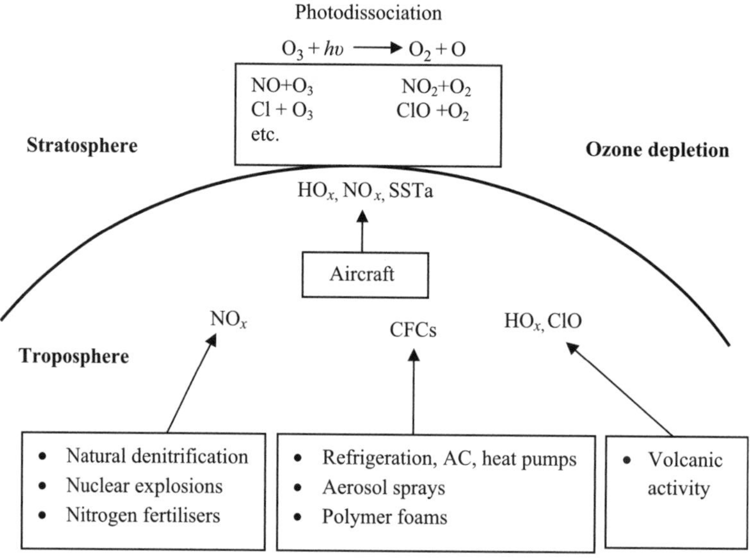

Figure 1.4 A schematic diagram representing sources of natural and anthropogenic ozone depleters.

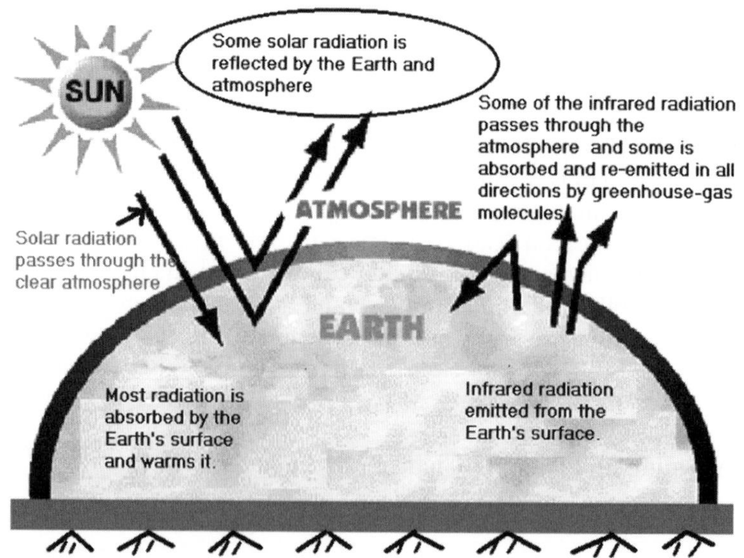

Figure 1.5 Greenhouse effect on planet Earth.

Water vapour is usually taken out of this list. Since it comes down as rain or snow as part of the water cycle, the man's activities in producing water do not make much difference on its concentration in the atmosphere. It is mostly due to evaporation from rivers, lakes, oceans *etc*. The CO_2 is different due to man's activities that do make a difference in CO_2 concentration in the atmosphere.

The greenhouse effect makes human life on the planet Earth feasible by keeping the Earth warm at about 30 °C. However, excessive amounts of greenhouse gases emitted by human being disturb the delicate balance by trapping too much energy. This causes the average temperature of the Earth to rise and the climate generally changes at some localities. These undesirable features of the greenhouse effect are generally referred to as **global warming** or **climate change**.

The excessive use of fossil fuels such as coal, petroleum products and natural gas in electric power generation, transportation and manufacturing processes is responsible for global climate change. The present concentration of CO_2 in the atmosphere is about 360 ppm (0.36 percent). This is 20 percent higher than the level a century ago. Further, it is projected to increase over 700 ppm by the year 2100. Under normal conditions, vegetables consume CO_2 and release CO_2 during the photosynthesis process, thus keeping the CO_2 concentration in the atmosphere in check. A mature growing tree consumes about 12 kg of CO_2 a year and exhales enough oxygen to support a family of four. However, deforestation and the huge increase in CO_2 production due to the fast growing industrialisation in recent decades has disturbed this balance. Also, a major source of greenhouse gas emissions is transportation. Each litre of petrol burned by a vehicle produces about 2.5 kg of CO_2. Also, a car emits about 6000 kg of CO_2 to the atmosphere in a year, which is nearly 4 times the weight of a car.

1.4.4 Solutions to Environmental Problems

There should be an effort to find ways to replace fossil fuels with more environmental friendly alternatives, particularly renewable energy resources, as potential solutions to the current environmental problems associated with the harmful pollutant emissions from fossil fuels. The use of renewable energy should be encouraged worldwide, with incentives. It is necessary to make the Earth a better place to live in. In recent times the advancements in thermodynamics have greatly contributed to improving conversion efficiencies of systems and thus to reduce pollution for clean environment. As individuals, we can also help in sustainable environment by using efficient energy conservation devices and by making energy efficiency a high priority in our purchases. Some of the potential solutions to environmental problems are as follows:

- clean renewable energy technologies;
- efficient energy conservation devices;
- clean alternative energy for transportation;
- energy source switching from fossil fuel to environmentally benign energy forms;
- energy storage technologies for better use;
- clean coal-based technologies;
- recycling method;
- encouraging forestation;
- use of locally available renewable energy resources;
- changing life style;
- increasing public awareness among users;
- educating and training for clean energy-based technologies.

1.5 SUSTAINABLE RENEWABLE ENERGY SOURCES

The continuing depletion of fossil fuels, and the environmental hazard problems posed by fast growing industrial development, are gradually shifting the path of devolvement towards

(i) environmental sustainability, (ii) better sociability and (iii) climate change. This in turn emphasises the need for use of renewable energy sources by human beings. The area of renewable energy sources is expanding rapidly. Numerous innovations as well as its applications based on renewable energy sources are taking place. The decentralised renewable energy systems has been recognised the world over as an answer to meeting the energy demands both in the household and in the agroindustrial sector. The exhaustion of natural sources and the accelerated demand of conventional energy have forced planners and policy makers to look for alternate sources, *i.e.* clean renewable energy sources.

A secure supply of energy resources is generally agreed to be a necessary but not sufficient requirement for development within a society. Furthermore, sustainable development demands a sustainable supply of energy sources that, in the long term, is readily and sustainably available at reasonable cost and can be utilised for all required tasks without causing negative social impacts. Supplies of energy resources like fossil fuels (coal, oil and natural gas) and uranium are generally acknowledged to be finite and other energy sources such as sunlight, wind and falling water (hydro) are generally considered renewable, and therefore sustainable over the relative long term. Waste and biomass fuels are also usually viewed as sustainable energy sources. In general, the implications of these statements are numerous and depend on how the word "sustainable" is defined.

Environmental concerns are an important factor in sustainable development. For a variety of reasons, activities that continually degrade the environment are not sustainable overtime, *i.e.* cumulative impact of such activities on the environment often leads over time to a variety of health, ecological and other problems. A large portion of the environmental impact in a society is associated with its utilisation. Ideally, a society seeking sustainable development utilises only energy resources that cause no environmental impact (e.g. that release no emissions to the environment). However, since all energy resources lead to some environmental impact, it is reasonable to suggest that some (not all) of the concerns regarding the limitations imposed on sustainable development by environmental emissions and their negative impacts, can be in part overcome through increased energy efficiency. Clearly, a strong relation exists between energy efficiency and environmental impact, since for the same services or products, less resource utilisation and pollution is normally associated with increased energy efficiency.

1.5.1 Global Scenario

Presently, even though commercial energy sources like coal, oil, and natural gas are being utilised to a large extent, renewable sources of energy are slowly gaining importance. Renewable energy plays a basic role in sustainable development. Such sources can supply the energy we need for indefinite periods of time polluting far less than fossil fuels. The advantages of renewables are well known, as far as they enhance diversity in energy supply markets; secure long-term sustainable energy supplies; reduce local and global atmospheric emissions; and create new employment opportunities, offering possibilities for local manufacturing.[1]

According to the International Energy Agency (IEA), renewable energy includes hydropower, biomass, wind, solar, geothermal and marine energy. Figure 1.6 shows the world share of total primary energy consumption (TPES) from 1997.[2]

Figure 1.6 shows that renewable energy represented 13.4% of total world combustion in 1973 which increased up to 18.7% in 2008 as shown by Figure 1.7.[3]

Support is also focused upon the so-called "new" renewables such as wind energy, modern biomass and solar photovoltaic. Modern biomass in the sustainable biomass resources used for the production of modern energy vectors such as electricity and a variety of liquid as well as gaseous fuels using advanced technologies. Similarly, traditional biomass is the biomass resource used for domestic heating and cooking, mainly in developing countries based on the collection and combustion of wood and dung in crude stoves/fireplaces. Hence, biomass produced in a sustainable way

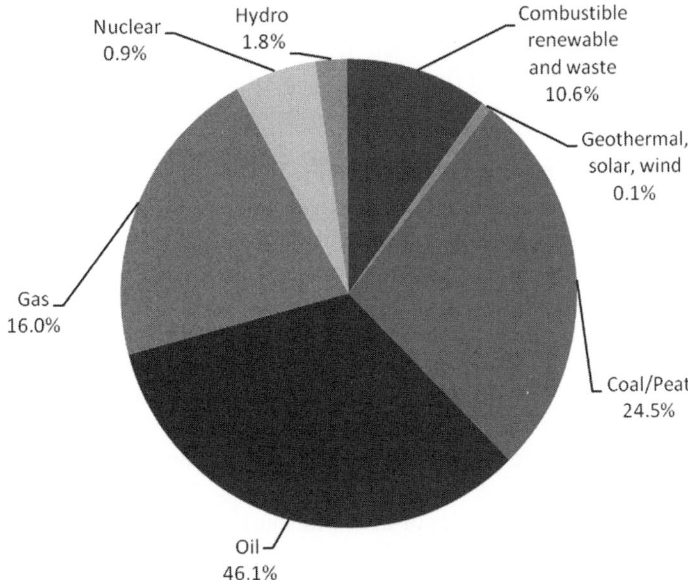

Figure 1.6 World shares of total primary energy supply (TPES) in 1973.

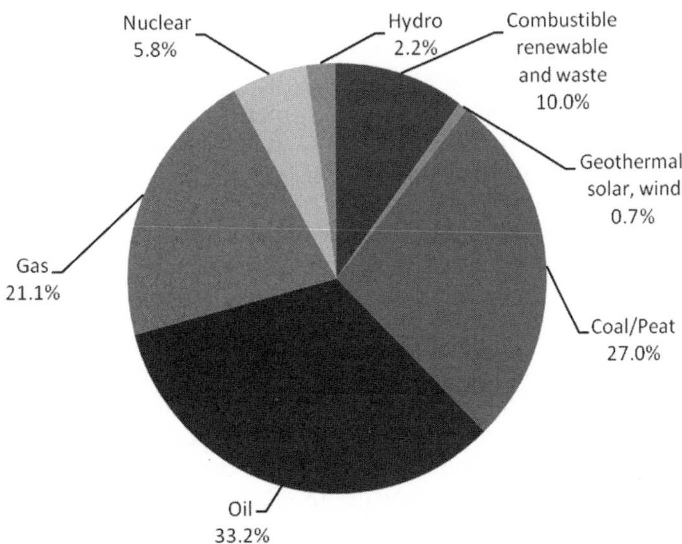

Figure 1.7 World shares of total primary energy supply (TPES) in 2008.

is called modern biomass, whereas biomass produced in an unsustainable way is called traditional biomass.

1.5.2 Indian Scenario

The economic development of the country is strongly depends on its energy utilisation. Presently, India ranks as the world's seventh largest producer and accounts for about 2.5% of the world's total annual energy production. This country is also the world's fifth largest energy consumer and accounts for about 3.5% of the world's total annual energy consumption.[4]

The 62nd report of the National Sample Survey Organisation (NSSO) states that 74% of households in rural area in India still depend on firewood as their cooking fuel, about 9% use dung cake and only 9% use LPG as cooking fuel. About 56% of households in rural India use electricity for lighting purpose, while 42% use kerosene. This reveals that a large section of rural India still depends on traditional biomass.[5]

In terms of the primary energy demand the country ranks fourth in the world and fifth when biomass is excluded. If it continues sustained economic growth, achieving 8–10% of GDP growth per annum till 2003, its primary energy supply will need to grow by three to four times and its electricity supply five to seven times.[6]

1.6 BASIC HEAT TRANSFER MECHANISMS

The energy transfer occurs by heat transfer through mechanical and electrical processes. Heat transfer is a well established topic. It does not require detailed discussions for small and moderate renewable energy applications due to small temperature difference (ΔT), less complicated geometrical configurations and lower energy fluxes.

An understanding of the basic principles of heat transfer is of vital importance in dealing with the renewable energy technologies. The object of this section is to review the fundamentals of basic heat transfer in connection with the renewable energy analysis. Heat is the form of energy that can be transferred from one system to another as a result of a temperature difference (ΔT). The science that deals with the determination of the rates of such energy transfer is heat transfer.

No net heart transfer between two media at the same temperature can take place as a temperature difference between two mediums is the first requirement. The rate of heat transfer in a certain direction depends on the magnitude of the temperature gradient in that direction. The higher the temperature gradient, the higher the rates of heat transfer.

Heat can be transferred as **(i) conduction**, **(ii) convection** and **(iii) radiation**. All modes of heat transfer are from the high temperature surface to a lower temperature one. Practical heat transfer problems are generally limited to a single method of heat transfer. Usually all three methods are involved in an overall heat transfer problems.[7]

1.6.1 Conduction

Conduction is the transfer of energy from the more energetic particles of a substance to the adjacent less energetic ones. Conduction can take place in solids, liquids or gases. Conduction is due to the collisions and diffusion of molecules during their random motion in gases and liquids. In solids, it is due to the combination of vibrations of the molecules in a lattice and the energy transport by free electrons. For example, a cold canned drink in a warm room eventually warms to the room air temperature as a result of heat transfer from the room to the drink through the aluminium by conduction.

The rate of heat conduction through a medium depends on (i) the geometry of the medium, (ii) its thickness (iii) the material of the medium and (iv) the temperature difference across the medium. For example, an insulated hot water tank reduces the rate of heat loss from the tank. The thicker the insulation, the smaller is the heat loss. A hot water tank will lose heat at a higher rate when the temperature of the room housing the tank is lowered. Further, tanks with larger surface area have a higher rate of heat loss.

1.6.1.1 Fourier's Law of Heat Conduction

Let us consider steady-state heat condition through a large plane wall of thickness $\Delta x = L$ and surface area A. For $\Delta T = T_2 - T_1$. On the basis of experiments it can be shown that the rate of heat transfer $\dot{Q}$ through the wall is doubled when the temperature difference ΔT across the wall or the area A normal to the direction of heat transfer is doubled. But it is halved when thickness L is doubled. Thus, one can conclude that the rate of heat conduction through a plane layer is proportional to ΔT across the layer and the heat transfer area (A), but is inversely proportional to the thickness of the layer (L). That is,

$$\text{Rate of heat conduction} \propto \frac{(\text{Area})\,(\text{temperature difference})}{\text{thickness}}$$

or,

$$\dot{Q} = K\,A\,\frac{\Delta T}{\Delta x} \tag{1.1}$$

where the constant K is the conductivity of the materials (Appendix V), which is a measure of the ability of a material to conduct heat. In the limiting case of $\Delta x \to 0$, the above measure equation reduces to the different form

$$\dot{Q} = -K\,A\,\frac{dT}{dx} \tag{1.2}$$

which is known as Fourier's law of heat conduction. Here, $\frac{dT}{dx}$ is the temperature gradient, which is the slope of the temperature curve on a T–x diagram at location x. The above relation indicates that "the rate of heat conduction in a direction is proportional to the temperature gradient in that direction". Heat is conducted in the direction of decreasing temperature and the temperature gradient becomes negative when temperature decreases with increasing x. Therefore, a negative sign is added to Eq. (1.2) to make heat transfer in the positive direction. The heat transfer area A is always normal to the direction of heat transfer.

1.6.1.2 Thermal Conductivity

The rate of conduction heat transfer under steady conditions (Eq. (1.1)) can also be viewed as the defining equation for thermal conductivity (K). Thus, the thermal conductivity of a material (K) can be defined as the rate of heat transfer through a unit thickness of the material ($\Delta x = 1$) per unit area per unit temperature difference ($\Delta T = 1$ °C). The thermal conductivity of a material is a measure of how fast heat flows through material. A large value of thermal conductivity indicates that the material is a good heat conductor. A low value indicates that the material is a poor heat conductor or insulator. The value of thermal conductivity of a few commonly used materials is given in Appendix V.

Thermal conductivity of gases, liquids and solids depends on temperature. Experimental studies have shown that for many materials, the dependence of thermal conductivity on temperature can be assumed to be linear.

$$K = K_0[1 + \beta(T - T_0)] \tag{1.3}$$

where K_0 is the thermal conductivity at temperature T_0 and β is a constant for the material. The values of K will increase for $T > T_0$ and decrease for $T < T_0$, however, the value of K is unaffected in the medium temperature range of renewable energy technologies in the present book.

1.6.1.3 Thermal Diffusivity

The thermal diffusivity is another material property that appears in heat conduction analysis. It represents how fast heat diffuses through a material and is defined as

$$\alpha = \frac{\text{Heat conducted}}{\text{Heat stored}} = \frac{K}{\rho C_p} \quad (\text{m}^2/\text{s}) \tag{1.4}$$

Here, the heat capacity (ρC_p) represents how much energy a material stores per unit volume. C_p and ρ are the specific heat and the density of the material, respectively. Therefore, the thermal diffusivity of a material can be viewed as the ratio of the heat conducted through the material to the heat stored per unit volume (Eq. (1.4)). A material that has a high thermal conductivity or a low heat capacity will obviously have a large thermal diffusivity (α). The larger the thermal diffusivity, the faster is the propagation of heat into the medium. A small value of thermal diffusivity means that heat is mostly absorbed by the material and a small amount of heat will be conducted further. Thermal diffusivities of some common materials are given in Appendix V.

Similarly, one dimensionless parameter is also used in heat conduction problems known as the Biot number (Bi) and is given by

$$\text{Bi} = \frac{hL}{K} = \frac{h}{K/L} \tag{1.5}$$

or,

$$\text{Bi} = \frac{\text{heat transfer coefficient at the surface of the solid}}{\text{internal conductance of solid across length L}}$$

Here, K is thermal conductivity of solid.

Therefore, the Biot number is defined as the ratio of the convective heat transfer coefficient at the surface to the conductive heat transfer coefficient within the body. When a solid body is being heated by the hotter fluid surrounding it, heat is first convected to the body and subsequently conducted within the body. The Biot number can also be defined as the ratio of the internal thermal resistance of a body to heat conducted to its external thermal resistance to heat convection. Hence, a small Biot number represents small resistance to heat conduction and thus small temperature gradients within the body.

1.6.1.4 Overall Heat Transfer

The practical problems in heat transfer involve a medium consisting of several different parallel layers each having different thermal conductivity or involve two or more of the heat transfer modes, namely, conduction, convection and radiation. In such a situation, the concept of an overall heat transfer coefficient (U) is applied to predict the one-dimensional steady-state heat transfer rate.

(A) Parallel Slabs[7]

Consider a composite wall as shown in Figure 1.8a through which heat is transferred from the hot fluid at temperature T_A to the cold fluid at temperature T_B. Assuming steady state, *i.e.* the heat-transfer rate, $\dot{Q}$ through the structure is the same through each layer and we can write,

$$\dot{Q} = Ah_a(T_A - T_0) = \frac{AK_1(T_0 - T_1)}{L_1} = \frac{AK_2(T_1 - T_2)}{L_2} = \frac{AK_3(T_2 - T_3)}{L_3} = Ah_b(T_3 - T_B) \tag{1.6a}$$

where terms like $h\,\Delta T$ represent heat transfer by convection and terms like $K\,(\Delta T/L)$ represent heat transfer by conduction through various layers.

Also, the rate of heat transfer per unit area is

$$\dot{q} = \frac{\dot{Q}}{A} = \frac{T_A - T_0}{R_a} = \frac{T_0 - T_1}{R_1} = \frac{T_1 - T_2}{R_2} = \frac{T_2 - T_3}{R_3} = \frac{T_3 - T_B}{R_b} \tag{1.6b}$$

where the Rs are thermal resistances at various surfaces and layers and are defined by,

$$R_a = \frac{1}{h_a}, R_1 = \frac{L_1}{K_1}, R_2 = \frac{L_2}{K_2}, R_3 = \frac{L_3}{K_3}, R_b = \frac{1}{h_b}$$

Equation (1.6b) can also be written the following form:

$$\begin{aligned} T_A - T_0 &= \dot{q}R_a \\ T_0 - T_1 &= \dot{q}R_1 \\ T_1 - T_2 &= \dot{q}R_2 \\ T_2 - T_3 &= \dot{q}R_3 \\ T_3 - T_B &= \dot{q}R_b \end{aligned}$$

or, summing up all the above equations we get

$$\dot{q} = \frac{T_A - T_B}{R} = U(T_A - T_B) \tag{1.6c}$$

where,

$$R = R_a + R_1 + R_2 + R_3 + R_b$$

and U is the overall heat transfer coefficient, W/m^2 K or W/m^2 K.

An overall heat transfer coefficient, U, is related to the total thermal resistance "R" of the composite wall by,

$$R = \frac{1}{U} = \frac{1}{h_a} + \frac{L_1}{K_1} + \frac{L_2}{K_2} + \frac{L_3}{K_3} + \frac{1}{h_b} \tag{1.7}$$

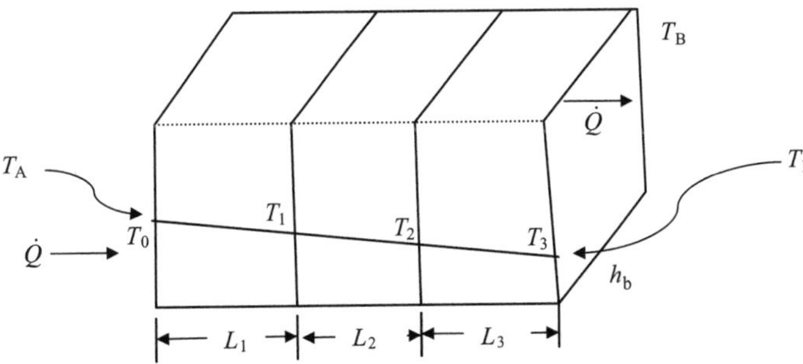

Figure 1.8a One-dimensional heat flow through parallel perfect contact slabs.

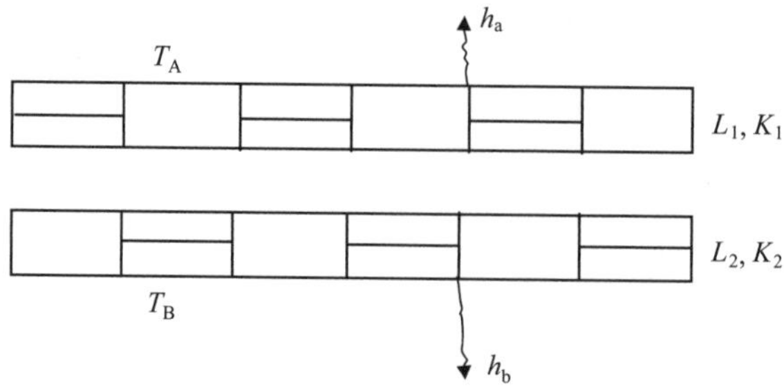

Figure 1.8b Configuration of parallel slabs with air cavity.

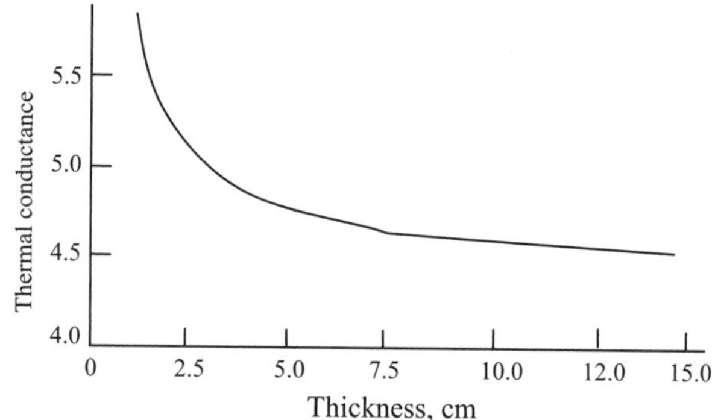

Figure 1.8c Variation of thermal air conductance with air gap thickness.

(B) Parallel Slabs with Air Cavity[7]

Consider a concrete greenhouse wall/roof with an air cavity of air conductance "C", as shown in Figure 1.8b.

The variation of thermal air conductance (C) with thickness of air gap (L) is given in Figure 1.8c.

The heat is transferred from the hot surface at temperature T_A to the cold surface temperature T_B (Figure 1.8b). For steady-state conditions, the rate of heat transfer per unit area of walls/roof will be the same at each layer boundary. Then, we can write,

$$\dot{Q} = Ah_a(T_A - T_0) = \frac{AK_1(T_0 - T_1)}{L_1} = AC(T_1 - T_2) = \frac{AK_2(T_2 - T_3)}{L_2} = Ah_b(T_3 - T_B) \quad (1.8)$$

The previous equation can be derived as done earlier and its expression is given below:

$$\dot{Q} = UA(T_A - T_B) \quad (1.9)$$

where,

$$U = \left[\frac{1}{h_a} + \frac{L_1}{K_1} + \frac{1}{C} + \frac{L_2}{K_2} + \frac{1}{h_b}\right]^{-1} = \frac{1}{R}$$

The expression for an overall heat transfer coefficient (U) for other configurations of parallel slabs with air cavities can be written as follows:

$$U = \left[\frac{1}{h_a} + \sum_i \frac{L_i}{K_i} + \sum_i \frac{1}{C_i} + \frac{1}{h_b} \right]^{-1} \quad (1.10)$$

1.6.2 Convection

The rate of heat transfer by convection between the fluid and the boundary surface may be evaluated by using the following expression,

$$\dot{Q} = h A \Delta T \quad (1.11)$$

where, h is the local convective heat transfer coefficient and the rate of heat flow at the fluid and body interface is related to the temperature difference between the surface of the body concerned and its surroundings.

1.6.2.1 Dimensionless Heat Convection Parameters

The convection equations contain the following dimensionless number depending on physical parameters of fluid:

$$\text{Nusselt number,} \quad \text{Nu} = \frac{h_c X}{K} \quad (1.12a)$$

$$\text{Reynolds number,} \quad \text{Re} = \frac{\rho\, v_0\, X}{\mu} = \frac{v_0\, X}{\nu} \quad (1.12b)$$

$$\text{Prandtl number,} \quad \text{Pr} = \frac{\mu C_p}{K} \quad (1.12c)$$

$$\text{Grashof number,} \quad \text{Gr} = \frac{g\beta' \rho^2 X^3 \Delta T}{\mu^2} = \frac{g\beta' X^3 \Delta T}{\nu^2} \quad (1.12d)$$

$$\text{Rayleigh number,} \quad \text{Ra} = \text{Gr Pr} = \frac{g\beta' X^3 C_p \Delta T}{\mu K} \quad (1.12e)$$

where

$$\nu = \text{kinematic viscosity} = \frac{\mu}{\rho}$$

and X is the characteristic dimension of the system.

The above dimensionless numbers given by Eqs. (1.12) can be obtained by using the physical properties of dry and moist air and water given in Appendix V.

The Reynolds (Re), Prandtl (Pr) and Grashof (Gr) numbers are calculated by using the physical properties of fluid at the average temperatures (T_f) of the hot surface (T_1) and surrounding air (T_2) *i.e.*

$$T_f = \frac{T_1 + T_2}{2} \quad (1.13a)$$

The thermal expansion coefficient (β') should be calculated at
(a) temperature of surrounding air (T_2) for exposed surface, *i.e.*

$$\beta' = \frac{1}{(T_2 + 273)} \tag{1.13b}$$

and (b) average temperature (T_f) for parallel plate, *i.e.*

$$\beta' = \frac{1}{(T_f + 273)} \tag{1.13c}$$

The characteristic dimension (X) of other shape is given by

$$X = \frac{A}{P} \tag{1.13d}$$

where A and P are the area and perimeter of the surface, which is generally used for irregular shapes.

Sometimes, for a rectangular horizontal surface ($L_0 \times B_0$), the characteristic dimension can also be calculated by

$$X = \left(\frac{L_0 + B_0}{2}\right) \tag{1.13e}$$

Nusselt number (Nu): This is the ratio of convective heat transfer to heat transfer by conduction in the fluid and is usually unknown in problems of convection, since it involves the heat transfer coefficient, h, which is an unknown parameter. Although the Nusselt number resembles the Biot number, the two are essentially different. The Biot number includes the thermal conductivity of a solid, the Nusselt number that of a fluid.

Reynolds number (Re): This is the ratio of the fluid dynamic force (ρv_0^2) to the viscous drag force ($\mu v_0/X$) where, ρ is the density and μ the dynamic viscosity. It indicates the flow behaviour in forced convection and serves as a criterion for the stability of laminar flow.

Prandtl number (Pr): This is the ratio of momentum diffusivity (μ/ρ) to the thermal diffusivity ($K/\rho C_p$); C_p is the specific heat at constant pressure. It gives the relation of heat transfer to fluid motion.

Grashof number (Gr): This is the ratio of the buoyancy force to the viscous force; β is the coefficient of volumetric thermal expansion, g is the gravitational acceleration and T is the temperature.

Rayleigh number (Ra): This is the ratio of the thermal buoyancy to viscous inertia.

1.6.2.2 Free Convection

For free convection, the terrestrial gravitational field, acting on the fluid with a nonuniform density distribution owing to the temperature difference between the fluid and the contacting surface, causes the fluid motion.

The coefficient of heat transfer, h, usually incorporated with Nusselt number, depends on whether the flow is laminar or turbulent, free or forced.

For free convection,

$$\text{Nu} = C'(\text{Gr}\,\text{Pr})^n\ K' \tag{1.14}$$

where the relationship is obtained by the method of dimensional analysis. The constants C' and n are determined by the correlation of experimental data of geometrically similar bodies. The correlation factor, K', is introduced to represent the entire physical behaviour of the problem.[8]

Some empirical relations used for free convention are given in Table 1.4.

EXAMPLE 1.2

Estimate Biot number (Bi) for a horizontal rectangular surface (1.0 m × 0.8 m) that is maintained at 134 °C. The hot surface is exposed to (a) water and (b) air at 20 °C.

Solution

The average film temperature, $T_f = (134+20)/2 = 77$ °C
From Appendix V, at $T_f = 77$ °C; $\nu = 20.8 \times 10^{-6}$ m^2/s; $K = 0.030$ W/m K, Pr = 0.697
$\beta' = 1/(20+273)$ due to exposure of surroundings.
For water
The average film temperature, $T_f = (134+20)/2 = 77$ °C
From Appendix V, water thermal properties at $T_f = 77$ °C;
$\mu = 3.72 \times 10^{-4}$ kg/m s; $K = 0.668$ W/m K, $\rho = 973.7$ kg/m^3, Pr = 2.33 and $\beta' = 1/(77+273) = 2.857 \times 10^{-3}$ K^{-1} and consider the characteristic dimension $(X = d) = (1.0+0.8)/2 = 0.90$ m.

The Grashof number is

$$Gr_L = \frac{g\beta'\rho^2(\Delta T)X^3}{\mu^2}$$
$$= \frac{(9.8)(2.857 \times 10^{-3})(973.7)^2(114)(0.9)^3}{(3.72 \times 10^{-4})^2} = 1.594 \times 10^{13}$$

This is a turbulent flow. For a heated plate facing upward for $X = L = (L_0 + B_0)/2$ and consider $C = 0.14$ and $n = 1/3$. Now, a convective heat transfer coefficient can be calculated as

$$h = \frac{K}{L}(0.14)(Gr_L\,\mathrm{Pr})^{1/3} = \frac{0.668}{0.9}(0.14)\left(1.594 \times 10^{13} \times 2.33\right)^{1/3}$$
$$= 3467 \text{ W/m}^2\text{K}$$

For the characteristic dimension of
(a) $L = A/P = 0.8/3.6 = 0.222$ m

$$Gr_L\,\mathrm{Pr} = \frac{(9.8)(134-20)(0.222)^3(0.697)}{(293)\left(2.08 \times 10^{-5}\right)^2} = 6.72 \times 10^7$$

Using Table 1.4 for a hot surface facing upward and turbulent flow conditions, the heat transfer coefficient can be calculated as

$$h = (K/L)\ 0.15\ (Gr_L\ \mathrm{Pr})^{0.333}$$
$$= (0.03/0.222)(0.15)(6.72 \times 10^7)^{0.333}$$
$$= 8.23 \text{ W/m}^2\ ^\circ\text{C}$$

Hence, the Biot number is

$$\mathrm{Bi} = \frac{hL}{K}$$

$$= (8.23 \times 0.222)/0.03 = 60.9$$

(b) $X = (L_0 + B_0)/2 = (1.0+0.8)/2 = 0.9$

$$Gr_L\ \mathrm{Pr} = 4.47 \times 10^7$$

and, $h = (0.03/0.9)(0.14)(4.47 \times 109)^{1/3} = 7.74$ W/m^2 °C

Table 1.4 Simplified equations for free convection from various surfaces to air at atmospheric pressure (Heat transfer, J. P. Holman, 1992).

Surface	*Laminar* $10^4 < Gr_f\, Pr_f < 10^9$	*Turbulent* $Gr_f\, Pr_f > 10^9$
Heated plate facing downward or cooled plate facing upward	$h = 0.59\,(\Delta T/L)^{1/4}$	
Horizontal plate: Heated plate facing upward or cooled plate facing downward	$h = 1.32\,(\Delta T/L)^{1/4}$	$h = 1.52(\Delta T)^{1/3}$
Horizontal cylinder	$h = 1.32\,(\Delta T/d)^{1/4}$	$h = 1.24\,(\Delta T)^{1/3}$
Vertical plane or cylinder	$h = 1.42\,(\Delta T/L)^{1/4}$	$h = 1.31(\Delta T)^{1/3}$

1.6.2.3 Forced Convection[9]

In forced convection, the fluid motion is artificially induced, say with a pump or a fan that forces the fluid flow over the surface. The external energy is supplied to maintain the process in which there are two types of forces (a) the fluid pressure related to flow velocity $(1/2)\rho v^2$ and (b) the frictional force produced by viscosity $(\mu.\mathrm{d}v/\mathrm{d}y)$. Their relative importance in heat transfer is signified by the nondimensional Reynolds number. It also controls the flow, laminar or turbulent, in the boundary layer with which the rate of heat transfer is closely connected. The heat transfer by forced convection is represented by the following Nusselt equation

$$\mathrm{Nu} = C(\mathrm{Re\ Pr})^n\, K \tag{1.15}$$

where, C and n are constants for a given type of flow and geometry. K is a correction factor (shape factor) added, to obtain a greater accuracy.

The empirical relation for forced convective heat transfer through cylindrical tubes may be represented as,

$$\overline{\mathrm{Nu}} = \frac{h\,D}{K_{\mathrm{th}}} = C\,\mathrm{Re}^m\,\mathrm{Pr}^n\,K \tag{1.16}$$

where $D = 4A/\mathrm{P}$, is the hydraulic diameter (m); P is the perimeter of the section (m) and K_{th} is the thermal conductivity (W/m K).

The values of C, m, n and K for various conditions are given in Table 1.5.

For fully developed laminar flow in tubes at constant wall temperature the following relation applies,

$$\mathrm{Nu_d} = 3.66 + \frac{0.0668(d/L)\mathrm{Re\ Pr}}{1 + 0.04[(d/L)\mathrm{Re\ Pr}]^{2/3}} \tag{1.17}$$

The heat transfer coefficient calculated from this relation is the average value over the entire length of the tube. When the tube is sufficiently long the Nusselt number approaches a constant value of 3.66. For the plate heated over its entire length, the Nusselt number can be obtained by integrating the equation given below over the length of the plate,

$$\mathrm{Nu}_x = 0.332(\mathrm{Pr})^{1/3}\,\mathrm{Re}_x^{1/2} \tag{1.18}$$

Table 1.5 The value of constants for forced convection.

Cross section	*D*	*C*	*m*	*n*	*K*	*Operating conditions*
(circle, diameter d)	*d*	1.86	1/3	1/3	$(d/l)^{1/3}\ (\mu/\mu_w)^{0.14}$	Laminar flow short tube for Re<2000, and Gr>10
	d	3.66	0	0	1	Laminar flow long tube Re< 2000 Gr<10
	d	0.023	0.8	0.4	1	Turbulent flow of gases Re>2000
	d	0.027	0.8	0.33	$(\mu/\mu_w)^{0.14}$	Turbulent flow of highly viscous liquids for 0.6<Pr<100

Now,

$$\bar{h} = \frac{1}{L}\int_0^L \frac{K}{x}(0.332)\left(\mathrm{Pr}^{1/3}\right)\left(\mathrm{Re}_x^{1/2}\right)\mathrm{d}x = \frac{K}{L}(0.332)\left(\mathrm{Pr}^{1/3}\right)\int_0^L \frac{1}{x}\left(\frac{v_0 x}{\nu}\right)^{1/2}\mathrm{d}x$$

$$= \frac{K}{L}(0.332)\left(\mathrm{Pr}^{1/3}\right)(2)\left(\frac{v_0 L}{\nu}\right)^{1/2} = \frac{K}{L}(0.664)\left(\mathrm{Pr}^{1/3}\right)\left(\mathrm{Re}_L^{1/2}\right)$$

Thus,

$$\overline{N}\mathrm{u}_L = 0.664\mathrm{Re}_L^{1/2}\mathrm{Pr}^{1/3} \tag{1.19}$$

1.6.2.4 Convective Heat Transfer Due to Wind

The heat transfer from a flat plate exposed to outside winds has been analysed by several workers. The following equation for convective heat transfer coefficient is generally used

$$h_c = 5.7 + 3.8V \quad \text{for } 0 \leq V \leq 5 \text{ m s}^{-1} \tag{1.20a}$$

where V is the wind speed, m/s.

The above equation for zero wind speed gives heat loss by natural convection. It may be mentioned here that the process taking place is not as simple as it appears, as the wind may not always be blowing parallel to the surface.

It is probable that in this equation the effects of free convection and radiation are included. For this reason, this equation should be,

$$h_c = 2.8 + 3.0V \quad \text{for } 0 \leq V \leq 7 \text{ m s}^{-1} \tag{1.20b}$$

The sensibility of these parameters is also demonstrated through a comparison, another relation for convective heat transfer coefficient is given by,

$$h_c = 7.2 + 3.8V \tag{1.20c}$$

Several other correlations are also available in the literature and generally, h_c is determined from an expression in the formed expressed as

$$h_c = a + bV_a^b \tag{1.20d}$$

where, $a=2.8$, $b=3$ and $n=1$ for $V_a < 5$ m/s and $a=0$, $b=6.15$ and $n=0.8$ for $V_a > 5$ m/s.

(The source and reference of Eqs. (1.20) can be obtained from Tiwari (2002).)

1.6.3 Radiation

Thermal radiation involves the transfer of heat from a body at a higher temperature to another at a lower temperature by electromagnetic waves (0.1 to 100 μm). Temperature is transmitted in the space in the form of electromagnetic waves. Thermal radiation is in the infrared range and obeys all the rules as that of light, namely, travels in straight lines through a homogenous medium, is converted into heat when it strikes any body that can absorb it and is reflected and refracted according to the same rules as those of light.

1.6.3.1 Radiation Involving Real Surfaces

When radiant energy falls on a body, a part of it is reflected, another part is absorbed and the rest is transmitted through it. The conservation of energy states that the total sum must be equal to the incident radiation, thus,

$$I_r + I_a + I_t = I_T \tag{1.21a}$$

or,

$$\rho' + \alpha' + \tau = 1 \tag{1.21b}$$

where, ρ', α' and τ are the reflectivity, absorptivity and transmissivity of the intercepting body, respectively. The ratio of the energy reflected to that which is incident is called the reflectivity. The ratio of the energy absorbed and the energy transmitted to that which is incident are the absorptivity and transmissivity, respectively.

For an opaque surface, $\tau = 0$, therefore $\rho' + \alpha' = 1$. However, when $\rho' = \tau = 0$; $\alpha = 1$, that is, the substance absorbs the whole of the energy incident on it. Such a substance is called a blackbody. Similarly, for a white body that reflects the whole of the radiation falling on it, $\alpha' = \tau = 0$, $\rho' = 1$.

The energy that is absorbed is converted into heat and this heated body, by virtue of its temperature, emits radiation. The radiant energy emitted per unit area of a surface in unit time is referred to as the emissive power (E_λ). However, if defined as the amount of energy emitted per second per unit area perpendicular to the radiating surface in a cone formed by a unit solid angle between the wavelengths lying in the range $d\lambda$, it is called spectral emissive power (e_λ). Further, emissivity, defined as the ratio of the emissive power of a surface to the emissive power of a blackbody of the same temperature, is the fundamental property of a surface.

1.6.3.2 Kirchhoff's Law

This states that for a body in thermal equilibrium, the ratio of its emissive power to that of a blackbody at the same temperature is equal to its absorptivity, *i.e.*

$$\frac{e}{e_b} = \alpha' \text{ or } \varepsilon = \alpha' \tag{1.21c}$$

Thus, a body can absorb as much incident radiation as it can emit at a given temperature. However, it may not be valid if the incident radiation comes from a source at different temperature. Further, it applies to surfaces bearing the grey surface characteristics, namely radiation intensity is taken to be a constant proportional to that of a blackbody. The radiative properties α_λ, ε_λ and ρ_λ are assumed to be uniform over the entire wavelength spectrum.

1.6.3.3 Laws of Thermal Radiation

Laws of thermal radiations have been obtained for black bodies and conditions of thermodynamic equilibrium.

(i) Planck's Law

The emission of energy with respect to wavelength is not uniform and depends on temperature. Planck's law establishes the relation of the spectral emissive power, wavelength and temperature and is written as,

$$E_{b\lambda} = \frac{C_1}{(\lambda)^5} \frac{1}{(\exp[C_2/(\lambda T)] - 1)} \tag{1.22a}$$

where, $C_1 = 3.742 \times 10^8$ W μm^4/m^2 (= 3.7405×10^{-6} W m^2) and $C_2 = 1.4387 \times 10^4$ μm K (= 0.01439 mK) are called Planck's first and second radiation constants, respectively. Planck's law has two limiting cases depending on the relative value of C_2 and λT:

a) when $\lambda T \gg C_2$

$$E_{b\lambda} = \frac{C_1}{(\lambda)^5} \frac{\lambda T}{C_2} \text{ Rayleigh} - \text{Jeans law} \tag{1.22b}$$

b) when $\lambda T \ll C_2$

$$E_{b\lambda} = \frac{C_1}{(\lambda)^5} \exp(-C_2/\lambda T) \tag{1.22c}$$

(ii) Wien's Displacement Law

The wavelength corresponding to the maximum intensity of blackbody radiation for a given temperature T is given by this law:

$$\lambda_{\max} T = C_3 \tag{1.22d}$$

where, $C_3 = 2897.6$ μm K. Hence, an increase in temperature shifts the maximum blackbody radiation intensity towards the shorter wavelength.

(iii) Stefan–Boltzmann Law

This law relates the hemispherical total emissive power, namely total energy and temperature. By integrating Planck's law over all wavelengths, the total energy emitted by a blackbody is found to be,

$$E_b = \int_0^\infty E_{b\lambda} = \sigma T^4 \tag{1.22e}$$

where $\sigma = 5.6697 \times 10^{-8}$ W/m^2 K^4 is the Stefan–Boltzmann constant.

(iv) Sky Radiation

In order to evaluate radiation exchange between a body and the sky, certain equivalent blackbody sky temperature is defined. This accounts for the fact that the atmosphere is not at a uniform temperature and that it radiates only in certain wavelength regions. Thus, the net radiation to a surface with emittance ε and temperature T is,

$$\dot{Q} = A\,\varepsilon\,\sigma\,({T_{sky}}^4 - T^4) \tag{1.23a}$$

In order to express the equivalent sky temperature T_{sky}, in terms of ambient air temperature, various expressions have been given. These relations, although simple to use, are only approximations. The sky temperature to the local air temperature can be given by the relation,

$$T_{sky} = 0.0552 T_a^{1.5} \tag{1.23b}$$

where, T_{sky} and T_a are both in Kelvin.

Another commonly used relation is given as

$$T_{sky} = T_a - 6 \tag{1.23c}$$

or,

$$T_{sky} = T_a - 12 \tag{1.23d}$$

1.6.3.4 Radiative Heat Transfer Coefficient

The radiant heat exchange between two infinite parallel surfaces per m^2 at temperatures T_1 and T_2 may be given as,

$$\dot{q}_r = \varepsilon\ \sigma\ (T_1^{\ 4} - T_2^{\ 4}) \tag{1.24a}$$

$$\dot{q}_r = h_r(T_1 - T_2) \tag{1.24b}$$

where,
$$h_r = \varepsilon\ \sigma\ (T_1^2 + T_2^2)(T_1 + T_2) = \varepsilon(4\sigma\overline{T})^3 \quad \text{for}\ \overline{T}_1 \cong \overline{T}_2$$

$$\text{and}\ \varepsilon = \frac{1}{\varepsilon_1} + \frac{1}{\varepsilon_2} - 1,\ \text{for two parallel surfaces}$$
$$= \varepsilon,\ \text{for surface exposed to atmosphere}$$

ε_1 and ε_2 are the emissivities of the two surfaces. When one of the surfaces is sky, Eq. (1.24a) becomes,

$$Q_r = \varepsilon\ \sigma\ (T_1^{\ 4} - T_{sky}^{\ \ 4}) \tag{1.25a}$$

The above equation may be rewritten as,

$$Q_r = \varepsilon\ \sigma\ (T_1^{\ 4} - T_a^{\ 4}) + \varepsilon\ \sigma(T_a^{\ 4} - T_{sky}^{\ \ 4}) \tag{1.25b}$$

or,
$$Q_r = h_r\,(T_1 - T_a) + \varepsilon\Delta\mathrm{R} \tag{1.25c}$$

where $\Delta R = \sigma[(T_a + 273)^4 - (T_{sky} + 273)^4]$ is the difference between the long-wavelength radiation incident on the surface from sky and surroundings and the radiation emitted by a blackbody at ambient temperature. The reduction of Q_r in the form of Eq. (1.25c) will enable one to find the exact closed form solution for T_1. It may be noted here that this solution is based on the assumption that T_a and T_{sky} are constant.

1.6.4 Simultaneous Heat and Mass Transfer

In a moving single-component medium, heat is transferred by conduction and convection; the process is known as convective heat transfer.

$$\dot{Q} = h\,(T_w - T_a) \tag{1.26a}$$

where T_w and T_a are the fluid (water) and the surrounding air temperatures, respectively. By analogy, the process of molecular and molar transport of matter; in a moving heterogeneous medium, is called convective mass transfer. The surface of the liquid phase plays a role similar to that of a solid wall in heat transfer process without accompanying diffusion. The process of heat and mass transfer are of practical interest in evaporation, condensation, *etc*. The heat transfer is based on Newton's law.

Mass transfer rate is based on a similar equation:

$$\dot{m} = h_D\,(\rho_w{}^0 - \rho_a{}^0) \tag{1.26b}$$

where m is the rate of mass flow per unit area, (kg/m^2/s), h_D the mass transfer coefficient [(kg/s)(sqm/kg/m^3)], $\rho_w{}^0$ the partial mass density of water vapour, kg/m^3, $\rho_a{}^0$ is the partial mass density of air (kg/m^3).

According to the Lewis relation for an air and water vapour mixture

$$h_{cw} = h_D \rho^\circ C_p \tag{1.26c}$$

From the perfect gas equation for 1 mole of air,

$$PM = \rho RT$$

where R is the universal gas constant.

By assuming $T_w = T_a = T$ at the water/air interface, Eq. (1.26b) becomes

$$\dot{m} = \frac{h_{cw}}{\rho_a^0\,C_{pa}}\frac{M_w}{RT}\,(P_w - P_a) \tag{1.26d}$$

The rate of heat transfer on account of mass transfer of water vapour is

$$\dot{Q}_{ew} = \dot{m}L \tag{1.26e}$$

where L is the latent heat of vapourisation and P_w and P_a are the partial pressures of water vapour and air respectively.

$$\dot{Q}_e = \frac{L\,h_{cw}}{\rho_a^0\,C_{pa}}\frac{M_w}{RT}\,(P_w - P_a) \tag{1.26f}$$

Let

$$H_o = \frac{L\,h_{cw}}{\rho_a^0\,C_{pa}}\frac{M_w}{RT}$$

then $\dot{Q}_e = H_o(P_w - P_a)$

Using the perfect gas equation $\rho_a^0 = \dfrac{P_a M_a}{RT}$ for air, (for 1 mole of air) and by substituting in the expression for H_0

$$\frac{H_0}{h_{cw}} = \frac{L}{C_{pa}}\frac{M_w}{M_a}\frac{1}{P_a} \tag{1.26g}$$

For small values of P_w, $P_T = P_a$ the above equation becomes

$$\frac{H_0}{h_{cw}} = \frac{L}{C_{pa}}\frac{M_w}{M_a}\frac{1}{P_T} \tag{1.26h}$$

where P_T is the total pressure of the air–vapour mixture. The values of different parameters used in Eq. (1.26h) are as follows:

L = Latent heat of vapourisation = 2200 kJ/kg
C_{pa} = Specific heat of air = 1.005 kJ/kg °C
M_w = 18 kg/mol (molar mass of water)
P_T = Total pressure of air–vapour mixture = 1 atm
1 atm = 101 325 N/m^2

Substituting all these values in Equation (1.26h) and solving

$$\frac{H_0}{h_{cw}} = 0.013$$

The best representation of heat and mass transfer phenomenon is obtained if the values of $\frac{H_0}{h_{cw}}$ is taken to be 16.27×10^{-3} instead of 0.013. Thus, the rate of heat transfer on account of mass transfer is written as

$$\dot{q}_{ew} = 16.273 \times 10^{-3}\, h_{cw}(P_w - P_a) \tag{1.27a}$$

If the surface is exposed to atmosphere, then the above equation reduces to

$$\dot{q}_{ew} = 16.273 \times 10^{-3}\, h_{cw}(P_w - \gamma P_a) \tag{1.27b}$$

where γ is the relative humidity of air.
Also,

$$\dot{q}_{ew} = h_{cw}(T_w - T_a)$$

Hence, the evaporation heat transfer coefficient can be given as,

$$h_{ew} = \frac{16.276 \times 10^{-3} h_{cw}\left[\overline{P_w} - \gamma \overline{P_a}\right]}{(T_w - T_a)} \tag{1.27c}$$

The values of P_w and P_a for the ranges of temperature (10–90 °C) can be obtained from the following expression (Fernanez and Chargoy, 1990).

$$P(T) = \exp\left[25.317 - \frac{5144}{T + 273}\right] \tag{1.27d}$$

1.7 LAWS OF THERMODYNAMICS

A physical system containing a large number of atoms or molecules is called the **thermodynamic system** if macroscopic properties, such as the temperature, pressure, mass density, heat capacity, *etc.*, are the properties of main interest. The number of atoms or molecules contained, and hence the volume of the system, must be sufficiently large so that the conditions on the surfaces of the system do not affect the macroscopic properties significantly. From the theoretical point of view, the size of the system must be infinitely large, and the mathematical limit in which the volume, and proportionately the number of atoms or molecules, of the system are taken to infinity is often called the **thermodynamic limit**.

The **thermodynamic process** is a process in which some of the macroscopic properties of the system change in the course of time, such as the flow of matter or heat and/or the change in the volume of the system. It is stated that the system is in **thermal equilibrium** if there is no thermodynamic process going on in the system, even though there would always be microscopic molecular motions taking place. The system in thermal equilibrium must be uniform in density, temperature, and other macroscopic properties.

1.7.1 The Zeroth Law of Thermodynamics

If two thermodynamic systems, A and B, each of which is in thermal equilibrium independently, are brought into thermal contact, one of two things will take place: either (i) a flow of heat from one system to the other or (ii) no thermodynamic process will result. In the latter case the two systems are said to be in thermal equilibrium with respect to each other.

According to the zeroth law of thermodynamics, "If two systems are in thermal equilibrium with each other and there is a physical property that is common to the two systems, this common property is called the temperature."

Let the condition of thermodynamic equilibrium between two physical systems A and B be symbolically represented by

$$A \Leftrightarrow B$$

Then, experimental observations confirm the statement

if $A \Leftrightarrow C$ and $B \Leftrightarrow C$, then $A \Leftrightarrow B$.

Based on preceding observations, some of the physical properties of the system C can be used as a measure of the temperature, such as the volume of a fixed amount of the chemical element mercury under some standard atmospheric pressure. The zeroth law of thermodynamics is the assurance of the existence of a property called the **temperature**.

1.7.2 The First Law of Thermodynamics

Let us consider a situation in which a macroscopic system has changed state from one equilibrium state P_1 to another equilibrium state P_2, after undergoing a succession of reversible processes. Here, the processes mean that a quantity of heat energy Q has cumulatively been absorbed by the system and an amount of mechanical work W has cumulatively been performed upon the system during these changes.

According to the first law of thermodynamics, there are many different ways or routes to bring the system from state P_1 to the state P_2; however, it turns out that the sum is independent of the ways or the routes as long as the two states P_1 and P_2 are fixed, even though the quantities W and Q may vary individually depending upon the different routes.

$$W + Q$$

Consider, now, the case in which P_1 and P_2 are very close to each other and both W and Q are very small. Let these values be d/W and d/Q. According to the first law of thermodynamics, the sum, $d/W + d/Q$, is independent of the path and depends only on the initial and final states, and hence is expressed as the difference of the values of a quantity called the *internal energy*, denoted by U, determined by the physical, or thermodynamic, state of the system, *i.e.*,

$$dU = U_2 - U_1 = d/W + d/Q$$

Mathematically speaking, d/W and d/Q are not *exact differentials* of state functions since both d/W and d/Q depend upon the path; however, the sum, $d/W + d/Q$, is an exact differential of the state function U. This is the reason for using primes on those quantities.

1.7.3 The Second Law of Thermodynamics

There are several different ways of expressing the second law of thermodynamics, and the following are three examples.

Clausius' principle: No cyclic process exists that has as its sole effect the transference of heat from a colder body to a hotter body.

Kelvin's principle: No cyclic process exists that produces no other effect than the extraction of heat from a body and its conversion into an equivalent amount of work.

Caratheodory's principle: There are states of a system, differing infinitesimally from a given state, which is unattainable from that state by any quasistatic adiabatic process.

The second law (especially in the Caratheodory form) allows one to order the states according to the direction that the system is allowed to evolve. A parameter, called empirical entropy, is ascribed to each state, such that this parameter never decreases spontaneously. In the absence of external agents doing work on the system, any change in the system is either reversible, which involves no change in entropy, or irreversible, which involves an increase in entropy.

A bucket of warm water with a warm cannon ball has higher entropy than the bucket of cold water and the red hot cannon ball, so that the evolution is irreversible and can go in only one direction. To reverse the change we need to do work on the system: mechanical work to lift the cannon ball, and thermodynamic work, using some form of refrigerator to decrease the temperature of the bucket of water, and some other process to heat the cannon ball.

1.7.4 The Third Law of Thermodynamics

The entropy of a system approaches a constant value as the temperature approaches absolute zero.

The third law is relatively easy to understand from a statistical point of view in which entropy is associated with disorder. As absolute zero is approached, all thermal motions cease, and any system must approach an ordered state in which the particles do not move. Hence, the entropy of a system is defined only to within an arbitrary constant and only changes in entropy have physical significance. The changes in entropy become negligibly small as absolute zero is approached.

1.8 CONSERVATION OF ENERGY AND MOMENTUM

1.8.1 Conservation of Energy

The moving fluid follows the fundamental laws of mechanics related to the conversation of mass, energy and momentum for its transfer of energy. The energy balance for solar, wind, hydro and wave energy devices are based on the energy transfer of moving fluids (both liquids and gases). Compressibility provides the important point for distinguishing a liquid from gas. A gas is more compressible than a liquid. The flow pattern in fluids are mostly steady, *i.e.* flow does not change with time. Here, we shall consider broadly the incompressible flow.

Bernoulli's equation: The Bernoulli equation is a relation between pressure, velocity and elevation in steady, incompressible, frictionless flow. Despite its simplicity, it has proven to be a very powerful tool in liquid mechanics. A key assumption in the derivation of the Bernoulli equation is to consider the viscous effects to be negligible and thus fluid to be inviscid. Such flows are usually designated as frictionless flow. There is no fluid with zero viscosity and thus this assumption is valid only when the viscous effect is small compared with other effects such as gravity and pressure.

Therefore, care should be exercised when making this assumption. In the absence of frictional effect and less common effects such as surface tension, the fluid motion is governed by the combined effects of pressure and gravity forces.

The motion of a particle and the path it follows are described by the vector velocity as a function of time and space coordinates and the initial position of the particle. When the flow is steady (no change with time at a specific location), all particles that pass through the same point will follow the same path (which is the streamline) and the velocity vectors remain tangent to the path at every point. The flow stays within well defined (though imaginary) stream tubes, *i.e.* tubes bounded by streamlines. Here, we assume no work is done.

Figure 1.9 shows a streamline that rises from a height of h_1 to a height of h_2. The tube is narrow enough that h will probably be consistent over each section of the tube. We consider a control volume bounded by the streamlines and two perpendicular slices across the stream tube at 1 and 2.

A mass $m = \rho A_1 v_1 \Delta t$ enters the control volume at 1 and an equal $m = \rho A_1 v_1 \Delta t$ leaves at 2 due to conservation of mass. For Figure 1.9, the energy balance on the fluid can be written as

Potential energy lost + work done by pressure forces
= kinetic energy gain + heat due to friction

or,

$$mg(h_1 - h_2) + [(p_1 A_1)(v_1 \Delta t) - (p_2 A_2)(v_2 \Delta t)] = \frac{1}{2} m\left(v_1^2 - v_2^2\right) + E_t \tag{1.28}$$

where the pressure forces $p_1 A_1$ and $p_2 A_2$ act through a distance $v_1 \Delta t$ and $v_2 \Delta t$, respectively. E_t is the thermal heat energy generated by the friction in the flow channel.

After neglecting the effect of fluid friction, Eq. (1.28) can be written as

$$\frac{p_1}{\rho} + gh_1 + \frac{1}{2} v_1^2 = \frac{p_2}{\rho} + gh_2 + \frac{1}{2} v_2^2 \tag{1.28a}$$

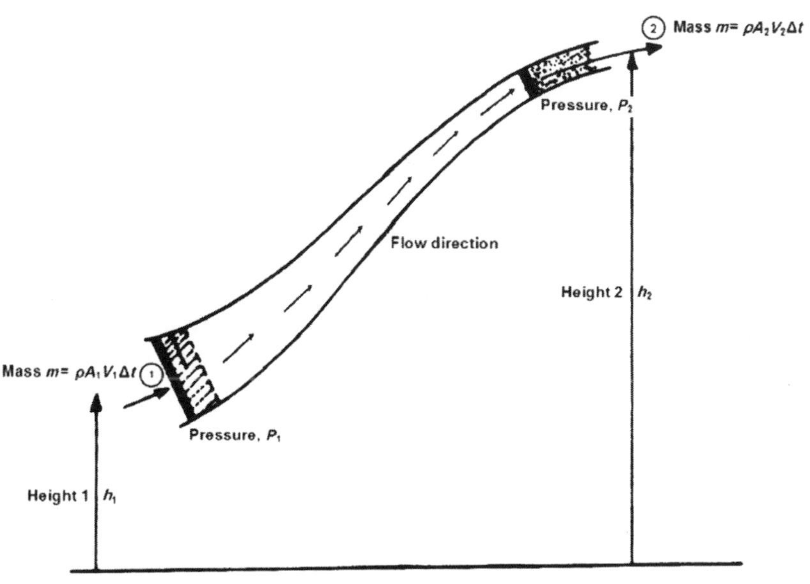

Figure 1.9 Conservation of energy (a stream tube rise from height h_1 to height h_2).

or,

$$\frac{p}{\rho g} + h + \frac{v^2}{2g} = \text{constant along a streamline flow through channel with no loss of energy of any kind}$$

Therefore,

$$\frac{p}{\rho g} + h + \frac{v^2}{2g} = \text{constant} \tag{1.28b}$$

The above equation is termed Bernoulli's equation. In Bernoulli's equation each term has the dimension of length and represents some kind of head. $\frac{p}{\rho g}$ is the pressure head, which represents the height of a fluid column that produces the static pressure p. h is the elevation head, which represents the potential energy of the fluid. $\frac{v^2}{2g}$ is the velocity head, which represents the elevation needed for a fluid to reach the velocity v during frictionless free fall.

Equation (1.28b) can also be written as,

$$\frac{p}{\rho g} + h + \frac{v^2}{2g} = H = \text{constant} \tag{1.28c}$$

where H denotes the total head for the flow. Therefore, Bernoulli's equation can also be expressed in terms of head. The sum of the pressure head, elevation head and velocity heads along with a streamline is constant during steady flow when the compressibility and frictional effects are negligible. Equation (1.28c) can be rewritten as

$$\frac{p}{\rho} + hg + \frac{v^2}{2} = H = \text{constant}$$

Here, $\frac{p}{\rho}$ is pressure energy, hg is potential energy and $\frac{v^2}{2}$ is kinetic energy per unit mass. Therefore, Bernoulli's equation can also be expressed in terms of energy as: the sum of the pressure, potential and kinetic energies of a fluid particle is constant along a streamline during steady flow when compressibility and frictional effects are negligible.

1.8.2 Conservation of Momentum

Newton's second law of motion for particles can be generalised for fluids as **"At any instant in steady flow, the resultant force acting on the moving fluid within a fixed volume of space, equals the next rate of outflow of momentum from the closed surface bounding that volume"**. This is known as the momentum theorem.

Let us consider the fluid passing across a turbine in a pipe as shown in Figure 1.10. The dotted line in the Figure 1.10 shows the control surface over which the momentum theorem is applied.

In Figure 1.10 fluid flowing at speed v_1 into the left of the control surface carries momentum $\rho v_1 \hat{x}$ per unit volume, where $\hat{x}$ is the unit vector in the direction of flow. In time Δt, the volume entering the surface is $A_1 v_1 \Delta t$. Therefore, the rate at which the momentum is entering the control surface is

$$(A_1 v_1 \Delta t)(\rho v_1 \hat{x})/\Delta t = \rho A_1 v_1^2 \hat{x} \tag{1.29a}$$

Similarly, the rate at which momentum is leaving the control volume is $\rho A_2 v_2^2 \hat{x}$. The next rate of outflow momentum can be given as,

$$F_1 = \rho(A_2 v_2^2 - A_1 v_1^2)\hat{x} = (\dot{m}v_2 - \dot{m}v_1)\hat{x} \tag{1.29b}$$

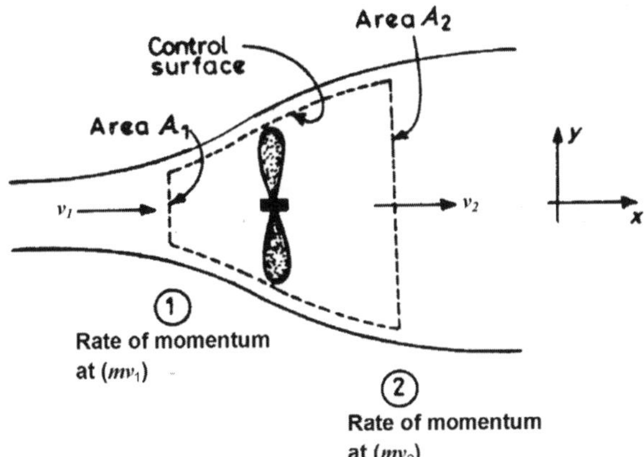

Figure 1.10 Conservation of momentum.

where $\dot{m} = \rho A_1 v_1 = \rho A_2 v_2$ is the mass flow. The momentum theorem tells us that F_1 is the force on the fluid and so, by Newton's third law, $-F_1$ is the force exerted on the turbine and pipe by the fluid. Normally, $v_2 < v_1$, so that F_1 points in the negative x direction and $-F$ (*i.e.* the force on the turbine) is in the direction of flow, as expected.

There are two points to note in applying the momentum theorem: (i) momentum is a vector and (ii) the expression for flow of momentum (e.g., $\rho A_1 v_1^2 \hat{x}$) typically involves products of velocity.

OBJECTIVE QUESTIONS

1.1 Energy is defined as
   (a) The rate of doing work
   (b) The rate of applying force
   (c) The rate of displacement
   (d) None of them

1.2 The unit of energy is
   (a) J
   (b) Wh
   (c) KWh
   (d) all

1.3 Renewable energy has
   (a) Infinite source
   (b) Zero source
   (c) Finite source
   (d) None

1.4 Nonrenewable energy has
   (a) Infinite source
   (b) Zero source
   (c) Finite source
   (d) None

1.5 Fossil fuel has
   (a) Infinite source

(b) Zero source
(c) Finite source
(d) None

1.6 Fossil fuel sources are
(a) Increasing
(b) Depleting
(c) Constant
(d) None

1.7 Chemical energy is the energy
(a) Stored in the chemical bonds of molecules
(b) Due to random motion of particles in solids
(c) Due to elevation of objects in gravitational field
(d) All of them

1.8 Potential energy is the energy
(a) Stored in the chemical bonds of molecules
(b) Due to random motion of particles in solids
(c) Due to elevation of objects in gravitational field
(d) All of them

1.9 Heat energy is the energy
(a) Stored in the chemical bonds of molecules
(b) Due to random motion of particles in solids
(c) Due to elevation of objects in gravitational field
(d) All of them

1.10 Kinetic energy is the energy
(a) Stored in the chemical bonds of molecules
(b) Due to random motion of particles in solids
(c) Due to elevation of objects in gravitational field
(d) Due to motion of an object

1.11 Nuclear energy is the energy
(a) Stored in the chemical bonds of molecules
(b) Due to random motion of particles in solids
(c) Due to elevation of objects in gravitational field
(d) Stored in the nucleus of an atom

1.12 The Sun is the source of
(a) All renewable energy
(b) All nonrenewable energy
(c) Both renewable and nonrenewable energy
(d) All of them

1.13 The cause of increase of CO_2 in environment is burning of
(a) Coal
(b) Oil
(c) Natural gas
(d) All

1.14 The acid rain is due to interaction between
(a) SO_2 and H_2O
(b) NO_2 and H_2O
(c) SO_2 and NO_2
(d) O_3 and H_2O

1.15 The increase of SO_2 in the atmosphere is due to
(a) Burning coal in thermal power plant
(b) Industrial processes

(c) Burning of biomass for domestic uses
(d) All of them

1.16 The main greenhouse gases are
(a) CO_2
(b) CH_4
(c) NO_x
(d) SO_x
(e) None

1.17 Most of infrared radiation is
(a) Blocked by greenhouse molecules
(b) Allowed to escaped by greenhouse molecules
(c) Absorbed by greenhouse molecules
(d) None of them

1.18 The thermal conductivity of material depends on
(a) Temperature
(b) Length
(c) Thickness
(d) None

1.19 The thermal conductivity of insulating material is
(a) Low
(b) High
(c) Infinity
(d) Zero

1.20 The thermal conductivity of conducting material is
(a) Infinite
(b) Zero
(c) Very high
(d) Low

1.21 The heat transfer coefficient is inversely proportional to
(a) Thermal resistance
(b) Thermal conductivity
(c) Thickness
(d) None

1.22 The rate of heat transfer from higher to lower temperature is due to
(a) Conduction
(b) Convection
(c) Radiation
(d) All

1.23 The conductive heat transfer is governed by
(a) Fourier's law
(b) Stefan–Boltzmann law
(c) Wien's displacement law
(d) None

1.24 The radiation heat transfer is governed by
(a) Stefan–Boltzmann's law
(b) Fourier's law
(c) Wien's displacement law
(d) None

1.25 The wavelength of radiation depends inversely to
(a) Temperature
(b) Temperature difference

(c) Area of surface and
(d) None

1.26 Expression for an overall heat transfer coefficient (U) is derived under
(a) Transient conditions
(b) Periodic conductions
(c) Quasisteady state
(d) Steady-state conditions

1.27 The unit of thermal conductance of air is
(a) Same as heat transfer coefficient
(b) Different from heat transfer coefficient
(c) Same as thermal conductivity of air
(d) None of them

1.28 The thermal conduction of air is unaffected for
(a) Larger air cavity
(b) Smaller air cavity
(c) Infinity air cavity
(d) Zero air cavity

1.29 The thermal conductance of air is very large for
(a) Smallest air cavity
(b) Largest air cavity
(c) Zero air cavity
(d) None of them

1.30 The conduction, convection and radiation losses are
(a) Dependent on each other
(b) Independent of each other
(c) Independent of temperature
(d) None of them

1.31 Thermal expansion coefficients depend on
(a) Temperature
(b) Thermal air conductance
(c) Thermal conductivity
(d) None of these

1.32 The convective heat transfer depends on
(a) Physical properties of fluid
(b) Physical properties of solid
(c) Characteristics dimension
(d) All

1.33 The forced convective heat transfer at higher temperature is higher then
(a) Free convective
(b) Conductive
(c) Radiative
(d) Evaporative

1.34 The radiation heat transfer between two surfaces is mainly due to
(a) Short wavelength radiation
(b) Infrared
(c) UV
(d) Long wavelength radiation

1.35 The evaporative heat transfer coefficient (h_{ew}) is
(a) Proportional to h_{cw} (convective heat transfer coefficient)
(b) Inversely proportional to h_{cw}

(c) Independent of convective heat transfer coefficient
(d) None

1.36 The evaporative heat transfer coefficient (h_{ew}) depends on convective heat transfer coefficient due to
(a) Lewis relation
(b) Newton's law
(c) Fourier's law
(d) None

1.37 The shape (geometrical) factor for parallel surfaces is
(a) One
(b) Ten
(c) Less than one
(d) Infinity

1.38 The shape (geometrical) factor for nonparallel surfaces is
(a) One
(b) Less than one
(c) Ten
(d) Infinity

1.39 The radiative heat transfer coefficient for parallel surfaces having a temperature difference by 1 °C is
(a) 6 W/m^2 K
(b) 60 W/m^2 K
(c) 0.6 W/m^2 K
(d) None

1.40 The expression for radiative heat transfer coefficient (h) for a surfaces having temperatures almost the same but different is
(a) $4\varepsilon\sigma T^4$
(b) $4\varepsilon\sigma T^3$
(c) $\frac{1}{4}\varepsilon\sigma T^3$
(d) $0.4\varepsilon\sigma T^3$

1.41 Expression for an overall heat transfer coefficient for single slab is
(a) $U = h_0 + \frac{k}{L} + h_i$
(b) $U^{-1} = \left[\frac{1}{h_0} + \frac{L}{k} + \frac{1}{h_i}\right]$
(c) $U = \frac{1}{h_0} + \frac{L}{k} + \frac{1}{h_i}$
(d) None

1.42 For an inclined surface, an expression for free convective heat transfer coefficient can be obtained from
(a) $\mathrm{Nu} = C(\mathrm{GrPr})^n$
(b) $\mathrm{Nu} = C(\mathrm{GrPr}\sin\theta)^n$
(c) $\mathrm{Nu} = \frac{1}{C}(\mathrm{GrPr})^{\frac{1}{n}}$
(d) $\mathrm{Nu} = C(\mathrm{GrPr}\cos\theta)^n$
Here, $C = 0.54$ and $n = \frac{1}{4}$

1.43 For horizontal surface facing upward, an expression for free convection is
(a) $\mathrm{Nu} = \frac{1}{C}(\mathrm{GrPr})^n$
(b) $\mathrm{Nu} = C(\mathrm{GrPr})^n$
(c) $\mathrm{Nu} = C(\mathrm{GrPr}\cos\theta)^n$
(d) None
Here, $C = 0.54$ and $n = \frac{1}{4}$

1.44 For forced convection, an expression for convective heat transfer coefficient is
(a) $\mathrm{Nu} = C\mathrm{Re}^m\mathrm{Pr}^n \cdot k$
(b) $\mathrm{Nu} = C(\mathrm{Re}\cos\theta)^m\mathrm{Pr}^n \cdot k$
(c) $\mathrm{Nu} = C(\mathrm{Re}^m \cdot (\mathrm{Pr}\cos\theta)^n) \cdot k$
(d) None

1.45 An expression for wind dependent convective heat transfer coefficient is
(a) $h = 3 + 2.8V$
(b) $h = 2.8 + 3V$
(c) $h = 3 + 3V$
(d) None

1.46 An expression for wind dependent convective and radiative heat transfer coefficient is
(a) $h = 3.8 + 5.7V$
(b) $h = 3.8 + 3V$
(c) $h = 5.7 + 3.8V$
(d) None

1.47 The properties of a selective surface are
(a) High value of absorptivity and emissivity of surface
(b) Low value of absorptivity and emissivity of surface
(c) High value of absorptivity and low value of emissivity
(d) None

1.48 The sky temperature with respect to ambient temperature is
(a) Less
(b) More
(c) Equal
(d) None

1.49 The partial vapour pressure depends on temperature
(a) Linearly
(b) Proportionally
(c) Exponentially
(d) None

1.50 The value of heat transfer is increased by
(a) Increasing the volume
(b) Increasing the mass
(c) Decreasing the surface area
(d) Increasing the surface area

1.51 The value of heat transfer of a given surface area is increased by using
(a) Black surface
(b) Reflected surface
(c) Fins
(d) None of these

1.52 The value of heat transfer for a hot surface facing upward is maximum for
(a) Horizontal surface
(b) Inclined surface
(c) Vertical surface
(d) None

1.53 The value of heat transfer is significantly effected by
(a) Streamline flow
(b) Turbulent flow

(c) Steady-state flow
(d) Constant flow rate

1.54 The convective heat transfer coefficient in the case of water in comparison with air as fluid for a hot surface facing upward is
(a) Significantly higher
(b) Equal value
(c) Less value
(d) None of these

1.55 The partial vapour pressure depends on temperature at low operating temperature range up to 40 °C
(a) Nonlinear
(b) Constant
(c) Linear
(d) None

1.56 According to Bernoulli's theorem
(a) Sum of pressure, potential and kinetic energy is constant
(b) Sum of pressure, potential and kinetic energy varies
(c) Sum of pressure, potential and kinetic energy is zero
(d) None of them

1.57 Conservation of momentum for fluids flowing through pipes depends on
(a) Cross-sectional area of pipe and flow velocity
(b) Radius of pipe and flow velocity
(c) Diameter of pipe and flow velocity
(d) All of them

1.58 The second law of thermodynamics states that heat transfer takes place
(a) From a colder body to a hotter body
(b) From a hotter body to a colder body
(c) From same-temperature bodies
(d) None of them

1.59 The second law of thermodynamics is
(a) Reversible
(b) Irreversible
(c) Both
(d) None

1.60 The first law of thermodynamics is
(a) Reversible
(b) Irreversible
(c) Both
(d) None

1.61 Energy conservation process depends on
(a) The first law of thermodynamics
(b) The second law of thermodynamics
(c) Zeroth law of thermodynamics
(d) Third law of thermodynamics

1.62 The existence of temperature depends on
(a) The first law of thermodynamics
(b) The second law of thermodynamics
(c) Zeroth law of thermodynamics
(d) Third law of thermodynamics

ANSWERS

1.1 **(a)**; 1.2 **(d)**; 1.3 **(a)**; 1.4 **(c)**; 1.5 **(c)**; 1.6 **(b)**; 1.7 **(a)**; 1.8 **(c)**; 1.9 **(b)**; 1.10 **(d)**; 1.11 **(d)**; 1.12 **(a)**; 1.13 **(d)**; 1.14 **(a)**; 1.15 **(d)**; 1.16 **(a)**; 1.17 **(a)**; 1.18 **(a)**; 1.19 **(a)**; 1.20 **(c)**; 1.21 **(a)**; 1.22 **(d)**; 1.23 **(a)**; 1.24 **(a)**; 1.25 **(a)**; 1.26 **(d)**; 1.27 **(a)**; 1.28 **(a)** & **(c)**; 1.29 **(a)**; 1.30 **(b)**; 1.31 **(a)**; 1.32 **(a)** & **(c)**; 1.33 **(a)** & **(b)**; 1.34 **(d)**; 1.35 **(a)**; 1.36 **(a)**; 1.37 **(a)**; 1.38 **(b)**; 1.39 **(a)**; 1.40 **(b)**; 1.41 **(b)**; 1.42 **(d)**; 1.43 **(b)**; 1.44 **(a)**; 1.45 **(b)**; 1.46 **(c)**; 1.47 **(c)**; 1.48 **(a)**; 1.49 **(c)**; 1.50 **(d)**; 1.51 **(c)**; 1.52 **(a)**; 1.53 **(b)**; 1.54 **(a)**; 1.55 **(c)**; 1.56 **(a)**; 1.57 **(d)**; 1.58 **(b)**; 1.59 **(b)**; 1.60 **(a)**; 1.61 **(a)**; 1.62 **(c)**.

References

1. Gupta, C. L., *Renew. Sustain. Energy Rev.*, 2003, **7**, 155–174.
2. Gross, R., Leach, M. and Baven, *A. Environ, Int.*, 2003, **29**, 105–122.
3. IEA, World Energy Outlook, International Energy Agency, Paris, 2008.
4. Buran, B., Butler, L., Currano, A. and Smith, E., *Appl. Energy*, 2003, **76**, 89–100.
5. Anon. The United States Central Intelligence Agency. The World Factbook: India, 2001. Available from http:/www.cia.gov/publication/factbook/index.
6. Lynch, R., Available from: http:/www.fe.doe.gov/enternational/indiover.html. An Energy Overview of India, 2001.
7. Tiwari, G. N., *Solar Energy: Fundamental, Design, Modelling and Applications. Narosa Publishing House*, New Delhi and CRC Press, New York, 2002.
8. W. C. Mc Adams, *Heat Transmission*, McGraw Hill, New York, 1954.
9. J. P. Holman, *Heat Transfer*, McGraw Hill Int. (UK) Ltd, 1992.

Solar Energy is a Direct Source of Energy for Renewable Sources and an Indirect Source of Energy for Nonrenewable Sources Too.

CHAPTER 2

Solar Energy

2.1 SOLAR RADIATION

2.1.1 Solar Constant

The orientation of the Earth's orbit around the Sun is such that the Sun–Earth distance varies only by 1.7% and since the solar radiation outside the Earth's atmosphere is nearly of fixed intensities, the radiant energy flux received per second by a surface of unit area held normal to the direction of Sun's rays at the mean Earth–Sun distance, outside the atmosphere, is practically constant throughout the year. This is termed as the solar constant I_{sc} and its value is now adopted to be 1367 W/m^2. However, this extraterrestrial radiation suffers variation due to the fact that the Earth revolves around the Sun not in a circular orbit but follows an elliptic path, with the Sun at one of the foci. The intensity of extraterrestrial radiation measured on a plane normal to the radiation on the nth day of the year (Figure 2.1) is given in terms of solar constant (I_{sc}) as follows:[1,2]

$$I_{ext} = I_{sc}\ [1.0 + 0.033\cos(360n/365)] \tag{2.1}$$

For June 22, 2010, $n = 173$, $I_{ext} = 1322.49\ W/m^2$ (Eq. (2.1))

For December 21, 2010, $n = 355$, $I_{ext} = 1411.43\ W/m^2$ (Eq. (2.1))

It is to be noted that for a leap year, February has 29 days instead of 28 days and accordingly the value of n increases by 1 in Eq. (2.1). For a normal year, the variation of extraterrestrial radiation on the Earth's surface for end of each month using Eq. (2.1) can be shown as follows:

2.1.2 Solar Time

There is not much difference between solar time (ST) and local apparent time (LAT) for Jaipur and New Delhi.

2.1.3 Sun–Earth Angles

The energy flux of beam radiation on a surface with arbitrary orientation can be obtained from the knowledge of flux either on a surface perpendicular to the Sun's rays or on a horizontal surface.

If θ_i is the angle of incidence of a beam of flux I, incident on a plane surface then the flux incident on the plane surface is $I\cos\theta_i$ (Figures 2.2–2.4).

Advanced Renewable Energy Sources
G. N. Tiwari and R. K. Mishra

Published by the Royal Society of Chemistry, www.rsc.org

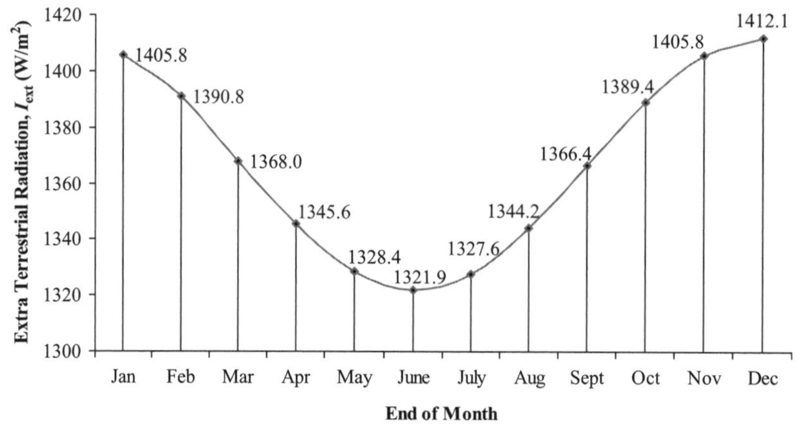

Figure 2.1 Variation of I_{ext} with month of the year.

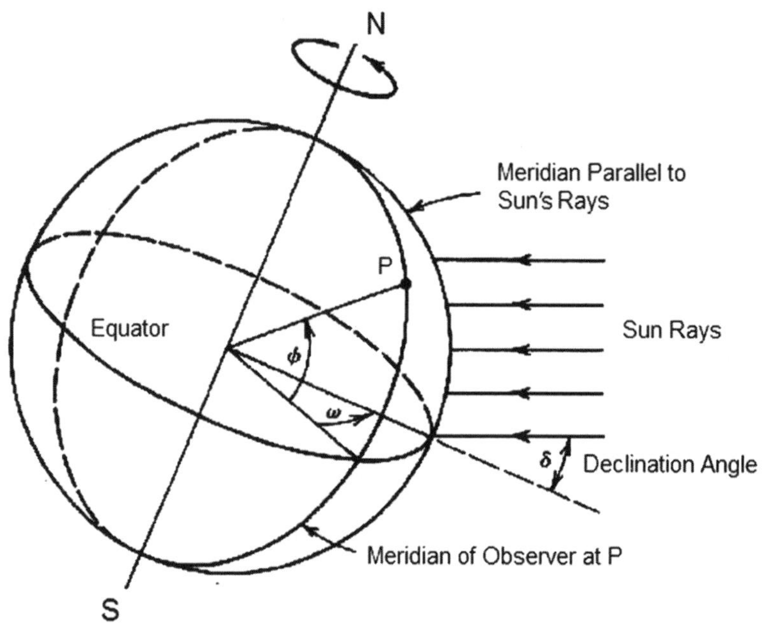

Figure 2.2 Earth always inclined at 23.5° rotating about self-axis.

Latitude (ϕ): The latitude of a location is the angle made by the radial line, joining the given location to the centre of the Earth, with its projection on the equatorial plane. The latitude is positive for the Northern hemisphere and negative for the Southern hemisphere. The latitude for some places in India is given in Table 2.1.

Declination (δ): Declination may be defined as the angle between the line joining the centres of the Sun and the Earth and its projection on the equatorial plane. Declination is due to the rotation of the Earth about an axis that makes an angle of 66.5° with the plane of its rotation around the Sun. The declination varies from a maximum value of 23.45° on June 21 to a minimum value of –23.45° on December, 21. It may be calculated by the following relation.[3]

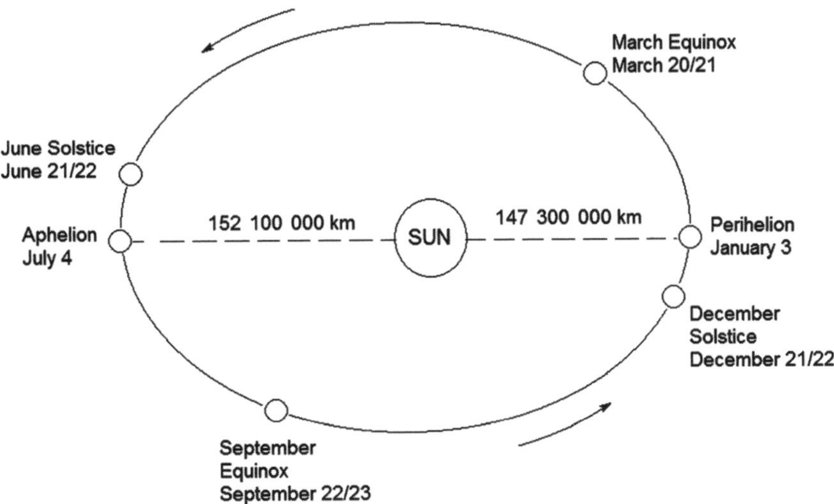

Figure 2.3 Earth orbit around Sun with different positions.

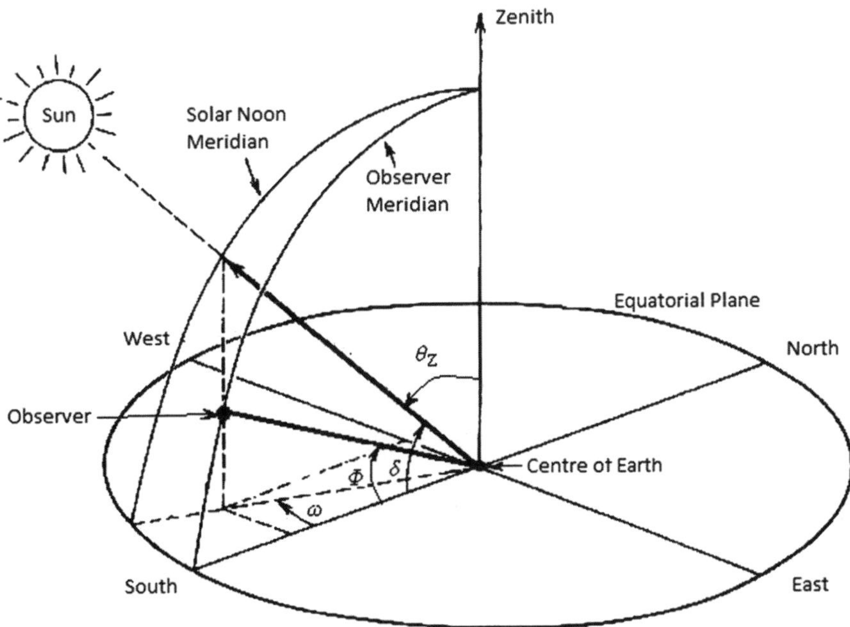

Figure 2.4 View of different Sun–Earth angles.

$$\delta = 23.45 \sin\left[\frac{360}{365}(284 + n)\right] \tag{2.2}$$

The variation of declination angle with the nth day of year is depicted in Figure 2.5.

Hour angle (ω): This is the angle through which the Earth must be rotated to bring the meridian of the plane directly under the Sun. In other words, it is the angular displacement of the Sun east or

Table 2.1 Latitude, longitude and elevation for different places in India.

Place	*Latitude* (Φ)	*Longitude* (L_{loc})	*Elevation* (E_0)
Bangalore	12°58′ N	77°35′ E	921 m above msl
Jaipur	26°55′ N	75°55′ E	431 m above msl
Jodhpur	26°18′ N	73°01′ E	224 m above msl
Mt. Abu	24°36′ N	72°43′ E	1195 m above msl
Mumbai	18°54′ N	72°49′ E	11 m above msl
New Delhi	28°35′ N	77°12′ E	216 m above msl
Simla	31°06′ N	77°10′ E	2202 m above msl
Srinagar	34°05′ N	74°50′ E	1586 m above msl
Kolkata	22°32′ N	88°20′ E	6 m above msl
Chennai	13°00′ N	80°11′ E	16 m above msl

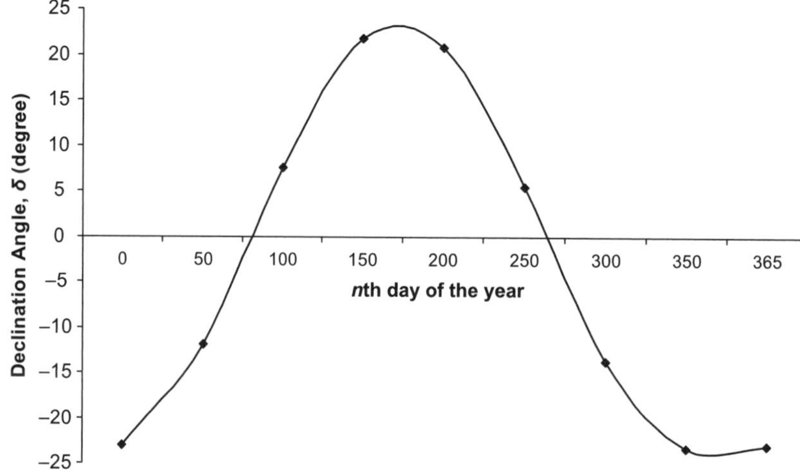

Figure 2.5 Variation of declination angle with *n*th day of the year.

west of the local meridian, due to the rotation of the Earth on its axis at 15° per hour. The hour angle is zero at solar noon, negative in the morning and positive in the afternoon, as shown in Table 2.2 for the Northern hemisphere (India) and *vice versa* for Southern hemisphere (Australia). The expression for the hour angle is

$$\omega = (\mathrm{ST} - 12)\ 15° \tag{2.3}$$

where ST is local solar time.

Zenith (θ_z): This is defined as the angle between the Sun's ray and the line perpendicular to the horizontal plane.

Altitude or solar altitude angle (α): This is defined as the angle between line Sun's rays and a horizontal plane. Also, $\alpha = 90 - \theta_z$.

Slope (β): This is the angle between the plane surface, under consideration, and the horizontal. It is taken to be positive for a surface sloping towards South and negative for surfaces sloping towards North.

Surface azimuth angle (γ): This is the angle in the horizontal plane, between the line due South and the projection of the normal to the surface (inclined plane) on the horizontal plane. By convention, the angle will be taken negative for the Northern hemisphere (India) and *vice versa* for the Southern hemisphere (Australia), if the projection is east of South and positive if west of South. The values of γ for some orientations are given in Table 2.3.

Table 2.2 The value of hour angle with time of the day (For northern hemisphere).

Time of the day (h)	6	7	8	9	10	11	12
Hour angle (°)	−90	−75	−0	−45	−30	−15	0
Time of the day (h)	12	13	14	15	16	17	18
Hour angle (°)	0	+15	+30	+45	+60	+75	+90

Table 2.3 Surface azimuth angle (γ) for various orientations in Northern hemisphere.

Surface orientation	γ
Sloped towards South	0°
Sloped towards North	180°
Sloped towards East	−90°
Sloped towards West	+90°
Sloped towards South-East	−45°
Sloped towards South-West	+45°

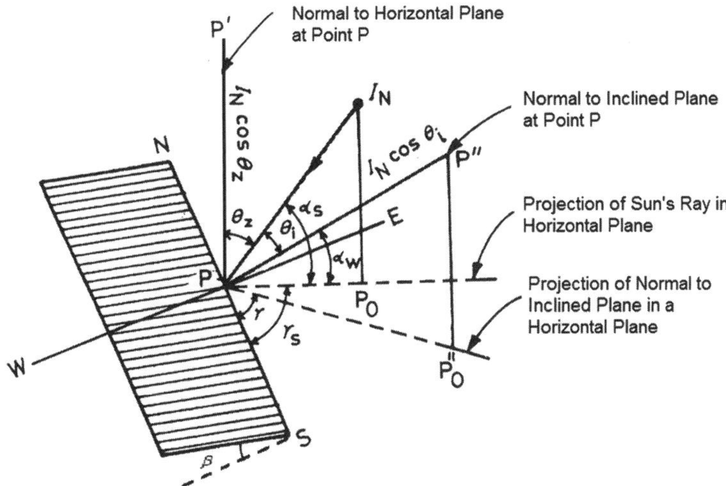

Figure 2.6 View of various Sun–Earth angles for an inclined surface.

Solar azimuth angle (γ_s): This is the angle in a horizontal plane, between the line due South and the projection of beam radiation on the horizontal plane. By convention, the angle is taken to be positive and negative, respectively, if the projection is east of South and west of South for the Northern hemisphere (India) and *vice versa* for the Southern hemisphere.

Angle of incidence (θ_i): This is the angle between beam radiation on a surface and the normal to that surface (Figure 2.6).

In general, the angle of incidence (θ_i) can be expressed as,

$$\begin{aligned}\cos\theta_i = {} & (\cos\varphi\cos\beta + \sin\varphi\sin\beta\,\cos\gamma)\cos\delta\cos\omega \\ & + \cos\delta\sin\omega\sin\beta\sin\gamma + \sin\delta\,(\sin\varphi\cos\beta - \cos\varphi\sin\beta\cos\gamma)\end{aligned} \tag{2.4}$$

For a horizontal plane facing due South, $\gamma = 0$, $\beta = 0$, $\theta = \theta_z$ (zenith angle)

$$\cos\theta_z = \cos\varphi\cos\delta\cos\omega + \sin\delta\sin\varphi \tag{2.5}$$

2.1.4 Solar Radiation on Horizontal Surface

(A) By Using Turbidity Factor (T_R) without Knowing Global and Diffuse Radiation

The value of direct solar radiation in the terrestrial region will depend on the turbidity factor of atmosphere (Table 2.4) and it can be expressed as follows:

$$I_N = I_{ext}\exp\left[\frac{-T_R}{(0.9 + 9.4 \sin\alpha)}\right] \tag{2.6}$$

where, $\alpha = 90 - \theta_Z$, therefore, $\sin\alpha = \cos\theta_z$.

The value of T_R for different weather conditions in India is given in Appendix II.

Normally, the beam radiation (I_b) and diffuse radiation (I_d) on a horizontal surface are recorded. In the case of nonavailability of data for beam and diffuse radiation, the following expression for beam and diffuse radiation on the horizontal surface can be used:

$$I_b = I_N \cos\theta_z \tag{2.7a}$$

$$I_d = \frac{1}{3}[I_{ext} - I_N]\cos\theta_z \tag{2.7b}$$

where expressions for I_{ext} and I_N are given by Eqs. (2.1) and (2.6), respectively.

The program for evaluating beam and diffuse radiation on a horizontal surface using the turbidity factor, in Jaipur is given in Table 2.5.

Table 2.4 The turbidity factor (T_R) for different months for blue sky condition.

Months Region	*1*	*2*	*3*	*4*	*5*	*6*	*7*	*8*	*9*	*10*	*11*	*12*
Mountain	1.8	1.9	2.1	2.2	2.4	2.7	2.7	2.7	2.5	2.1	1.9	1.8
Flat land	2.2	2.2	2.5	2.9	3.2	3.4	3.5	3.3	2.9	2.6	2.3	2.2
City	3.1	3.2	3.5	3.9	4.1	4.2	4.3	4.2	3.9	3.6	3.3	3.1

For cloudy condition the values of T_R depends on sky condition.

Table 2.5 Calculation of Beam and Diffuse Radiation from Turbidity Factor.

```
Phi=26.916;                                              % for Jaipur
n=173;                                                   % for June 22, 2010
TR=2.7                                                   % Turbidity Factor
Isc=1367;                                                % solar constant
Iext=Isc*(1+0.033*cos((360*n/365)*(pi/180)));            % Extra Terrestrial solar radiation
Del=23.45*sin((360/365)*(284+n));                        % Hour angle from 7:00 AM to 5:00 PM
Omega=[-75 -60 -45 -30 -15 0 15 30 45 60 75];
t=7:1:17;                                                % Time of day from 7:00 AM to 5:00 PM
CTZ=cos(phi*pi/180)*cos(del*pi/180)*cos(omega*pi/180)
  +sin(phi*pi/180)*sin(del*pi/180);
for i=1:11;
In(i)=Iext*exp((-TR/(0.9+9.4.*CTZ(i))));
end
for j=1:11;
Ib(j)=In(j).*CTZ(j);
Id(j)=(1/3)*(Iext-In(j)).*(CTZ(j));
End
plot(t,Ib)                                               %Plot of Beam Radiation vs. Time
```

(B) Known Data of Beam and Diffuse Radiation

Figures 2.7a and 2.7b show the hourly variation of beam and diffuse radiation. The hourly and monthly data for beam and diffuse radiation have been generated on the basis of ten years data obtained from IMD, Pune. The results are shown in Figures 2.8a–f.

There is good agreement between the generated data of beam and solar radiation in subsection (A) from the turbidity factor and known solar radiation data from Indian Meteorological Data (IMD), Pune. Low percentage error with high correlation coefficient was found between predicted and measured data for the month of December due to clear sky conditions in winter months as compared to cloudy sky conditions in June (*i.e.* with varying turbidity factor), which can be taken from Figure 1.7a and Figure 1.8a for the measuring location at Jaipur (Rajasthan, India).

2.2 SOLAR RADIATION ON AN INCLINED SURFACE

2.2.1 Hourly Variation

The total solar radiation incident on a surface consists of (i) beam solar radiation (ii) diffuse solar radiation and (iii) solar radiation reflected from the ground and the surroundings. After

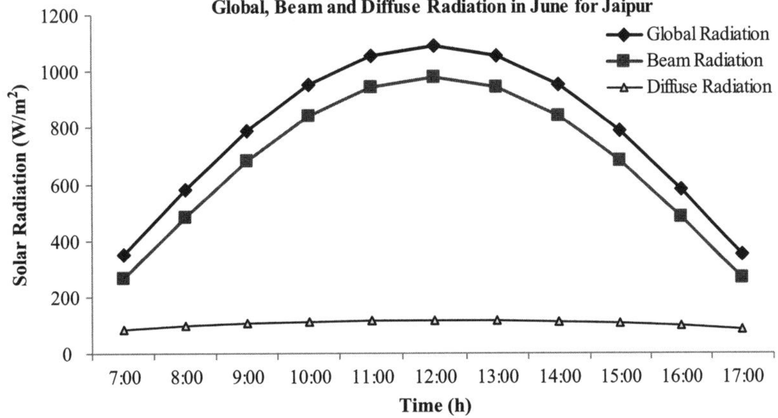

Figure 2.7a Hourly variation of beam and diffuse radiation (June 22, 2010).

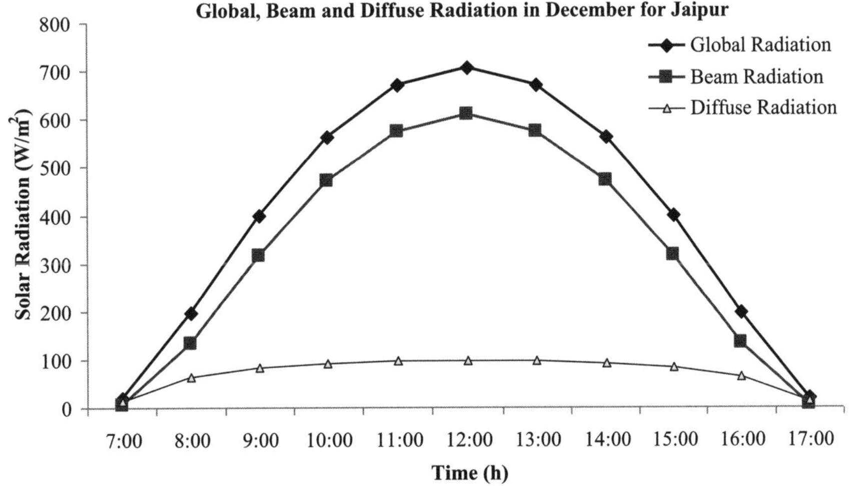

Figure 2.7b Hourly variation of beam and diffuse radiation (December 21, 2010).

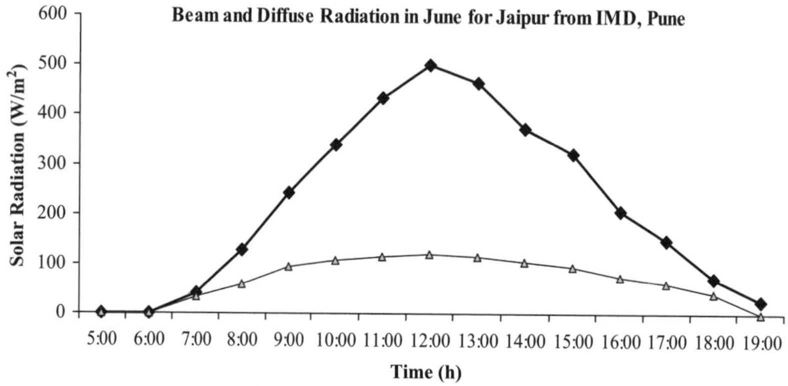

Figure 2.8a Hourly variation of beam and diffuse radiation (June, 2010) (Source, IMD).

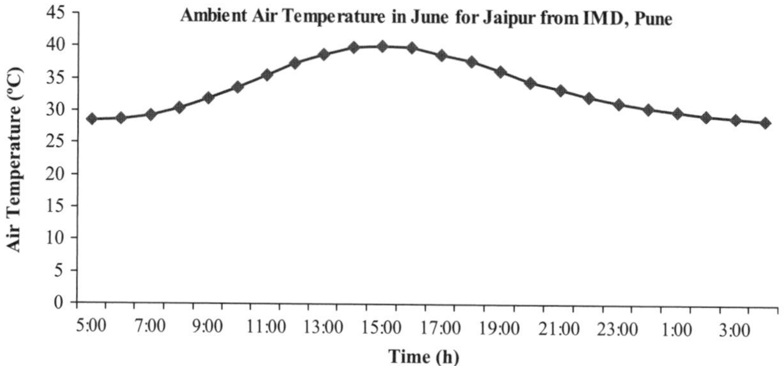

Figure 2.8b Hourly variation of ambient air temperature (June, 2010) (Source, IMD).

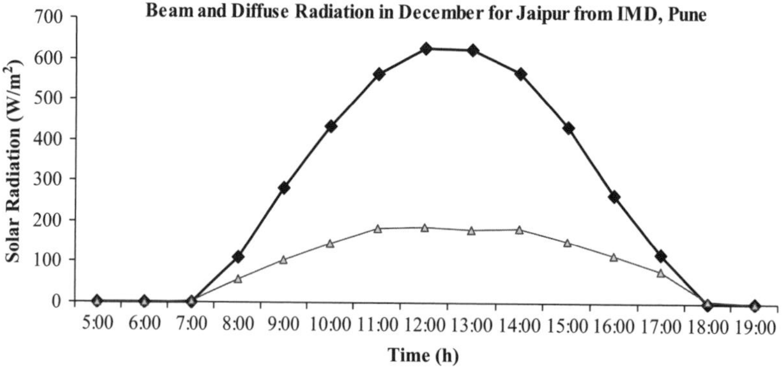

Figure 2.8c Hourly variation of beam and diffuse radiation and ambient air temperature (December 21, 2010) (Source, IMD).

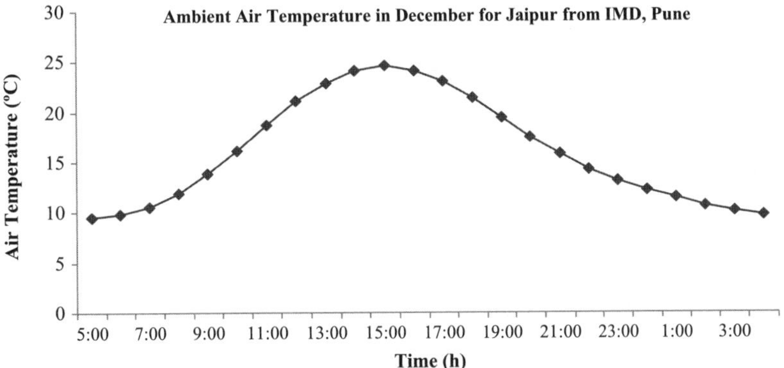

Figure 2.8d Hourly variation of ambient air temperature (December 21, 2010) (Source, IMD).

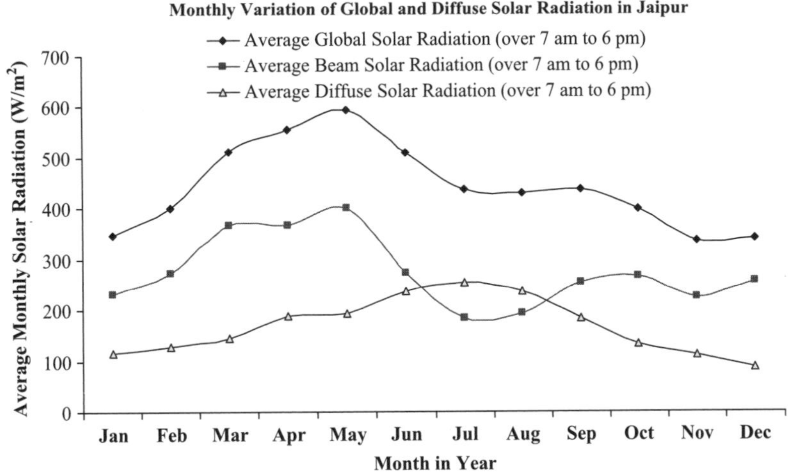

Figure 2.8e Monthly data of beam and diffuse radiation.

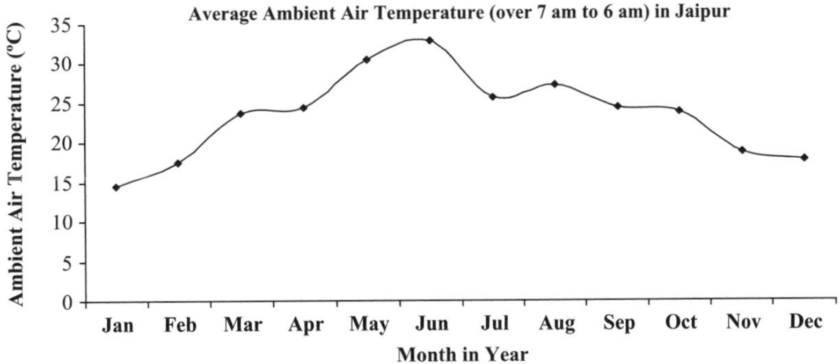

Figure 2.8f Monthly data of ambient air temperature.

determining the beam and diffuse radiation on horizontal surface, Liu and Jordan gave a formula to evaluate total radiation on a surface of arbitrary orientation.[3]

$$I = I_b R_b + I_d R_d + \rho R_r (I_b + I_d) \qquad (2.8)$$

where R_b, R_d and R_r are known as conversion factors for beam, diffuse and reflected components respectively, and ρ is the reflection coefficient of the ground (=0.2 and 0.6 for ordinary and snow covered ground, respectively). The expressions for R_b, R_d and R_r are given below:

i) $\boldsymbol{R_b}$: This is defined as the ratio of flux of beam radiation incident on an inclined surface (Eq. (2.4)) to that on a horizontal surface (Eq. (2.5)). Now, R_b, for beam radiation can be obtained as,

$$R_b = \frac{I'_b}{I_b} = \frac{\cos\theta_i}{\cos\theta_z} \qquad (2.9)$$

Depending on the orientation of the inclined surface, the expression for $\cos\theta_i$ and $\cos\theta_z$ can be obtained from Eqs. (2.4) and (2.5).

ii) $\boldsymbol{R_d}$: This is defined as the ratio of the flux of diffuse radiation falling on the tilted surface to that on the horizontal surface.

This conversion factor depends on the distribution of diffuse radiation over the sky and on the portion of sky seen by the surface. But a satisfactory method of estimating the distribution of diffuse radiation over the sky is yet to be found. It is, however, widely accepted that sky is an isotropic source of diffuse radiation. If $(1+\cos\beta)/2$ is the radiation shape factor for a tilted surface with respect to sky, then

$$R_d = \frac{1+\cos\beta}{2} \qquad (2.10)$$

iii) $\boldsymbol{R_r}$: The reflected component comes mainly from the ground and other surrounding objects. If the considered reflected radiation is diffuse and isotropic, then the situation is opposite to that in the above case.

$$R_r = \left(\frac{1-\cos\beta}{2}\right) \qquad (2.11)$$

It may be mentioned here that both the beam and diffuse components of radiation undergo reflection from the ground and the surroundings.

2.2.2 Solar Radiation and Sol Air Temperature on Building Surfaces

Figure 2.9 shows a schematic view of a building with a single sloped roof and a horizontal roof (dotted line) with dimensions of walls and roofs for Jaipur. In order to determine the total solar radiation on building cover, it is necessary to determine the total solar radiation on each of the walls and roof of the building by using Eq. (2.12). There are four vertical walls and one roof ($i=5$). The roof is also made horizontal for some cases, as shown by the dotted line in Figure 2.9. Hence, the total solar radiation on building surfaces can be obtained by using the following expression:

$$\text{Total Solar Radiation} = \sum_{i=1}^{i=5} A_i I_i \qquad (2.12)$$

where A_i and I_i are an area of the ith section, and total solar radiation available on the ith section are shown in Figures 2.10a and b.

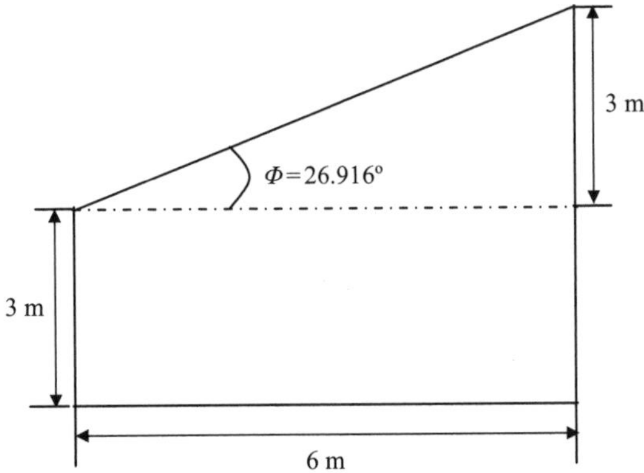

Figure 2.9 Schematic views of single slope and horizontal roof building.

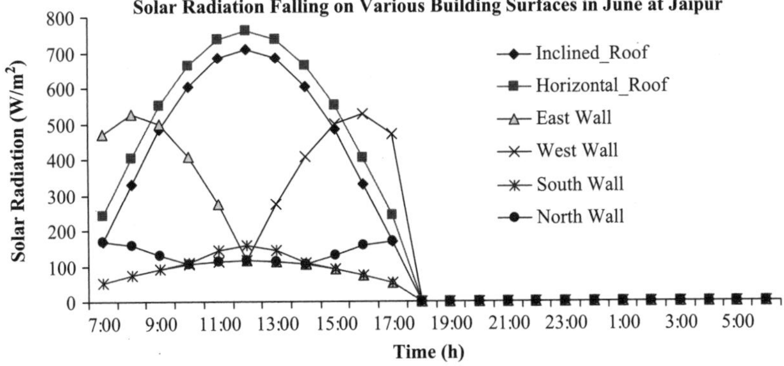

Figure 2.10a Solar radiation on different building surfaces in June.

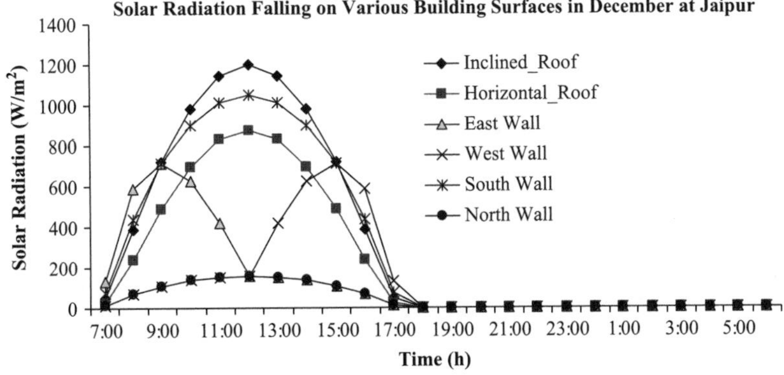

Figure 2.10b Solar radiation on different building surfaces in December.

In this section, T_{sa} is a sol air temperature and an expression for this can be written as,

$$T_{sa} = \frac{\alpha}{h_o} I(t) + T_a - \frac{\varepsilon \Delta R}{h_o} \quad \text{for bare surface of wall or roof} \tag{2.13}$$

where α is the absorptivity of the wall or the roof surface, is the ambient air temperature and T_a is the solar radiation on the building surface.

The term $\varepsilon\Delta R$ represents longwave radiation from the roof or wall surface of the building and its value for a vertical wall is taken as 60 W/m^2 and 0 W/m^2 for a horizontal roof. The longwave radiation for inclined roof shown in Figure 2.9 is taken as

$$\varepsilon\Delta R \times \cos\beta$$

where β = slope of inclined surface = φ

$h_o = h_r + h_c$,

h_o = Outside heat transfer coefficient of building

h_r = Radiative heat transfer coefficient of building

h_c = Convective heat transfer coefficient of building

The variation of sol air temperature on outside building surfaces are plotted and shown in Figures 2.11. The value of sol air is found to be directly dependent on the solar radiation falling on the outside surface of a building.

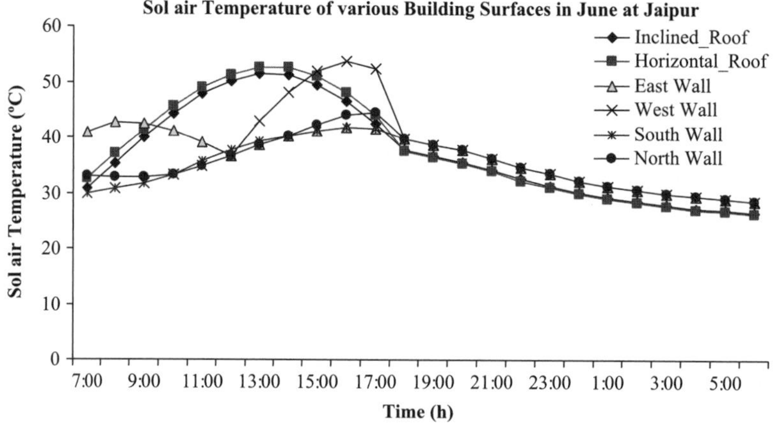

Figure 2.11a Sol air temperatures on different building surfaces in June.

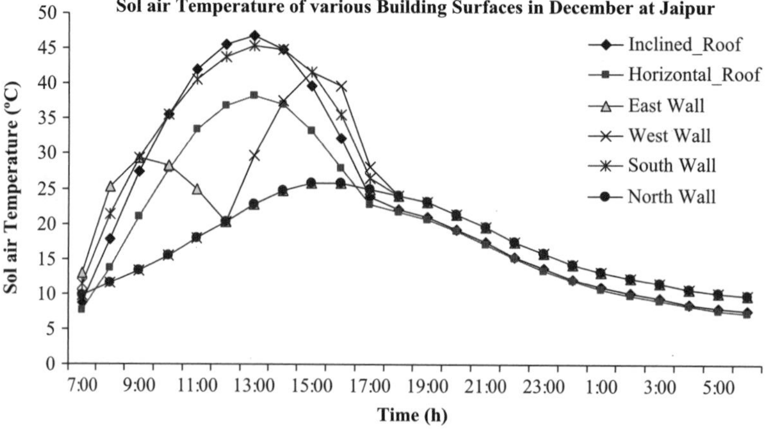

Figure 2.11b Sol air temperatures on different building surfaces in December.

The program for evaluating solar radiation and sol air temperature on a building surface is given in Table 2.6, which is used for obtaining Figures 2.10 and 2.11.

Table 2.6 Calculation of Solar Radiation and Sol air Temperatures on Building Surfaces in Matlab.

```
%Jaipur
phi = 26.916*(pi/180);
Isc = 1367;
omega = [−75 −60 −45 −30 −15 0 15 30 45 60 75 105 105 105 105 105 105 105 105 105 105 105 105 105]*pi/180;;
% 7:00 AM to 6:00 AM
t = 7:1:17;
n = 173; %nth day of the year (June 22, 2010, n = 173, December 21, 2010, n = 355)
rho = 0.2; % Reflectivity of surface
TR = 3.1; % Turbidity of surface TR = 3.1 in June,
TR = 2.7 in December
delta = (23.45*sin((360/365)*(n + 284)*pi/180))*pi/180;
Iext = Isc*(1 + 0.033*cos(2*pi*n/365));
for j = 1:1:24
CThz(1,j) = cos(phi)*cos(delta).*cos(omega(1,j)) + sin(delta)*sin(phi);
if CThz(1,j) < 0;
CThz(1,j) = 0;
end
if omega(1,j) > 90;
CThz(1,j) = 0;
end
end
CThz1 = repmat(CThz,24,1);
for j = 1:1:24
In(1,j) = Iext.*exp(-TR/((0.9 + 9.4.*CThz(1,j))));
Ib(1,j) = In(1,j).*CThz1(j,j);
Id(1,j) = (1/3)*(Iext-In(1,j)).*CThz1(j,j);
end
Ib1 = repmat(Ib,24,1);
Id1 = repmat(Id,24,1);
Ta = [28.5 28.7 29.3 30.4 31.9 33.6 35.6 37.4 38.8 39.8 40.1 39.8 38.7 37.7 36.2 34.6 33.4 32.2 31.3 30.6 30 29.4 29
  28.6];% Jaipur in June
%Ta = [9.5 9.8 10.6 11.9 13.9 16.1 18.7 21.1 22.9 24.1 24.6 24.1 23.1 21.4 19.5 17.5 15.8 14.3 13.1 12.2 11.5 10.7
  10.1 9.7];% Jaipur in December
gamma = [0 0 −90 90 0 −180]*pi/180;
%[InclinedRoof Horizontal_Roof East West South North];
beta = [26.916 0 90 90 90 90]*pi/180;
%[InclinedRoof Horizontal_Roof East West South North];
for k = 1:1:6
for j = 1:1:24
Thz(j,k) = cos(phi)*cos(delta).*cos(omega(j)) + sin(delta)*sin(phi);
Thi(j,k) = (cos(phi)*cos(beta(k)) + sin(phi)*sin(beta(k))*cos(gamma(k)))*cos(delta).*cos(omega(j)) + cos(delta).*sin
  (omega(j))*sin(beta(k))*sin(gamma(k)) + sin(delta)*(sin(phi)*cos(beta(k))−cos(phi)*sin(beta(k))*cos(gamma(k)));
Rb(j,k) = Thi(j,k)./Thz(j,k);
num = size(Rb(j,k));
for i = 1:1:num;
if Rb(j,k) < 0;
Rb(j,k) = 0;
end
Rd(j,k) = (1 + cos(beta(k)))/2;
Rr(j,k) = rho*(1−cos(beta(k)))/2;
It(j,k) = Ib1(j,j).*Rb(j,k) + Id1(j,j).*Rd(j,k) + (Ib1(j,j) + Id1(j,j)).*Rr(j,k);
end
end
end
ho = 22.78; % Outside Heat Transfer Coefficient = 5.7 + 3.8v
```

Table 2.6 (*Continued*).

```
alpha = 0.6; % Absorptivity of surface
em = 0.9; % emissivity of surface
DeltaR_H = 60; % radiative value for horizontal surface
DeltaR_V = 60; % radiative value for verticle surface
Taa(1:24) = transpose(Ta);
% Sol-Air temperature calculations
Tsr1(1:24) = (alpha*It(1:24,1))*(1/ho) + Taa(1:24,1)-(em*DeltaR_V)/ho; %InclinedRoof Surface
Tsr(1:24) = (alpha*It(1:24,2))*(1/ho) + Taa(1:24,1)-(em*DeltaR_V)/ho;
%Horizontal Roof Surface
Tse(1:24) = (alpha*It(1:24,3))*(1/ho) + Taa(1:24,1)-(em*DeltaR_H)/ho;
%East wall
Tsw(1:24) = (alpha*It(1:24,4))*(1/ho) + Taa(1:24,1)-(em*DeltaR_H)/ho;
%West wall
Tss(1:24) = (alpha*It(1:24,5))*(1/ho) + Taa(1:24,1)-(em*DeltaR_H)/ho;
%South wall
Tsn(1:24) = (alpha*It(1:24,6))*(1/ho) + Taa(1:24,1)-(em*DeltaR_H)/ho;
%North wall
plot(t,Tsr) %Horizontal Roof Surface vs. Time
plot(t,Tse) %East wall vs. Time
plot(t,Tsw) %West wall vs. Time
plot(t,Tss) %South wall vs. Time
plot(t,Tsn) %North wall vs. Time
```

2.3 FLAT PLATE COLLECTOR (FPC)

The flat plate collector (FPC) (Figure 2.12) is the heart of any solar energy collection system designed for operation in the low (ambient 60 °C) or medium temperature range (ambient 100 °C). It is used to absorb solar energy, convert it into heat and then transfer that heat to a stream of liquid/gas. It absorbs both the beam and the diffuse radiation. It is usually placed on the top of a building or any structure with optimum inclination to receive the maximum solar radiation. It does not require tracking of the Sun. It requires little maintenance.

A FPC usually consists of the following components:

i) **Glazing:** This may be one or more sheets of glass or some other radiation-transmitting material.
ii) **Tubes/fins:** These are made of conducting material. It directs the heat transfer fluid from the inlet to the outlet.
iii) **Absorber plate:** This may be flat, corrugated or grooved with tubes/fins.
iv) **Header or manifolds:** This admits and discharges the fluid.
v) **Insulation:** This minimises heat loss from the back and sides of the FPC.
vi) **Container or casing:** This surrounds the various components and protects them from dust and moisture, *etc.*

2.3.1 Collector/Absorber Plate

The most important part of the FPC is the absorber plate made up of either copper, aluminium or steel integrated with the pipe or duct. The liquid or air passes through the pipe/duct that is in thermal contact with the plate for transferring thermal energy from the absorber to the fluid/gas. The function of the collector/absorber plate is to absorb the maximum possible solar radiation incident on it through the glazing for minimum heat loss to the atmosphere from the top surface. It is important to mention that the areas of absorber and glass cover are the same in the case of a FPC (*i.e.* $A_c = A_g$). The coating of the absorber plate should be such that it has high absorptivity and poor emissivity for the required temperature range. Selective surfaces are particularly important when the collector surface temperature is much higher than the ambient air temperature. The

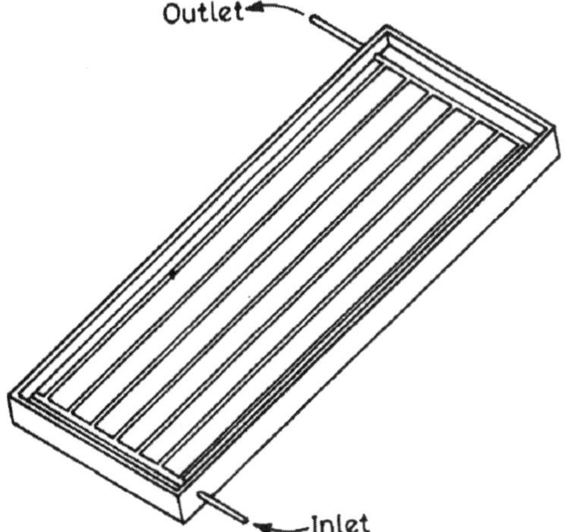

Figure 2.12 Complete view of FPC.

bottom and sides of the collector are covered with insulation to reduce the bottom/side conductive heat losses.

2.3.2 Basic Energy Balance Equation

The useful thermal energy output per unit time of a FPC of area A_c is the difference between the absorbed solar radiation, q_{ab}, and the thermal loss. It is given by,

$$\dot{Q}_u = A_c\,\dot{q}_u = A_c[\dot{q}_{ab} - U_L(T_p - T_a)] \tag{2.14}$$

$$\dot{q}_{ab} = (\tau_0\,\alpha_0)I(t)$$

where $I(t)$ is the solar intensity incident on a FPC, τ_0 and α_0 are the transmissivity and absorptivity, respectively. The expression for U_L, the overall heat loss coefficient.

The thermal instantaneous efficiency (η) of a FPC is given as,

$$\eta = \frac{\dot{Q}_u}{A_c\,I(t)} = \frac{\dot{q}_{ab}}{I(t)} - \frac{U_L(T_p - T_a)}{I(t)} \tag{2.15}$$

The overall thermal collection efficiency of a FPC (the ratio of the daily useful gain to the daily incident solar energy) is given by,

$$\eta_c = \frac{\int \dot{Q}_u \mathrm{d}t}{A_c \int I(t)\mathrm{d}t} \tag{2.16}$$

2.3.3 Collector Efficiency Factor, F'

The FPC efficiency factor, F', is defined as the ratio of actual rate of useful heat collection to the rate of useful heat collection rate when the collector absorbing plate (T_p) is placed at the local fluid temperature (T_f), *i.e.*

$$F' = \frac{\dot{Q}_{useful}}{\dot{Q}_u(T_p = T_f)} = \frac{\dot{Q}_{useful}}{A_c[\dot{q}_{ab} - U_L(T_f - T_a)]}$$

or,

$$\dot{Q}_{useful} = F'\,A_c[\dot{q}_{ab} - U_L(T_f - T_a)] \tag{2.17}$$

2.3.4 The Outlet Fluid Temperature of FPC (T_{fo})

The outlet fluid temperatures (T_{fo} at $x = L_r$) from a FPC can be obtained as,

$$T_{fo} = T_f(x = L_r) = [(\dot{q}_{ab}/U_L) + I_a] + [T_{fi} - T_a - (\dot{q}_{ab}/U_L)]\ \exp[-A_c\, U_L\, F'/(\dot{m}\, C_f)] \tag{2.18}$$

where A_c ($= n_o\, W\, L_r$) is the collector area and L_r the length of riser in the flow direction.

2.3.5 Collector Heat Removal Factor, F_R

The FPC heat removal factor (F_R) is defined as the ratio of the actual useful energy gain to the useful energy gain if the entire FPC were at the fluid inlet temperature (T_{fi}) (forced circulation flow). It can be expressed as,

$$F_R = \frac{\dot{m}\, C_f\, (T_{fo} - T_{fi})}{A_c\, [\,\dot{q}_{ab} - U_L\, (T_{fi} - T_a)]} \tag{2.19a}$$

The above equation can further be written, Tiwari (2002), as,

$$F_R = [\dot{m}\, C_f\, /(A_c\, U_L)][1 - \exp(-\, A_c\, U_L\, F'/(\dot{m} C_f)] \tag{2.19b}$$

From Eq. (2.19a), the actual rate of useful energy collected by fluid is given by,

$$\dot{Q}_{useful} = A_c\, F_R[\dot{q}_{ab} - U_L(T_{fi} - T_a)] \tag{2.19c}$$

From Eq. (2.19c), an instantaneous thermal efficiency (η_i) of a FPC can be defined as follows:

$$\eta_i = \frac{\dot{Q}_{useful}}{A_c\, I(t)} = F_R\left[(\alpha\tau) - U_L\frac{T_{fi} - T_a}{I(t)}\right] \tag{2.20}$$

2.3.6 Outlet Fluid Temperature at the *N*th Collector

For a number of collectors (FPC) connected in series as shown in Figure 2.13, the outlet fluid temperature of the first FPC is the inlet temperature of the second FPC; the outlet temperature of the second is the inlet temperature of the third FPC and so on. Hence, for a system of N collectors connected in series, the outlet fluid temperature from the Nth collector can be derived with the help of Eq. (2.18) in terms of the inlet temperature of the first collector. An expression for the outlet fluid temperature is given below.

$$T_{foN} = \left(\frac{\dot{q}_{ab}}{U_L} + T_a\right)\left\{1 - \exp\left(-\frac{NA_cU_LF'}{\dot{m}C_f}\right)\right\} + T_{fi}\ \exp\left\{-\frac{NA_cU_LF'}{\dot{m}C_f}\right\} \tag{2.21}$$

The rate of useful energy at the end of Nth collector is given by,

$$\dot{Q}_{UN} = \dot{m}C_f(T_{foN} - T_{fi}) \tag{2.22}$$

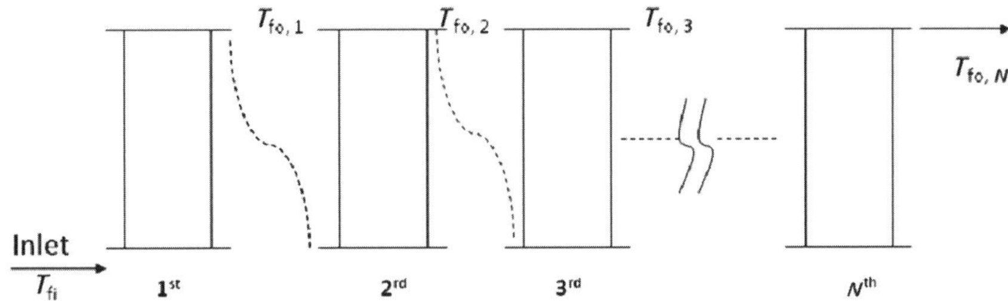

Figure 2.13 N-FPC connected in series.

2.3.7 Optimum Inclination of a Flat Plate Collector

As shown in Figures 2.10 (Eq. 2.8), there is a variation of solar radiation with inclination of the surface for given latitude, orientation, the time of the day and day of the year. Hence, there should be an optimum inclination to receive maximum radiation. On the basis of a literature survey an optimum inclination of the surface receiving maximum radiation for winter/summer condition is given by,

$$\beta_{\text{optimum}} = \varphi \pm 15^\circ \tag{2.23}$$

where the +ve sign refers to winter conditions and the −ve sign refers to summer conditions.

For the year-round performance, the optimum tilt is 0.9 times the latitude of the location.

2.4 EVACUATED TUBULAR COLLECTOR (ETC)

In an evacuated tubular collector, the vacuum is created between the absorber and transparent glass cover. This is briefly discussed in the following sections.

2.4.1 Solaron Collector

Figure 2.14a shows a cross-sectional view of a Solaron collector. The tube cover above the selective surface is evacuated. In this case, evacuated tubes are arranged above the selective surface/absorber so that there should not be any space left between consecutive tubes. The evacuated tubes provide a vacuum layer above the absorber to reduce the top loss coefficient. The vacuum layer suppresses the convection heat loss from the absorber to glass cover. Similar to the FPC, an incident solar radiation is absorbed by a selectively coated surface/absorber after transmission through the glass cover and transparent evacuated tubes, as shown in Figure 2.14a. After absorption, most of the available thermal energy at the selective surface/absorber will be first conducted and then convected to the working fluid below the surface/absorber. The rest of the absorbed thermal energy is lost to the upper portion of the evacuated tubes by radiative heat loss. Further, there will be convective and radiative heat losses from the upper portion of the evacuated tubes to the glass cover. Since, the temperature of the upper portion of the evacuated tubes will be small; there will be small heat losses. The working fluid may be either a liquid fluid (say water) or air. The temperature of the working fluid in this case will be more in comparison to the fluid temperature of a conventional FPC due to reduced upward heat loss.

2.4.2 Phillips (Germany) Collector

The performance of a Solaron collector with an evacuated cover can be further increased by increasing the convective heat transfer from the absorber to the working fluid through the surface

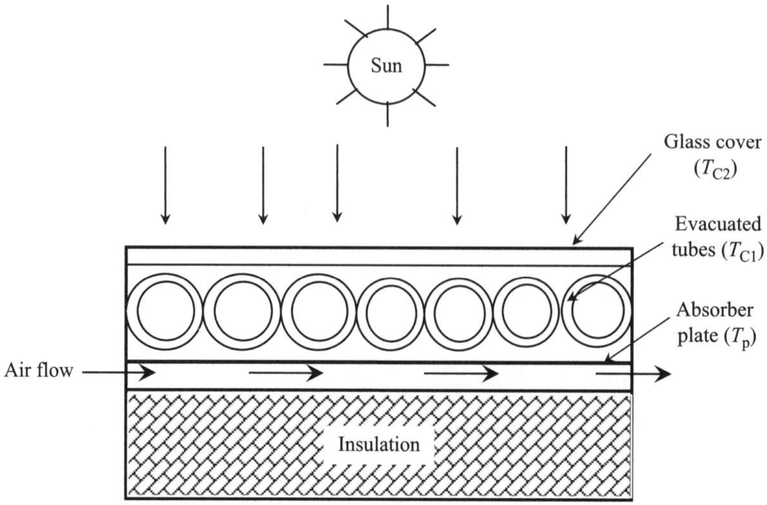

Figure 2.14a View of solaron collector with evacuated tube cover.

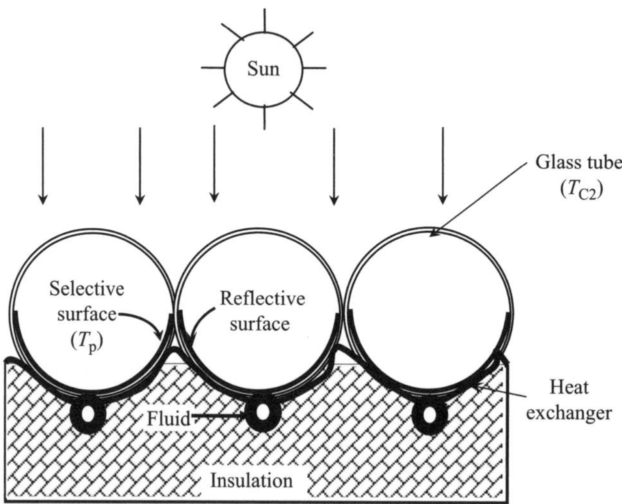

Figure 2.14b View of Phillips collector with evacuated tube cover.

area of the absorber, as shown in Figure 2.14b. In this case, the working fluid passes through the tubes attached at the bottom of semicircular absorber. The rise in the temperature of the working fluid becomes more due to the lower value of the heat capacity of the working fluid. The curved selective surface absorber acts as a heat exchanger. The top surface of the evacuated tubes is directly exposed to solar radiation, unlike the Solaron collector.

Solar radiation is transmitted after reflection from the outer curved portion of the glass tube to the inside vacuum space. It is finally absorbed by the curved selective surface after reflection. The reflected radiation is further transmitted to atmosphere through the curved outer portion of the tube. Most of the absorbed thermal energy is transferred to the working fluid through conduction and convection. The rest is lost to atmosphere through radiation, conduction and convection.

2.4.3 Thermal Efficiency

The efficiency of the evacuated cover collector (Figure 2.14b) can be written as

$$\eta = \alpha(1-\rho) - \frac{\varepsilon\sigma\left(T_p^4 - T_c^4\right)A_a}{IA_c} - \frac{q_f}{IA_c} \tag{2.24}$$

and, the rate of energy balance for cover will be

$$\varepsilon\,\sigma\left({T_p}^4 - {T_c}^4\right)A_a = h_c(T_c - T_a)A_a \tag{2.25}$$

where

α = absorptance of the absorbing surface,
ρ = fraction of incident radiation reflected by the glass tube,
σ = Stefan–Boltzmann constant (5.67×10^{-8} W/m^2 K),
ε = emmittance of the absorbing surface at the collector temperature
A_a = area of the absorber surface (m^2),
A_c = aperture area of the collector (m^2),
$h_c = 5.7 + 3.8\ V$,
I = solar radiation on the collector (W/m^2),
q_f = rate of nonradiant heat loss (W/m^2),
T_p = temperature of the absorbing surface (K),
T_c = temperature of the glass tube (K),
T_a = temperature of ambient air (K),
V = wind velocity (m/s).

If the evacuated tubular collector is covered with a glass cover as shown in Figure 2.14a, then there will be another energy balance equation for the cover that can be written as follows:

For cover I

$$\varepsilon\sigma\left(T_p^4 - T_{c1}^4\right)A_a = h_1(T_{c1} - T_{c2})A_c \tag{2.26}$$

For cover II

$$h_1(T_{c1} - T_{c2})A_c = h_c(T_{c2} - T_a)A_c \tag{2.27}$$

where h_1 is the sum of the convective and radiative heat transfer coefficients between cover I and cover II, respectively.

In the remaining collector, each tube acts as a collector and then connected either in parallel or in series to make a module with an effective area equal to a conventional FPC for comparison.

2.4.4 An Instantaneous Thermal Efficiency (η) for Evacuated Tubular Collector (ETC)

By linearising the right-hand side of Eq. (2.25), one gets

$$\varepsilon\,\sigma\left({T_p}^4 - {T_c}^4\right)A_a = h_r\left(T_p - T_c\right)A_a = h_c(T_c - T_a)A_a \tag{2.28}$$

The above equation can further be written as

$$h_r\left(T_p - T_c\right) = U_t\left(T_p - T_a\right) \tag{2.29}$$

where

$$h_r = \varepsilon\sigma\left(T_p + T_c\right)\left(T_p^2 + T_c^2\right) \text{ and } U_t = \frac{h_r h_c}{h_r + h_c}$$

Substitute the value of $h_r\left(T_p - T_c\right)$ in Eq. (2.24), one gets

$$\eta = \alpha(1-\rho) - \frac{U_t(T_p - T_a)A_a}{IA_c} - \frac{q_f}{IA_c} \tag{2.30}$$

If $q_f = U_b(T_p - T_a)$, the overall bottom heat loss, then the above equation becomes

$$\begin{aligned} \eta &= \alpha\,(1-\rho) - \frac{U_L\left(T_p - T_a\right)A_a}{I\,A_c} \\ &= (\alpha\tau) - \left(U_L\frac{A_a}{A_c}\right)\left[\frac{T_p - T_a}{I}\right] \end{aligned} \tag{2.31}$$

The above equation is similar to the characteristic equation of a FPC (Eq. 2.15) with intercept of $(\alpha\tau) = \alpha\,(1-\rho)$ and slope of $(-U_L\frac{A_a}{A_c})$. Here, $U_L = U_t + U_b$.

2.5 SOLAR AIR COLLECTOR

The basic description and analysis of some solar air heaters used for space heating and crop drying has been discussed. The solar air heaters have certain advantages and disadvantages over the liquid FPC. These are as follows:

(A) Advantages

(i) **Simple system:** It is compact, cheap and less complicated.
(ii) **Corrosion:** It can cause serious problems in a solar water heater. It is completely eliminated in a solar collector.
(iii) **Leakage:** It does not pose any major problem in a solar air collector.
(iv) **Freezing:** The freezing of working fluid virtually never occurs.
(v) **Pressure:** The pressure inside the collector does not become very high.

(B) Disadvantages

(i) **Heat transfer:** There is a poor heat transfer property of air.
(ii) **Handling:** The need for handling large volumes of air due to its low density.
(iii) **Storage:** It cannot be used as a storage fluid due to low heat capacity.
(iv) **High cost:** In the absence of proper design the cost of an air heater can be very high.

2.5.1 Description and Classification of Solar Air Collector

A solar air collector is basically a FPC with an absorber plate, a transparent cover at the top and insulation at the bottom and on the sides. The whole assembly is encased in a sheet-metal container. The working fluid is air. The material for construction of air heaters is similar to those of liquid FPCs. The transmission of solar radiation through the cover and its subsequent absorption in the absorber plate can be given by expressions identical to those of liquid FPCs. A selective coating on the absorber plate can be used to improve the collection efficiency.

The solar air collector can be nonporous or porous, as shown in Figures 2.15.

(A) Nonporous Type

In this case, air flows above or below the absorber plate.

Case (i): Above absorber:

No separate passage is required. The air flows between the transparent cover system and the absorber plate (Figure 2.15a). In this case, as the hot air flows above the absorber that is in contact

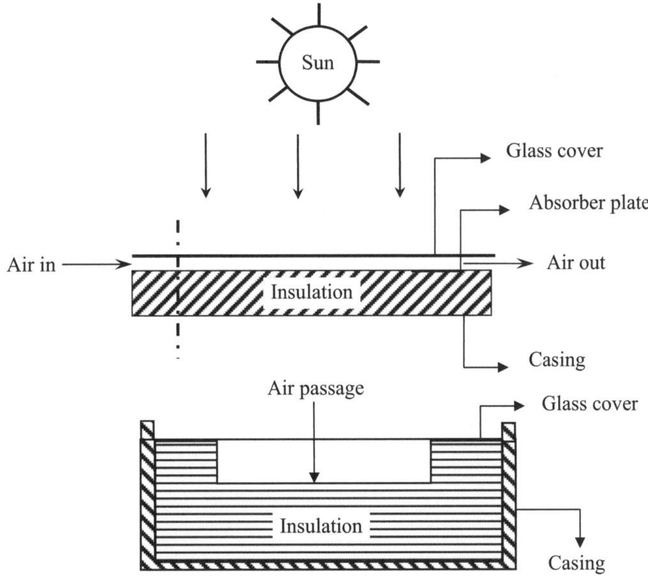

Figure 2.15a Schematic of nonporous absorber-type air heaters.

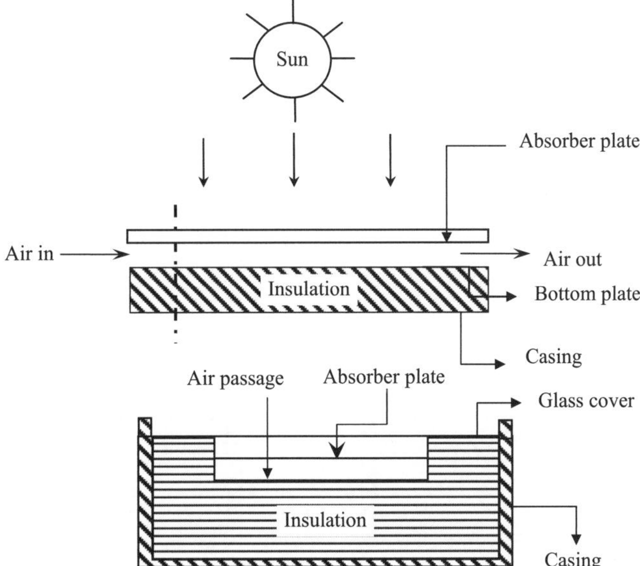

Figure 2.15b Schematic of nonporous absorber-type air heaters.

with the glass cover more heat losses take place to the ambient air and hence this solar air heater is not recommended.

Case (ii): Below absorber

The air passage below the absorber is the most commonly used solar air collector. A plate parallel to the absorber plate is provided in between the absorber and the insulation, thus forming a passage of high aspect ratio (Figure 2.15b) for the air flow.

Case (iii): Both sides of absorber

In another variety of nonporous type of air heater, the absorber plate is cooled by an air stream flowing on both sides of the plate (Figure 2.15c).

It may be noted that the heat transfer between the absorber plate and the flowing air being low, the efficiency of air heaters is less. The performance, however, can be improved by roughening the absorber surface or by using a vee-corrugated plate as the absorber. The heat transfer can also be increased by adding fins to the absorber plate (Figure 2.15d). Turbulence induced to the air flow helps increase the convective heat transfer.

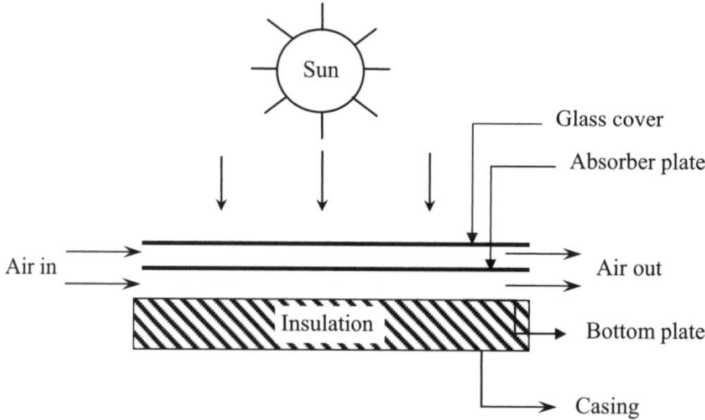

Figure 2.15c Schematic of nonporous absorber-type air heaters.

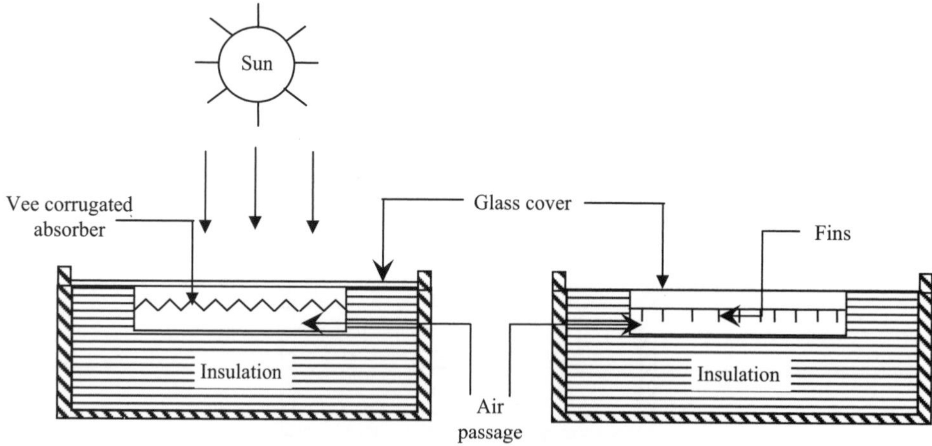

Figure 2.15d Schematic of nonporous absorber-type air heaters.

The radiative losses from the absorber plate are significant, unless selective coatings are used, decreasing the collection efficiency. Also, the use of fins may result in a prohibitive pressure drop, thus limiting the applicability of the nonporous type.

(B) Porous Type

The second type of air heaters has a porous absorber that may include slit and expanded metal, overlapped glass plate absorber and transpired honeycomb. Figure 2.16a shows the schematics of porous absorber type air heaters.

The air heater with a porous type of absorber has the following advantages:

(i) Solar radiation penetrates to a greater depth inside a porous type of absorber. It is absorbed along the path of penetration. Thus, the solar radiation loss decreases. The air stream heats up as it passes through the matrix.
(ii) The pressure drop is usually lower.

It may be noted that an improper choice of matrix porosity and thickness may cause a reduction in efficiencies as beyond an optimum thickness; the matrix may not be hot enough to transfer the heat to the air stream.

Wire mesh (Figure 2.16b) that is made up of porous broken bottles and overlapped glass plate (Figure 2.16c) are some examples of porous-type absorbers.

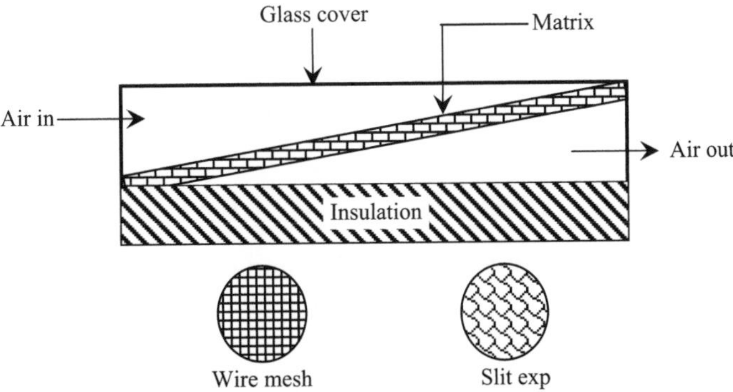

Figure 2.16a Schematics of porous absorber-type air heaters (Selcuk, 1977).

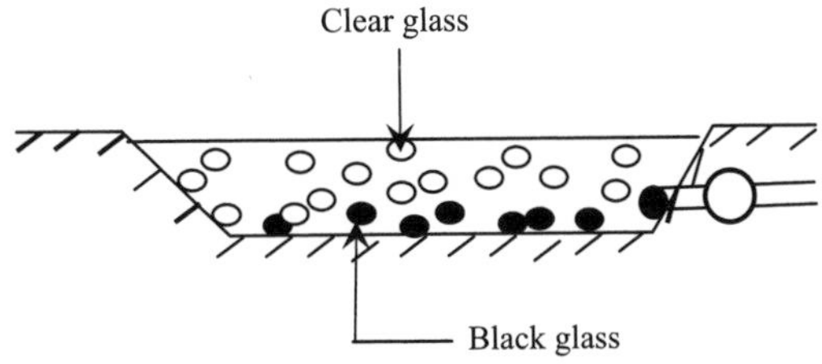

Figure 2.16b Schematics of porous absorber-type air heaters (Selcuk, 1977).

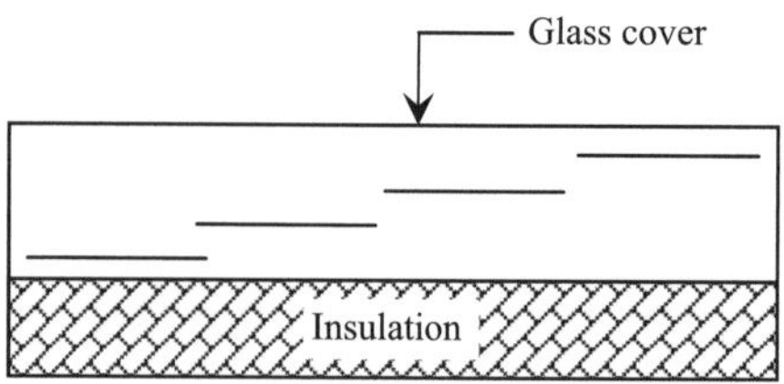

Figure 2.16c Schematics of porous absorber-type air heaters (Selcuk, 1977).

2.5.2 Thermal Analysis of Conventional Air Heater

The schematic of a conventional air heater is shown in Figure 2.17. The air flows in a parallel passage below the absorber as discussed in case (ii). The performance analysis of a solar air collector does not include the fin effect or the tube-to-plate bond conductance as mentioned in the case of liquid FPCs.

The thermal performance of such a solar air collector was first investigated analytically by Whillier.[4] The analysis under steady-state conditions is discussed below:

Let the length and the width of the absorber plate be L_1 and L_2, respectively. Let us consider an element of area $L_2 dx$ at a distance x from the inlet, as shown in Figure 2.17. The energy balance equations for the absorber plate, bottom plate and air stream can be written as:

$$I(t) = U_t (T_{pm} - T_a) + h_{pf} (T_{pm} - T_f) + h_{rpb} (T_{pm} - T_{bm}) \tag{2.32}$$

and,

$$\dot{m} C_{air} \, d T_f = h_{pf} L_2 \, dx(T_{pm} - T_f) + h_{bf} L_2 \, dx(T_{bm} - T_f) \tag{2.33}$$

Equations (2.32)–(2.33) can be solved for the air temperature in a way similar to that of a liquid FPC.

$$h_{rpb}(T_{pm} - T_{bm}) = h_{bf}(T_{bm} - T_f) + U_b(T_{bm} - T_a) \tag{2.34}$$

Then, the rise in the temperature of the air as a function of the length of the solar air collector through the duct can be obtained. The expression for the rate of useful heat gain from the solar air collector (q_u) is given in the following form:

$$\dot{q}_u = F_R A_P[I - U_L(T_{fi} - T_a)] \tag{2.35}$$

where F_R is the collector heat removal factor and is given by,

$$F_R = \dot{m} C_{air}/(U_L A_P)[1 - \exp[-F' U_L A_P /(\dot{m} C_{air})]$$

where,

$$U_L = U' + (1/F')[U_b h_{bf}/(h_{rpb} + h_{bf} + U_b)]$$

$$U' = U_t + [h_{rpb} U_b /(h_{rpb} + h_{bf} + U_b)]$$

and the collector efficiency factor (F') is given by,

$$F' = [1 + U'/ h_e]^{-1} \text{ with } h_e = h_{pf} + [h_{bf} h_{rab} /(h_{rpb} + h_{bf} + U_b)]$$

where A_P is the area of the absorber plate (m^2), I is the solar intensity (W/m^2), U_b is the bottom loss coefficient (W/m^2 °C), U_t is the top loss coefficient (W/m^2 °C), h_{bf} is the heat transfer coefficient

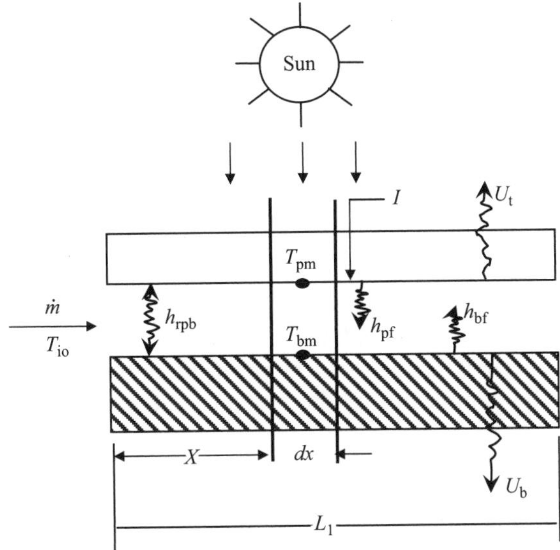

Figure 2.17 Heat transfer process in a conventional air heater.

between bottom and fluid (W/m^2 °C), h_{pf} is the heat transfer coefficient between plate and fluid (W/m^2 °C), h_{rpf} is the heat transfer coefficient between the plate and the bottom (W/m^2 °C), T_a, T_f, T_{fi}, T_{pm} and T_{bm} are, respectively, the ambient, the fluid, the inlet air, the mean plate and the mean bottom plate temperature (°C), $\dot{m}$ is the flow rate (kg/s) and C_{air} the specific heat of air (J/kg °C).

Equation (2.35) is in a form similar to that of liquid FPCs (Eq. (2.19c)). The effect of various parameters on its performance and hence the instantaneous thermal efficiency can also be obtained just like Eq. (2.20).

2.6 SOLAR CONCENTRATOR

A solar concentrator is a device that concentrates the incident solar energy over a larger surface area onto a smaller surface area, unlike a FPC. **Further, it is important to mention too that the area of absorber and glass cover is not the same as in the case of a FPC (*i.e.* $A_c < A_a$).** In this case, the concentration is achieved by the use of suitable reflecting or refracting elements. This results in an increased flux density on the absorber surface area. In order to get a maximum concentration, an arrangement for tracking the Sun's virtual motion is required. An accurate focusing device is also required. Thus, a solar concentrator consists of (i) a focusing device, (ii) a receiver system and (iii) a tracking arrangement. The temperature from a solar concentrator can be obtained up to the order of 3000 °C. Hence, they have the potential applications in both thermal and power generation (electrical power) at high delivery temperatures.

The advantages of a concentrator are as follows:

(i) It increases the solar intensity by concentrating the solar energy available over a larger surface area onto a smaller surface area (absorber).
(ii) Due to the concentration on a smaller surface area, the heat loss area is reduced. Further, the thermal mass is much smaller than that of a FPC. Hence, transient effects due to small water mass are small.

(iii) The delivery temperatures being high, a thermodynamic match between the temperature level and the task occurs.
(iv) It helps in reducing the cost by replacing an expensive large receiver by a less expensive reflecting or refracting surface area.

However, a solar concentrator is an optical system and hence the optical loss terms become important. Further, it operates only on the beam component of solar radiation. This results in the loss of the diffuse component. The basic concepts of FPCs are applicable to concentrating collectors as well. A number of complications arise because of nonuniform flux on absorbers, wide variations in shape, temperature and heat loss behaviour of absorbers. Finally the optical considerations are important in the energy balance conditions. It may be noted that the higher the concentration of a solar collector, the higher is the precision of the optics and the higher the cost of the unit. In addition to the complexity of the systems, the maintenance requirements are also increased.

2.6.1 Characteristic Parameters

Definitions of the terms that characterise concentrating solar collectors are given below:

2.6.1.1 Aperture Area (A_a)

This is the plane opening of the solar concentrator through which the incident solar flux is received. It may be defined by the physical extremities of the concentrator.

2.6.1.2 Acceptance Angle ($2\ \theta_c$)

This is the limiting angle for a beam incident ray path that may deviate from the normal to the aperture plane and still reach the absorber. Solar concentrators with large acceptance angle need to be moved only seasonally. Solar concentrators with small acceptance angles must be moved continuously to track the Sun.

2.6.1.3 Absorber Area (A_{abs})

This is the total surface area receiving the concentrated solar beam radiation. This is the area to deliver the useful thermal energy at high temperature.

2.6.1.4 Geometric Concentration Ratio (C)

This is defined as the ratio of the collecting aperture surface area to the area of the absorber surface. The values of C vary from unity (FPC, $C=1$) to several thousand (parabolic dish).

As shown in Figure 2.18, let us consider the circular concentrator with aperture area A_a and receiver area A_r. It views the Sun of radius r at a distance R. The half angle made by the Sun is θ_s. If the concentrator is perfect, the beam radiation from the Sun on the aperture (and consequently on the receiver) is the fraction of the radiation emitted by the Sun. It is intercepted by the aperture.

$$Q_{s\to r} = A_a \frac{r^2}{R^2} \sigma T_s^4$$

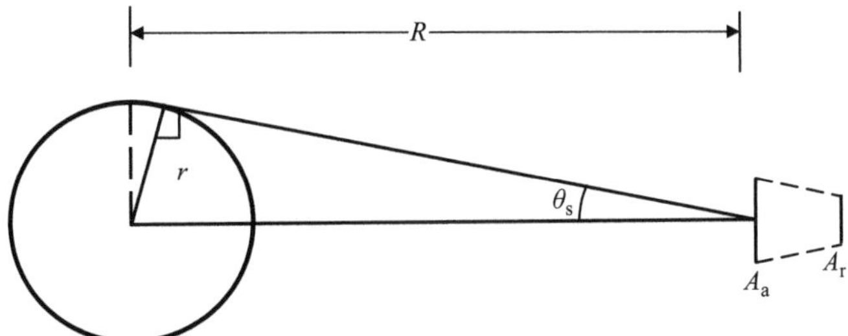

Figure 2.18 Schematic of Sun at T_s at a distance R from a concentrator with aperture area A_a and receiver area A_r.

A perfect receiver (blackbody) emits radiation in the from of thermal energy equal to $A_r T_r^4$, and a fraction of this, E_{r-s}, reaches the Sun.

$$Q_{r\rightarrow s} = A_r \sigma T_r^4 E_{r-s}$$

When T_r and T_s are the same, $Q_{s\rightarrow r}$ will be equal to $Q_{r\rightarrow s}$ by the second law of thermodynamics. Thus,

$$A_a \frac{r^2}{R^2} \sigma T_s^4 = A_r \sigma T_r^4 E_{r-s}$$

The maximum possible value of E_{r-s} is unity, the maximum concentration ratio for circular concentrators is given by,

$$\left(\frac{A_a}{A_r}\right)_{max} = \frac{R^2}{r^2} = \frac{1}{\sin^2 \theta_s} \tag{2.36a}$$

A similar relation can be obtained for linear concentrators as it is given by.

$$C = A_a / A_{abs} \tag{2.36b}$$

2.6.1.5 *Intercept Factor* (γ)

This is the fraction of focused energy intercepted by the absorber of a given size.

$$\gamma = \frac{\int_A^B I(x)\mathrm{d}x}{\int_{-\text{infinity}}^{\text{infinity}} I(x)\mathrm{d}x} \tag{2.37}$$

For a typical concentrator–receiver design, the value of γ depends on the size of the absorber, γ usually has a value greater than 0.9. If the solar radiation is normal to the aperture, the value of γ is 1.

2.6.1.6 *Optical Efficiency* (η_0)

This is defined as the ratio of the solar energy absorbed by the absorber to that incident on the solar concentrator. It includes the effect of mirror surface shape and reflection, transmission losses, tracking accuracy, shading by the receiver; cover transmission, absorptance of the absorber and solar beam incident angle. It is given by

$$\eta_0 = \dot{q}_u / (\alpha\tau I_b) \tag{2.38}$$

2.6.1.7 Thermal Efficiency (η_c)

This is the ratio of the available useful thermal energy to the incident beam radiation on the aperture of a solar concentrator. It is given by

$$\eta_0 = \dot{q}_u / I_b \tag{2.39}$$

2.6.2 Classification of Concentrators

Solar concentrators may be classified as (i) tracking type and (ii) nontracking type. Tracking may be continuous or intermittent. It may be of one-axis or two-axes design. The Sun may be followed by moving either the focusing part or the receiver or both. Solar concentrators may also be classified on the basis of optical components. They may be (i) reflecting or refracting type, (ii) imaging or nonimaging type, and (iii) line focusing or point focusing type.

There are a number of methods by which the flux of radiation on receivers can be increased. A few of them are described below:

(i) Cylindrical Parabolic Concentrator

A cylindrical parabolic trough is a conventional optical imaging device used as a solar concentrator. It consists of a cylindrical parabolic reflector and a metal tube receiver at a focal plane (Figure 2.19). The receiver is selective/blackened at the outside surface. It is covered by a cylindrical glass tube. It is rotated about one axis to track the Sun's diurnal motion. The working fluid flows through the absorber tube, gets heated and thus carries away the thermal energy.

The aperture diameter, rim angle, absorber size and shape may be used to define the solar concentrator. The absorber tube may be made of mild steel/copper. Depending on the temperature requirement different heat transfer liquids may be used. Reflectors may be of anodised aluminium sheet, aluminised Mylar or curved silvered glass. Since it is difficult to curve a very large glass, mirror strips are sometimes used in the shape of a parabolic cylinder. The reflecting part having high reflectance is fixed on a lightweight structure. A cylindrical parabolic trough may be oriented in any of three directions: East–West, North–South or polar. The concentration ratio for a cylindrical absorber varies from 5 to 30. Such solar concentrators have been in use for many years.

(ii) Paraboloidal Dish Concentrator

This is used to concentrate solar beam radiation. It has a high concentration ratio. Due to the compound curvature with perfect optics and a point source of light, beam radiation (light) is focused at a point in a paraboloid. A degraded image is obtained if the object is off-axis. The rays from the central region of the paraboloid travel a shorter distance in arriving at the focus. The rays from the edges travel a larger distance, which results in a spread of the image. Thus, a three-dimensional image of the Sun in the shape of an ellipsoid is formed, as shown Figure 2.20.

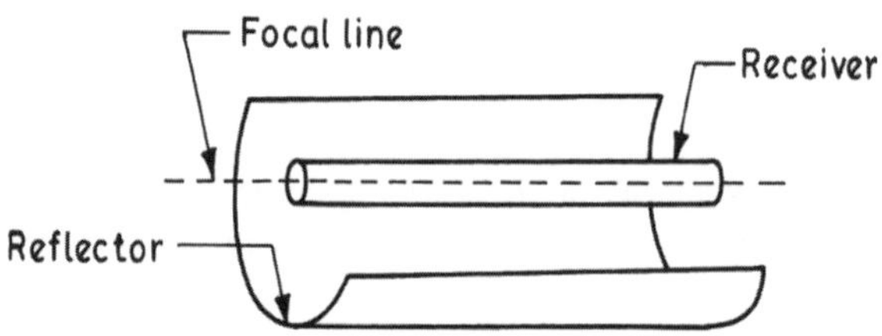

Figure 2.19 Cylindrical parabolic concentrator.

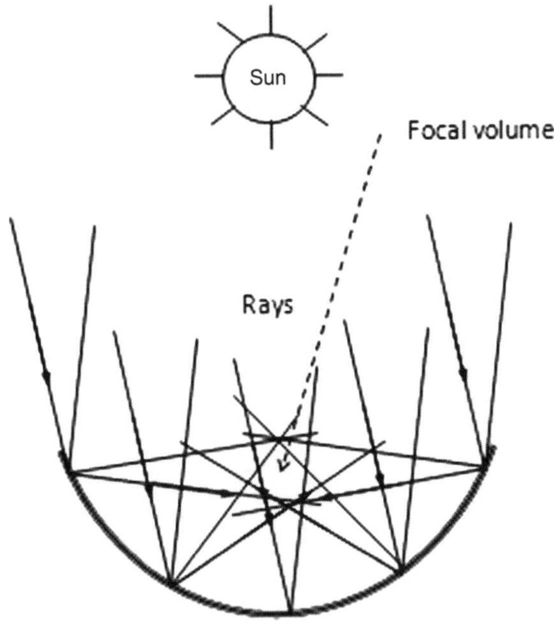

Figure 2.20 Illustration of formation of ellipsoid image in paraboloid.

The thermal losses from a paraboloid are primarily radiative. It can be reduced by decreasing the absorber aperture area. This, however, results in a smaller intercept factor. The optimum intercept factor is about 0.95–0.98. The larger the surface errors, the larger must be the absorber size to achieve the optimum beam radiation intercept. High collection efficiency and high quality thermal energy are the features of a paraboloid or parabolic dish type of concentrator. The delivery temperatures are very high. These devices can be used as sources for a variety of purposes.

(iii) Flat Receiver with Booster Mirror

Figure 2.21 shows a flat receiver with plane reflectors at the edges. Reflectors reflect total radiation in addition to beam radiation incidence on the receiver. Mirrors are also called booster mirrors. The concentration ratio of such solar concentrators is relatively low. It has a maximum value less than four. As the solar incidence angle increases, the mirrors become less effective. For a single collector, booster mirrors can be used on all the four sides. When the Sun angle exceeds the semiangle of the booster mirrors, the mirror actually starts casting a shadow on the absorber. In the case of an array of collectors, booster mirrors can be used only on two sides.

The efficiency of a boosted flat plate system can be increased if the angle of the flat mirrors can be changed several times during the year. The advantage of such a system is that it makes use of the diffuse radiation in addition to the beam radiation. The attainable temperature and collection efficiency will be higher than that of a FPC of the same collection area.

(iv) Compound Parabolic Concentrator (CPC)

The CPC is a nonimaging one. It belongs to a family of concentrator that has highest possible concentration permissible by the thermodynamic limit for a given acceptance angle. Further, it has a large acceptance angle. It needs to be intermittently turned towards the Sun.

The compound parabolic concentrator (CPC) consists of two parabolic segments. It is oriented such that the focus of one is located at the bottom end point of the other and *vice versa*, as shown

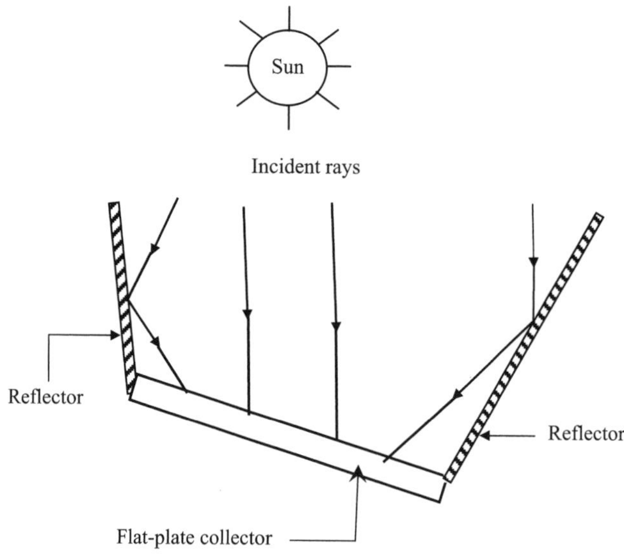

Figure 2.21 Flat plate collector with booster mirrors.

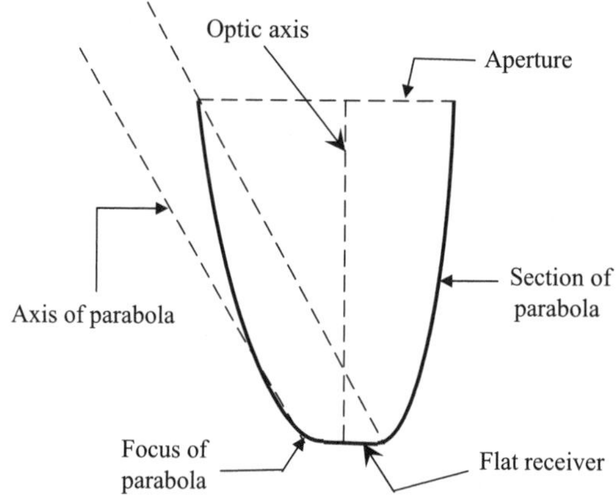

Figure 2.22 Schematic cross section of a CPC.

Figure 2.22. The axes of the parabolic segments subtend an angle that is equal to the acceptance angle with the CPC axis. The slope of the reflector surfaces at the aperture plane is parallel to the CPC axis. The receiver is a flat surface parallel to the aperture joining two foci of the reflecting surfaces.[5,6]

Beam rays incident in the central region of the aperture undergo no reflection, whereas those near the edges undergo one or more reflections. The number of reflections depends on the incident angle of beam radiation, collector depth and concentration ratio.[7] To reduce the cost of the unit, the CPC can be truncated in height to half without any significant change in concentration.

Extensive investigations on this concentrator have led to several modified designs of the ideal CPC. The salient modifications can be listed as follows:

(i) The use of receiver shapes such as fins, circular pipes is for better optical and thermal performance.
(ii) Truncation of the CPC height reduces the system's to physical size and cost.

(iii) Asymmetric orientation of source and aperture deliver seasonally varying outputs.
(iv) Design of CPC as a second-stage concentrator.

The CPC can be used in a nontracking mode for concentration ratios of about 6. However, for higher ratios, the reflector surface area becomes very large and hence cannot be used.

2.6.3 Thermal Performance

Due to the following reasons in solar concentrators, namely,

(i) receiver shapes are variable;
(ii) the beam radiation intensity at the receiver is not uniform;
(iii) the temperature is high and hence edge losses and conduction effects are significant;

complications occur in the calculation of thermal losses. Thus, it is not possible to give a general analysis for the estimation of thermal losses of concentrators. Each receiver has to be analysed separately. However, from a basic knowledge of FPC, we can derive the expression for collection efficiency or thermal efficiency in terms of inlet fluid temperature, fluid flow rate, ambient temperature and solar intensity.

The thermal efficiency, η_c, is given by,

$$\eta_c = (\dot{q}_u / I_b) \tag{2.40a}$$

where I_b is the incident beam solar radiation and q_u the rate of useful energy per unit aperture area. In some of the solar concentrators, the acceptance angle may be large so that in addition to beam components, some fraction of the diffuse radiation also comes in. In the case of high concentration systems, even the direct component is curtailed. I_b is, thus, to be corrected accordingly.

Case (i) Without thermal energy losses

With no losses

$$\dot{q}_{loss} = 0 \text{ and } \dot{q}_u = \dot{q}_{abs} = A_a S \tag{2.40b}$$

$$\eta_c = \eta_0 = A_a S / I_b$$

where $S = I_b\ \rho(\gamma\tau\alpha)_n\ K_{\gamma\tau\alpha}$ is the absorbed radiation per unit area of unshaded aperture, I_b is the beam component of the incident radiation, ρ is the specular reflectance of the concentrator, γ, τ and α are functions of the angle of incidence of radiation on the aperture and $K_{\gamma\tau\alpha}$ is an incidence angle modifier.

Case (ii) With thermal energy losses

Expression for q_u is given by

$$\dot{q}_u = \dot{q}_{abs} - \dot{q}_{loss} \tag{2.41}$$

where q_{abs} and q_{loss} are, respectively, the rate of energy absorbed and lost per unit aperture area. As has been mentioned earlier, the losses depend on the geometry of the system. Let us consider linear concentrating systems with cylindrical receivers (Figure 2.19), the overall heat transfer coefficient from the surroundings to the fluid in the tube is,

$$U_0 = \left[\frac{1}{U_L} + \frac{D_0}{h_{fi}\,D_i} + \frac{D_0 \ln(D_0/D_i)}{2K}\right]^{-1} \tag{2.42}$$

where U_L is the heat transfer coefficient from the receiver to the ambient; D_i and D_0 are the inside and outside tube diameters; h_{fi} is the heat transfer coefficient inside the tube, and K the thermal conductivity of the tube.

The useful energy gain, q_u, in terms of the absorbed solar radiation per unit length is,

$$\dot{q}_{u'} = \frac{\dot{q}_u}{L} = \frac{A_a\,S}{L} - \frac{A_r\,U_L}{L}(T_r - T_a) \tag{2.43a}$$

where A_a is the aperture area and A_r is the area of the receiver.

In a steady-state condition, the useful energy gain per unit collector length in terms of the energy transfer to the fluid at local fluid temperature, T_f, is,

$$\dot{q}_{u'} = \frac{(A_r/L)(T_r - T_f)}{\dfrac{D_0}{h_{fi}\,D_i} + \left(\dfrac{D_0}{2K}\ln\dfrac{D_o}{D_i}\right)} \tag{2.43b}$$

On elimination of T_r from Eqs. (2.43a) and (2.43b), we get the following expression for the rate of useful thermal energy as

$$\dot{q}_{u'} = F'\frac{A_a}{L}\left[S - \frac{A_r}{A_a}U_L(T_f - T_a)\right] \tag{2.44}$$

where, the collector efficiency factor, F', is given as

$$F' = \frac{1/U_L}{\dfrac{1}{U_L}\dfrac{D_0}{h_{fi}\,D_i} + \left(\dfrac{D_0}{2K}\ln\dfrac{D_o}{D_i}\right)}$$

or,

$$F' = U_0/U_L$$

Using Eqs. (2.40) and (2.41), we get,

$$\eta_c = (\dot{q}_{abs}/I_b) - (\dot{q}_{loss}/I_b) = \eta_0 - (\dot{q}_{loss}/I_b) \tag{2.45}$$

The collector heat loss factor (U_L) with respect to collector aperture area is defined as,

$$U_L = \dot{q}_{loss}/(T_r - T_a) \tag{2.46}$$

where T_r and T_a are, respectively, the receiver plate and ambient temperature.

As in the case of a FPC, the collection efficiency in terms of average fluid temperature can be written as,

Hence,

$$\eta_c = F'\left[\eta_0 - \{U_L(\overline{T}_f - T_a)/I_b\}\right] \tag{2.47a}$$

$$\eta_c = F'\left[\eta_0 - \{U_L(\overline{T}_f - T_a)/I_b\}\right] \tag{2.47b}$$

where the collector efficiency factor, F', is given by the ratio between the thermal resistance from the receiver surface to ambient and the thermal resistance from the fluid to ambient.

In terms of inlet fluid temperature (T_{fi}), Eq. (2.47b) becomes

$$\eta_c = F_R\left[\eta_0 - \{U_L(\overline{T}_{fi} - T_a)/I_b\}\right] \tag{2.48}$$

where $\dot{m}$ the heat removal factor, F_R is,

$$F_R = \frac{\dot{m}\,C_p}{U_L}\left[1 - \exp\left(-\frac{U'F}{\dot{m}C_p}\right)\right] \tag{2.49}$$

Here, is the flow rate per unit area and C_p is the fluid heat capacity.

2.7 SOLAR CROP DRYING

Food is a basic need of a human being after air and water. Food holds a key position in the development of a country. The drying of food is necessary to avoid food losses between harvesting and consumption. High moisture content is one of the reasons for its spoilage during the course of storage at the time of harvesting. Crops with high moisture are prone to fungus infection. Crops are attacked by insects, pests and the increased respiration of agricultural produce. To solve this problem drying of a crop with optimum moisture content is the answer. Dried crops can be used for longer periods after storage. Drying has the following advantages:

i) facilitates early harvest;
ii) permits planning the harvest season;
iii) helps in long-term storage;
iv) helps farmers to achieve better returns;
v) helps farmers to sell a better quality product;
vi) reduces the requirement of storage space;
vii) helps in handling, transport and distribution of crops; and
viii) permits maintaining viability of seeds.

Different crops have different level of safe moisture content, as given in the Table 2.7.

2.7.1 Working Principle

The following methods are used for solar energy drying:

(i) Open Sun Drying (OSD)

Figure 2.23 gives information about the working principle of open Sun drying (OSD). The short-wavelength solar radiation [$I(t)$] falls on the uneven crop surface having an area of A_t. A part of this energy is reflected back. The remaining part is absorbed [$\alpha_c\, I(t)$] by the crop surface depending upon the colour of the crops. The absorbed radiation is converted into thermal energy. The temperature of the crop [T_c] starts increasing. This results in long wavelength radiation and convective heat loss too, due to the blowing wind losses from the surface of crop to ambient air. Evaporation of moisture takes place in the form of evaporative losses and so the crop is dried. Further, a part of the absorbed thermal energy is conducted into the interior of the product. This causes a rise in the temperature of the crop. There is formation of water vapour inside the crop that diffuses towards

Table 2.7 Initial and final moisture content and maximum allowable temperature for drying for some crops (Brooker *et al.*, 1992; Sharma *et al.*, 1993).

Sl.No.	*Crop*	*Initial moisture content (%,w.b.)*	*Final moisture content (%,w.b.)*	*Max. allowable Temp.(°C)*
1	Paddy, raw	22 – 24	11	50
2	Paddy, Parboiled	30 – 35	13	50
3	Maize	35	15	60
4	Wheat	20	16	45
5	Corn	24	14	50
6	Rice	24	11	50
7	Pulses	20 – 22	9 – 10	40 – 60
8	Oil seed	20 – 25	7 – 9	40 – 60
9	Gren peas	80	5	65
10	Cauliflower	80	6	65
11	Carrots	70	5	75
12	Green beans	70	5	75
13	Onions	80	4	55
14	Garlic	80	4	55
15	Cabbage	80	4	55
16	Sweet potato	75	7	75
17	Potatos	75	13	75
18	Chillies	80	5	65
19	Apples	80	24	70
20	Apricot	85	18	65
21	Grapes	80	15 – 20	70
22	Bananas	80	15	70
23	Gauvas	80	7	65
24	Okra	80	20	65
25	Pineapples	80	10	65
26	Tomatoes	96	10	60
27	Brinjal	95	6	60

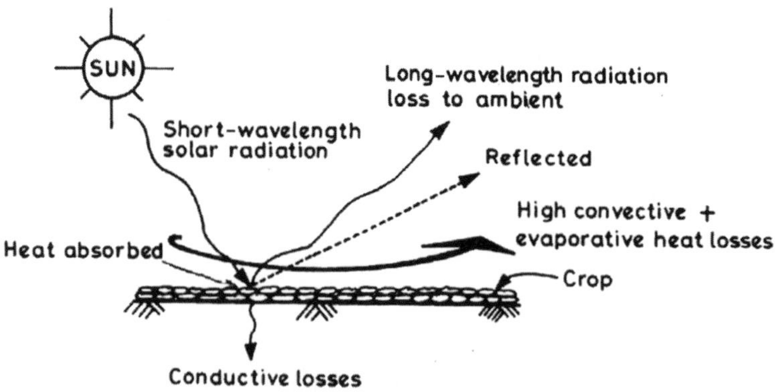

Figure 2.23 Working principle of open Sun drying.

the surface of the crop. Finally, there are thermal energy losses in the form of evaporation from the surface of crop. In the initial stages, the moisture removal is rapid since the excess moisture on the surface of the product presents a wet surface to the drying air. Subsequently, drying depends upon the rate at which the moisture within the product moves to the surface by a diffusion process depending upon the type of product.[8]

In open Sun drying, there is a considerable loss due to various reasons such as rodents, birds, insects and micro-organisms. The unexpected rain or storm further worsens the situation. Further, overdrying, insufficient drying, contamination by foreign materials like dust, dirt, insects and micro-organisms as well as discolouring by UV radiation are characteristic of open Sun drying. In general, open Sun drying does not fulfil the international quality standards and therefore it cannot be sold on the international market.

(ii) Direct Solar Drying (DSD)

The working principle of direct solar crop drying has been shown in Figure 2.24. It is referred to as a cabinet dryer. A part of the incident solar radiation on the glass cover is reflected back to the atmosphere. Remaining part of solar radiation is transmitted through the glass cover inside the cabinet dryer. Further, a part of transmitted radiation is reflected back from the surface of the crop. The rest part is absorbed by the surface of the crop. Due to the absorption of solar radiation, the crop temperature increases and the crop starts emitting long-wavelength radiation that is not allowed to escape to atmosphere due to the presence of the glass cover, unlike open Sun drying. Thus, the temperature above the crop inside chamber becomes higher. The glass cover serves one more purpose of reducing direct convective losses to the ambient that further becomes beneficial for a rise in crop production. However, convective and evaporative losses occur inside the chamber from the heated crop. The moisture content is taken away by the air entering into the chamber from below and escaping through another opening provided at the top, as shown in Figure 2.24.

A cabinet dryer has the following limitations:

* it has small capacity;
* the discolouration of crops occurs due to direct exposure to solar radiation;
* the moisture condensation takes place inside the glass cover that reduces its transmittivity.

To solve the above problems, indirect solar drying is preferred.

(iii) Indirect Solar Drying (ISD)

In this case, the crop is not directly exposed to solar radiation to minimise discolouration and cracking on the surface of the crop. Figure 2.25 describes another principle of indirect solar drying that is generally known as conventional dryer. In indirect crop drying, a separate unit termed a solar air heater is used

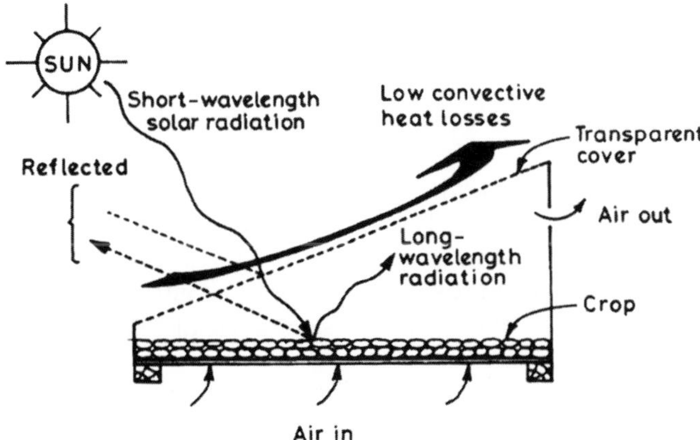

Figure 2.24 Working principle of direct solar drying.

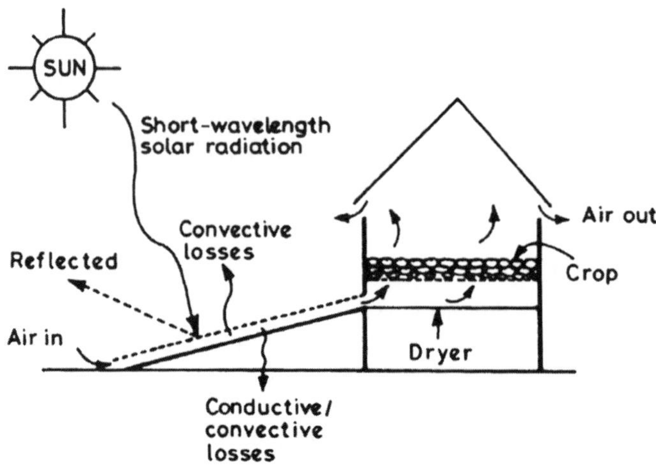

Figure 2.25 Working principle of indirect solar.

for crop heating by allowing hot air into the drying chamber. The hot air is allowed to flow through the wet crop. The drying is basically achieved by the difference in moisture concentration between the drying air and the air in the vicinity of crop surface. A better control over drying is achieved in a indirect type of solar drying system and the product obtained is of good quality.

2.8 SOLAR DISTILLATION

The supply of potable water is a major problem in underdeveloped as well as in some developing countries due to pollution developed in underground water by global industrialisation. It has caused a scarcity of fresh water in many towns and villages near lakes and rivers. Along with food and air, good quality water is a basic necessity for man. Man has been dependent on rivers, lakes and underground water reservoirs for fresh water. Surveys show that about 79% of water available on the Earth is salty, only 1% is fresh. The remaining 20% is brackish.

Distillation of brackish/saline water is a good method to obtain fresh water (potable water). However, the conventional distillation processes such as multieffect evaporation, multistage fresh evaporation, thin-film distillation, reverse osmosis and electrodialysis are energy intensive techniques. Therefore, solar distillation is an attractive alternative due to its simple technology and nonrequirement of highly skilled labour for maintenance work and low energy consumption. As such, it can be used at any place without much problem.

2.8.1 Working Principle of Solar Distillation

Figure 2.26 shows the various components of energy balance and thermal energy losses in a conventional double slope solar distiller unit. It is an airtight basin. It is usually constructed out of concrete/cement, galvanised iron sheet (GI) or fibre-reinforced plastic (FRP) with a top cover of transparent material like glass/plastic, *etc*. The inner surface of the rectangular base is blackened to absorb solar radiation incident on the transparent surface. There is a provision to collect the distillate at lower end of the glass cover. The brackish/saline water is fed into the basin for purification.

The working principle of a conventional double slope solar distiller unit is as follows:

"The solar radiation is transmitted inside an enclosure of the distiller unit after reflection and absorption by the glass cover. The transmitted radiation [$\tau_g I(t)$] is further partially reflected [$R'_w I(t)$] and absorbed [$\alpha'_w I(t)$] by the water mass. The attenuation of the solar flux in the water mass depends on its absorptivity and depth. The solar radiation finally reaches the blackened surface

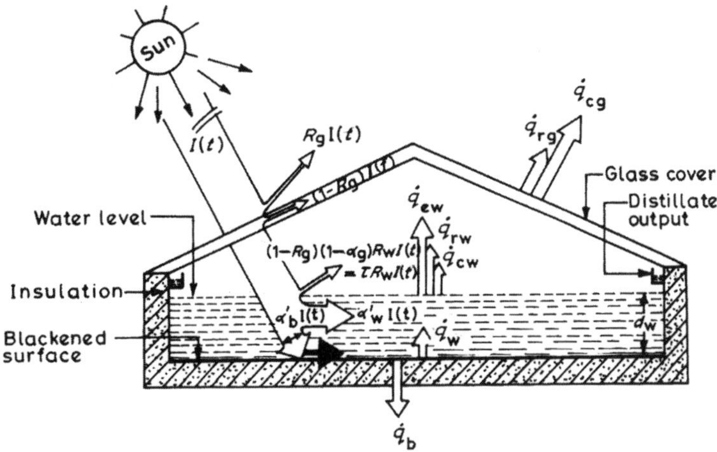

Figure 2.26 Energy flow diagram in a conventional solar still.

where it is mostly absorbed. After absorption of solar radiation at the basin liner (blackened surface), most of the thermal energy is convected to the water mass. The rest is lost to the atmosphere by conduction through the insulating bottom. Consequently, the basin water gets heated, leading to an increased difference of water and glass cover temperatures. There are basically three modes of heat transfer namely radiation (q_{rw})*, convection* (q_{cw}) *and evaporation* (q_{ew}) *from the water surface to the glass cover. The evaporated water gets condensed on the inner surface of the glass cover after releasing the latent heat. The condensed water trickles into the channels provided at the lower ends of the glass cover under gravity. The collected water in the channel is taken out of the system for further use. The thermal energy received by the glass cover through radiation, convection and latent heat from water surface is lost to the ambient by radiation and convection."*

The fraction of solar flux, at different components of the distiller unit is shown in Figure 2.26.

2.8.2 Thermal Efficiency

(A) Instantaneous Efficiency

An energy balance for the steady state for the water basin can be written as:[9]

$$[\alpha'_{\mathrm{w}} + \alpha'_{\mathrm{b}}]I(t)A_{\mathrm{S}} = \dot{q}_{\mathrm{ev}} + \dot{q}_{\mathrm{losses}} \tag{2.50}$$

[Rate of energy in] = [Rate of energy out]

where $\dot{q}_{\mathrm{ev}} = \dot{m}_{\mathrm{w}}\, L$ and $\dot{q}_{\mathrm{losses}} = U_{\mathrm{L'}}(T_{\mathrm{w}} - T_{\mathrm{a}})\, A_{\mathrm{s}}$, U_L is the overall heat transfer coefficient from water to the ambient through the top, bottom and sides of the distiller unit. It is assumed that,

$$(\alpha'_{\mathrm{w}} + \alpha'_{\mathrm{b}}) = (\alpha\tau)_{\mathrm{w}}$$

Here, in the Tamini model, q_{losses} does not include the evaporative heat loss. The analysis has been compared with analysis of a FPC. However, it may be noted that in a FPC, the upward heat/thermal losses should be minimum unlike in the case of a solar still. In the case of a solar still, the upward heat/thermal losses should be maximum in order to get a higher yield. In the conventional solar still, the radiative, convective and evaporative thermal losses from the water to the condensing cover (glass) are grouped together and it is taken as the total internal heat transfer coefficient from the water to the glass. These are temperature-dependent heat transfer coefficients.

Equation (2.50) can be rewritten as,

$$\dot{q}_{ev} = \dot{m}_w L = (\alpha\tau)_w I(t) A_s - U_{L'}(T_w - T_a) A_s \tag{2.51}$$

The expression for instantaneous efficiency (η_i), can be given as,

$$\eta_i = \frac{\dot{m}_w L}{I(t)A_s} = (\alpha\tau)_w - U_{L'} \frac{(T_w - T_a)}{I(t)} \tag{2.52}$$

The plot of η_i versus $(T_w - T_a)/I$ will represent a straight line with $(\alpha\tau)_w$ and $-U_L{}'$ as the intercept and the slope respectively. $U_L{}'$ can be taken as a constant. The expression for η_i is similar to that for a FPC except for the heat removal factor. Thus, the distiller unit can be considered as a special type of FPC that collects the solar energy and produces distilled water.

(B) Overall Thermal Efficiency

The overall thermal efficiency of the solar distiller system in the passive and active modes of operation can be mathematically expressed as,

$$\eta_{passive} = \frac{\Sigma \dot{m}_w L}{A_s \int I(t)dt} \times 100 \tag{2.53a}$$

$$\eta_{active} = \frac{\Sigma \dot{m}_w L}{\left[A_s \int I(t)dt + \eta A_c \int I'(t)dt\right]} \times 100 \tag{2.53b}$$

Here, the latent heat of vaporisation (L) in Joule/kg can be considered to be temperature dependent. It can be given as,[10,11]

$$L = 3.1615 \times 10^6 \ [1 - 7.6160 \times 10^{-4}\, T] \tag{2.54a}$$

for temperatures higher than 70 °C and,

$$L = 2.4935 \times 10^6 [1 - 9.4779 \times 10^{-4}\, T + 1.3132 \times 10^{-7}\, T^2 - 4.7974 \times 10^{-9}\, T^3] \tag{2.54b}$$

for operating temperatures less than 70 °C.

2.8.3 Thermal Analysis of Conventional Solar Still

The conventional single basin solar still is the most practical design for an installation to provide distilled water. The distilled water can be used as a potable water after treatment for daily requirement.

The basin of the distillation unit is made watertight to avoid any water leakage. The surface inside is blackened so as to absorb maximum solar radiation. The bottom and the sides of the basin are insulated to reduce the heat losses through conduction. The basin may be constructed of (i) concrete/bricks/cement (ii) galvanised iron sheet (G.I. sheet) or (iii) fibre-reinforced plastic (FRP).

The schematic diagram of a single basin solar still is shown in Figure 2.27.

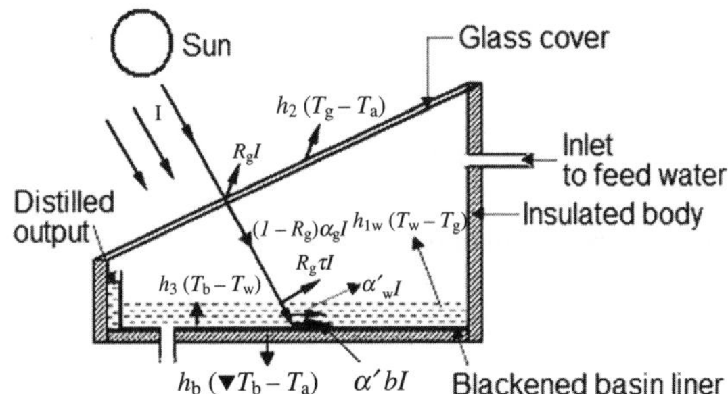

Figure 2.27 Schematic diagram of a single slope basin solar still.

The basin heat-flux components at various points are shown in Figure 2.28. The following assumptions have been made in writing the energy balance in terms of Joules per s per m^2.

(i) inclination of the glass cover is very small;
(ii) the heat capacity of the glass cover, the absorbing material and the insulation (bottom and sides) is negligible;
(iii) the solar distiller unit is vapour-leakage proof;
(iv) the analysis is for quasisteady state.

The energy balance for different components of the still is as follows:

$$\alpha'_g\ I(t) + h_{1w}(T_w - T_g) = h_{1g}(T_g - T_a) \tag{2.55a}$$

$$\alpha_{w'}\ I(t) + h_w(T_b - T_w) = (MC)_w \frac{dT_w}{dt} + (T_w - T_g) \tag{2.55b}$$

$$\alpha'_b I(t) = h_w(T_b - T_w) + h_b(T_b - T_a) \tag{2.55c}$$

Substituting the values of T_g and T_b from Eqs. (2.55a) and (2.55c) in Eq. (2.55b) and simplifying, we get,

$$\frac{dT_w}{dt} + a\,T_w = f(t) \tag{2.56}$$

where

$$f(t) = \frac{(\alpha\tau)_{eff}\, I(t) + U_L\, T_a}{(MC)_w}$$

$$a = \frac{U_L}{(MC)_w}$$

$$(\alpha\tau)_{eff} = \alpha'_b \frac{h_w}{h_w + h_b} + \alpha'_w + \alpha'_g \frac{h_{1w}}{h_{1w} + h_{1g}}$$

$$U_L = U_b + U_t;\ U_b = \frac{h_w\, h_b}{h_w + h_b},\ U_t = \frac{h_{1w}\, h_{1g}}{h_{1w} + h_{1g}}$$

In order to obtain an approximate solution of Eq. (2.56) with the above initial conditions, the following assumptions have been made:

(i) the time interval Δt $(0<t<\Delta t)$ is small;
(ii) the function $f(t)$ is constant, *i.e.* $f(t)=\bar{f}(t)$ for the time interval Δt; and
(iii) a is constant during the time, interval Δt.

The value of h_{1w} can be determined by considering known values of water and glass temperatures at

$$t=0, i.e.\ T_{w\text{-}t=0}=T_w 0 \quad \text{and} \quad T_{g\text{-}t=0}=T_g 0$$

The solution of Eq. (2.56) can be written as,

$$T_w=\frac{\bar{f}(t)}{a}[1-\exp(-a\Delta t)]+T_w 0 \exp(-a\Delta t) \tag{2.57}$$

where T_{w0} is the temperature of basin water at $t=0$ and $\bar{f}(t)$ is the average value of $f(t)$ for the time interval between 0 and t.

2.9 SOLAR WATER HEATING SYSTEMS

2.9.1 Collection-Cum-Storage Water Heater

The collection-cum-storage water heater can be classified as (i) a built-in storage water heater and (ii) a shallow solar pond water heater.

Detailed descriptions of the two are given below:

(A) Built-In Storage Water Heater

A collection-cum storage water heater combines both collection and storage in the same unit. This eliminates the need for a separate insulated tank for storage of hot water. These water heaters employ water as a transport fluid for energy transfer from the collection unit to the storage one. Water, is a low cost and easily available fluid throughout. It has many advantageous thermo-chemical properties such as nontoxicity, nonflammability, high specific heat, good heat transfer and fluid dynamic characteristics.

The built-in storage water heaters are compact in design. In addition, they have a good collection efficiency and satisfactory overnight thermal storage.

A built-in storage water heater as shown in Figure 2.28a consists of a rectangular metallic tank having the dimension of 1.12 m × 0.08 m × 0.1 m. It is covered with 5-cm thick glass-wool insulation or an air gap with reflecting sheet on sides and bottom. It is encased in a mild iron or wooden box (typically 1.22 m × 0.9 m × 0.2 m). The top surface of the tank is sprayed with black board paint. It is covered with one or two window glass covers (3 mm thick); with about 4 cm air gap between the absorbing surface and the glass cover. In order to avoid bulging of the tank due to enormous water pressure, braces are provided.

Cold water is fed through the inlet and hot water is withdrawn from the outlet.

During off-sunshine hours, the major thermal losses in a built-in storage water heater are radiative and convective losses from the top surface. A simple method to improve the night time performance is to cover the glass with an adequate thickness of glass-wool insulation.

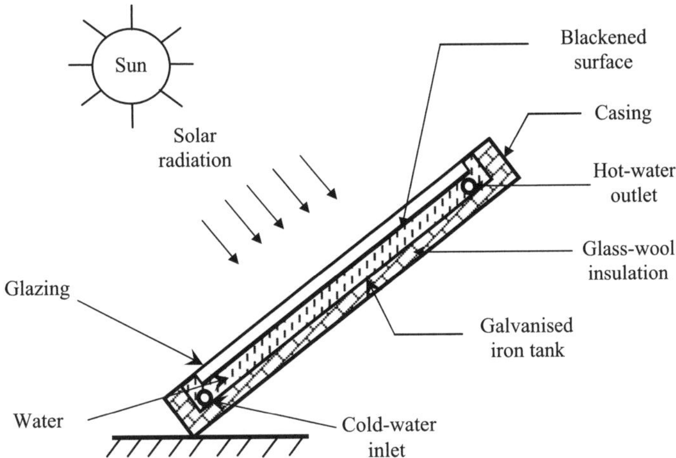

Figure 2.28a Built-in storage water heater.

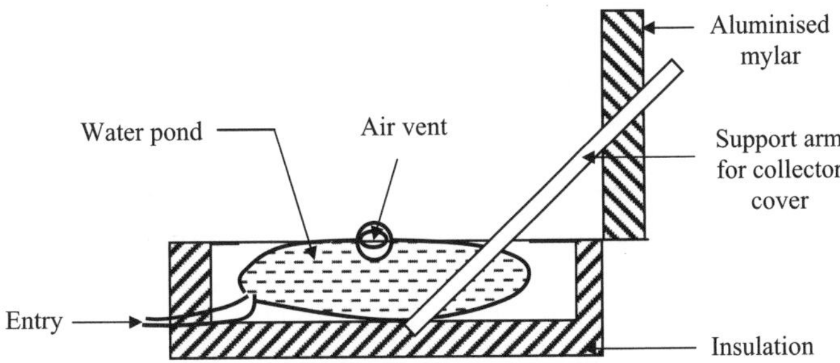

Figure 2.28b Schematic diagram of a compact SSP.

(B) Shallow Solar Pond Water Heater (SSP)

A shallow solar pond (SSP) water heater is a simple and cost effective device to harness the solar energy for domestic applications.

A SSP water heater consists of a blackened tray holding some water in it, the depth being very small. The evaporative cooling in a SSP is suppressed by covering the water by means of a transparent plastic film in such a way that the film comes in contact with the top surface of water. The incident energy is absorbed at the blackened bottom and transferred to the water column due to convection. It can, thus, heat large quantities of water to appreciable temperature.

The solar energy collection efficiency of a SSP is directly proportional to the water depth inside. The grade of thermal collection (*i.e.* the temperature) is inversely proportional to the water depth. In a typical design, a compact shallow solar pond consists of a pillow type of water bag, encased in a wooden box, the top being transparent, the bottom is painted black and the walls and bottom are insulated. The system has a thermally insulated cover at the top. It can be used as a booster mirror during the day, and as an insulation cover during off-sunshine hours providing a means of overnight storage, as shown in Figure 2.28b.

An efficient performance of the SSP water heater demands continuous withdrawal of hot water from the pond, to reduce the thermal losses.

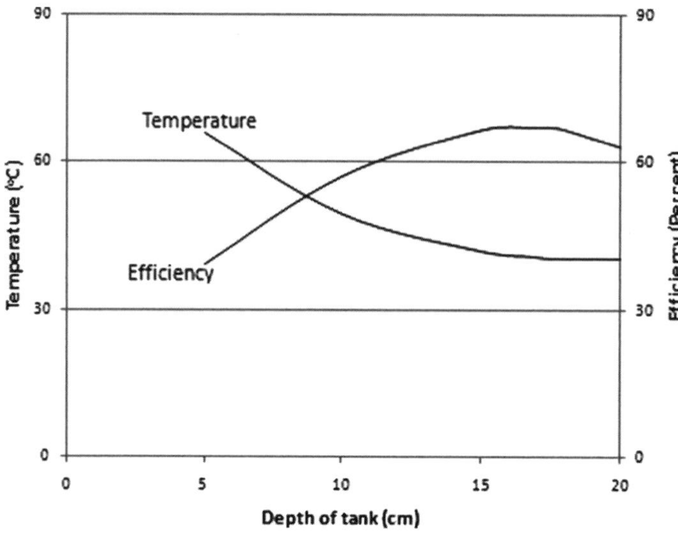

Figure 2.29 Effect of storage tank depth on storage temperature and efficiency.

Three modes of hot water withdrawal from a SSP water heater have been suggested. These modes are as follows:

(i) **Batch withdrawal**: In the batch withdrawal mode, the pond is filled early in the morning and is emptied into an insulated storage reservoir in the afternoon, when the temperature is maximum.
(ii) **Closed-cycle continuous flow**: In the closed-cycle mode, water is circulated continuously at a constant rate between the pond and the storage. In this case also, the pond water is transferred to storage in the afternoon.
(iii) **Open-cycle continuous flow**: In the open-cycle mode, water at an initial temperature enters the pond and the hot water flowing through the pond is drained out continuously into storage or for some other end use application. The pond is emptied in the evening when the heat collection reduces to zero.

As seen from Figure 2.29, the collection efficiency increases due to the decrease of the thermal losses to the outside air with increase of water depth. However, the rise in efficiency is significant upto a depth of 10 cm and then becomes stable. It can, thus, be concluded that a 10-cm depth gives the optimal performance for both types of collection-cum-storage water heater.

2.9.2 Solar Water Heating System

The solar water heating system consists of two units; namely the collection unit and the storage unit. The collection unit is a FPC to absorb the solar radiation incident on it. The flowing water in thermal contact with the absorber gets heated and then it is transferred to the storage unit. The storage unit is a well insulated tank to reduce the possible heat losses. The transportation of the heated liquid from the collector to the storage unit takes place by two modes;

(i) **Thermosyphon mode (Natural circulation)**: The circulation of heated water is accomplished by the natural convection; and
(ii) **Forced circulation mode**: A small pump is required for the flow of water between FPC and storage tank.

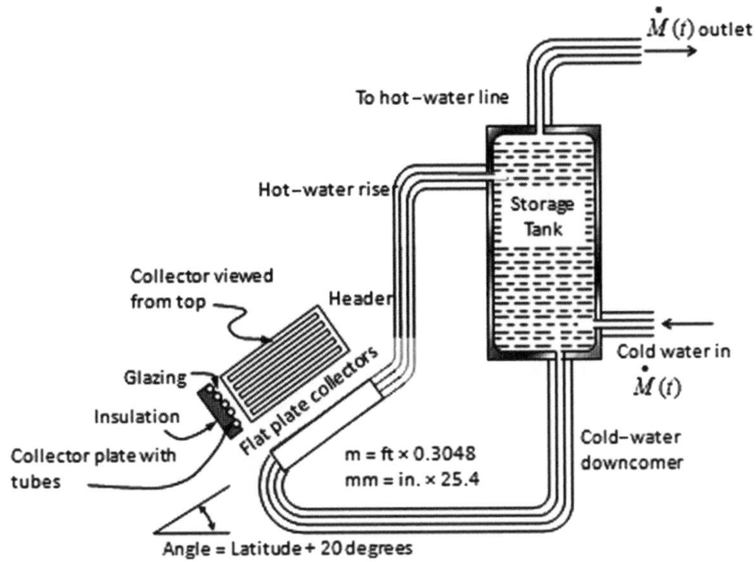

Figure 2.30 Schematic diagram of solar water heater with natural circulation.

(i) Natural Circulation

The schematic diagram of a solar water heater with natural circulation mode is shown in Figure 2.30. The absorber is a set of N collectors connected in parallel. The storage tank is an insulated one with two inlets, one for the hot water from the collector and the other one to allow the cold water from the mains to reach the bottom of the tank without mixing with hot water. There are two outlets; namely one for the withdrawal of hot water and the other one is used to feed cold water to the collector inlet. The entire length of the connecting pipes is covered with glass-wool insulation to reduce the heat loss. Solar radiation incident on the FPC heats the water inside the tube. The hot water with low density rises up to the upper header of FPC. The vacuum created by this flow is filled up by the cold water from the storage tank. Thus, the upper header end of the collector has hot water, while the lower header end has cold one. This hot water then enters the storage tank, from inlet 3, from where it can be withdrawn for further use.

With reference to Figure 2.30, the energy balance for the system and the tank can be written with the following assumptions:

(i) The connecting pipes are well insulated so that the heat loss from the pipes to the ambient is negligible.
(ii) The mean body temperature of the collector plate and tubes, the storage tank and the connecting pipes is equal to the mean temperature of water within them.
(iii) The mean water temperature in the collector tubes, the storage tank and the connecting pipes is equal and denoted by the mean system temperature, T_m.

i.e. $$T_c \sim T_{wc};\ T_p \sim T_{wp};\ T_t \sim T_{wt} \text{ and } T_c = T_p = T_t = T_m$$

$$W\frac{dT_m}{dt} + U_{LN}(T_m - T_a) + \dot{M}(t)\,C_w(T_m - T_{in}) = (\alpha\tau)I(t)F'A_{CN}$$

The rate of heat stored	The rate of heat lost by the absorber	The rate of heat lost due to withdrawal of hot water from tank	The rate of solar radiation absorbed

where

$$W = (mC)_c + (mC)_{wc} + (mC)_p + (mC)_{wp} + (mC)_t + (mC)_{wt} \tag{2.57}$$

$$U_{LN} = A_{CN} F' U_L + U_p A_P + U_i A_i \quad \text{and,} \quad A_{CN} = N A_C \tag{2.58}$$

(ii) Forced Circulation Water Heater

In this case, a water pump at the inlet of collector is used to transfer the hot water available at the upper header of the collector to the insulated storage tank. The collectors can also be connected in series for higher water temperature, if required. A single-loop water heating system will be discussed in the present section.

In this case the flow of the cold water from the storage tank to the collector is maintained by a pump (Figure 2.31). The energy balance can be written as,

$$\dot{Q}_{useful} = W_t \frac{dT_m}{dt} + U_T A_t (T_m - T_a) + \dot{M}(t) C_w (T_m - T_a) \tag{2.59}$$

2.10 SOLAR COOKER

Box-type solar cookers are suitable mainly for cooking of food, as shown in Figure 2.32. The cooking temperature is around 100 °C. The quantities of heat required for physical and chemical

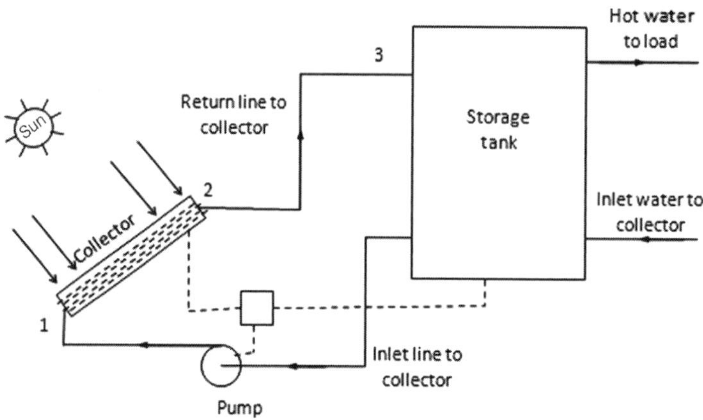

Figure 2.31 Schematic diagram of a single-loop water heating system.

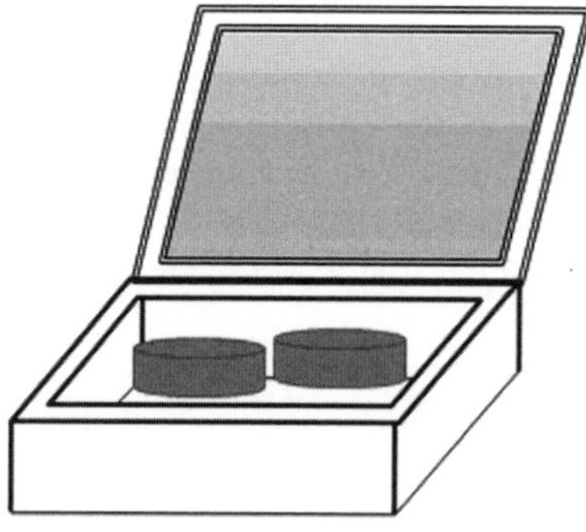

Figure 2.32 Box-type solar cooker.

changes involved in cooking are small. It is compared to the sensible heat of increasing food temperature and energy required for meeting heat losses that normally occur in cooking.

The complete thermal analysis of the cooker is complex due to the 3-dimensional transient heat transfers involved. In a solar cooker, there is no control over the temperature. Hence, the operation is transient unlike FPC. A quasisteady state is achieved when the stagnation temperature is attained.

The components of solar cooker are as follows:

(i) double glass lid to reduce the top loss coefficient;
(ii) aluminium cooking pots for fast heat transfer;
(iii) inner tray formed out of aluminium sheet, which is conductive;
(iv) outer box of teakwood with glass-wool insulation; and
(v) Booster mirror to increase the solar radiation on double-glazed surface.

There are two types of radiation incident on the double glass cover of the container of the solar cooker. One is the direct incident $I(t)$ and the other is that reflected from the mirror $\rho I'(t)$. The resultant incident radiation $I_T(t)$ can be given as the sum of the two.

For thermal analysis of the solar cooker, the following assumptions have been made:

(i) There is no stratification in water in the cooking pot.
(ii) The cooking pot is in contact with the inner surface of the cooker and hence their temperatures are the same.
(iii) The physical properties of the material to be cooked are the same as that of water.

The rate of thermal energy stored in the cooking pot of solar cooker is given by

$$Q_u = (MC)_w (dT_w/dt) \tag{2.60a}$$

The rate of net thermal energy available to the cooking pot of solar cooker is given by

$$\dot{Q}_u = F' A_p[(\alpha\tau) I_T(t) - U_L(T_p - T_a)] \tag{2.60b}$$

Now as per assumption (ii) $T_p \sim T_w$, and then the above two equations (2.60) can be equated to get water temperature as:

$$(MC)_w \frac{dT_w}{dt} = F' A_p[(\alpha\tau) I_T(t) - U_L(T_w - T_a)] \tag{2.61a}$$

or,

$$\frac{dT_w}{[(\alpha\tau) I_T(t) - U_L(T_w - T_a)]} = \frac{F' A_p}{(MC)_w} dt$$

After integration of above equation, one gets,

$$-\frac{1}{U_L} \ln[(\alpha\tau) I_T(t) - U_L(T_w - T_a)] = \frac{F' A_p}{(MC)_w} t + C \tag{2.61b}$$

At $t = 0$, $T_w = T_{w0}$ (initial water temperature in cooking pot) and hence from the above equation, one gets,

$$C = -(1/U_L)\ln[(\alpha\tau) I_T(t) - U_L(T_{w0} - T_a)]$$

On substituting of expression of C in Eq. (2.61b), we get,

$$\frac{1}{U_L}\ln\left[\frac{(\alpha\tau)I_T(t)-U_L(T_w-T_a)}{(\alpha\tau)I_T(t)-U_L(T_{w0}-T_a)}\right]=\frac{-F'A_p}{(MC)_w}t$$

The above equation can or,

$$t=-t_0\ln\left[\frac{(\alpha\tau)-\dfrac{U_L(T_w-T_a)}{I_T}}{(\alpha\tau)-\dfrac{U_L(T_{w0}-T_a)}{I_T}}\right] \quad (2.61c)$$

where t_0 is the time constant and is given by,

$$t_0=\frac{MC}{U_L F' A_p}$$

also be written as,

$$e^{-t/t_0}=\frac{(\alpha\tau)-\dfrac{U_L(T_w-T_a)}{I(t)}}{(\alpha\tau)-\dfrac{U_L(T_{w0}-T_a)}{I_T}} \quad (2.61d)$$

EXAMPLE 2.1

Calculate the time taken for the water at 40 °C in a cooking pot to boil in a solar cooker with the following specifications:

$(\alpha\tau)=0.7$; $F'=0.85$, $U_L=6$ W/m^2 °C, $A_p=0.36$ m^2, $MC=4\times 4190$ J/°C;
$T_a=15$ °C $T_{w0}=40$ °C, $T_w=100$ °C
I_T (with + without reflector) = (400 + 600) = 1000 W/m^2

Solution

From Eq. (2.61c)

$$t_0=(MC/U_L F'A_p)=9129\text{ s}$$

and

$$t=-9129\ln\left\{\frac{\left[0.7-\dfrac{6(100-15)}{1000}\right]}{\left[0.7-\dfrac{6(40-15)}{1000}\right]}\right\}=9676.7\text{ s}=161\text{ min}$$

or, $t=2.69$ h

EXAMPLE 2.2

If the water in a solar cooker with the specification given in Example 2.1 takes 2 h to boil; calculate the value of initial water temperature (T_{w0}) required.

Solution

From Example 2.1,

$$t_0 = 9129 \text{ s}, \; T = 2 \text{ h (given)} = 7200 \text{ s}$$

Using above values in Eq. (2.61c), we have,

$$7200 = -9129 \ln \left\{ \frac{\left[0.7 - \frac{6(100-15)}{1000}\right]}{\left[0.7 - \frac{6(T_{w0}-15)}{1000}\right]} \right\}$$

or,

$$0.45 = \frac{0.19}{0.7 - 6\frac{(T_{wo}-15)}{1000}}$$

Solving the above equation, we get

$$T_{w0} = 61.3\,^{\circ}\text{C} \sim 61\,^{\circ}\text{C}$$

EXAMPLE 2.3

Calculate the minimum value of the total intensity (I_T) incident on a solar cooker with the specifications given in Example 2.1, so as to heat the water at 50 °C in 3 h.

Solution

From Example 2.1, $t_0 = 9129$ s, $t = 3$ h $= 10\,800$ s (given), $T_{w0} = 50$ °C

Substitute the above value in Eq. (2.61c), we have

$$10\,800 = -9129 \ln \left[\frac{0.7 - \frac{6 \times 85}{I_T}}{0.7 - \frac{6 \times 35}{I_T}} \right]$$

or,

$$0.306 = \frac{0.7 - \frac{510}{I_T}}{0.7 - \frac{210}{I_T}}$$

Solving the above equation, we get,

$$I_T = 917.5 \sim 917 \text{ W/m}^2$$

That is the minimum intensity required is 917 W/m^2

2.11 APPLICATIONS OF FPC

There are many thermal applications of FPC, however, only some of them will be discussed briefly.

2.11.1 Swimming Pool

Solar energy can be used for the low temperature heating of fluids. It can be done either by direct or by indirect method. The direct method is generally referred to as a passive system while the indirect method is known as an active system. One of the applications of solar energy is in heating swimming pools. Outdoor swimming pools can be heated by both passive and active systems, while indoor swimming pools can only be heated by an active system.

(A) Passive Heating of Swimming Pools

A transparent floatable plastic cover is used over the surface of the water in the swimming pool during sunshine hours. The bottom and side inner surface of the swimming pool is preferably blackened to absorb solar radiation. The solar radiation is transmitted through the transparent plastic cover. It reaches the bottom of the pool and is finally absorbed by the blackened surface. Some of the radiation is absorbed by the water of the pool. Most of the absorbed solar radiation is convected to the water mass in the swimming pool and the rest is lost to the inside ground. The thermal loss can be minimised by using a layer of insulating material beneath the bottom surface. On receiving heat from the blackened bottom surface, the water gets heated and moves in an upward direction due to its low density. As the water temperature is higher than the ambient air temperature, particularly at night, convective and radiative heat losses occur from the water surface. In order to minimise these losses, the pool surface is covered with a waterproof material that acts as an insulating material during off-sunshine hours/night hours. As the transparent sheet must be removed when the pool is in use, there are unavoidable evaporative heat losses during this period.

(B) Active Heating of Swimming Pool

As explained above, the temperature of swimming pool water can be increased marginally by use of a passive system. But under extreme cold climatic conditions, passive heating is not sufficient. For further increasing of swimming pool water temperature, extra thermal energy is needed. This can be achieved by integrating FPC at the bottom of the swimming pool that can maintain the temperature of the swimming pool in a comfortable range ($\sim$20°C). This can be done by connecting a panel of FPC to the pool either directly or through a heat exchanger. The area of flat-plate collectors (FPC) depends on the capacity of the swimming pool and the climatic conditions.

An indoor swimming pool has certain advantages over an outdoor swimming pool namely,

(i) it is protected from dust, birds, climate, *etc.*;
(ii) it becomes an integral part of the building;
(iii) it is easy to clean;
(iv) it requires less maintenance; and
(v) it can be used in extreme cold climatic conditions.

A schematic diagram of an indoor solar swimming pool active heating is shown in Figure 2.33. There are three major components of the system, namely (i) panel of FPCs, (ii) circulation system consisting of pumps, valves and connecting pipes, (iii) control system integrated with a panel of collectors and the pool water.

The cost of an active system mainly depends on the type of collectors used for heating the swimming pool water. It is governed by the atmospheric conditions and the availability of solar radiation. The types of collectors generally used for swimming pool heating are as follows:

(i) Unglazed collectors:

a) plastic panels;
b) strip collectors;
c) plastic pipe collectors; and
d) permanent collectors.

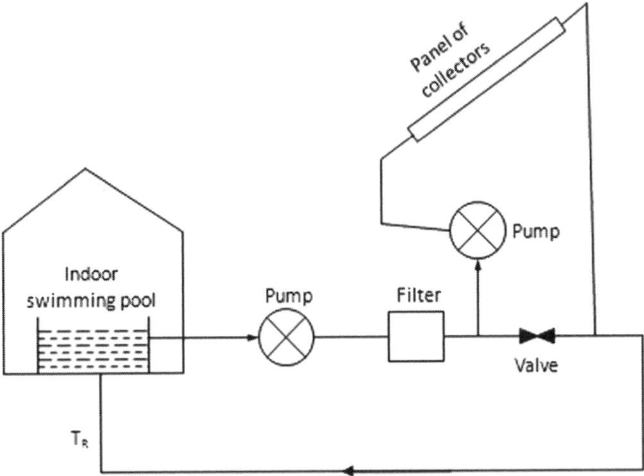

Figure 2.33 Schematic diagram of indoor swimming pool heated by an active system.

ii) Glazed collectors:

a) boxed collectors; and
b) integrated collector, *i.e.* an integral part of the roof of a building.

Unglazed collectors are preferred for thermal heating of indoor/outdoor swimming pool for a smaller temperature range. For a higher temperature requirement, glazed FPCs are preferred. Unglazed collectors are more cost effective than any conventional FPCs.

To write down the energy balance of an active indoor swimming pool integrated with a panel of FPCs, the following assumptions have been made:

i) The rate of useful energy available from the FPCs is uniformly fed at the bottom of swimming pool basin.
ii) There is no stratification along the depth of swimming pool.
iii) Heat losses through connecting pipes are negligible through proper insulation.
iv) The panel of FPCs is disconnected during off-sunshine hours.
v) The pool surface is covered with a floating, waterproof, insulating cover to minimise upward convective and radiative heat losses.
vi) The evaporative heat loss is considered only when the pool is in use and it is taken out by means of an exhaust fan.
vii) The bottom heat loss is considered under a steady-state mode.
viii) The enclosure temperature is almost equal to ambient air temperature.
ix) Proper insulating material is used in the basin of swimming pool to avoid downward heat losses.

The energy balance of swimming pool without and with evaporation (not in use) is given by,

Without evaporation:

$$M_w \frac{dT_w}{dt} \quad + \quad A_p U_t(T_w - T_a) \quad + \quad A_p h_b(T_w - T_R) \quad = \quad \dot{Q}_u \tag{2.62a}$$

Rate of thermal energy stored in the pool water	Rate of thermal energy lost from the pool surface to ambient	Rate of thermal energy lost through sides/bottom	Rate of thermal energy available from collector panel

where we have,

$$\dot{Q}_u = A_c F_R [(\alpha\tau)I - U_L(T_w - T_a)] \quad \text{without heat exchanger}$$
$$= 0 \quad \text{when collectors are disconnected from the pool}$$

h_e $(= h_{ew})$ is given by Eq. (1.27c).

With evaporation:

$$M_w \frac{dT_w}{dt} + A_p h_c(T_w - T_a) + 0.013\, h_c(P_w - \gamma P_a) A_p + A_p h_b(T_w - T_R) = \dot{Q}_u \quad (2.62b)$$

or,

$$M_w \frac{dT_w}{dt} + A_p h_r(T_w - T_a) + h_e(T_w - T_a) + A_p h_b(T_w - T_R) = \dot{Q}_u$$

2.11.2 Biogas Plant

Biogas plants have gained importance in the rural areas of developing and underdeveloped countries. Animal dung is available in plenty in these countries. The temperature requirement for biogas production is normally met by the usual atmospheric temperature. In addition to providing gas (for cooking, lighting and small-scale industries), biogas plant also produce good quality manure.

Biogas (70% methane (CH_4) and 30% carbon dioxide (CO_2)) has a calorific value of 20 MJ/m^3. Biogas is produced from the slurry (50% water + 50% dung) at an average optimum temperature of about 35–40 °C by chemical and biological processes. This is known as anaerobic fermentation.

The quantity of gas production depends on the nature of dung and temperature. The gas production rate for different type of dung is shown in Table 2.8. The optimum temperature for maximum production is achieved after a number of days. It is referred to as the retention period. The slurry is fed into the digester of the system. The production of biogas starts only after the retention period. The length of the retention period can be reduced by supplying additional thermal energy to the digester of the biogas system by external means, *i.e.* by heating slurry using either passive or active methods.

On the basis of the mean atmospheric air temperature during winter condition, India is divided into various zones as shown in Table 2.9.

Biogas is produced by the decomposition of decaying biomass and animal wastes by decomposer organisms namely fungi and bacteria. The process is favored by wet, warm and dark conditions. The final stages are achieved by different species of bacteria classified as (i) aerobic and (ii)

Table 2.8 Potential gas production from different feed stocks.

Type of Feedstock (dung)	*Gas yield per kg (m^3)*	*Normal manure availability per animal per day*	*Gas yield per day (m^3)*
Cattle	0.036	10.00	0.36
Buffalo	0.036	15.00	0.54
Pig	0.078	2.25	0.18
Chicken	0.062	0.18	0.01
Human (adult)	0.070	0.40	0.028

Table 2.9 Different zones in India.

	Retention period (days)	*Temperature*	*States*
Zone I	30	>25 °C	Andaman & Nicobar Islands; Andhra Pradesh, Goa, Karnataka
Zone II	35	20–25 °C	Kerala; Maharastra, Pondicherey, Tamil Nadu
Zone III	40	15–20 °C	Bihar, Gujarat, Haryana Jammu, Madhya Pradesh, Orissa, Punjab, Rajasthan, Uttar Pradesh and West Bengal
Zone IV	55	10–15 °C	Himachal, N-E States, Sikkim, Kashmir, Hill Station of UP
Zone V		not suitable for <10 °C	Srinagar, Leh, Simla biogas plant etc.

anaerobic. Aerobic bacteria flourish in the presence of oxygen. Anaerobic bacteria survive in closed conditions with no oxygen available from the environment. Being accomplished by micro-organisms, the reactions are called fermentations. The term "digestion" is often used in the anaerobic conditions that lead to methane. The energy available from the combustion of biogas ranges between 60 and 90% of the dry matter heat of combustion of the input material. The biogas is obtained from slurries of 50% water. An advantage of biogas is that the digested effluent is a significantly less health hazard than the input material. Economic benefits of biogas are achieved when the digester is placed in a flow of waste materials already present (sewage systems, cattle shed slurries, abattoir wastes, food-processing residues and municipal refuse landfill dumps). Biogas generation is used for both small- and large-scale operations, being particularly attractive for integrated farming.

The general equation for anaerobic digestion can be given as:

$$C_x H_y O_z + \left(x - \frac{y}{4} - \frac{z}{2}\right) H_2O \rightarrow \left(\frac{x}{2} - \frac{y}{8} + \frac{z}{4}\right) CO_2 + \left(\frac{x}{2} + \frac{y}{8} - \frac{z}{4}\right) CH_4$$

For cellulose this becomes,

$$(C_6 H_{10} O_5)_n + nH_2O \rightarrow 3nCO_2 + 3n\,CH_4$$

It is seen that digestion at higher temperature proceeds at a faster pace than at lower temperature.

The biochemical processes occur in three stages, each stage being facilitated by distinct sets of anaerobic bacteria.

(i) Insoluble biodegradable materials (cellulose, fats) are reduced to soluble carbohydrates and fatty acids. This takes about a day at 25 °C.
(ii) Acid-forming bacteria produce mainly acetic and propionic acid. This stage also takes about a day at 25 °C.
(iii) Methane-forming bacteria complete the digestion to ~70% CH_4 and ~30% CO_2. This takes about 14 days at 25 °C.

The methane-forming bacteria work well in mildly acidic conditions. For successful operation of the digester the two basic requirements are the maintenance of constant temperature and suitable input material.

The biogas system may be classified as follows:

(A) Floating Type

The slurry from the inlet settlement tank reaches the digester and gets heated by absorbing the solar radiation, as shown in Figure 2.34a. Biochemical process takes place. The biogas thus produced comes out through the pipe provided for gas supply. The digested fluid comes out to the outlet tank. A partition is provided in the digester for faster circulation of slurry. The floating gas holder provided at the top of the digester helps to keep the pressure constant. It rises when the pressure is increased to let out the biogas. It lowers when the pressure is decreased to stop the supply of biogas.

(B) Fixed-Dome Type

A fixed-dome digester is shown in Figure 2.34b. The dome is fixed and thus the pressure keeps on increasing. The process is identical to the floating-dome type except for the fact that in this case the pressure in the fermentation compartment is not fixed.

A comparison of the two biogas system is given in Table 2.10.

(C) Heating Process

In order to achieve the desired temperature (say about 37 °C) for greater fermentation for short retention periods, the following heating methods may be used:

(i) Passive Methods

(a) use of movable insulation and erection of solar canopy (greenhouse) over the dome, which is most applicable to floating-type biogas plant;
(b) use of water heater/solar still over the dome;

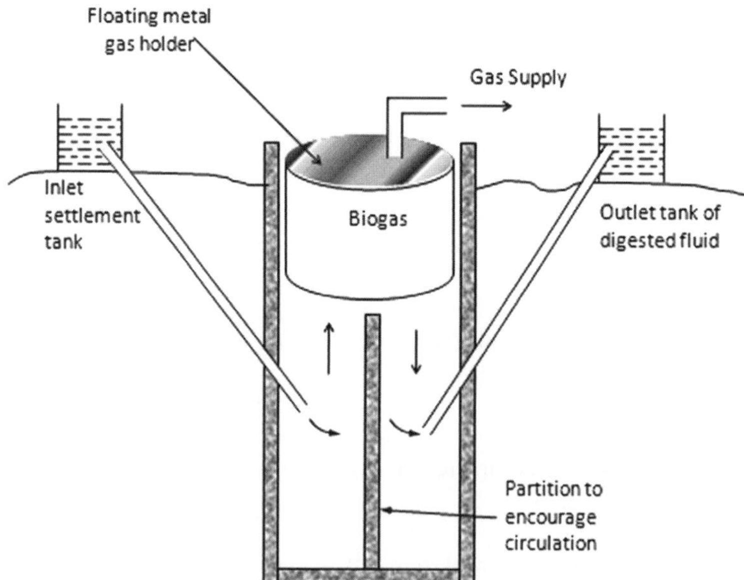

Figure 2.34a Floating gas holder digester.

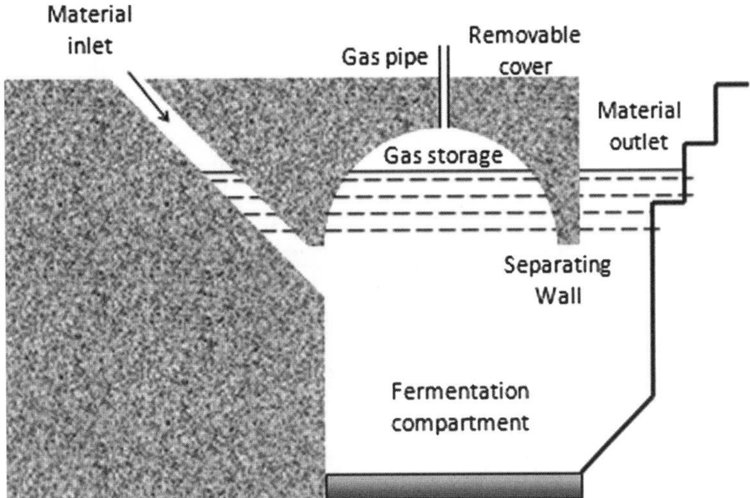

Figure 2.34b Fixed-dome digester.

Table 2.10 Comparison of two type of biogas plant.

Floating gas holder type	*Fixed dome type*
Gas is released at constant pressure	Gas is released at variable pressure
Location of defects in gas holder easy	Location of defects in dome difficult
Cost of maintenance is high	Cost of maintenance is low
Capital cost is high	Capital cost is low (for same capacity)
Movable drum does not allow the use of space for other purpose.	Space above the drum can be used
Temperature is low during winter	Temperature is high during winter
Life span is short	Life span is comparatively long
Requires relatively less excavation	Requires more excavation work

(c) erection of solar canopy (green house) over the dome with a provision for movable insulation;
(d) constructing the digester with insulating material to reduce the bottom and side losses.

(ii) Active Method

(a) **Hot charging of the slurry**: This method is most suitable for zone III and IV, where large heating is required due to the low mean ambient air temperature.
(b) **Integration of collectors**: It is integrated with the digester through a heat exchanger (Figure 2.35).

Biogas plant is integrated with a panel of FPCs through a heat exchanger placed inside the digester. When the heated water from the FPC passes through the heat exchanger, heat is transferred to the slurry by conduction and convection. It thus raises the temperature of the slurry in the digester. An excessive rise in the slurry temperature may lead to the death of anaerobic bacteria. Hence, the slurry should be heated slowly so that anaerobic bacteria exist for microbiological processes.

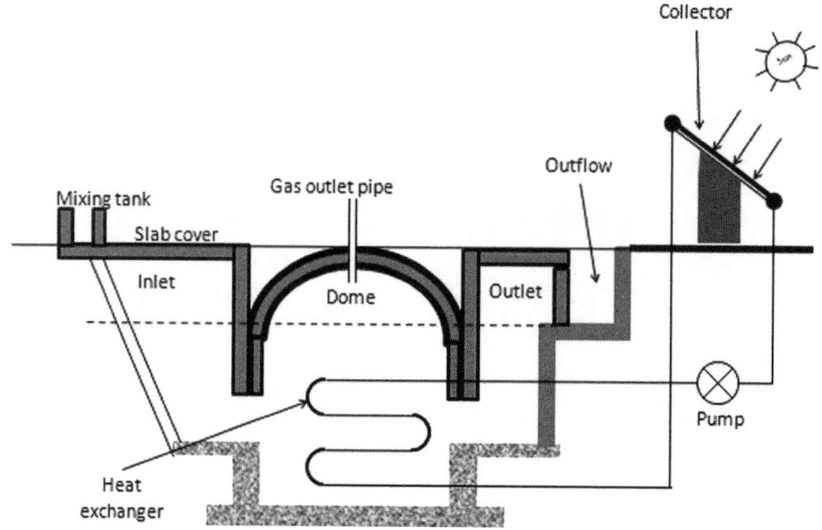

Figure 2.35 Fixed-dome-type biogas plant integrated with panel of collectors.

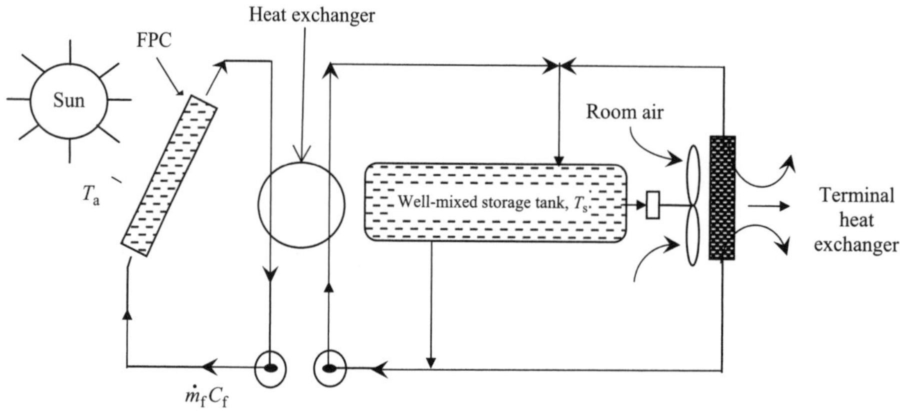

Figure 2.36 Schematic of solar water heating system with terminal heat exchanger.

2.11.3 Space Heating

The effect of thermal heating load to a room is shown in Figure 2.36.

If l_T is the demand expressed as the head load per unit collector area and it is defined as

$$l_T = \frac{(T_b - T_a)^+ (UA)_b}{A_c} \tag{2.63}$$

where the positive sign (+) indicates that only positive value in the bracket should be considered for thermal heating of a building. The $(UA)_b$ is the overall heat loss coefficient from the room of a building to ambient air through different walls/roof/floor/doors/windows, *etc.* The typical value of $[(UA)_b/A_c] = 0.01386$ MJ/m^2 °C h. This can vary depending upon design of a building. The base temperature of room air (T_b) can be considered as 25 °C for the present case.

Now, the expression for the net heat collected (q_N) can be written as

$$q_N = q_T - l_T \tag{2.64}$$

The average temperature of a well mixed storage tank can be written as

$$\bar{T}_s = T_{so} + \frac{q_N}{2m_s C_s} \tag{2.65}$$

With the help of Eqs. (2.63) to (2.65), an expression for the net heat collection can be derived and it is given by

$$q_N = \frac{F_x'[(\alpha\tau)I_T - U_L(T_{si} - \bar{T}_a)t_T] - l_T}{1 + \dfrac{F_x' U_L t_T}{2m_s C_s}} \tag{2.66}$$

The rise in temperature of the storage tank during the day can be obtained from

$$T_{so} = T_{si} + \frac{q_N}{m_s C_s} = T_{si} + \frac{q_T - l_T}{m_s C_s} \tag{2.67}$$

Further, an expression for the total heat collected can be obtained by substituting the above expression in Eq. (2.64) as

$$q_T = \frac{F_x'[(\alpha\tau)I_T - U_L(T_{si} - \bar{T}_a)t_T]}{1 + \dfrac{F_x' U_L t_T}{2m_s C_s}} + \frac{l_T}{1 + \dfrac{2m_s C_s}{F_x' U_L t_T}} \tag{2.68}$$

EXAMPLE 2.4

Calculate the heating load for a climatic data of a table given below. Also calculate the total heat collected and its corresponding storage temperature by using hourly steps.

Time, h	*Solar intensity (I), W/m²*	*Ambient air temperature (T_a), °C*	l_T, *MJ/m²*	q_T, *MJ/m², using Eq. (5.26)*	T_{so}, °C
9	424	11.4	0.1890	0.6046	46.30
10	558	13.5	0.1599	1.0159	48.98
11	641	15.8	0.1279	1.2879	52.61
12	669	18.1	0.0959	1.4057	56.70
13	641	19.8	0.0723	1.3564	60.71
14	558	20.9	0.0570	1.1427	64.10
15	424	21.3	0.0514	0.7742	66.36
Final					
Average					
Total	3915	17.23	0.7534	7.5874	

Solution

From Eq. (2.63), first the calculation of heating load will be carried out for the data given above

For the first time step:

$$l_T = \frac{(T_b - T_a)^+ (UA)_b}{A_c} = (25 - 11.4) \times 0.0139 = 0.1890 \text{ MJ/m}^2$$

For second time step:

$$l_T = \frac{(T_b - T_a)^+ (UA)_b}{A_c} = (25 - 13.5) \times 0.0139 = 0.1599 \text{ MJ/m}^2$$

Similar calculations can be carried out for other time steps. The results have been shown in the same table.

After calculating the heating load, the total heat collected can be obtained from Eq. (2.68) as

$$q_T = \frac{F_x'[(\alpha\tau)I_T - U_L(T_{so} - \bar{T}_a)t_T]}{1 + \dfrac{F_x' U_L t_T}{2m_s C_s}} + \frac{l_T}{1 + \dfrac{2m_s C_s}{F_x' U_L t_T}}$$

$$= \frac{1 \times [0.8 \times 424 - 5 \times (45 - 11.4)] \times 0.0036}{1 + \dfrac{1 \times 5 \times 0.0036}{2 \times 0.32}}$$

$$+ \frac{0.1890}{1 + \dfrac{2 \times 0.32}{1 \times 5 \times 0.0036}} = \frac{0.6163}{1.0281} + \frac{0.1890}{36.5872} = 0.6046 \text{ MJ/m}^2$$

For the first time step:
From Eq. (2.67),

$$T_{so} = T_{si} + \frac{q_T - l_T}{m_s C_s} = 45 + \frac{0.6046 - 0.1890}{0.32} = 46.30\ °\text{C}$$

For the second time step:

$$q_T = \frac{1 \times [0.8 \times 558 - 5 \times (45 - 13.5)] \times 0.0036}{1 + \dfrac{1 \times 5 \times 0.0036}{2 \times 0.32}} + \frac{0.1599}{1 + \dfrac{2 \times 0.32}{1 \times 5 \times 0.0036}} = \frac{1.0400}{1.0281} + \frac{0.1599}{36.5872}$$

$$= 1.0159 \text{ MJ/m}^2$$

$$T_{so} = 46.30 + \frac{1.0159 - 0.1599}{0.32} = 48.98\ °\text{C}$$

A similar calculation can be carried out for other time step. The results are given in the same table.

ADDITIONAL EXERCISES

Exercise 2.1: Derive an expression for air mass in terms of zenith angle (θ_z) and altitude angle (α).

Solution

We know

$$\text{air mass} = \sec\theta_z = \frac{1}{\cos\theta_z}$$

Since $\alpha = 90 - \theta_z$, hence

$$\text{air mass} = \text{cosec}\ \alpha$$

Exercise 2.2: Calculate the angle of incidence of beam radiation on a surface located at New Delhi at 1:30

(Solar time) on 16 February 1995, if the surface is tilted 45′ from the horizontal and pointed 30′ west of South (1° = 60′).

Solution

In the given problem, the value of n is 47.

$\delta = -13.0'$ (from Eq. (2.2)); $\omega = +22.5''$ (from Eq. (2.3)).

$\gamma = 30'$; $\beta = 45'$; $\phi = +28'35'$ (Table 2.1)

Now, the angle of incidence can be calculated by using Eq. (2.4) as follows:

$$\begin{aligned}\cos\theta_i = &(\cos(28°35') \cos45° + \sin(28°35') \sin45° \cos30°) \cos(-13°) \cos(22.5°)\\ &+ \cos(-13°) \sin(22.5°) \sin45° \sin30°\\ &+ \sin(-13°) (\text{in}(28°35') \cos45° - \cos(28°35') \sin45° \cos30°)\end{aligned}$$

$$\theta_i = \cos^{-1}(0.999) = 2.56°$$

Exercise 2.3: Calculate the number of daylight hours (sunshine hours) in Delhi on 22 December and 22 June 1995.

Solution

Here,

$\Phi = 28°35'$ (Table 2.1); For 22 December 1995, $n = 356$, and $\delta = -23.44°$

we have,

$$N = \frac{2}{15}\cos^{-1}[-\tan(-23.44°)\tan(28°35')]$$

$$N = \frac{2}{15}\cos^{-1}[-(-0.434)(0.545)]$$

$$= (2/15)\cos^{-1}[0.237] = 10.18\,\text{h}$$

Similarly, for 22 June 1995, $n = 173$; $\delta = 23.45°$ (from Eq. (2.2))

Further, we can have,

$$N = \frac{2}{15}\cos^{-1}(-\tan 23.45° \tan 28°35') = 13.82\,\text{h}$$

Exercise 2.4: Calculate the zenith angle of the Sun at New Delhi at 2.30 pm on 20 February 1995.

Solution

We have, $n = 51$; $\Phi = 28°35'$ (Table 2.1)

$\delta = -11.58°$ (Eq. (2.2)); $\omega = 37.5°$ (Eq. (2.3))

From Eq. (1.13), we have,

$$\begin{aligned}\cos\theta_z &= \cos(28°35') \cos(-11.58°) \cos(37.5°) + \sin(-11.58°) \sin(28°35')\\ &= 0.587\end{aligned}$$

$$\theta_z = \cos^{-1}(0.587) = 54.03°$$

Exercise 2.5: Calculate the net rate of useful energy per m^2 for the following parameters:

(i) The overall heat loss coefficient $(U_L) = 6.0$ W/m^2 °C and $F' = 0.8$
(ii) $\dot{m} = 0.35$ kg/s and $C_f = 4190$ J/kg °C
(iii) $I(t) = 500$ W/m^2 and $\alpha_0\tau_0 = 0.8$
(iv) $T_{fi} = 60$ °C and $T_a = 40$ °C.

Solution

The flow rate factor is given by Eq. (2.19b) as

$$F_R = [\dot{m}\, C_f\, /(\, A_c\, U_L\,)][1 - \exp\,(\, - \,A_c\, U_L\, F'/(\dot{m}C_f))]$$
$$= [0.35 \times 4190/(1 \times 6)][1 - \exp(-6 \times 1 \times 0.8/(0.35 \times 4190))] = 0.7986$$

The net rate of useful energy per m^2 can be calculated from Eq. (2.19c) as

$$\dot{q}_u = F_R\, [\alpha_0\tau_0 I(t) - U_L\, (\, T_{fi} - T_a\,)]$$
$$= 0.7986[0.8 \times 500 - 6(60 - 40)] = 223.6\ \mathrm{W/m^2}$$

Exercise 2.6: Determine the rate of useful energy per m^2 for Example 3.7(a) with the mass flow rate of 0.035 kg/s.

Solution

The flow rate factor, Eq. (2.19b), can be evaluated as

$$F_R = \frac{\dot{m}C_f}{A_c U_L}\left[1 - \exp\left(-\frac{A_c U_L F'}{\dot{m}C_f}\right)\right]$$

$$F_R = \frac{(0.035 \times 4190)}{(1 \times 6)}\left[1 - \exp\left(-\frac{1 \times 6 \times 0.8}{0.035 \times 4190}\right)\right] = 0.787$$

The net rate of useful energy per m^2 will be

$$\dot{q}_u = 0787[0.8 \times 500 - 6(60 - 40)] = 220.36\ \mathrm{W/m^2}$$

It is clear that the change in flow rate have no effect on $\dot{q}_u$ for a given design and climatic parameters of a collector.

Exercise 2.7: Find the threshold radiation flux for Example 5, for $\alpha_0\tau_0 = 0.80$, 0.60, 0.40 and 0.20.

Solution

From Example 3.1 (b), $T_p = 100$ °C, $T_a = 16$ °C and $U_L = 6$ W/m^2 °C

Therefore, the threshold radiation flux levels from Eq. (3.73b) are

$I_{th} = \frac{6(100-16)}{0.8} = 630$ W/m^2 for $\alpha_0\tau_0 = 0.8$
$= 840$ W/m^2 for $\alpha_0\tau_0 = 0.6$
$= 1260$ W/m^2 for $\alpha_0\tau_0 = 0.4$
$= 2520$ W/m^2 for $\alpha_0\tau_0 = 0.2$

This indicates that solar radiation can't be used for thermal heating for $\alpha_0\tau_0 = 0.2$ and 0.4 due to the higher value of I_{th}.

Exercise 2.8: A linear parabolic concentrator with aperture, $a = 2.00$ m and focal length, $f = 1.00$ m is continuously adjusted about a horizontal East–West axis. It is to be fitted with a liquid heating receiver unit whose length is 10.0 m. A strip of the reflector 0.21 m wide is shaded by the receiver. The receiver is designed to be just large enough to intercept all of the specularly reflected beam radiation when the incident beam radiation is normal to the aperture. The normal-beam radiation is 950 W/m^2. Here, $\alpha\tau$ for the receiver is 0.78 with radiation normal to the aperture, ρ is 0.84. The inlet fluid temperature is 180 °C, the ambient temperature is 20 °C and F_R is 0.85.

Solution

Since the radiation is normal to the aperture $\gamma = 1$. A fraction of the reflector 0.21/2.00, or 0.10, is shaded by the receiver, so (1 – 0.10) or 0.90 of the reflector is effective.

The product $\rho\gamma\tau\,\alpha$ is, $\rho\gamma\tau\,\alpha = 0.84 \times 1 \times 0.78 = 0.655$ (Eq. (2.40b))

Based on the area of the unshaded aperture,

$$\dot{q}_{ab} = S = 950 \times 0.655 = 622.25\ \mathrm{W/m}$$

At an estimated mean receiver surface temperature of 200 °C, the overall heat transfer coefficient, U_L is 13.7 W/m^2 °C.

At a concentration ratio A_a/A_r of $(2.00 - 0.21) \times 10/(0.21 \times 10) = 8.5$

With the help of Eq. (2.48), we can have

$$\begin{aligned}\dot{Q}_u &= 0.85 \times 10.0\ (2.00 - 0.21)\ [622 - 13.7/8.5\ (180 - 20)]\\ &= 0.85 \times 10.0 \times 1.79\ [364] = 5540\ \mathrm{W} = 5.54\ \mathrm{kW}.\end{aligned}$$

Exercise 2.9: A cylindrical parabolic concentrator with width 2.0 m and length 8 m has an absorbed radiation, per unit area of aperture, of 400 W/m^2. The receiver is a cylinder painted flat black and surrounded by an evacuated glass cylindrical envelope. The absorber has a diameter of 55 mm, and the transparent envelope has a diameter of 85 mm. The collector is designed to heat a fluid entering the absorber at 220 °C at a flow rate of 0.04 kg/s. The value of C_p for the fluid is 3.26 kJ/kg °C. The heat transfer coefficient inside the tube is 280 W/m^2 °C and the overall loss coefficient is 12 W/m^2 °C. The tube is made of stainless steel ($K = 16$ W/m °C) with a wall thickness of 5 mm. If the ambient temperature is 22 °C, calculate the useful gain and exit fluid temperature.

Solution

The area of the receiver $A_r = \pi\, D.L = \pi \times 0.055 \times 8 = 1.382\ \mathrm{m}^2$

Taking into account the shading of the central part of the collector by receiver,

$$A_a = (2.0 - 0.085)\ 8 = 15.32\ \mathrm{m}^2$$

From Eq. (2.44),

$$F' = \frac{1/U_L}{\dfrac{1}{U_L} + \dfrac{D_0}{h_{fi} D_i} + \left(\dfrac{D_0}{2K}\ln\dfrac{D_o}{D_i}\right)} = \frac{1/12}{\dfrac{1}{12} + \dfrac{0.055}{280 \times 0.045} + \dfrac{0.055}{2 \times 16}\ln\left(\dfrac{0.055}{0.045}\right)} = 0.946$$

F_R is given by Eq. (8.17), and its value is

$$F_R = \frac{\dot{m}\,C_p}{A_r\,U_L}\left[1 - \exp\left(-\frac{A_r U_L F'}{\dot{m} C_p}\right)\right]$$

Here,

$$\frac{\dot{m}C_p}{A_r U_L F'} = \frac{0.04 \times 3.26 \times 10^3}{1.382 \times 12 \times F'} = \frac{7.87}{F'} = \frac{7.87}{0.946} = 8.31$$

Substituting the above values in expression for F_R, we get

$$F_R = \frac{\dot{m}\,C_p}{A_r\,U_L}\left[1 - \exp\left(-\frac{A_r U_L F'}{\dot{m} C_p}\right)\right] = 7.87\left[1 - \exp\left(-\frac{1}{8.31}\right)\right] = 0.89$$

Now, the useful gain is,

$$\dot{Q}_u = F_R A_a\left[S - \frac{A_r}{A_a} U_L (T_{fi} - T_a)\right] = 0.89 \times 15.32\left[400 - \frac{1.382}{15.32} \times 12 \times (220 - 22)\right] = 2787.5\ \mathrm{W}$$

Now, the expression for $\dot{Q}_u$ can be written as:

$$\dot{Q}_u = \dot{m}C_p(T_{fo} - T_{fi})$$

or

$$T_{fo} = T_{fi} + \frac{\dot{Q}_u}{\dot{m}C_p}$$

Thus, the exit fluid temperature is

$$T_{fo} = T_{fi} + \frac{\dot{Q}_u}{\dot{m}C_p} = 220 + \frac{2787.5}{0.04 \times 3260} = 241.4\,^\circ\mathrm{C}$$

OBJECTIVE QUESTIONS

2.1 The diffuse radiation on the extraterrestrial region is
   (a) Maximum
   (b) Minimum
   (c) Zero
   (d) None of these

2.2 The solar constant is measured
   (a) Near Sun
   (b) Near Earth
   (c) Extraterrestrial region
   (d) Terrestrial region

2.3 The atmosphere containing greenhouse gases transmits
   (a) Ultraviolet radiation
   (b) Short-wavelength radiation
   (c) Long-wavelength radiation
   (d) Infrared radiation

2.4 The latitude of North/South pole is
(a) $\pm 90°$
(b) $\pm 45°$
(c) zero
(d) $\pm 60°$

2.5 The short-wavelength radiation reaching on the Earth is
(a) 0.03–0.30 μm
(b) 3–30 μm
(c) 0.3–3 μm
(d) None of these

2.6 The black body temperature of the Sun is
(a) 6000 K
(b) 600 K
(c) 16 000 K
(d) 3000 K

2.7 The value of solar radiation in summer is maximum on the surface having inclination equal to
(a) Latitude
(b) Zero
(c) 45°
(d) 90°

2.8 The optical properties of atmosphere and transparent materials are
(a) Different
(b) Differ by 30%
(c) Same
(d) Opposite

2.9 The minimum solar radiation on the Earth occurs at
(a) Early morning
(b) Late evening
(c) Night time
(d) Noon time

2.10 The one hour angle (ω) is
(a) 30°
(b) 95°
(c) 45°
(d) 15°

2.11 The sunshine hour (N) at the equator
(a) Varies with "n"
(b) Constant
(c) Zero
(d) 24 h

2.12 The values of sunshine hour (N) at the equator
(a) 12 h
(b) 24 h
(c) Zero
(d) None of these

2.13 The reflected solar radiation on horizontal surface is
(a) Zero
(b) Minimum
(c) Maximum
(d) 1367 W/m^2

2.14 The values of solar constant is
   (a) 1000 W/m^2
   (b) Zero
   (c) 1367 W/m^2
   (d) 500 W/m^2
2.15 The value of extraterrestrial solar radiation is
   (a) Constant
   (b) Varying
   (c) 1320–1420 W/m^2
   (d) None of these
2.16 The blackbody temperature of planet Earth is
   (a) 300 K
   (b) 3000 K
   (c) 600 K
   (d) 30 K
2.17 The relation between zenith (θz) and solar altitude (α) angles is
   (a) $\theta z + \alpha = 60°$
   (b) $\theta z + \alpha = 90°$
   (c) $\theta z + \alpha = -90°$
   (d) $\theta z + \alpha = 0$
2.18 The energy generated at the core of the Sun is due to
   (a) Fission reaction
   (b) Fusion reaction
   (c) Conduction
   (d) Radiation
2.19 The radiation from the Sun is governed by
   (a) Stefan–Boltzmann law
   (b) Fourier's law of conduction
   (c) Faraday's law
   (d) Wien's displacement law
2.20 The radiation from the Earth is governed by
   (a) Fourier's law of conduction
   (b) Stefan–Boltzmann law
   (c) Wien's displacement law
   (d) None of these
2.21 The unit of thermal conductance of air is
   (a) Same as heat transfer coefficient
   (b) Different from heat transfer coefficient
   (c) Same as thermal conductivity of air
   (d) None of them
2.22 The thermal conduction of air is unaffected for
   (a) Larger air cavity
   (b) Smaller air cavity
   (c) Infinity air cavity
   (d) Zero air cavity
2.23 The thermal conductance of air is very large for
   (a) Smallest air cavity
   (b) Largest air cavity
   (c) Zero air cavity
   (d) None of them

2.24 The conduction, convection and radiation losses are
(a) Dependent on each other
(b) Independent of each other
(c) Independent of temperature
(d) None of them
2.25 Thermal expansion coefficients depend on
(a) Temperature
(b) Thermal air conductance
(c) Thermal conductivity
(d) None of these
2.26 The convective heat transfer depends on
(a) Physical properties of fluid
(b) Physical properties of solid
(c) Characteristics dimension
(d) All
2.27 The forced convective heat transfer at higher temperature is higher than
(a) Free convective
(b) Conductive
(c) Radiative
(d) Evaporative
2.28 The radiation heat transfer between two surfaces is mainly due to
(a) Short wavelength
(b) Infrared
(c) UV
(d) Long-wavelength radiations
2.29 The evaporative heat transfer coefficient (h_{cw}) is
(a) Proportional to h_{cw} (convective heat transfer coefficient)
(b) Inversely proportional to h_{cw}
(c) Independent of convective heat transfer coefficient
(d) None
2.30 The evaporative heat transfer coefficient (hew) depends on convective heat transfer coefficient due to
(a) Lewis relation
(b) Newton's law
(c) Fourier's law
(d) None
2.31 The shape (Geometrical) factor for nonparallel surfaces is
(a) One
(b) Ten
(c) Less than one
(d) Infinity
2.32 The shape (geometrical) factor for nonparallel surfaces is
(a) One
(b) Less than one
(c) Ten
(d) Infinity
2.33 The radiative heat transfer coefficient for parallel surfaces having a temperature difference of 1 °C is
(a) 6 W/m^2 K
(b) 60 W/m^2 K

(c) 0.6 W/m^2 K
(d) None

2.34 The expression for radiative heat transfer coefficient (h) for surfaces having temperatures almost the same but different is
(a) $4\varepsilon\sigma T^4$
(b) $4\varepsilon\sigma T^3$
(c) $1\varepsilon\sigma T^3$
(d) $0.4\varepsilon\sigma T^3$

2.35 An expression for wind dependent convective heat transfer coefficient is
(a) $h = 3 + 2.8V$
(b) $h = 2.8 + 3V$
(c) $h = 3 + 3V$
(d) None

2.36 An expression for wind dependent convective and radiative heat transfer coefficient is
(a) $h = 3.8 + 5.7V$
(b) $h = 3.8 + 3V$
(c) $h = 5.7 + 3.8V$
(d) None

2.37 The flow rate factor (F_R) is
(a) More than flat plate collector efficiency (F')
(b) Less than flat plate collector efficiency (F')
(c) Equal to the flat plate collector efficiency (F')
(d) None

2.38 The flow rate factor (FR) depends on
(a) Mass flow rate
(b) An overall heat transfer coefficient (U_L)
(c) Area of collector (A_c)
(d) All

2.39 The flow rate factor (FR) is derived under
(a) Natural flow
(b) Turbulent flow
(c) Streamline flow
(d) None

2.40 The vacuum between absorber and glass cover can be maintained only in
(a) Flat plate collector
(b) Concentrating collector
(c) Tubular collector
(d) None

2.41 The outlet fluid temperature of a collector is maximum in
(a) Flat plate collector
(b) Evacuated tubular collector
(c) Fin-type flat plate collector
(d) None

2.42 The instantaneous efficiency of evacuated tubular collector at higher operating temperature is
(a) Higher than the flat plate collector
(b) Lower than the flat plate collector
(c) Equal to flat plate collector
(d) None

2.43 The overall top loss coefficient for evacuated tubular collector (U_t) is
(a) Much less than the value for flat plate collector
(b) More than the value for flat plate collector
(c) Equal to the value for flat plate collector
(d) None

2.44 The Phillips (Germany) evacuated tube collectors are connected in
(a) Series
(b) Parallel
(c) Series and parallel
(d) None

2.45 The Sanyo evacuated tube collectors are connected in
(a) Parallel
(b) Series
(c) Series and parallel
(d) None

2.46 In solaron air collector, the air passes through
(a) Evacuated tube
(b) Below evacuated tube
(c) Above evacuated tube
(d) None

2.47 In Phillips (Germany) water collector, the water passes through
(a) Below evacuated tube
(b) Above evacuated tube
(c) Through evacuated tube
(d) None

ANSWERS

2.1 **(c)**; 2.2 **(c)**; 2.3 **(b)**; 2.4 **(a)**; 2.5 **(c)**; 2.6 **(a)**; 2.7 **(b)**; 2.8 **(c)**; 2.9 **(c)**; 2.10 **(d)**; 2.11 **(b)**; 2.12 **(a)**; 2.13 **(c)**; 2.14 **(c)**; 2.15 **(c)**; 2.16 **(a)**; 2.17 **(b)**; 2.18 **(b)**; 2.19 **(a)**; 2.20 **(b)**; 2.21 **(a)**; 2.22 **(a)** & **(c)**; 2.23 **(a)**; 2.24 **(b)**; 2.25 **(a)**; 2.26 **(a)** & **(c)**; 2.27 **(a)** & **(b)**; 2.28 **(d)**; 2.29 **(a)**; 2.30 **(a)**; 2.31 **(a)**; 2.32 **(b)**; 2.32 **(a)**; 2.33 **(b)**; 2.34 **(b)**; 2.35 **(c)**; 2.36 **(b)**; 2.37 **(d)**; 2.38 **(c)**; 2.39 **(c)**; 2.40 **(b)**; 2.41 **(b)**; 2.42 **(a)**; 2.43 **(a)**; 2.44 **(a)**; 2.46 **(b)**; 2.47 **(a)**.

REFERENCES

1. Tiwari, G. N. 2002. *Solar Energy, Fundamentals, Design, Modeling and Applications*, Narosa Publishing House, New Delhi, India.
2. Duffie, J. A. and Beckman, W. A. 1991. *Solar Engineering of Thermal Processes*, John Wiley and Sons Inc., New York.
3. Liu, B. Y. H. and Jordan, R. C. 1962. *ASHRAE J.*, **3**(10), 53–59.
4. Whillier, A. 1964. *Solar Energy* **8**(3), 95–598.
5. Winston, R. 1974. *Solar Energy* **16** p. 89.
6. Baranov, V. K. 1975. *Gelioteckhika* II (3–4) pp. 45–52 (English Translation pp. 36–41).
7. Rabl, A. 1976. *Solar Energy* **18**, 497–511.
8. Sodha, M. S., Bansal, N. K., Kumar, K., Bansal, P. K., and Malik, M. S. S. 1987. *Solar Crop Drying*. Volume II. CRC Press. Boca Raton, Florida.
9. Tamini, A. 1987. *Int. J. Solar Wind Technol.* **4**, 443.
10. Fernadez, J. L. and Chargoy, N. 1990. *Solar Energy* **44**(4), 215–223.
11. Toyama, S., Aragaki, T., Murase, K. and Tsumura, K. 1983. *Desalination* **45**, 101–108.

The Ecofriendly Photovoltaic Thermal (PVT) System will be the Future Electrical and Thermal Energy Power to Achieve Energy Security for Developing/Underdeveloped Countries.

CHAPTER 3

Photovoltaic and Photovoltaic Thermal Systems

3.1 INTRODUCTION

A solar cell or photovoltaic (PV) cell is a device that converts solar energy into electricity by the photovoltaic effect. Photovoltaics is the field of technology and research related to the application of solar cells as solar energy. Sometimes the term solar cell is reserved for devices intended specifically to capture energy from sunlight, while the term photovoltaic cell is used when the source is unspecified. Photovoltaic generation of power is caused by radiation that separate positive and negative charge carriers in absorbing material. In the presence of an electric field, these charges can produce a current for use in an external circuit. Such fields exist permanently at junctions or inhomogeneities in materials as "built-in" electric fields and provide the required EMF for useful power production.

Junction devices are usually known as photovoltaic cells or solar cells, although, the term is a misnomer in the sense that it is the current that is produced by the radiation photons and not the "voltage". The cell itself provides the source of electromagnetic field (EMF). It is to be noted that photoelectric devices are electrical current sources driven by a flux of radiation. A majority of photovoltaic cells are silicon semiconductor junction devices. Thus, in order to study the photovoltaic cells we should have an understanding of the basics of the semiconductors; a brief description of which follows in the subsequent sections.

A solar cell constitutes the basic unit of a PV generator that, in turn, is the main component of a solar generator. A PV generator is the total system consisting of all PV modules that are connected in series or parallel or combination of both series and parallel with each other.

Solids can be divided into three categories, on the basis of electrical conduction through them. They are: conductors, semiconductors and insulators. The gap between the valence band and the conduction band (forbidden energy band) in the case of insulators ($h\upsilon < E_g$, h is Planck's constant and υ is the frequency), is very large. Thus, it is not possible for the electrons in the valence band to reach the conduction band; hence there is no conduction of current. In the case of semiconductors ($h\upsilon > E_g$), the gap is moderate and the electrons in the valence band may acquire sufficient energy for them to cross the forbidden region (Figure 3.1). While, in the case of conductors ($E_g \approx 0$), no forbidden gap exists and electron can easily move to the conduction band.

The semiconductor can again be divided into two categories: intrinsic and extrinsic. Intrinsic (pure) semiconductors have a Fermi level in the middle of the conduction and valence bands. In this case, the densities of free electrons in conduction and free holes in the valence band are equal $n = p = n_i$ and each is proportional to exp $(-E_g/2kT)$.

Advanced Renewable Energy Sources
G. N. Tiwari and R. K. Mishra

Published by the Royal Society of Chemistry, www.rsc.org

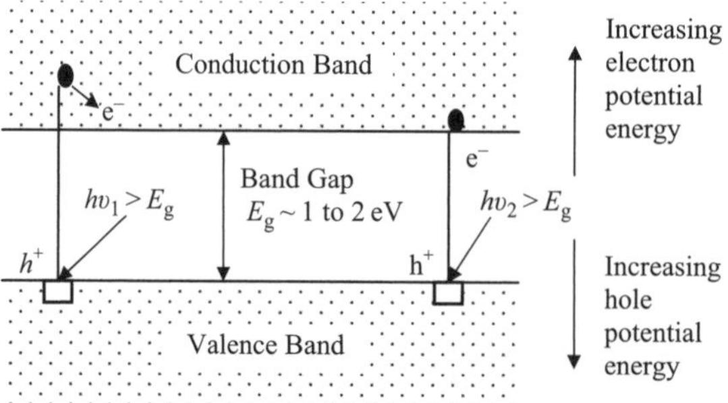

Figure 3.1 Semiconductor band structure of intrinsic material. (i) Photon absorption $h\upsilon < E_g$, no photo-electric absorption. (ii)($h\upsilon_1$–E_g) excess energy dissipated as heat and (iii) $h\upsilon_2 = E_g$, photon energy equals bandgap.

EXAMPLE 3.1

Determine the bandgap in a silicon crystal at 40 °C.

Solution

The variation of bandgap with temperature is given by the relation:

$$E_g(T) = E_g(0) - \frac{aT^2}{T+b}$$

where, a and b for different materials are as follows:

Material	$E_g(0)$	a	b
Silicon (Si)	1.16 eV	7×10^{-4} eV K^{-1}	1100 K
Gallium arsenide (GaAs)	1.52 eV	5.8×10^{-4} eV K^{-1}	300 K

Substituting the appropriate values in the above equation, we get

$$E_g(T) = 1.16 - \frac{7 \times 10^{-4} \times (313)^2}{313 + 1100} = 1.11 \text{ eV}$$

Solar cells are classified into three generations that indicates the order in which each became prominent. At present there is concurrent research into all three generations, while the first-generation technologies are most highly represented in commercial production, accounting for 89.6% of 2007 production.[1]

3.1.1 First Generation

First generation cells consist of large area, high quality and single junction devices. First generation technologies involve high energy and labour inputs that prevent any significant progress in reducing production costs. Single junction silicon devices are approaching the theoretical limiting efficiency of 33%[2] and achieve cost parity with fossil fuel energy generation after a payback period of 5–7 years.

3.1.2 Second Generation

Second generation materials have been developed to address energy requirements and production costs of solar cells. Alternative manufacturing techniques such as vapour deposition and electroplating are advantageous as they reduce high temperature processing significantly. It is commonly accepted that as manufacturing techniques evolve production costs will be dominated by constituent material requirements,[2] whether this be a silicon substrate, or glass cover. Second generation technologies are expected to gain market share in 2008.[1]

The most successful second generation materials are cadmium telluride (CdTe), copper indium gallium selenide, amorphous silicon and micromorphous silicon.[1] These materials are applied in a thin film to a supporting substrate such as glass or ceramics reducing material mass and therefore costs. These technologies do hold promise of higher conversion efficiencies and offer significantly lower production costs.

3.1.3 Third Generation

Third generation technologies aim to enhance poor electrical performance of second generation thin-film technologies, while maintaining very low production costs. Current research is targeting conversion efficiencies of 30–60%, while retaining low cost materials and manufacturing techniques.[2] There are a few approaches to achieving these high efficiencies:[3]

1. multijunction photovoltaic cell;
2. modifying incident spectrum (concentration);
3. use of excess thermal generation to enhance voltages or carrier collection.

3.2 PHOTOVOLTAIC FUNDAMENTALS

3.2.1 Doping

In order to increase the conductivity of intrinsic semiconductors, controlled quantities of specific impurity ions are added to the intrinsic semiconductor to produce doped (extrinsic) semiconductors. Impurity ions of valency less than the semiconductor enter the semiconductor lattice and become electron-acceptor sites that trap free electrons. These traps have an energy level within the bandgap, but near the valence band. The absence of free electrons produce positively charged states called holes that also move through the material as free carriers. Such a material is called p-type material, having holes as majority carriers and electrons as minority carriers. If impurity ions of a valency greater than that of the semiconductor are added then n-type material results that has electrons as majority carriers and holes as minority carriers. Both p- and n-type extrinsic semiconductors have higher electrical conductivity than the intrinsic basic material.

3.2.2 Fermi Level

The Fermi level is the apparent energy level within the forbidden bandgap from which majority carriers (electrons in n-type and holes in p-type) are excited to become charge carriers. The probability for the majority-carrier excitation varies as $\exp\,[-e\varphi/(kT)]$, where e is the charge of the electron and hole, and φ is the electric potential difference between the Fermi level and the valence or conduction band, T the temperature (K) and k the Boltzmann constant, 1.38×10^{-23} J/K.

For *n*-type material,

$$E_F = E_c + kT \ln \frac{N_0}{N_c} \tag{3.1}$$

where E_F is the Fermi-energy level, E_c the conduction band energy; k the Boltzmann constant; N_0 the donor concentration and N_c the effective density of states in conduction band, and is constant at fixed temperature T.

For p-type material,

$$E_F = E_V - kT \ln \frac{N_A}{N_V} \tag{3.2}$$

where E_V is the valence band energy, N_A is the acceptor ion concentration, N_V is the effective density of states in the valence band.

EXAMPLE 3.2

Calculate the shift in Fermi energy level in a silicon crystal doped with a V group impurity of concentration 10^{15} cm^3. Given: the effective density of states in the conduction band $= 2.82 \times 10^{19}$ cm^3 and the bandgap $= 1.1$eV; room temperature $= 27$ °C.

Solution

From Equation 3.1, we have,

$$E_F = E_C + kT \ln (N_D/N_C)$$

If the valence band is taken as the reference level, then $E_C = 1.1$ eV. Substitution of the values gives,

$$E_F = 1.1 + (1.38 \times 10^{-23}/1.6 \times 10^{-19}) \times 300 \ln (10^{15}/2.82 \times 10^{19})$$
$$= 1.1 + 2.00 = 3.1$$

The shift is 3.1 – 0.55 = 2.55 eV.

3.2.3 p-n Junction

The basic requirement for photovoltaic energy conversion is an electronic asymmetry in the semiconductor structure known as a junction. When n-and p-type semiconductors are brought into contact, then electrons from the n-region near the junction would flow to the p-type semiconductor leaving behind a layer that is positively charged. Similarly, holes will flow in the opposite direction leaving behind a negatively charged layer. A steady state is finally reached, resulting in a junction, which contains practically no mobile charges, hence the name depletion region.

The p-n junction (Figures 3.2a and b) may be connected to a battery in two ways (i) in forward bias as in Figure 3.3a, the positive conventional circuit current passes from p to n material across a reduced band potential difference V_B (ii) in reverse bias as in Figure 3.3b, the conventional positive current has an increased band potential difference V_B to overcome. Thermally or otherwise generated electrons and holes recombine after a typical relaxation time τ, having moved a typical diffusion length L through the lattice. In intrinsic material the relaxation time can be long, $\tau \sim 1$ s, but for commercial doped materials relaxation time are much shorter, $\tau \sim 10^{-2}$ to 10^{-8} s.

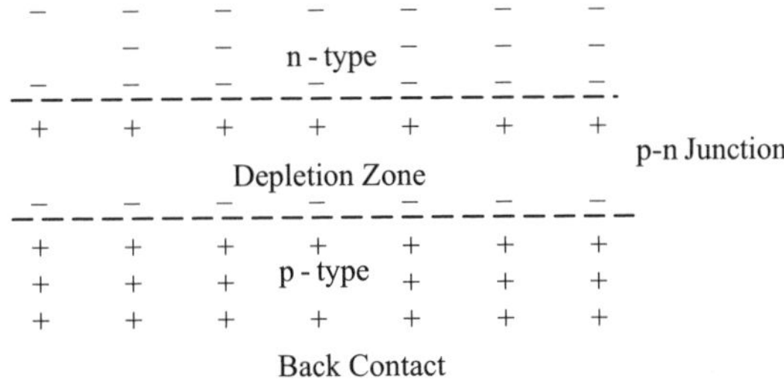

Figure 3.2a Energy levels in a p-n junction.

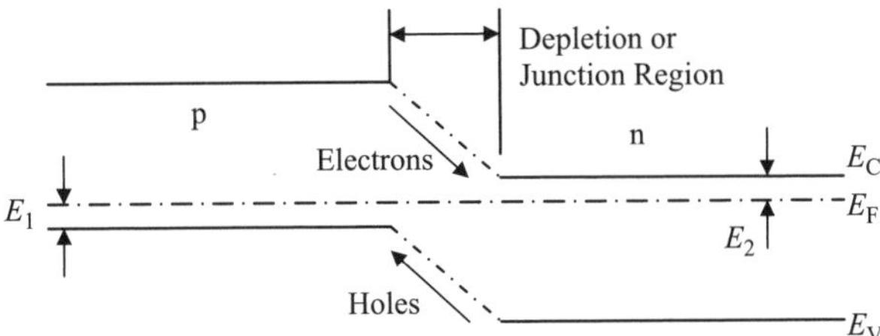

Figure 3.2b Energy levels for a p-n junction.

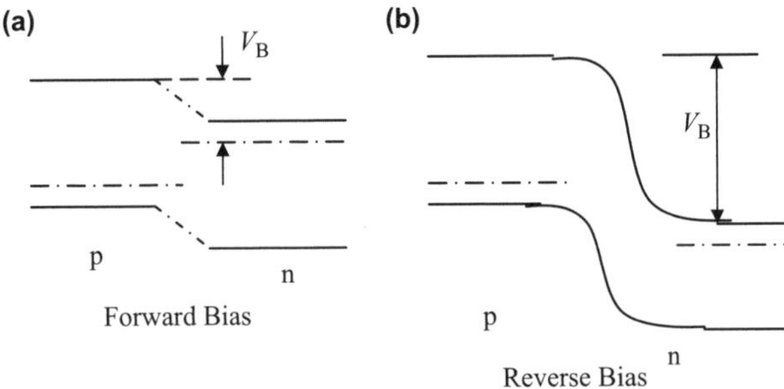

Figure 3.3 Energy levels for p-n junction with (a) forward bias and (b) reverse bias.

Electrons and holes may be generated thermally or by light, and become carriers in the material (Figure 3.4). Minority carriers in the depletion region are pulled across electrostatically down their respective potential gradients. The minority carriers that cross the region become majority carriers in the adjacent layer. The passage of these carriers causes the generation current, I_g, which is mainly controlled by temperature in a given junction without illumination.

In an isolated junction, there can be no overall imbalance of current across the depletion region. Thus, a reverse recombination current I_r of equal magnitude occurs from the bulk material, this restores the normal internal electric field. The band potential V_B is slightly reduced by I_r. The recombination current, I_r can be varied by external bias as explained earlier (Figure 3.5).

3.2.4 p-n Junction Characteristics

The p-n junction characteristics are given in Figure 3.6.

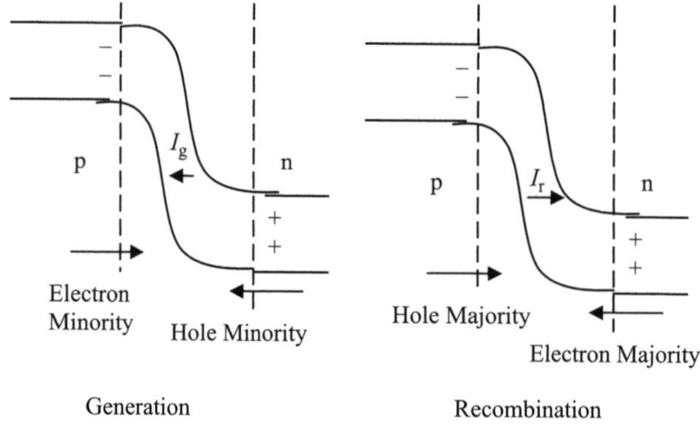

Figure 3.4 Generation and recombination currents at p-n junction.

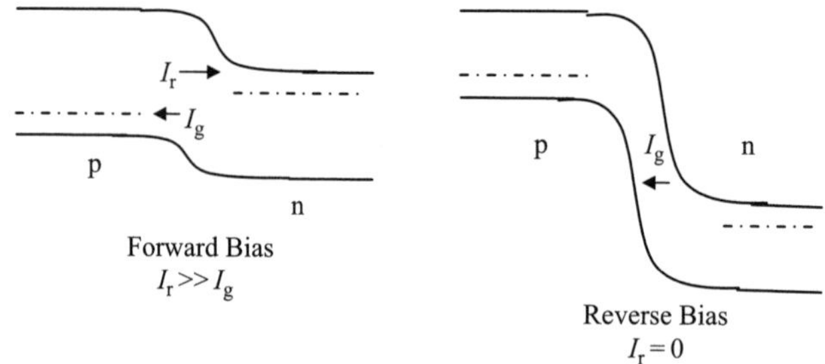

Figure 3.5 Generation and recombination currents with external bias.

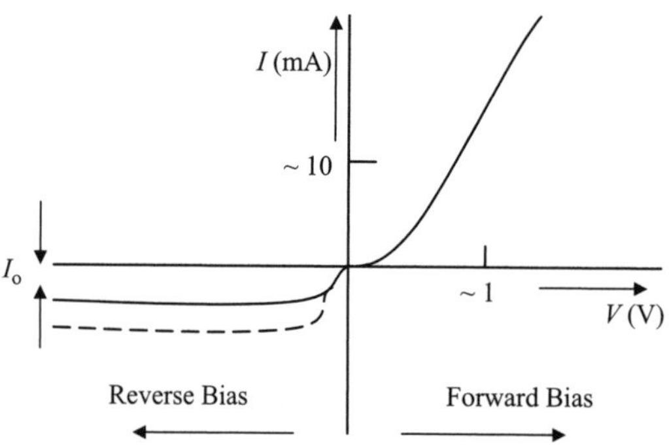

Figure 3.6 Dark characteristics for p-n junction.

With no external bias ($V = 0$).

$$I_r = I_g \tag{3.3}$$

with a forward bias of voltage V, the recombination current becomes an increased forward current.

$$I_r = I_g \exp(eV/kT) \tag{3.4}$$

The total current (with no illumination) is,

$$I_D = I_r - I_g = I_g[\exp(eV/kT) - 1] \tag{3.5}$$

The above equation is the Shockley equation and can be written as,

$$I_D = I_0[\exp(eV/kT) - 1] \tag{3.6}$$

where I_D (= I_g) is the saturation current under reverse bias, before avalanche breakdown occurs. It is also known as the leakage or diffusion current. For good solar cells $I_0 \sim 10^{-8}$ A m^{-2}. Its value increases with temperature (Figure 3.6, dotted curve).

EXAMPLE 3.3

Determine the value of saturation current for silicon at 40 °C.

Solution

The dependence of saturation current on temperature is given by the relation:

$$I_0 = AT^3 \exp(E_g/kT)$$

Here, A is the nonideality factor and its value is taken as 1,

$$E_g = 1.11 \text{ eV} = 1.11 \times 1.6 \times 10^{-19} \text{ J}.$$

Substituting the known values in the above equation, we get,

$$I_0 = (40 + 273)^3 \exp\left(-\frac{1.11 \times 1.6 \times 10^{-19}}{1.38 \times 10^{-23} \times 313}\right) = 4.26 \times 10^{-11} \text{ A m}^{-2}$$

EXAMPLE 3.4

Determine the value of dark current in the limiting case $V \to 0$.

Solution

From Eq. (3.6),
as $V \to 0$, $\exp(eV/kT) \to 1$ and hence dark current $I_D \to 0$.

3.2.5 Photovoltaic Effect

When the solar cell (p-n junction) is illuminated, electron–hole pairs are generated, and acted upon by the internal electric fields, resulting in a photocurrent (I_L). The generated photocurrent flows in a direction opposite to the forward dark current. Even in the absence of an external applied voltage, this photocurrent continues to flow, and is measured as the short-circuit current (I_{sc}). This current depends linearly on the light intensity, because absorption of more light results in additional electrons to flow in the internal electric field force.

The overall cell current I is determined by subtracting the light induced current I_L from the diode dark current I_D.

$$I = I_D - I_L \tag{3.7}$$

Then,

$$I = I_0\left[\exp\left(\frac{eV}{kT}\right) - 1\right] - I_L \tag{3.8}$$

This phenomenon is called the photovoltaic effect.

EXAMPLE 3.5a

Determine the value of the overall cell current in the limiting case $V \to 0$.

Solution

From Eq. (3.8),
as $V \to 0$, $\exp(eV/kT) \to 1$ and hence, $I \to -I_L$.

EXAMPLE 3.5b

Determine the voltage for zero overall cell current.

Solution

Substitute $I=0$ in Eq. (3.8), we get

$$I_0\left[\exp\left(\frac{eV}{kT}\right) - 1\right] - I_L = 0$$

$$\exp\left(\frac{eV}{kT}\right) = \frac{I_L}{I_0} + 1$$

$$V = \frac{kT}{e}\ln\left[\frac{I_L}{I_0} + 1\right]$$

3.2.6 Photovoltaic Material

The solar cells are made of various materials and with different structure in order to reduce the cost and achieve maximum efficiency. There are various types of solar cell material, single-crystal, polycrystalline and amorphous silicon, compound thin-film material, and other semiconductor absorbing layers that give highly efficient cells for specialised applications.

Crystalline silicon cells are the most popular, though they are expensive. Amorphous silicon thin-film solar cells are less expensive. The amorphous silicon layer is used with both hydrogen and fluorine incorporated in the structure. These a-Si: F: H alloy have been produced by the glow-discharge decomposition of SiF_4 in the presence of hydrogen. The efficiency of an a-Si module is about 6–8%.

A variety of compound semiconductor can also be used to manufacture thin-film solar cells. These compound materials are $CuInSe_2$, CdS, CdTe, Cu_2S, InP. The $CuInSe_2$ solar cell stability appears to be excellent. The combinations of different bandgap material in the tandem configurations lead to photovoltaic generators of much higher efficiencies.

3.2.6.1 Silicon (Si)

The most prevalent bulk material for solar cells is crystalline silicon (c-Si), also known as "solar grade silicon". Bulk silicon is separated into multiple categories according to crystallinity and crystal size in the resulting ingot, ribbon, or wafer.

1. **Monocrystalline silicon (c-Si)**: often made using the Czochralski process. Single-crystal wafer cells tend to be expensive, and because they are cut from cylindrical ingots, do not completely cover a square solar cell module without a substantial waste of refined silicon. Hence, most c-Si panels have uncovered gaps at the corners of four cells.
2. **Poly- or multicrystalline silicon (poly-Si or mc-Si)**: made from cast square ingots – large blocks of molten silicon carefully cooled and solidified. These cells are less expensive to produce than single-crystal cells but are less efficient. Polycrystalline silicon wafers are made by wire sawing block-cast silicon ingots into very thin (180 to 350 μm) slices or wafers. The wafers are usually lightly p-type doped. To make a solar cell from the wafer, a surface diffusion of n-type dopants is performed on the front side of the wafer. This forms a p-n junction a few hundred nanometers below the surface.
3. **Ribbon silicon**: formed by drawing flat thin films from molten silicon and having a multicrystalline structure. These cells have lower efficiencies than poly-Si, but save on production costs due to a great reduction in silicon waste, as this approach does not require sawing from ingots.

Antireflection coatings, which increase the amount of light coupled into the solar cell, are typically applied next. Over the past decade, silicon nitride has gradually replaced titanium dioxide as the antireflection coating of choice because of its excellent surface passivation qualities. It is typically applied in a layer several hundred nanometers thick using plasma-enhanced chemical vapour deposition (PE-CVD). Some solar cells have textured front surfaces that, like antireflection coatings, serve to increase the amount of light coupled into the cell. Such surfaces can usually only be formed on single-crystal silicon, though in recent years methods of forming them on multicrystalline silicon have been developed.

Silicon thin films are mainly deposited by chemical vapour deposition (typically plasma-enhanced (PE-CVD)) from silane gas and hydrogen gas. Depending on the deposition's parameters, this can yield:

1. amorphous silicon (a-Si or a-Si:H);
2. protocrystalline silicon; or
3. nanocrystalline silicon (nc-Si or nc-Si:H).

These types of silicon present dangling and twisted bonds, which results in deep defects (energy levels in the bandgap) as well as deformation of the valence and conduction bands. The solar cells made from these materials tend to have lower energy conversion efficiency than bulk silicon, but are also less expensive to produce. The quantum efficiency of thin-film solar cells is also lower due to the reduced number of collected charge carriers per incident photon.

Amorphous silicon has a higher bandgap (1.7 eV) than crystalline silicon (c-Si) (1.1 eV), which means it absorbs the visible part of the solar spectrum more strongly than the infrared portion of the spectrum. As nc-Si has about the same bandgap as c-Si, the two materials can be combined in thin layers, creating a layered cell called a tandem cell. The top cell in a-Si absorbs the visible light and leaves the infrared part of the spectrum for the bottom cell in nanocrystalline Si.

Recently, solutions to overcome the limitations of thin-film crystalline silicon have been developed. Light trapping schemes where the incoming light is obliquely coupled into the silicon and the light traverses the film several times enhance the absorption of sunlight in the films. Thermal processing techniques enhance the crystallinity of the silicon and passivate electronic defects. The result is a new technology – thin-film crystalline silicon on glass (CSG).[4] CSG solar devices represent a balance between the low cost of thin films and the high efficiency of bulk silicon.

A silicon thin-film technology is being developed for building integrated photovoltaics (BIPV) in the form of semitransparent solar cells that can be applied as window glazing. These cells function as window tinting while generating electricity. Despite the numerous attempts at making better solar cells by using new and exotic materials, the reality is that the photovoltaics market is still dominated by silicon wafer-based solar cells (first generation solar cells). The aim of the research is to achieve the lowest $/watt solar cell design that is suitable for commercial production.

3.2.6.2 Cadmium Telluride (CdTe)

Cadmium telluride is an efficient light-absorbing material for thin-film solar cells. Compared to other thin-film materials, CdTe is easier to deposit and more suitable for large-scale production. Despite much discussion of the toxicity of CdTe-based solar cells, this is the only technology (apart from amorphous silicon) that can be delivered on a large scale, as shown by First Solar and Antec Solar. Other companies such as Primestar Solar, AVA Technologies as well as Arendi SRL have also started CdTe divisions, respectively. There is a 40 megawatt plant in Ohio (USA) and a 10 megawatt plant in Germany. First Solar is scaling up to a 100-MW plant in Germany and has started building another 100-MW plant in Malaysia (2007).

The perception of the toxicity of CdTe is based on the toxicity of elemental cadmium, a heavy metal that is a cumulative poison. Scientific work, particularly by researchers of the National Renewable Energy Laboratories (NREL) in the USA, has shown that the release of cadmium to the atmosphere is lower with CdTe-based solar cells than with silicon photovoltaics and other thin-film solar cell technologies.[5]

3.2.6.3 Copper-Indium Selenide ($CuInSe_2$)

The materials based on $CuInSe_2$ that are of interest for photovoltaic applications include several elements from groups I, III and VI in the periodic table. These semiconductors are especially attractive for thin-film solar cell application because of their high optical absorption coefficients and versatile optical and electrical characteristics.

3.2.6.4 Gallium Arsenide (GaAs) Multijunction

High efficiency cells have been developed for special applications such as satellites and space exploration. These multijunction cells consist of multiple thin films produced using molecular beam epitaxy. A triple-junction cell, for example, may consist of the semiconductors: GaAs, Ge, and $GaInP_2$.[6] Each type of semiconductor will have a characteristic bandgap energy that causes it to absorb light most efficiently at a certain colour, or more precisely, to absorb electromagnetic radiation over a portion of the spectrum. The semiconductors are carefully chosen to absorb nearly the entire solar spectrum, thus generating electricity from as much of the solar energy as possible.

GaAs multijunction devices are the most efficient solar cells to date, reaching a record high of 40.7% efficiency under solar concentration and laboratory conditions.[7] These devices use 20 to 30 different semiconductors layered in series.

3.2.6.5 Single-Crystal Solar Cell

Single-crystalline solar cells made from high purity material (solar grade) show excellent efficiencies and long-term stability but they are generally considered to be too expensive for large-scale mass production.

Figure 3.7 shows the diagram of a silicon solar cell structure and mechanism. The electric current generated in the semiconductor is extracted by contacts to the front and rear of the cell. The cell is covered with a thin layer of dielectric material, the antireflecting coating or ARC (to minimise the reflection from the top surface).

The total series resistance of the cell can be expressed as:

$$R_s = R_{cp} + R_{bp} + R_{cn} + R_{bn} \tag{3.9}$$

where R_{cp} is the metal contact to p-type semiconductor resistance; R_{bp} the bulk p-type resistance (bulk of p-type region is where most of electron–hole pairs are generated by the absorption of light and where minority carriers (electron) are transported by diffusion and partially lost by recombination); R_{cn} the contact to n-type semiconductor resistance and R_{bn} the bulk n-type resistance.

The idealised junction current is given as,

$$I = I_0 \left[\exp \frac{e(V \times IR_s)}{kT} - 1 \right] \tag{3.10}$$

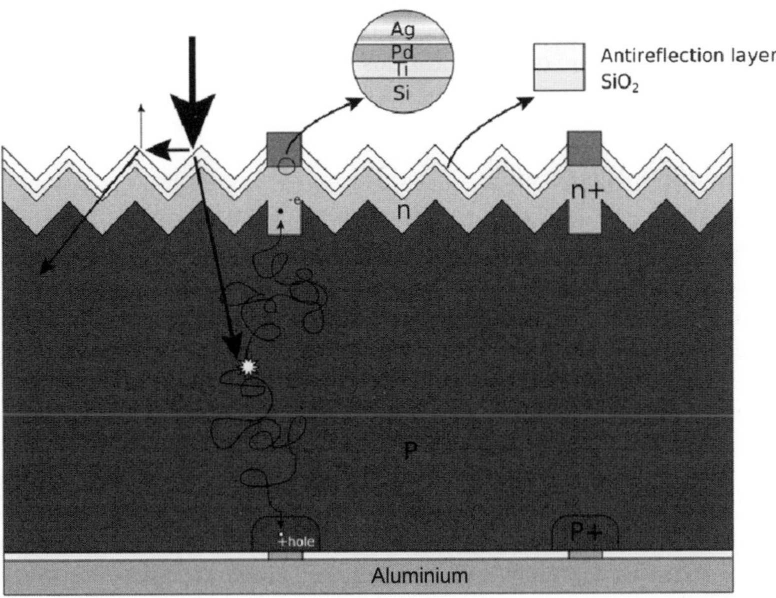

Figure 3.7 The structure of silicon solar cell and working mechanism.

In addition, a shunt path may exist for current flow across the junction due to surface effects or a poor junction region. This alternate path for current constitutes a shunt resistance R_p across the junction. Then,

$$I = I_L - I_0\left[\exp\left(\frac{e(V - IR_s)}{AkT}\right) - 1\right] - \left(\frac{V - IR_s}{R_p}\right) \qquad (3.11)$$

where A is an empirical nonidealist factor and is usually 1.

3.2.6.6 Light-Absorbing Dyes (DSSC)

Typically, a ruthenium metalorganic dye (Ru-centred) is used as a monolayer of light-absorbing material. The dye-sensitised solar cell depends on a mesoporous layer of nanoparticulate titanium dioxide to greatly amplify the surface area (200–300 m^2/g TiO_2, as compared to approximately 10 m^2/g of a flat single crystal). The photogenerated electrons from the light-absorbing dye are passed onto the n-type TiO_2, and the holes are passed to an electrolyte on the other side of the dye. The circuit is completed by a redox couple in the electrolyte, which can be liquid or solid.[8]

This type of cell allows a more flexible use of materials, and is typically manufactured by screen printing, with the potential for lower processing costs than those used for bulk solar cells. However, the dyes in these cells also suffer from degradation under heat and UV light, and the cell casing is difficult to seal due to the solvents used in assembly. In spite of the above, this is a popular emerging technology with some commercial impact forecast within this decade.

3.2.6.7 Organic/Polymer Solar Cells

Organic solar cells and polymer solar cells are built from thin films (typically 100 nm) of organic semiconductors such as polymers and small molecule compounds like polyphenylene vinylene, copper phthalocyanine (a blue or green organic pigment) and carbon fullerenes.[9] Energy conversion efficiencies achieved to date using conductive polymers are low compared to inorganic materials, with the highest reported efficiency of 6.5% for a tandem-cell architecture. However, these cells could be beneficial for some applications where mechanical flexibility and disposability are important.

3.2.6.8 Nanocrystalline Solar Cells

These structures make use of some of the same thin-film light absorbing materials but are overlain as an extremely thin absorber on a supporting matrix of conductive polymer or mesoporous metal oxide having a very high surface area to increase internal reflections (and hence increase the probability of light absorption). Using nanocrystals allows one to design architectures on the length scale of nanometers, the typical exciton diffusion length. In particular, single-nanocrystal (channel) devices, an array of single p-n junctions between the electrodes and separated by a period of about a diffusion length, represent a new architecture for solar cells and potentially high efficiency.

3.2.6.9 Low Cost Solar Cells

The dye-sensitised solar cell is considered as a low cost solar cell. This cell is extremely promising because it is made of low cost materials and does not need elaborate apparatus to manufacture, so it can be made in a DIY way allowing more players to produce it than any other type of solar cell. In bulk, it should be significantly less expensive than older solid state cell designs. It can be engineered into flexible sheets. Although its conversion efficiency is less than the best thin-film cells, its price/performance ratio should be high enough to allow them to compete with fossil fuel electrical generation.

EXAMPLE 3.5c

What is the condition for zero idealised junction current ($I=0$).

Solution

Substitute $I=0$ in Eq. (3.10), we get $exp\left(\frac{eV}{kT}\right) = 1 \Rightarrow V = 0$

3.2.7 Basic Parameters of Solar Cell

There are certain parameters to be mentioned in the I–V characteristics of a solar cell. They are:

3.2.7.1 Overall Current (I)

This is determined by subtracting the light-induced current from the diode dark current and can be expressed as:

$$\text{Overall current } (I) = \text{Diode dark current } (I_D) - \text{light induced current } (I_L) \tag{3.12}$$

or,

$$I = I_0\left[\exp\left(\frac{eV}{kT}\right) - 1\right] - I_L \tag{3.13}$$

where, I_0 is saturation current which is also known as leakage or diffusion current ($I_0 \approx 10^{-8}$ Am^{-2} for good solar cells); e is charge on electron and hole and k is the Boltzmann constant.

Both I_L and I_0 depend on the structure of solar cells.

3.2.7.2 Short-Circuit Current (I_{sc})

This is the light-generated current or photocurrent, I_L. It is the current in the circuit when the load is zero in the circuit. It can be achieved by connecting the positive and negative terminals by a copper wire.

3.2.7.3 Open-Circuit Voltage (V_{oc})

It is obtained by setting $I=0$ in expression for overall current, *i.e.* $I=0$ when $V=V_{oc}$.

$$V_{oc} = \frac{kT}{e}\ln\left(\frac{I_L}{I_0} + 1\right) \tag{3.14}$$

The open-circuit voltage is the voltage for maximum load in the circuit.

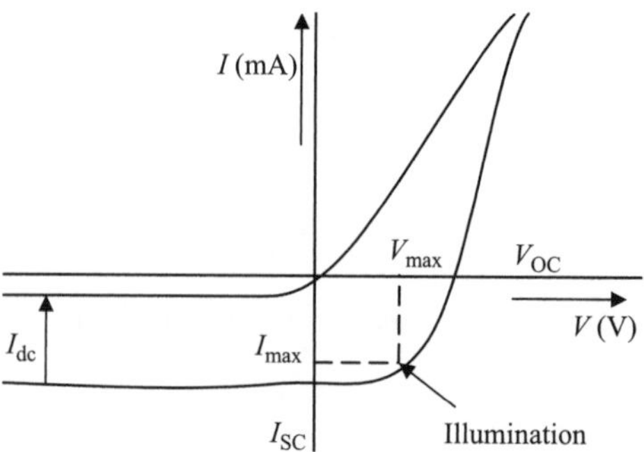

Figure 3.8 *I–V* characteristics of a solar cell with and without illumination.

3.2.7.4 I–V *Characteristics*

The current equation for a solar cell is given by,[10] $I = I_0\left[\exp\frac{e(V-IR_s)}{kT} - 1\right]$ and is shown in Figure 3.8. For a good solar cell, the series resistance, R_s, should be very small and the shunt (parallel) resistance, R_p, should be very large. For commercial solar cells, R_p is much greater than the forward resistance of a diode so that it can be neglected and only R_s is of interest. The following are a few of the characteristics parameters that have been discussed.

The optimum load resistance R_L $(P_{max}) = R_{pmax}$ is connected, if the PV generator is able to deliver maximum power.

$$P_{max} = V_{pmax} I_{pmax} \tag{3.15}$$

and,

$$R_{p\,max} = \frac{V_{p\,max}}{I_{p\,max}} \tag{3.16}$$

The efficiency is defined as,

$$\eta = P/\Phi \tag{3.17}$$

where, $P = V \times I$, is the power delivered by the PV generator.

$\Phi = I_T \times A$, is the solar radiation falling on the PV generator.

I_T is the solar intensity and A is the surface area irradiated.

3.2.7.5 *Fill Factor (FF)*

The fill factor, also known as the curve factor (Figure 3.9), is a measure of the sharpness of the knee in the *I–V* curve. It indicates how well a junction was made in the cell and how low the series resistance has been made. It can be lowered by the presence of series resistance and tends to be higher whenever the open-circuit voltage is high. The maximum value of fill factor is one, which is not possible. Its maximum value in Si is 0.88.

$$\text{FF} = \frac{P_{max}}{V_{oc} \times I_{sc}} = \frac{I_{max} \times V_{max}}{V_{oc} \times I_{sc}} \tag{3.18}$$

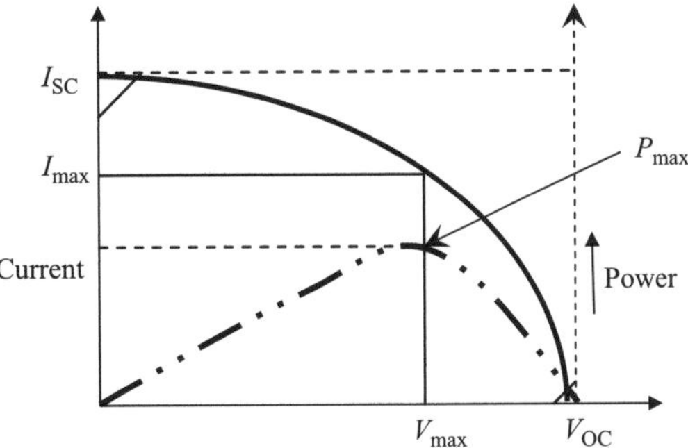

Figure 3.9 Characteristic and power curve for determining the fill factor (FF).

3.2.7.6 *Maximum Power* (P_{max})

No power is generated under short or open circuit. The power output is defined as,

$$P_{out} = V_{out} \times I_{out} \tag{3.19}$$

The maximum power P_{max} provided by the device is achieved at a point on the characteristics, where the product IV is maximum. Thus,

$$P_{max} = I_{max} \times V_{max} \tag{3.20}$$

The maximum possible output can also be given as,

$$P_{max} = V_{oc} \times I_{sc} \times \text{FF} \tag{3.21}$$

where, FF is the fill factor given by Eq. (3.18).

3.2.7.7 *Solar Cell Efficiency* (η_{ec})

The solar cell power conversion efficiency can be given as,

$$\eta_{ec} = \frac{P_{max}}{P_{in}} = \frac{I_{max} \times V_{max}}{\text{Incident solar radiation} \times \text{Area of solar cell}} = \frac{V_{OC} \times I_{SC} \times \text{FF}}{I(t) \times A_C} \tag{3.22}$$

where, I_{max} and V_{max} are the current and voltage for maximum power, corresponding to solar intensity ($I(t)$).

EXAMPLE 3.6

Calculate the fill factor for a solar cell that has the following parameters:

$$V_{oc} = 0.2\ \text{V},\ I_{sc} = -5.5\ \text{mA},\ V_{max} = 0.125\ \text{V},\ I_{max} = -3\ \text{mA}$$

Solution

Substituting the appropriate values in Eq. (3.18), we get,

$$\text{Fill factor} = \frac{V_{max} I_{max}}{V_{oc} I_{sc}} = \frac{0.125 \times 3}{0.2 \times 5.5} = 0.34$$

EXAMPLE 3.7

Calculate the maximum power and cell efficiency of the cell at an intensity of 200 W/m^2.

Given: $V_{oc} = 0.24$ V, $I_{sc} = -9$ mA, $V_{max} = 0.14$ V and $I_{max} = -6$ mA, $A_C = 4$ cm^2

Solution

From Eq. (3.20), we have,

$$P_{max} = V_{max} \times I_{max} = 0.14 \times (-6) = -0.84 \text{ mW}$$

and from Eq. (3.22), we have,

$$\text{Cell efficiency} = \text{output/input} = (0.14 \times 6 \times 10^{-3})/(200 \times 4 \times 10^{-4}) = 0.0105 = 1.05\%$$

EXAMPLE 3.8

Calculate the power output from a solar cell at standard test condition ($I(t) = 1000$ W/m^2 and $T_c = 25$ °C), when $\eta = 16\%$, FF $= 0.782$, aperture area $= 4.02 \times 10^{-4}$ m^2.

Solution

Power output $= 0.16 \times 1000 \times 4.02 \times 10^{-4} \times 0.782 = 0.05$ W

3.2.7.8 Thin-Film Solar Cell

Thin-film solar cells are efficient for large-scale photovoltaic energy conversion. This not only reduces the semiconductor material required but is also beneficial for production of large-area modules.

Semiconductor materials for thin-film solar cells should have a high absorption coefficient ($\alpha > 10^4$ cm^{-1}). Two groups of material meet this requirement.

(i) compound semiconductors with direct bandgap and polycrystalline structure;
(ii) amorphous semiconductors.

3.2.7.9 Amorphous Si Solar Cells (a-SiH)

Hydrogenated amorphous silicon film represents extremely suitable material for the solar cell mainly due to its optical properties. Only a thin film of about 0.7 μm thickness absorbs a large fraction of the incident solar radiation due to high absorption coefficient. The optical bandgap of pure a-SiH is well matched with the solar spectrum.

3.2.7.10 Tandem Solar Cell

Tandem systems can be realised as stack of cells with decreasing bandgap in the direction of the light path.

3.2.7.11 Concentrating Solar Cell

The most advanced solar cells, for concentrator application, are based on the crystalline silicon and AlGaAs/GaAs single junction cells. The most successful Si concentrator cells are p^+-n-n^+ or n^+-p^+ configurations.

3.2.7.12 *Effect of Cell Temperature on Cell Efficiency*

The temperature of operation of a PV module can be determined by an energy balance. The solar energy absorbed by a module is converted partly into thermal energy and partly into electrical energy. The electrical energy is removed from the cell through the external circuit. The thermal energy is dissipated by a combination of heat transfer mechanisms; the upward losses and back losses.[10] Back losses, in this case, are more important, as the heat transfer from the module should be maximised so that the cell operates at the lowest possible temperature.

An energy balance on a unit area of module, cooled by losses to the surroundings can be written as,

$$\tau\alpha I_{\mathrm{T}} = \eta_{\mathrm{c}} I_{\mathrm{T}} + U_{\mathrm{L}}(T_{\mathrm{c}} - T_{\mathrm{a}}) \tag{3.23}$$

where τ is the transmittance of any cover that may be over the cells, α is the fraction of the radiation incident on the surface of the cells, that is absorbed, and η_{c} is the efficiency, of the module, of conversion of incident radiation into electrical energy. The efficiency will vary from zero to a maximum, depending on how close to the maximum power point, the module is operating. The loss coefficient, U_{L}, will include losses by convection and radiation from top and bottom and by conduction through any mounting framework that may be present, to the ambient temperature T_{a}.

The nominal operating cell temperature (NOCT) is defined as that cell or module temperature that is reached when the cells are mounted in their normal way at a solar radiation level of 800 W/m^2, a wind speed of 1 m/s, an ambient temperature of 20 °C, and no load operation (*i.e.* with $\eta_{\mathrm{c}}=0$).

From Eq. (3.23), $\tau\alpha/U_{\mathrm{L}}$ is given as,

$$\alpha\tau/U_{\mathrm{L}} = (T_{\mathrm{C,NOCT}} - T_{\mathrm{a}})/I_{\mathrm{T,NOCT}} \tag{3.24}$$

Knowing T_{a}, $I_{\mathrm{T,NOCT}}$, $T_{\mathrm{C,NOCT}}$, $\tau\alpha/U_{\mathrm{L}}$ can be calculated. Then, treating $\tau\alpha/U_{\mathrm{L}}$ as a constant, the temperature at any other condition can be found from the relation:

$$T_{\mathrm{c}} = T_{\mathrm{a}} + (I_{\mathrm{T}}\,\tau\alpha/U_{\mathrm{L}})(1 - \eta_{\mathrm{c}}/\tau\alpha) \tag{3.25}$$

The electrical efficiency (η_{el}), as a function of temperature, is given by:[11]

$$\eta_{\mathrm{el}} = \eta_0[1 - \beta_0(T_{\mathrm{c}} - 298)] \tag{3.26}$$

where η_0 is efficiency of PV module at temperature of 298 K; β_0 silicon efficiency temperature coefficient (0.0045 K^{-1} or 0.0064 K^{-1}) and T_{c} cell temperature (K).

3.3 PHOTOVOLTAIC MODULE AND ARRAY

A photovoltaic array is a linked collection of photovoltaic modules, which are in turn made of multiple interconnected solar cells. The cells convert solar energy into direct current electricity *via* the photovoltaic effect. The power that one module can produce is seldom enough to meet requirements of a home or a business, so the modules are linked together to form an array. Most PV arrays use an inverter to convert the DC power produced by the modules into alternating current that can plug into the existing infrastructure to power lights, motors, and other loads. The modules in a PV array are usually first connected in series to obtain the desired voltage; the individual strings are then connected in parallel to allow the system to produce more current.

Solar arrays are typically measured by the electrical power they produce, in watts, kilowatts, or even megawatts.

The electrical output of the module depends on the size and number of cells, their electrical interconnection, and, of course, on the environmental conditions to which the module is exposed. Solar electric panels come in all shapes and sizes, and may be made from different materials. However, the most commonly used module is a "glass-plate-sandwich" that has 36 PV cells connected in series to produce enough voltage to charge a 12-volt battery. The purpose of the structure is to provide a rigid package and protect the intercell connections from the environment. Plus (+) and minus (−) connectors are located on the back of the module for interconnection. The modules may have an individual metal frame or be protected by a rubber gasket and intended for installation in a larger mounting system designed to hold several modules.

There are four factors that determine any solar electric panel's output – efficiency of the photovoltaic cells, the load resistance, solar irradiance, and cell temperature. The solar cell efficiency is set by the manufacturing process – today's commercially available modules are from 9% to 17% efficient at converting the solar energy to electrical energy. The load resistance determines where, on the current and voltage (*I–V*) curve, the module will operate. The obvious preferred operating point is where maximum power (power is calculated by multiplying the current times the voltage) is generated – called the peak power point.

For a given solar cell area, the current generated is directly proportional to solar irradiance (*S*) and is almost independent of temperature (*T*). Thus, as the Sun's brightness increases the output voltage and power decreases as temperature increases. The voltage of crystalline cells decreases about 0.5 percent per degree centigrade temperature increase. Therefore, arrays should be mounted in the sunniest place (no shading) and kept as cool as possible by ensuring air can move over and behind the array.

A photovoltaic module is a packaged interconnected assembly of photovoltaic cells, also known as solar cells. An installation of photovoltaic modules or panels is known as a photovoltaic array or a solar panel. Photovoltaic cells typically require protection from the environment. For cost and practicality reasons a number of cells are connected electrically and packaged in a photovoltaic module, while a collection of these modules that are mechanically fastened together, wired, and designed to be a field-installable unit, sometimes with a glass covering and a frame and backing made of metal, plastic or fibreglass, are known as a photovoltaic panel or simply solar panel. A photovoltaic installation typically includes an array of photovoltaic modules or panels, an inverter, batteries (for off-grid) and interconnection wiring.

Most solar PV panels have 30 to 36 cells connected in series. Each cell produces about 0.5 V in sunlight, so a panel produces 15 V to 18 V. These panels are designed to charge 12 V batteries. A 30-cell panel (15 V) can be used to charge the battery without a controller, but it may fail to charge the battery completely. A 36-cell panel (18 V) will do better, but needs a controller to prevent over-charging. The current depends on the size of each cell, and the solar radiation intensity. Most cells produce a current of 2 A to 3 A in bright sunlight. The current is the same in every cell because the cells are connected in series.

Panels are rated in peak watts (Wp), namely the power produced in an optimally matched load with incident solar radiation 1000 W/m^2. A typical panel rating is 40 Wp. In a tropical climate 40 Wp may produce an average of 150 Wh of electricity per day, but as the weather changes the energy varies, typically between 100 Wh to 200 Wh per day.

If two 40-Wp panels, each giving 2.5 A at 16 V in bright sunlight, are connected in parallel they give 5 A at 16 V. If they are connected in series they give 2.5 A at 32 V. In both cases the power is the same: 80 W.

Since the intensity of sunlight is rarely at the peak value, the power output from a panel is usually much less than the peak rating. At low solar radiation intensities the voltage remains almost the same, but the current is low.

Panels should normally be mounted facing the point where the celestial equator crosses the meridian, but should be tilted at least 5° to allow rain to drain off. Since the power output of solar cells is reduced by high temperatures there should be at least 100 mm clearance for ventilation under the panels. There must be no shading of the panels by obstructions, and the panels should be kept clean. Even partial shading of one or more panels can create a resistance in the circuit and reduce the performance of the system.

The majority of modules use wafer-based crystalline silicon cells or a thin-film cell based on cadmium telluride or crystalline silicon, which is commonly used in the wafer form in photovoltaic (PV) modules, is derived from silicon, a relatively multifaceted element.

In order to use the cells in practical applications, they must be:

- connected electrically to one another and to the rest of the system;
- protected from mechanical damage during manufacture, transport and installation and use (in particular against hail impact, wind and snow loads). This is especially important for wafer-based silicon cells that are brittle;
- protected from moisture, which corrodes metal contacts and interconnects, (and for thin-film cells the transparent conductive oxide layer) thus decreasing performance and lifetime;
- electrically insulated including under rainy conditions; and
- mountable on a substructure.

Most modules are rigid, but there are some flexible modules available, based on thin-film cells. Electrical connections are made in series to achieve a desired output voltage and/or in parallel to provide a desired amount of current-source capability. Diodes are included to avoid overheating of cells in case of partial shading. Since cell heating reduces the operating efficiency it is desirable to minimise the heating. Very few modules incorporate any design features to decrease temperature; however, installers try to provide good ventilation behind the module. New designs of module include concentrator modules in which the light is concentrated by an array of lenses or mirrors onto an array of small cells. This allows the use of cells with a very high cost per unit area (such as gallium arsenide) in a cost competitive way. Depending on construction the photovoltaic can cover a range of frequencies of light and can produce electricity from them, but cannot cover the entire solar spectrum. Hence, much of the incident sunlight energy is wasted when used for solar panels, although they can give far higher efficiencies if illuminated with monochromatic light. Another design concept is to split the light into different wavelength ranges and direct the beams onto different cells tuned to the appropriate wavelength ranges. This is projected to raise efficiency to 50%.[1] Sunlight conversion rates (module efficiencies) can vary from 5–18% in commercial production.

3.3.1 Single-Crystal Solar Cell Module

After testing solar cells under test conditions and sorting to match current and voltage, about 36 solar cells are interconnected and encapsulated to form a module (Figure 3.10). A module consists of the following components: (i) front cover low-iron tempered glass (ii) encapsulating, transparent, insulating, thermoplastic polymer, the most widely used one is EVA (ethylene vinyl acetate) (iii) the solar cell and metal interconnected (iv) back cover usually a foil of tedlar or Mylar.

Cells are usually mounted in modules and multiple modules are used in arrays. Individual modules may have cells connected in series and parallel combinations to obtain the desired voltage. Arrays of modules may also be arranged in series and parallel depending upon the requirement of current and voltage.

There are two types of crystalline PV modules namely opaque PV modules and semitransparent PV modules, as shown in Figures 3.11a and b, respectively. Figures 3.11b and c show the

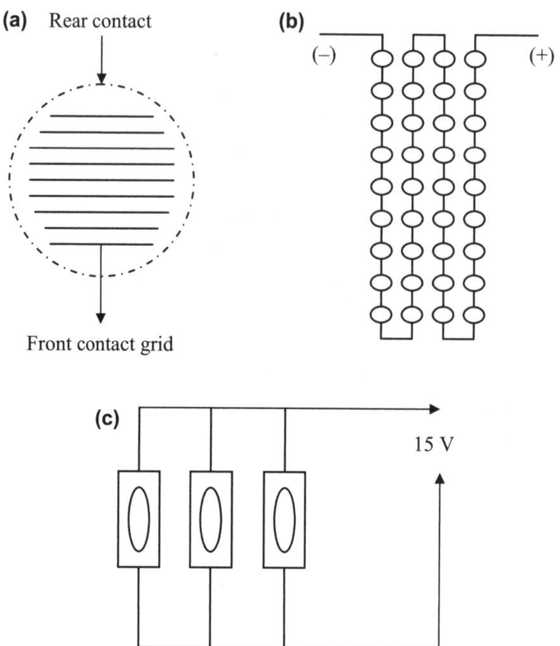

Figure 3.10 Typical arrangements of commercial Si solar cells (a) cell (b) module of 36 cells (c) array of PV module.

Figure 3.11a Opaque PV module of 75 Wp.

semitransparent PV modules with different packing factors. In opaque PV module, there is a white tedlar layer at the back surface of the PV module (Figure 3.11a), whereas in semitransparent PV modules, a transparent glass covers is used (Figures 3.11b and c). Such semitransparent PV modules have been applicable in building such as façades and roofs integrated for (i) electric generation and (ii) thermal and day lighting.

Figure 3.11b Semitransparent PV module of 75 Wp.

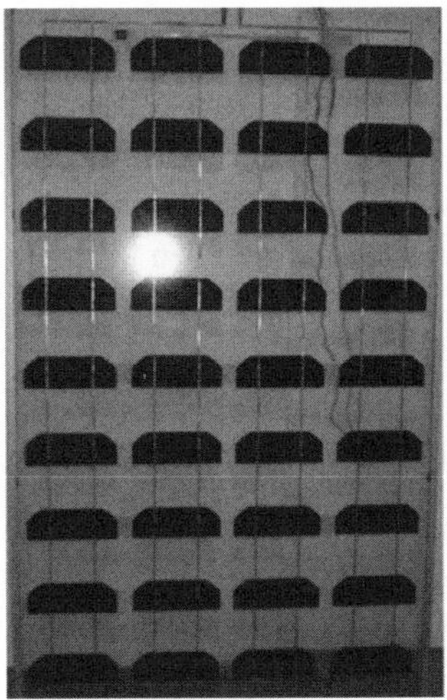

Figure 3.11c Semitransparent PV module of 37 Wp.

3.3.2 Thin-Film PV Modules

The advantages of thin films over crystalline cells have been the important driver to initiate a PV market in this area. Thin-film module production manufacturing processes operate at a much lower temperature than that of crystalline silicon and this reduces the embodied energy per watt-peak. Another manufacturing advantage is the fact that PV films can be easily deposited on a wide variety of both rigid and flexible substrates including glass, steel and plastics. However, thin-film technologies show significant initial performance degradation when deployed outdoors (Staebler–Wronski effect) and the most important challenge of thin-film technologies remains the production improvement of the technology so as to increase the efficiency of industrially produced cells.[12] Initial products of this technology were made from very thin films of silicon in a form known as amorphous silicon (a-Si) and there has been a growing interest in this technology due to its promise of low production costs and the fact that conditions for preparing a-Si are even less critical, in principle, than those for preparing polycrystalline silicon.[13] Amorphous silicon cells should be cheaper to produce than those made from crystalline silicon and are better light absorbers, facilitating in this way thinner and therefore cheaper cells. However, stabilised amorphous silicon efficiencies of the best commercial modules remain low at 6–7%.[14]

Amorphous silicon is by no means the only material suited to thin-film technologies. Amongst the many other possible thin-film technologies some of the most promising are those based on compound semiconductors in particular copper indium–diselenide (CIS), copper gallium diselenide (CGS), copper indium gallium diselenide (CIGS), and cadmium telluride (CdTe) with the major technical progress particularly for those based on CdTe and CIGS. More specifically, several II-VI chalcopyrite compounds and their mixed compounds were found to be very suitable for photovoltaic applications. The most important of them are CIS, CGS and CIGS. Currently, the highest efficiency has been reported to be 19.9% with CIGS solar cells that were developed at NREL, USA, while the highest aperture area conversion efficiency of 13% for a CIGS power module has been achieved by Wurth Solar in Germany. There are a variety of different production deposition methods for the fabrication of absorber films. The main production techniques for this technology are coevaporation and reactive annealing of metal precursor films. Other methods include non-vacuum techniques like electrodeposition. Manufacturing plants with a total production capacity of about 20MW are already in operation in Germany.[15]

Another front runner of thin-film solar PV technology is CdTe, due to its stability and chemical simplicity. CdTe is a direct bandgap II-VI semiconductor with an optical bandgap of 1.44 eV, which is close to the optimum for photoconversion. A disadvantage of using CdTe and most of the II-VI semiconductors is the difficulty in electronic doping and specifically controlling the doping concentration in p-type CdTe. Furthermore, the toxic nature of cadmium and the environmental consequences of deploying large solar systems based on toxic materials have caused serious concerns that are currently being carefully examined, even though the trace amounts of this material in a thin-film PV module do not approach toxic limits. The most efficient CdTe cell has been reported with an efficiency of 16.5%, demonstrated and confirmed by NREL.

3.3.3 III-V Single and Multijunction PV Modules

A promising approach for achieving better efficiencies is through utilising greater portions of the solar spectrum. One method for achieving this is by combining cells of different bandgaps in a tandem arrangement. A single junction cell can provide theoretical efficiencies of 30% while as the number of junctions increases from 1 to infinity, the thermal loss due to absorption of light with energy greater than the bandgap goes to zero, resulting in a thermodynamic performance limit of 68% and for sunlight of full concentration, the new limit is above 85%. Ultrahigh efficiency multijunction solar cells have therefore attracted a lot of attention. More specifically, III-V solar cells based PV modules have become the standard modules for space power generation, mainly due to their high efficiency, reliability and ability to be integrated into very lightweight panels.

A number of techniques have been developed to produce multijunction solar cells. These cells are mechanically stacked, monolithically integrated or created through a combination of both techniques. Mechanically stacked techniques have led to dual junction cells based on gallium arsenide (GaAs) reaching efficiencies well above 31% and recently researchers at the Fraunhofer Institute for Solar Energy Systems (ISE) have achieved a record efficiency of 41.1% for the conversion of sunlight into electricity by concentrating sunlight by a factor of 454 onto a small 5 mm^2 multijunction solar cell of GaInP/GaInAs/Ge (gallium indium phosphide, gallium indium arsenide on a germanium substrate). The use of GaAs as a solar cell material has also the disadvantage of the limited gallium resources that dictates that GaAs will always be an expensive material. This is offset by the fact that GaAs cells are ideal for use in systems that concentrate light, thus the amount of material required for a given power output is reduced.[16]

3.3.4 Emerging and New PV Systems

The development of new PV modules is the subject of numerous research activities worldwide; the target of which is the lowering of costs and the increase of conversion efficiency. New device technologies include organic, dye-sensitised, quantum well solar cells and in general nanostructured materials for solar energy conversion. PV modules in this category can be distinguished mainly through the approaches taken to tailor the properties of the active layer to better match the solar spectrum and approaches that modify the incoming solar spectrum and function at the periphery of the active device.

A different concept to the existing solar cell approach has emerged with the inclusion of organic solar cells in the field of PV. Organic photovoltaics comprise of electron-donor and electron-acceptor materials rather than semiconductor p-n junctions and are characterised either as hybrid when organic solar cells retain an inorganic component or fully organic. It is essential that both hybrid and fully organic solar cells are made more stable and that their efficiency increases to the 15% target for laboratory cells by 2015, if this technology is to have potential in the future. Nanostructure use in photovoltaic devices has attracted major interest with dye-sensitised photo-electrochemical solar cells (DSSC) based on nanoporous titanium dioxide which is the best representative of the family of nanostructured PV devices. These solar cells have been widely investigated during the past decade and an efficiency of 10.4% was achieved by O'Regan and Gratzel.[17] This technology may prove extremely important in the establishment of new designs that reduce the production cost of photovoltaic devices, because it is made of low cost materials and does not need elaborate apparatus to manufacture.

Another important novel alternative in PV system is the quantum well solar cell (QWSC). Unlike multijunction solar cells, these are created by growing simple quantum wells of a smaller bandgap material within the space charge region of p-n or p-i-n structures. The idea behind this technology is to facilitate the ability to absorb light of energy below the bulk bandgap energy. These solar cells are fabricated by either molecular beam epitaxy or metalorganic chemical vapour deposition. QWSC tends to increase short-circuit current, open-circuit voltage and in general the efficiency of solar devices. The specifications of different solar cell materials (at a solar intensity of 1000 W/m^2 and a cell temperature of 25 °C) are shown in Appendix III.[18]

The highest PV module efficiencies obtained from a survey undertaken by the National Renewable Energy Laboratory (NREL), USA, on commercial flat plate manufacturers' websites in April 2008 are listed in Table 3.1.

Photovoltaic generators, Figure 3.12, may be used to drive machines such as electric pumps, refrigerators, and other devices. A PV array mounted on the roof-top offer the possibility of large-scale power generations in decentralised medium-size grid-connected units. The PV system supplies the electricity needs of the building, feeds the surplus electricity needs of the building to the grid to earn revenue, and draws electricity from the grid at low insolation.

3.3.5 Packing Factor (β_c) of PV Module

This is defined as the ratio of total solar cell area to the total module area and can be expressed as:

$$\beta_c = \frac{\text{area of solar cells}}{\text{area of PV module}} \tag{3.27}$$

It is clear that β_c is less than unity (pseudosolar cell) and it has maximum value of one when all the area is covered by solar cells (rectangular solar cells).

Table 3.1 Module efficiencies from National Renewable Energy Laboratories (NREL) survey of manufacturers' websites.

Module	*Technology*	*Efficiency (%)*
SunPower 315	Mono-Si, special junction (sp. j.)	19.3
Sanyo HIP-205BAE	CZ-Si, "HIT," sp. j.	17.4
BP7190	CZ-Si, sp. j.	15.1
Kyocera KC200GHT-2	MC-Si, standard junction (std.j.)	14.2
Solar World SW 185	CZ-Si, std. j.	14.2
BP SX3200	MC-Si, std. j.	14.2
Suntech STP 260S-24V/b	MC or CZ-Si, std. j.	13.4
Solar World SW 225	MC-Si, std. j.	13.4
Evergreen Solar ES 195	String-ribbon-Si std. j.	13.1
Würth Solar WS11007/80	CIGS	11.0
First Solar FS-275	CdTe	10.4
Sharp NA-901-WP	a-Si/nc-Si	8.5
GSE Solar GSE120-W	CIGS	8.1
Mitsubishi Heavy MA100	a-Si, single junction	6.3
Uni-Solar PVL136	a-Si, triple junction	6.3
Kaneka T-SC(EC)-120	a-Si single junction	6.3
Schott Solar ASI-TM86	a-Si/a-Si same bandgap tandem	5.9
EPV EPV-42	a-Si/a-Si same bandgap tandem	5.3

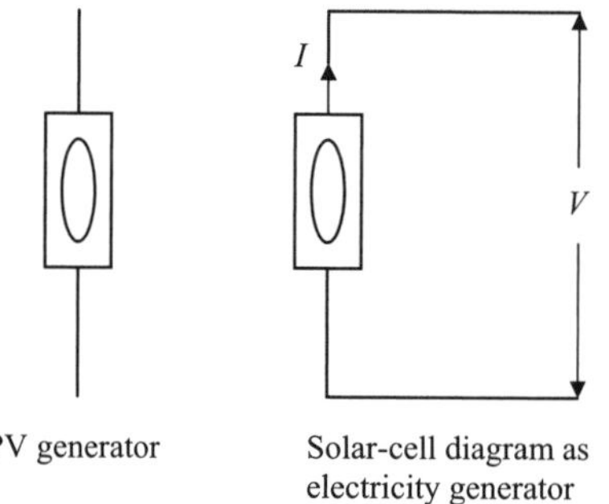

Figure 3.12 Technical signs for various unit of PV generator.

3.3.6 Efficiency of PV Module

The electrical efficiency of PV module can be expressed as:

$$\eta_{em} = (\eta_{ec} \times \beta_c) \times 100 \tag{3.28}$$

It can also be expressed as

$$\eta_{em} = \left(\frac{FF \times I_{sc} \times V_{oc}}{A_{\mathrm{m}} \times I_{\mathrm{p}}}\right) \times 100 \tag{3.29}$$

where, A_{m} = area of PV module; I_{p} = incident solar intensity on PV module.

The maximum value of fill factor (FF) in Si is 0.88.

The equivalent thermal efficiency of the PV module may be expressed as:

$$\eta_{eth} = \left(\frac{\eta_e}{0.38}\right) \times 100 \tag{3.30}$$

The electrical load efficiency may be expressed as:

$$\eta_{load} = \left(\frac{I_L \times V_L}{A_m \times I_p}\right) \times 100 \tag{3.31}$$

The overall thermal efficiency of the hybrid PVT system may be written as:

$$\eta_{ov,th} = \eta_{th} + \frac{\eta_e}{0.38} \tag{3.32}$$

where η_{th} is thermal efficiency.

The overall exergy efficiency of the hybrid PVT system may be written as:

$$\eta_{ov,ex} = \eta_{ex} + \eta_e \tag{3.33}$$

where η_{ex} is exergy efficiency $= \eta_{th}\left(1 - \frac{T_{\sin k}}{T_{source}}\right)$, T is temperature in Kelvin.

EXAMPLE 3.9

Calculate the packing factor of PV module (36 solar cells) of area 0.605 m^2, each pseudosolar cell having an area of 0.015 m^2.

Solution

From Eq. (3.27), we get

$$\beta_c = \frac{0.54}{0.605} \times 100 = 89.2\%$$

EXAMPLE 3.10

Calculate the efficiency of a PV module at an intensity of 400 W/m^2. Given:

FF = 0.8, I_{SC} = 3.2 A, V_{oc} = 16 V, I_L = 1 A, V_L = 14 V, area of module = 1 m^2.

Solution

From Eq. (3.29), we have

$$\eta_{em} = \frac{0.8 \times 3.2 \times 16}{400 \times 1} \times 100 = 10.24\%$$

EXAMPLE 3.11

Using Example 3.9, calculate the load efficiency of the PV module.

Solution

From Eq. (3.29), we have

$$\eta_{em} = \frac{1 \times 14}{400 \times 1} \times 100 = 3.5\%$$

3.3.7 Series and Parallel Combination of PV Modules

PV modules are connected in series or parallel to increase the current and voltage ratings. When modules are connected in series, the voltages of each module are added up. When modules are connected in parallel the currents of each module are added up. Thus, while interconnecting the modules; the installer should have this information available for each module. A solar panel is a group of several modules connected in series–parallel combination in a frame that can be mounted on a structure.

Series and parallel connection of modules in a panel is shown in Figure 3.13. In parallel connection, blocking diodes are connected in series with each series string of modules, so that if any string should fail, the power output of the remaining series string will not be absorbed by the failed string. Also, bypass diodes are installed across each module, so that if one module should fail, the power output of the remaining modules in a string will bypass the failed module. Some modern PV modules come with such internally embedded bypass diodes. A large number of interconnected solar panels is known as a solar PV array.

EXAMPLE 3.12

Calculate the daily load for domestic use and how much number of 40-Wp PV panels required in the array.

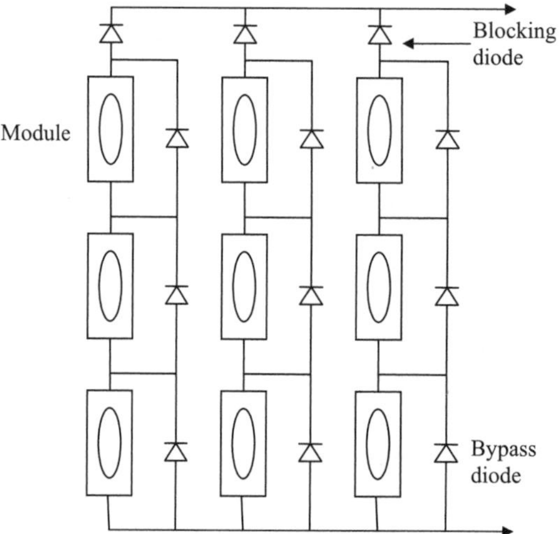

Figure 3.13 Series and parallel connection of modules in a panel.

Solution

Four 40-W lamps used 4 h per day: 640 Wh
One 15-W television used 4 h per day: 60 Wh
Two 35-W fans used 6 h per day: 420 Wh
One 60-W refrigerator used all day, compressor on 50% of the time: 720 Wh
Total daily load = 1840 Wh
Assuming each panel produces 150 Wh per day, then

$$= 1840 \text{ Wh}/150 \text{ Wh} = 12.3$$

Therefore, a 12 V system needs 13 panels connected in parallel.

3.3.8 Applications of Photovoltaic (PV) Modules

In urban and suburban areas, photovoltaic arrays are commonly used on rooftops to measure power use; often the building will have a pre-existing connection to the power grid, in which case the energy produced by the PV array will be sold back to the utility in some sort of net metering agreement. In more rural areas, ground-mounted PV systems are more common. The systems may also be equipped with a battery backup system to compensate for a potentially unreliable power grid. In agricultural settings, the array may be used to directly power DC pumps, without the need for an inverter. In remote settings such as mountainous areas, islands, or other places where a power grid is unavailable, solar arrays can be used as the sole source of electricity, usually by charging a storage battery. Satellites use solar arrays for their power. In particular, the International Space Station uses multiple solar arrays to power all the equipment on board. Solar photovoltaic panels are frequently applied in satellite power. However, the costs of production have been reduced in recent years for more widespread use through production and technological advances. For example, single-crystal silicon solar cells have largely been replaced by less expensive multicrystalline silicon solar cells, and thin-film silicon solar cells have also been developed recently at lower costs of production. Although they are reduced in energy conversion efficiency from single-crystalline Si wafers, they are also much easier to produce at comparably lower costs. Together with a storage battery, photovoltaics have become commonplace for certain low power applications, such as signal buoys or devices in remote areas or simply where connection to the electricity mains would be impractical.

3.3.8.1 PV in Buildings (Building Integrated Photovoltaic Systems)

The recent rapid expansion in installed photovoltaic capacity is largely due to the increase in grid-connected photovoltaic systems mounted on buildings. The term "building integrated" refers to PV systems that constitute part of a building envelope, but has also been used to describe systems that are simply mounted on the rooftop of buildings. For this reason, it is best to describe BIPV as systems that are readily integrated with the physical building or with the building's grid connection. The integration of such systems usually requires the advice of professional civil engineers, architects and PV system designers during the design of the system and the building. In this case, a good evaluation of the installation site is required so as to maximise solar coverage and electricity output. BIPV are usually installed on facades, building window systems and as flexible rolls on roofs. Consequently, BIPV systems often have restricted views of the Sun, and their orientation must be optimised for the particular circumstances of their installation site.[19]

In order for the BIPV systems to achieve multifunctional roles, various factors need to be taken into account, such as the PV's module temperature, shading, installation angle, and orientation. Among these factors, the irradiance and PV module temperature should be regarded as one of the most important factors, since it affects both the electrical efficiency of the BIPV system and the energy performance of buildings where BIPV systems are installed. The results of basic studies regarding irradiance and energy output of PV system have been reported by some researchers, while there have been other studies regarding the temperature and generation performance of PV modules. It is necessary to develop more diverse BIPV modules, like semitransparent ones, with various configurations and designs. These developments will provide more options for architects and building industries on how to apply PVs in buildings.

BIPV systems can achieve significant cost reductions when they are used as part of the building envelope and thereby offset the cost of the building materials they replace. Many modern exterior claddings can have costs per square meter comparable to the price of PV modules. At the leading edge of BIPV are the three main thin-film photovoltaic technologies (a-Si, CdTe and CIGS) that are at present commercially available. The most important issues for the successful integration of thin-film BIPV systems include gaining experience on the design and operation of such systems as well as acquiring knowledge of their lifecycle costs. Apart from thin-film BIPV, concentrator photovoltaic systems designed for building integration have also been gaining ground. These systems often work at low and medium concentration levels if installed on the rooftop of a building. Low concentration levels are preferred for integration into facades since direct views of the Sun are restricted and the diffuse component of light represents a larger proportion of the total irradiation available in such cases. Indeed, restrictions in the availability of direct light has led to many designs of building integrated concentrators that utilise the passive benefits of building integration, such as solar gain control, interior light distribution and collection of thermal energy for preheating of water in order to increase total system efficiency and cost effectiveness. A photograph of a building-integrated PVT roof module is shown in Figure 3.14.

3.3.8.2 PV in Transport

PV has traditionally been used for auxiliary power in space. PV is rarely used to provide motive power in transport applications, but is being used increasingly to provide auxiliary power in boats

Figure 3.14 Photograph of building-integrated PVT roof module. (Curtsey: http://www.iea-shc.org/countries/reports/report.aspx?Country = Canada).

and cars. Recent advances in solar cell technology, however, ha[illegible] administer significant hydrogen production, making it one of th[illegible] energy for automobiles.

3.3.8.3 PV in Standalone Devices

PV has been used for many years to power calculators and [illegible] integrated circuits and low power LCD displays make it poss[illegible] years between battery changes, making solar calculators less comm[illegible] remote fixed devices have seen increasing use recently, due to increasing co[illegible] nection of mains electricity or a regular maintenance programme. In particular, it is us[illegible] meters, emergency telephones, and temporary traffic signs.

3.3.8.4 PV in Agriculture

PV systems are used effectively worldwide to pump water for livestock, plants or humans. Water pumping appears to be most suitable for solar PV applications as water demand increases during dry days when plenty of sunshine is available. A SPV water pumping system is expected to deliver a minimum of 15 000 liters per day for 200 Wp and 170 000 liters per day for a 2250-Wp panel from a suction of 7 m and/or total head of 10 m on clear sunny day. PV is also used to power remote electric fences on farms.

3.3.8.5 Medical Refrigeration

For life-saving vaccines World Health Organisation (WHO) has laid down ground rules to maintain the cold chain from the point of their manufacturer to their application. WHO has the specified technical details for PV-based refrigeration. This has resulted in the success of WHO-sponsored immunisation programmes in those countries/remote areas where electricity is not available.

3.3.8.6 PV in Street Lights

Solar PV street lights can be used as yard lighting, peripheral lighting for industries, street lights in layout, compound lights, *etc*. The photovoltaic modules charge the batteries during the day time. At dusk, an automotive sensor switches on a powerful high efficiency light and at dawn the lamp is switched off automatically. A photograph of a solar PV street light is shown in Figure 3.15.

3.4 PHOTOVOLTAIC THERMAL (PVT) SYSTEMS

Photovoltaic thermal (PVT) technology refers to the integration of a PV module and a conventional solar thermal system in a single piece of equipment. The rationale behind the hybrid concept is that a solar cell converts solar radiation to electrical energy with a peak efficiency in the range of 9 to 12%, depending on the specific solar cell type and thermal energy through water heating. More than 80% of the solar radiation falling on photovoltaic (PV) cells is not converted to electricity, but is either reflected or converted to thermal energy. This leads to an increase in the PV cell's working temperature and consequently, a drop of electricity conversion efficiency. In view of this, hybrid photovoltaic thermal (PVT) systems are introduced to simultaneously generate electricity and thermal power.

Figure 3.15 PV module in street light application.

A collector is the heart of any solar energy collection system designed for operation in low or medium temperature ranges. It is used to absorb solar energy, convert it into heat and transfer it into a stream of liquid or air. In a conventional solar thermal collector, electrical energy is required to circulate the working fluid through the collector and the required electrical energy is usually supplied by grid electricity or a DC battery as a power source. In the case of a hybrid photovoltaic thermal (PVT) system (also called a PVT system), the electrical power source is not required as the PVT collector produces both electrical and thermal energy.

Thermal energy has wider applications in human life. It can be generally utilised in the form of either low grade (low temperature) or high grade (high temperature). Jones and Underwood[20] have studied the temperature profile of the photovoltaic (PV) module in a nonsteady-state condition with respect to time. They conducted experiment for cloudy as well clear day conditions. They observed that the PV module temperature varies in the range of 300–325 K (27–52 °C) for an ambient air temperature of 297.5 K ($\sim$24.5 °C). The main reasons for reduction of the electrical efficiency of the PV module is the packing factor (PF) of PV module, ohmic losses between two consecutive solar cells and the temperature of the module. The overall electrical efficiency of the PV module can be increased by increasing the packing factor (PF) and reducing the temperature of the PV module by withdrawing the thermal energy associated with the PV module. The packing factor is the ratio of total area of solar cells to the area of a PV module. The carrier of thermal energy associated with the PV module may be either air or water. Once thermal energy withdrawal is integrated with the photovoltaic (PV) module, it is referred to as a hybrid PVT system.

3.4.1 Energy Balance Equations for PV Modules

In order to write the energy balance equations for PV modules, the following assumptions have been made:

- one dimensional heat conduction;
- the system is in quasisteady state;
- the ohmic losses in the solar cell and PV module are negligible.

(a) For Opaque (Glass to Tedlar) PV Module (Figure 3.12a):[21]

$$\tau_g[\alpha_c\beta_c I(t) + (1-\beta_c)\alpha_T I(t)] = \left[U_{tc,a}(T_c - T_a) + h_{c,p}(T_c - T_a)\right] + \tau_g\eta_c\beta_c I(t) \tag{3.34}$$

or,

$$\tau_g[\alpha_c\beta_c I(t) + (1-\beta_c)\alpha_T I(t)] = \left(U_{tc,a} + h_{c,p}\right)(T_c - T_a) + \tau_g\eta_c\beta_c I(t)$$

or,

$$\tau_g[\alpha_c\beta_c I(t) + (1-\beta_c)\alpha_T I(t)] = U_{Lm}(T_c - T_a) + \eta_m I(t) \tag{3.35}$$

where,

$$U_{Lm} = U_{t,ca} + h_{cp} \quad \text{and} \quad \eta_m = \eta_c\tau_g\beta_c$$

From Eq. (3.35)

$$T_c - T_a = \frac{\left[\tau_g\{\alpha_c\beta_c + (1-\beta_c)\alpha_T - \eta_c\beta_c\}\right] I(t)}{U_{Lm}} \tag{3.36}$$

$$T_c - T_{ref} = (T_a - T_{ref}) + \frac{\left[\tau_g\{\alpha_c\beta_c + (1-\beta_c)\alpha_T - \eta_c\beta_c\}\right] I(t)}{U_{Lm}} \tag{3.37}$$

The temperature dependent electrical efficiency of the cell is given as,

$$\eta_c = \eta_{ref}[1 - \beta_{ref}(T_c - T_{ref})] \tag{3.38}$$

where, η_{ref} is the module's electrical efficiency at the reference temperature, T_{ref} and at solar radiation of 1000 W/m^2. β_{ref} is the temperature coefficient. These quantities are normally given by the PV manufacturer. The values of η_{ref} and β_{ref} are given in Table 3.2.

Putting the value of ($T_c - T_{ref}$) from Eq. (3.37) in Eq. (3.38), we can get the temperature-dependent electrical efficiency of the cell as:

$$\eta_c = \eta_{ref}\left[1 - \beta_{ref}\left\{(T_a - T_{ref}) + \frac{\left[\tau_g\{\alpha_c\beta_c + (1-\beta_c)\alpha_T - \eta_c\beta_c\}\right] I(t)}{U_{Lm}}\right\}\right]$$

or,

$$\eta_c = \frac{\eta_{ref}\left[1 - \beta_{ref}\left\{(T_a - T_{ref}) + \frac{\tau_g\{\alpha_c\beta_c + (1-\beta_c)\alpha_T\}}{U_{Lm}} I(t)\right\}\right]}{\left[1 - \frac{\eta_{ref}\beta_{ref}\tau_g\beta_c}{U_{Lm}} I(t)\right]} \tag{3.39}$$

(b) For Semitransparent (Glass to Glass) PV Module (Figures 3.12b and c):

$$\alpha_c\tau_g\beta_c I(t) = \left[U_{tc,a}(T_c - T_a) + h_{c,p}(T_c - T_a)\right] + \tau_g\eta_c\beta_c I(t) \tag{3.40}$$

or,

$$\alpha_c\tau_g\beta_c I(t) = (U_{tc,a} + h_{c,p})(T_c - T_a) + \tau_g\eta_c\beta_c I(t) \tag{3.41}$$

Table 3.2 Values of module electrical efficiencies and temperature coefficients.

T_{ref} (°C)	η_{Tref}	β_{ref}	*Comments*	*References*
25	0.15	0.0041	Mono-Si	Evans and Florschuetz[22]
28	0.117 (average) (0.104–0.124)	0.0038 (average) (0.0032–0.0046)	Average of Sandia and commercial cells	OTA[23]
25	0.11	0.003	Mono-Si	Truncellito and Sattolo[24]
25	0.13	0.0041	PVT system	Mertens[25]
		0.005		Barra and Coiante[26]
20	0.10	0.004	PVT system	Prakash[27]
25	0.10	0.0041	PVT system	Garg and Agarwal[28]
20	0.125	0.004	PVT system	Hegazy[29]
25		0.0026	a-Si	Yamawaki et al.[30]
25	0.13	0.004	Mono-Si	RETScreen[31]
	0.11	0.004	Poly-Si	
	0.05	0.0011	a-Si	
25	0.178	0.00375	PVT system	Nagano et al.[32]
25	0.12	0.0045	Mono-Si	Chow[33]
25	0.097	0.0045	PVT system	Zondag et al.[34]
25	0.09	0.0045	PVT system	Tiwari and Sodha[35]
25	0.12	0.0045	PVT system	Tiwari and Sodha[35]
25	0.12	0.0045	PVT system	Assoa et al.[36]
25	0.127	0.0063	PVT system	Tonui and Tripanagnostopoulos[37]
25	0.127 unglazed 0.117 glazed	0.006	PVT system	Tonui and Tripanagnostopoulos[37]
25		0.0054	PVT system	Othman et al.[38]

or,

$$\alpha_c \tau_g \beta_c I(t) = U_{Lm}(T_c - T_a) + \eta_m I(t) \tag{3.42}$$

where, $U_{Lm} = U_{t,ca} + h_{cp}$ and $\eta_m = \eta_c \tau_g \beta_c$.

From Eq. (3.42)

$$T_c - T_a = \frac{(\alpha_c \tau_g \beta_c - \eta_m) I(t)}{U_{Lm}} \tag{3.43}$$

or,

$$T_c - T_{ref} = (T_a - T_{ref}) + \frac{(\alpha_c \tau_g \beta_c - \eta_m) I(t)}{U_{Lm}} \tag{3.44}$$

With the help of Eq. (3.44), Eq. (3.38) becomes,

$$\eta_c = \eta_{ref}\left[1 - \beta_{ref}\left\{(T_a - T_{ref}) + \frac{(\alpha_c \tau_g \beta_c - \eta_c \tau_g \beta_c) I(t)}{U_{Lm}}\right\}\right]$$

or,

$$\eta_c = \frac{\eta_{ref}\left[1 - \beta_{ref}\left\{(T_a - T_{ref}) + \frac{\alpha_c \tau_g \beta_c}{U_{Lm}} I(t)\right\}\right]}{\left[1 - \frac{\eta_{ref}\beta_{ref}\tau_g \beta_c}{U_{Lm}} I(t)\right]} \tag{3.45}$$

The hourly variation of cell temperature and cell efficiency for a typical day of summer is shown in Figure 3.16. The figure shows that the increase in cell temperature decreases the cell efficiency and at the end of the day it will again increase due to a decrease in cell temperature, as conducted by Evans.[22]

An expression for the temperature-dependent electrical efficiency for a PVT air collector covered by opaque PV modules can be given as,

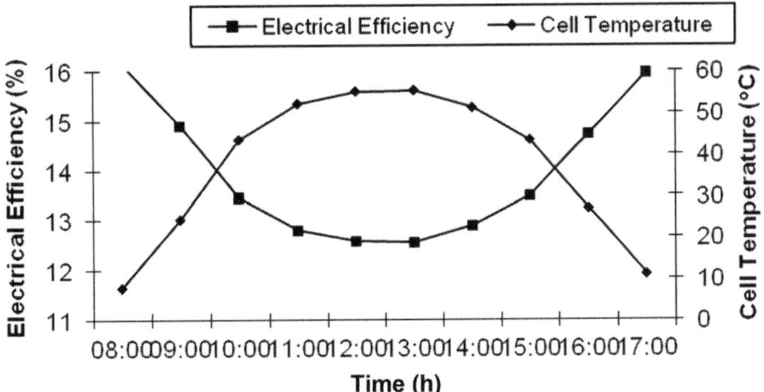

Figure 3.16 Hourly variation of cell temperature and cell efficiency for a typical day in summer.

$$\eta = \frac{\eta_0\left[1 - \frac{\beta_o\tau_g[\alpha_c\beta_c + \alpha_T(1-\beta_c)]I(t)}{U_T + h_T}\left\{1 + \frac{h_T h_{p1}}{h_t + U_{tT}} + \frac{h_T h_t h_{p1} h_{p2}}{(h_T + U_{tT})U_L}\left(1 - \frac{1-\exp(-X_o)}{X_o}\right)\right\}\right]}{1 - \frac{\beta_o\eta_0\tau_g\alpha_c\beta_c I(t)}{U_T + h_T}\left[1 + \frac{h_T h_{p1}}{h_t + U_{tT}} + \frac{h_T h_t h_{p1} h_{p2}}{(h_T + U_{tT})U_L}\left(1 - \frac{1-\exp(-X_o)}{X_o}\right)\right]} \tag{3.46}$$

and for a PVT air collector covered by semitransparent PV modules,

$$\eta = \frac{\eta_0\left[1 - \frac{\tau_g\beta_o}{U_{tc,a} + U_{Tc,f}}\left\{\alpha_c\beta_c + \frac{U_{Tc,f}}{U_L}(h_{p1}\alpha_c\beta_c + h_{p2}\alpha_p(1-\beta_c)\tau_g)\left(1 - \frac{1-\exp(-X_o)}{X_o}\right)\right\}I(t)\right]}{1 - \frac{\eta_0\beta_o\tau_g\beta_c\alpha_c I(t)}{U_{tc,a} + U_{Tc,f}}\left(1 + \frac{U_{Tc,f}h_{p1}}{U_L}\left(1 - \frac{1-\exp(-X_o)}{X_o}\right)\right)} \tag{3.47}$$

where, $X_o = \frac{bU_L L}{\dot{m}_a C_a}$,

$$(\alpha\tau)_{eff} = h_{p1}(\alpha\tau)_{1,eff} + h_{p2}(\alpha\tau)_{2,eff}$$
$$(\alpha\tau)_{1,eff} = \tau_g\alpha_c\beta_c(1-\eta) \text{ and } (\alpha\tau)_{2,eff} = \alpha_p(1-\beta_c)\tau_g^2$$

h_{p1} and h_{p2} are the penalty factors due to the glass cover of the PV module, which are

$$h_{p1} = \frac{U_{Tc,f}}{U_{tc,a} + U_{Tc,f}} \text{ and } h_{p2} = \frac{h_{p.f}}{U_{p,a} + h_{p,f}}$$

defined as,

$$U_{tc,a} = \left[\frac{L_g}{K_g} + \frac{1}{h_o}\right]^{-1}$$
$$h_o = 5.7 + 3.8\,V, \quad V = 0.5\,\text{m/s}$$
$$U_{Tc,f} = \left[\frac{L_g}{K_g} + \frac{1}{h_i}\right]^{-1}$$
$$h_{p,f} = h_i = 2.8 + 3v, \; v = 2\,\text{m/s}$$

$$U_{tT} = \frac{U_{Tc,f}\cdot U_{tc,a}}{U_{Tc,f} + U_{tc,a}}$$
$$U_T = \frac{U_{bp,a}\cdot h_{p,f}}{U_{bp,a} + h_{p,f}}$$
$$U_L = U_{tT} + U_T$$

$$U_{tT} = \frac{U_{Tc,f}\cdot U_{tc,a}}{U_{Tc,f} + U_{tc,a}}$$
$$U_T = \frac{U_{bp,a}\cdot h_{p,f}}{U_{bp,a} + h_{p,f}}$$
$$U_L = U_{tT} + U_T$$

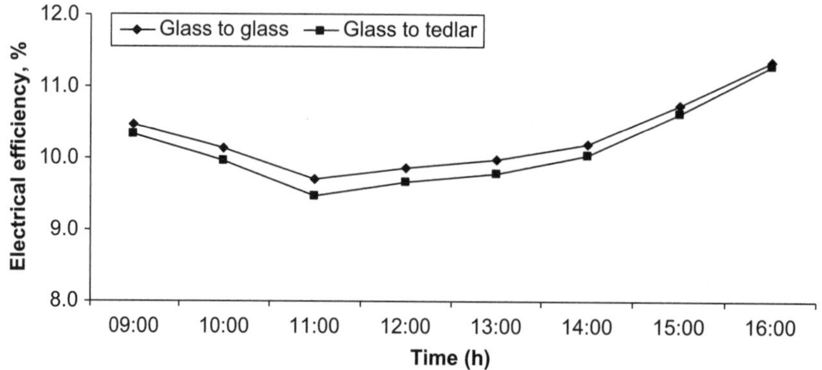

Figure 3.17 Hourly variation of electrical efficiency of glass to glass and glass to tedlar type PV module.

The hourly variation of electrical efficiency of PV modules is shown in Figure 3.17. Figure 3.17 shows that the higher efficiency is obtained by using glass-to-glass type PV modules due to the solar radiation falling on the nonpacking area of a glass-to-glass module is transmitted through the glass and absorbed by the blackened plate. Hence, the heat is convected to the flowing air by two ways from back surface of PV module as well as from top surface of the blackened plate. However, in the case of glass-to-tedlar all the radiation is absorbed by the tedlar and then carried away by conduction. This increases the temperature of the solar cell and its efficiency decreases.

3.4.2 Hybrid Photovoltaic Thermal (HPVT) Water Collector

3.4.2.1 Energy Balance Equations

The Lower Portion of FPC is Partially Covered by a PV Module

In this case, the lower portion of the absorber of the FPC is covered by the PV module and the upper portion is covered by the glass cover. The energy balance equations for each component of PV integrated collector are as follows:

(i) For Solar Cells of a PV Module (Glass–Glass):

$$\alpha_c\tau_g\beta_c I(t)W\mathrm{d}x = \left[U_{\mathrm{tc,a}}(T_c - T_a) + h_{\mathrm{c,p}}(T_c - T_p)\right]W\mathrm{d}x + \eta_c\tau_g\beta_c I(t)\cdot W\mathrm{d}x \tag{3.48a}$$

Here, it is important to note that the module electrical efficiency, $\eta_m = \eta_c\tau_g\beta_c$.
From Eq. (3.48a), the expression for cell temperature is From

$$T_c = \frac{(\alpha\tau)_{1,\mathrm{eff}}I(t) + U_{\mathrm{tc,a}}T_a + h_{\mathrm{c,p}}T_p}{U_{\mathrm{tc,a}} + h_{\mathrm{c,p}}} \tag{3.48b}$$

$$(\alpha\tau)_{1,\mathrm{eff}} = (\tau_g\alpha_c - \eta_c)\beta_c$$

where α_c is the absorptivity of solar cell, τ_g is the transmissivity of glass of PV module, β_c is the packing factor of the solar cell, $I(t)$ is the solar intensity, $W\mathrm{d}x$ is elementary section, η_c is the solar cell efficiency, $U_{\mathrm{tc,a}}$ is the overall heat transfer coefficient between the solar cell to ambient through the glass cover, $h_{\mathrm{c,p}}$ is the overall heat transfer coefficient from the solar cell to the blackened absorber plate through the glass cover and air gap, T_c is the temperature of the solar cell, T_a is the ambient temperature, T_p is the temperature of the blackened absorber plate.

(ii) For Blackened Absorber Plate Temperature below the PV Module (Glass–Glass):

$$\alpha_p(1-\beta_c)\tau_g^2 I(t)W\mathrm{d}x + h_{c,p}(T_c - T_p)W\mathrm{d}x = h_{p,f}(T_p - T_f)W\mathrm{d}x \tag{3.49}$$

From Eq. (3.36), the expression for the plate temperature is

$$T_p = \frac{(\alpha\tau)_{2,\mathrm{eff}}I(t) + PF_1(\alpha\tau)_{1,\mathrm{eff}}I(t) + U_{L1}T_a + h_{p,f}T_f}{U_{L1} + h_{p,f}} \tag{3.50}$$

where, $(\alpha\tau)_{2,\mathrm{eff}} = \alpha_p(1-\beta_c)\tau_g^2$, $\mathrm{PF}_1 = \frac{h_{c,p}}{U_{tc,b}+h_{c,p}}$, α_p is the absorptivity of blackened plate, $h_{p,f}$ is the conductive heat transfer coefficient from the plate to the flowing fluid, T_f is the temperature of the fluid, PF_1 is the penalty factor first due to the glass cover of the PV module, U_{L1} is an overall heat transfer coefficient from the blackened surface to the ambient.

(iii) For Water Flowing Through an Absorber Pipe below the PV Module (Glass–Glass):
The flow pattern of water when the PV is integrated at the bottom of the collector is shown in Figure 3.18.

The energy balance of flowing water through the absorber pipe (Figure 3.18) is given by,

$$\dot{m}_f C_f \frac{\mathrm{d}T_f}{\mathrm{d}x}\mathrm{d}x = F' h_{p,f}(T_p - T_f)W\mathrm{d}x \tag{3.51}$$

The solution of Eq. (3.51) with the help of Eqs. (3.48b) and (3.50) can be obtained as,

$$\dot{m}_f C_f \frac{\mathrm{d}T_f}{\mathrm{d}x}\mathrm{d}x = F'\left[\mathrm{PF}_2(\alpha\tau)_{m,\mathrm{eff}}I(t) - U_{L,m}(T_f - T_a)\right]W\mathrm{d}x \tag{3.52}$$

where, $(\alpha\tau)_{m,\mathrm{eff}} = \mathrm{PF}_1(\alpha\tau)_{1,\mathrm{eff}} + (\alpha\tau)_{2,\mathrm{eff}}$, $\mathrm{PF}_2 = \frac{h_{p,f}}{U_{L1}+h_{p,f}}$, $\dot{m}_f$ is mass of fluid, C_f is specific heat of fluid, F' is the collector efficiency factor, PF_2 is the penalty factor second due to the absorber below the PV module, $(\alpha\tau)_{m,\mathrm{eff}}$ is the effective absorptivity–transmissivity of the PV module, U_{Lm} is an overall heat transfer coefficient from the PV module.

By rearranging and integrating both side of Eq. (3.52) and using boundary conditions, namely, at $T_f|_{x=0} = T_{fi}$ and at $T_f|_{x=L} = T_{fo}$, one gets,

$$\frac{T_{fo} - T_a - \left(\dfrac{\mathrm{PF}_2(\alpha\tau)_{m,\mathrm{eff}}I(t)}{U_{L,m}}\right)}{T_{fi} - T_a - \left(\dfrac{\mathrm{PF}_2(\alpha\tau)_{m,\mathrm{eff}}I(t)}{U_{L,m}}\right)} = \exp\left(-\frac{F' A_m U_{L,m}}{\dot{m}_f C_f}\right)$$

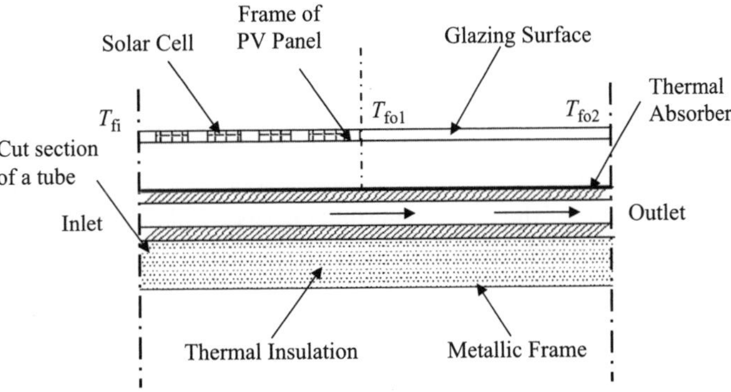

Figure 3.18 Cross-sectional side view of a PV integrated (bottom portion) single-glazed flat plate collector (FPC).

or,

$$T_{fo} = \left[\frac{PF_2(\alpha\tau)_{m,eff}I(t)}{U_{L,m}} + T_a\right]\left[1 - \exp\left(-\frac{F'A_m U_{L,m}}{\dot{m}_f C_f}\right)\right] + T_{fi}\exp\left(-\frac{F'A_m U_{L,m}}{\dot{m}_f C_f}\right) \quad (3.53)$$

Here, the outlet of water at the end of the PV module–absorber combination becomes the inlet to the glass–absorber combination. Such a collector is referred to as a photovoltaic thermal (PVT) water collector and T_{fo1} is the final outlet temperature of water from the PVT water collector.

The rate of thermal energy available at the end of absorber covered by the PV module–absorber combination

$$\dot{Q}_{u,m} = \dot{m}_f C_f (T_{fo} - T_{fi}) \quad (3.54)$$

After substituting the expression for T_{fo} from Eq. (3.53) into Eq. (3.54), we obtain,

$$\dot{Q}_{u,m} = A_m F_{Rm}\left(PF_2(\alpha\tau)_{m,eff}I(t) - U_{L,m}(T_{fi} - T_a)\right) \quad (3.55)$$

where, $A_m F_{Rm} = \frac{\dot{m}_f C_f}{U_{L,m}}\left[1 - \exp\left(-\frac{F'A_m U_{L,m}}{\dot{m}_f C_f}\right)\right]$

(iv) The Outlet Water Temperature at the End of a Collector:

An expression for the outlet water temperature at the end of a conventional FPC can be given as,[23]

$$T_{fo1} = \left[\frac{(\alpha\tau)_{c1,eff}I(t)}{U_{L,c1}} + T_a\right]\left[1 - \exp\left(-\frac{F'A_{c1}U_{L,c1}}{\dot{m}_f C_f}\right)\right] + T_{fi1}\exp\left(-\frac{F'A_{c1}U_{L,c1}}{\dot{m}_f C_f}\right) \quad (3.56)$$

Here, $T_{fi1} = T_{fo}$, the expression for final outlet temperature from the first PVT hybrid flat plate water collector can be given as,

$$\begin{aligned} T_{fo1} = {} & \left[\frac{(\alpha\tau)_{c1,eff}I(t)}{U_{L,c1}} + T_a\right]\left[1 - \exp\left(-\frac{F'A_{c1}U_{L,c1}}{\dot{m}_f C_f}\right)\right] \\ & + \left[\begin{array}{l} \left[\frac{PF_2(\alpha\tau)_{m,eff}I(t)}{U_{L,m}} + T_a\right]\left[1 - \exp\left(-\frac{F'A_m U_{L,m}}{\dot{m}_f C_f}\right)\right] \\ + T_{fi}\exp\left(-\frac{F'A_m U_{L,m}}{\dot{m}_f C_f}\right) \end{array}\right]\exp\left(-\frac{F'A_{c1}U_{L,c1}}{\dot{m}_f C_f}\right) \end{aligned} \quad (3.57)$$

An expression for the rate of thermal energy available from the first PVT hybrid flat plate water collector can be evaluated as,

$$\dot{Q}_{u1} = \dot{m}_f C_f (T_{fo1} - T_{fi}) \quad (3.58)$$

Further, an expression for the rate of thermal energy available from the first PVT hybrid flat-plate water collector will be sum of the rate of thermal energy available from the PV module and the conventional FPC, which can also be evaluated as,

$$\begin{aligned} \dot{Q}_{u1} &= \dot{Q}_{um} + \dot{Q}_{uc} \\ &= A_m F_{Rm}\left[PF_2(\alpha\tau)_{m,eff}I(t) - U_{L,m}(T_{fi} - T_a)\right] + A_c F_{Rc}\left[(\alpha\tau)_{c,eff}I(t) - U_{L,c}(T_{fo} - T_a)\right] \end{aligned} \quad (3.59)$$

Here, $T_{fo} = T_{fi} + \frac{\dot{Q}_{u,m}}{\dot{m}_f C_f}$.

On simplifying the above equation we get,

$$\dot{Q}_{\text{ul}} = \left[A_{\text{m}}F_{\text{Rm}}PF_2(\alpha\tau)_{\text{m,eff}}\left(1 - \frac{A_{\text{c}}F_{\text{Rc}}U_{\text{L,c}}}{\dot{m}_{\text{f}}C_{\text{f}}}\right) + A_{\text{c}}F_{\text{Rc}}(\alpha\tau)_{\text{c,eff}}\right]I(t) - \left[A_{\text{m}}F_{\text{Rm}}U_{\text{L,m}}\left(1 - \frac{A_{\text{c}}F_{\text{Rc}}U_{\text{L,c}}}{\dot{m}_{\text{f}}C_{\text{f}}}\right) + A_{\text{c}}F_{\text{Rc}}U_{\text{L,c}}\right](T_{\text{fi}} - T_{\text{a}}) \tag{3.60}$$

Equation (3.60) can also be obtained by substituting Eq. (3.57) into Eq. (3.58).
Equation (3.60) can be rewritten as

$$\dot{Q}_{\text{ul}} = (A\,F_{\text{R}}(\alpha\tau))_1 I(t) - (A\,F_{\text{R}}U_{\text{L}})_1(T_{\text{fi}} - T_{\text{a}}) \tag{3.61}$$

where

$$(A\,F_{\text{R}}(\alpha\tau))_1 = \left[A_{\text{m}}F_{\text{Rm}}\text{PF}_2(\alpha\tau)_{\text{m,eff}}\left(1 - \frac{A_{\text{c}}F_{\text{Rc}}U_{\text{L,c}}}{\dot{m}_{\text{f}}C_{\text{f}}}\right) + A_{\text{c}}F_{\text{Rc}}(\alpha\tau)_{\text{c,eff}}\right]$$

and

$$(A\,F_{\text{R}}U_{\text{L}})_1 = \left[A_{\text{m}}F_{\text{Rm}}U_{\text{L,m}}\left(1 - \frac{A_{\text{c}}F_{\text{Rc}}U_{\text{L,c}}}{\dot{m}_{\text{f}}C_{\text{f}}}\right) + A_{\text{c}}F_{\text{Rc}}U_{\text{L,c}}\right]$$

An instantaneous efficiency can be obtained from the above equation as.

$$\eta_{\text{i}} = (A\,F_{\text{R}}(\alpha\tau))_1 - (A\,F_{\text{R}}U_{\text{L}})_1\frac{T_{\text{fi}} - T_{\text{a}}}{I(t)} \tag{3.62}$$

From the above equation, expressions for the gain factor and loss coefficient can be obtained as

$$\text{Gain factor} = (A\,F_{\text{R}}(\alpha\tau))_1 \tag{3.63a}$$

and,

$$\text{Loss coefficient} = (A\,F_{\text{R}}U_{\text{L}})_1 \tag{3.63b}$$

3.4.2.2 Upper Portion of Absorber is Partially Covered by PV Module

In this case, the upper portion of the absorber is covered by the PV module and the lower portion is covered by the glass cover. The outlet of water at the end of the glass–absorber combination becomes the inlet to the PV module–absorber combination.

An expression for the rate of thermal energy available from the flat plate collector can be evaluated as,

$$\dot{Q}_{u,(c+m)} = \dot{m}_f C_f (T_{fo2} - T_{fi}) \tag{3.64}$$

An expression for the total thermal energy available from the PV integrated (upper side) flat plate collector can be evaluated as,

$$\dot{Q}_{u,(c+m)} = A_c F_{Rc}\left[(\alpha\tau)_{c,eff} I(t) - U_{L,c}(T_{fi} - T_a)\right] + A_m F_{Rm}\left[h_{p2}(\alpha\tau)_{m,eff} I(t) - U_{L,m}(T_{fo1} - T_a)\right] \tag{3.65}$$

Here,

$$T_{fo1} = T_{fi} + \frac{\dot{Q}_{u,c}}{\dot{m}_f C_f}$$

On simplifying the above equation we get,

$$\dot{Q}_{u,(c+m)} = \left[A_c F_{Rc}(\alpha\tau)_{c,eff}\left(1 - \frac{A_m F_{Rm} U_{L,m}}{\dot{m}_f C_f}\right) + A_m F_{Rm} h_{p2}(\alpha\tau)_{m,eff}\right] I(t) - \left[A_c F_{Rc} U_{Lc}\left(1 - \frac{A_m F_{Rm} U_{L,m}}{\dot{m}_f C_f}\right) + A_m F_{Rm} U_{Lm}\right](T_{fi} - T_a) \tag{3.66}$$

The hourly variation of solar cell temperature for two cases is shown in Figure 3.19a. Higher cell temperature is obtained when the PV is on the upper position than the case when the PV is on the lower position because in the latter case the water gets preheated from the glass–absorber combination of the collector area and then heated water goes into the PV–absorber combination of the collector area, resulting in less heat transfer from PV module. Higher cell temperatures decrease the cell efficiency, which is shown in Figure 3.19b. The cell efficiency varies from 0.111 to 0.099 and 0.108 to 0.096 for the two cases.

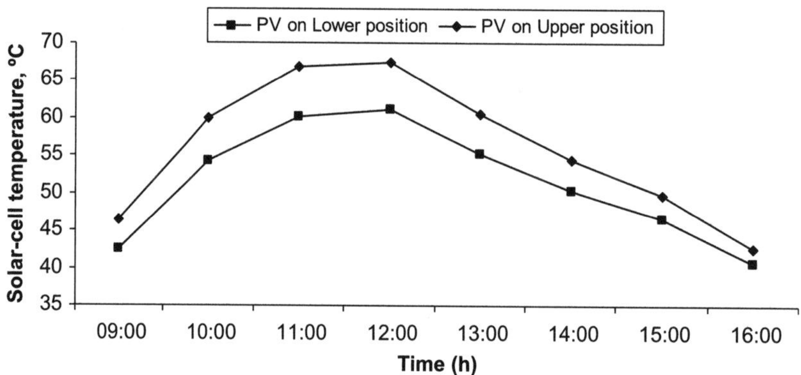

Figure 3.19a Hourly variation of solar cell temperature when the lower and upper portion of absorber is partially covered by PV module.

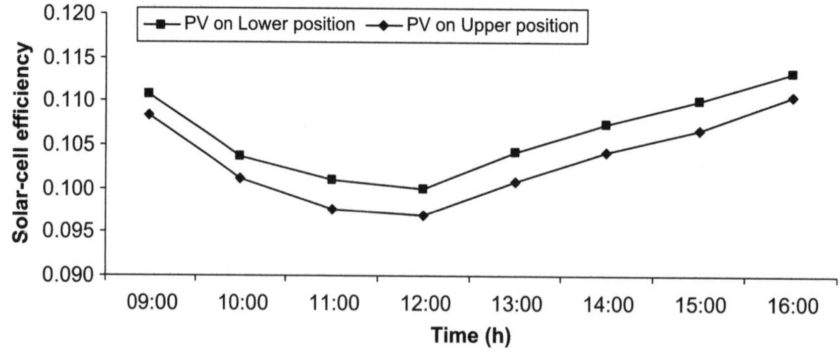

Figure 3.19b Hourly variation of cell efficiency when the lower and upper portion of absorber is partially covered by PV module.

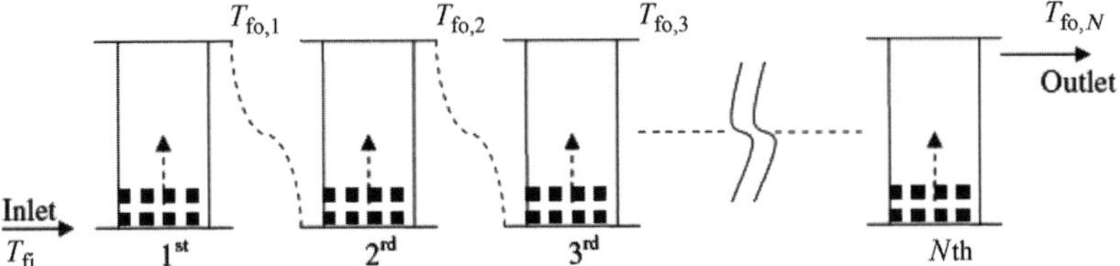

Figure 3.20 Collectors partially covered by PV modules and connected in series.

3.4.2.3 N-Hybrid Photovoltaic Thermal Water Collectors Connected in Series

The series connection of N-hybrid photovoltaic thermal water collectors is shown in Figure 3.20.

Similar to Eq. (3.61), the useful thermal output of the second PVT hybrid flat plate water collector can be written as,

$$\dot{Q}_{u2} = (A\,F_R(\alpha\tau))_2 I(t) - (A\,F_R U_L)_2 (T_{fo1} - T_a) \tag{3.67}$$

Here, $T_{fo1} = T_{fi} + \frac{\dot{Q}_{u1}}{\dot{m}_f C_f}$.

For the combination of two identical PVT hybrid flat plate collectors connected in series, the rate of useful thermal out can be given as,

$$\begin{aligned}\dot{Q}_{u1+2} = \dot{Q}_{u1} + \dot{Q}_{u2} = &\left[(A\,F_R(\alpha\tau))_1\left(1 - \frac{(A\,F_R U_L)_2}{\dot{m}_f C_f}\right) + (A\,F_R(\alpha\tau))_2\right] I(t) \\ &- \left[(A\,F_R U_L)_1\left(1 - \frac{(A\,F_R U_L)_2}{\dot{m}_f C_f}\right) + (A\,F_R U_L)_2\right](T_{fi} - T_a)\end{aligned} \tag{3.68}$$

In the above equation one can assume that

$$(A\,F_R(\alpha\tau))_1 = (A\,F_R(\alpha\tau))_2 \quad \text{and} \quad (A\,F_R U_L)_1 = (A\,F_R U_L)_2.$$

Similarly, the rate of useful thermal output from N identical PVT hybrid flat connected in series can be given as,

$$\dot{Q}_{u,N} = N.A_c\left[(\alpha\tau)_{eff,N} I(t) - U_{L,N}(T_{fi} - T_a)\right] \tag{3.69}$$

Here,

$$(\alpha\tau)_{eff,N} = (F_R(\alpha\tau))_1\left[\frac{1 - (1 - K_{K,A})^N}{N\,K_{K,A}}\right]$$

and

$$U_{L,N} = (F_R U_L)_1\left[\frac{1 - (1 - K_{K,A})^N}{N\,K_{K,A}}\right]$$

where, $K_{K,A} = \left[\frac{(A\,F_R U_L)_1}{\dot{m}_f C_f}\right]$.

The outlet fluid temperature at the end of the first PVT hybrid collector can also be evaluated from Eqs. (3.58) and (3.61) as,

$$T_{\mathrm{fo1}} = \frac{(A F_{\mathrm{R}}(\alpha\tau))_1}{\dot{m}_{\mathrm{f}} C_{\mathrm{f}}} I(t) + \frac{(A F_{\mathrm{R}} U_L)_1}{\dot{m}_{\mathrm{f}} C_{\mathrm{f}}} T_{\mathrm{a}} + T_{\mathrm{fi}} \left(1 - \frac{(A F_{\mathrm{R}} U_{\mathrm{L}})_1}{\dot{m}_{\mathrm{f}} C_{\mathrm{f}}}\right) \tag{3.70}$$

Similarly, the outlet fluid temperature at the end of the second PVT hybrid collector can be written as,

$$T_{\mathrm{fo2}} = \frac{(A F_{\mathrm{R}}(\alpha\tau))_2}{\dot{m}_{\mathrm{f}} C_{\mathrm{f}}} I(t) + \frac{(A F_{\mathrm{R}} U_{\mathrm{L}})_2}{\dot{m}_{\mathrm{f}} C_{\mathrm{f}}} T_{\mathrm{a}} + T_{\mathrm{fi2}} \left(1 - \frac{(A F_{\mathrm{R}} U_{\mathrm{L}})_2}{\dot{m}_{\mathrm{f}} C_{\mathrm{f}}}\right) \tag{3.71}$$

As, $T_{\mathrm{fi2}} = T_{\mathrm{fo1}}$, the outlet temperature of the first collector will be the inlet for the second collector, the outlet temperature of the second will be the inlet for the third collector and so on. Hence, for a system of N collectors connected in series, the outlet fluid temperature (T_{foN}) from the Nth PVT hybrid collector can be expressed in terms of the inlet temperature of the first collector.

For N identical set of PVT hybrid collectors connected in series, the outlet fluid temperature at the end of Nth collector can be derived as,

$$T_{\mathrm{fo\,N}} = \frac{(A F_{\mathrm{R}}(\alpha\tau))_1}{\dot{m}_{\mathrm{f}} C_{\mathrm{f}}} \left(\frac{1 - K_{\mathrm{K}}{}^N}{1 - K_K}\right) I(t) + \frac{(A F_R U_L)_1}{\dot{m}_{\mathrm{f}} C_{\mathrm{f}}} \left(\frac{1 - K_{\mathrm{K}}{}^N}{1 - K_{\mathrm{K}}}\right) T_{\mathrm{a}} + T_{\mathrm{fi}} K_{\mathrm{K}}{}^N \tag{3.72}$$

where, $K_{\mathrm{K}} = \left[1 - \frac{(A F_{\mathrm{R}} U_{\mathrm{L}})_1}{\dot{m}_{\mathrm{f}} C_{\mathrm{f}}}\right]$.

The hourly variation of temperature-dependent electrical efficiency by varying the number of collectors at constant flow rate ($\dot{m} = 0.04$ kg/s) for the above case is shown in Figure 3.21. As the number of collectors increases, the cell temperature increases due to the increase in water temperature and hence the cell efficiency decreases (0.091 to 0.086).

3.4.2.4 *Different Cases*

(a) For Fully Covered with Glass Cover (Conventional FPC)

Expressions for the rate of thermal energy available and the outlet temperature for N FPC collectors connected in series can be obtained from Eqs. (3.69) and (3.72) by substituting $A_{\mathrm{m}} = 0$. For FPC collectors connected in series.

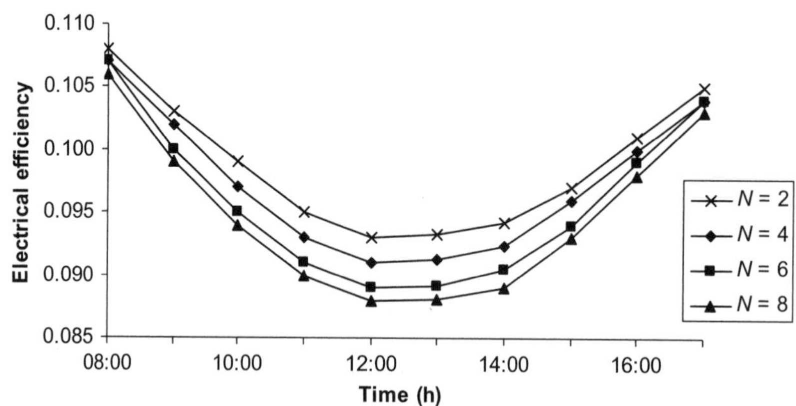

Figure 3.21 Hourly variation of temperature dependent electrical efficiency by varying the number of collectors at constant flow rate ($\dot{m} = 0.04$ kg/s).

The useful heat output,

$$\dot{Q}_{u,N} = \left[N\,A\,F_R(\alpha\tau)_{c,eff}\left\{\frac{1-(1-K_K)^N}{N\,K_K}\right\}\right]I(t) - \left[N\,A\,F_R\,U_{L,c}\left\{\frac{1-(1-K_K)^N}{N\,K_K}\right\}\right](T_{fi}-T_a) \quad (3.73)$$

where, $K_K = \left[\frac{A\,F_R\,U_{L,c}}{\dot{m}_f C_f}\right]$.

The gain factor and loss factor can be defined as,

$$(\alpha\tau)_{eff} = F_R(\alpha\tau)_{c,eff}\left[\frac{1-(1-K_K)^N}{N\,K_K}\right]$$

$$U_L = F_R\,U_{L,c}\left[\frac{1-(1-K_K)^N}{N\,K_K}\right]$$

The outlet fluid temperature

$$T_{foN} = \left[\frac{(\alpha\tau)_{c,eff}I(t)}{U_{L,c}} + T_a\right]\left[1-\exp\left(-\frac{NF'A_cU_{L,c}}{\dot{m}_f C_f}\right)\right] + T_{fi1}\exp\left(-\frac{NF'A_cU_{L,c}}{\dot{m}_f C_f}\right) \quad (3.74)$$

(b) For Fully Covered with PV Modules

Expressions for the rate of thermal energy available and the outlet temperature for N FPC collectors connected in series can be obtained from Eqs. (3.69) and (3.72) by substituting $A_c = 0$. For fully covered with PV modules:

$$T_{fo\,N} = \left[\frac{PF_2(\alpha\tau)_{m,eff}I(t)}{U_{L,m}} + T_a\right]\left[1-\exp\left(-\frac{N\,F'AU_{L,m}}{\dot{m}_f C_f}\right)\right] + T_{fi}\exp\left(-\frac{N\,F'AU_{L,m}}{\dot{m}_f C_f}\right) \quad (3.75)$$

(c) For Glazed PVT Hybrid Water Collectors

τ_g, τ_g^2 will be replaced by τ_g^2 and τ_g^3 in Eqs. (3.48a) and (3.49), respectively. Also, $U_{t,ca}$ will be the replaced by $U_{t,ca}$ for double-glazed surfaces.

(d) For Constant Collection Temperature

In this case, the outlet water temperature (Eq. (3.72)) is kept constant by changing mass flow rate by using the temperature sensor. For this, T_{foN} becomes T_o. For given T_o, the different mass flow rate can be achieved by varying the values of solar intensity and ambient air temperature. This is required by industrial applications.

(iv) Hybrid Photovoltaic Thermal (HPVT) Air Collector

The PVT air collectors are used for heating the air and electricity generation simultaneously. The hot air is used for space heating and drying purposes.

Similar to the thermal analysis as in the case of hybrid water collector the outlet fluid (air) temperature at the Nth collector fully covered with the PV module,

$$T_{fo\,N} = \left[\frac{(\alpha\tau)_{eff}I(t)}{U_L} + T_a\right]\left[1-\exp\left(-\frac{N\,bU_L L}{\dot{m}_a C_a}\right)\right] + T_{fi}\exp\left(-\frac{N\,bU_L L}{\dot{m}_a C_a}\right) \quad (3.76)$$

and the useful heat output of the Nth collector is derived as,

$$\dot{Q}_{u,N} = \left[F_R(\alpha\tau)_{eff}\left\{\frac{1-(1-K_K)^N}{N\,K_K}\right\}\right]I(t) - \left[F_R U_L\left\{\frac{1-(1-K_K)^N}{N\,K_K}\right\}\right](T_{fi} - T_a) \tag{3.77}$$

where,

$$K_K = \left[\frac{b.\,L.\,F_R U_L}{\dot{m}_a C_a}\right] \text{ and } F_R = \frac{\dot{m}_a C_a}{A_c U_L}\left[1 - \exp\left(-\frac{A_c U_L F'}{\dot{m}_a C_a}\right)\right]$$

3.4.2.5 Overall Thermal and Exergy Efficiency

(a) Thermal Efficiency

For the thermal analysis, an overall thermal efficiency of the system has been calculated. In this, electrical efficiency has been converted to the equivalent of thermal efficiency using the electric power generation efficiency η_{cp} for a conventional power plant. An expression for the overall thermal efficiency of a PVT system can be obtained as,

$$\eta_{overall,\,thermal} = \eta_{thermal} + \frac{\eta_{electrical}}{\eta_{cp}} \tag{3.78}$$

The value of η_{cp} depends on the quality of the coal. $\eta_{cp} = 0.38$ for a good quality of coal with low ash content. The value of η_{cp} varies between 0.20 to 0.40.

(b) Exergy Efficiency

For the exergy analysis, an overall exergy efficiency of the system has been calculated. In this, thermal efficiency has been converted into an equivalent electrical efficiency using the Carnot efficiency factor. The overall exergy efficiency of a PVT system is defined as,

$$\eta_{overall,\,exergy} = \eta_{electrical} + \eta_{thermal}\left(1 - \frac{T_a + 273}{T_f + 273}\right) \tag{3.79}$$

where, T_a and T_f are ambient and fluid temperature, respectively, and both are time dependent.

It was observed that the daily overall thermal efficiency of a IPVTS system increases with increase constant flow rate and decreases with increase of constant collection temperature. Figures 3.22a and b show the variation of overall thermal and exergy efficiency with hot water withdrawal rate. From the figures it is observed that the overall exergy and thermal efficiency of an integrated photovoltaic thermal solar system (IPVTS) is maximum at a hot water withdrawal flow rate of 0.006 kg/s.

(c) Energy Matrices and Cost Analysis

(i) Energy Payback Time (EPBT)

The EPBT depends on the energy spent to prepare the materials used for fabrication of the system and its components, *i.e.* embodied energy and the annual energy yield (output) obtained from such a system. To evaluate the embodied energy of various components of the system, the energy densities of different materials are required. It is the total time period required to recover the total energy spent to prepare the materials (embodied energy) used for fabrication of the hybrid PVT

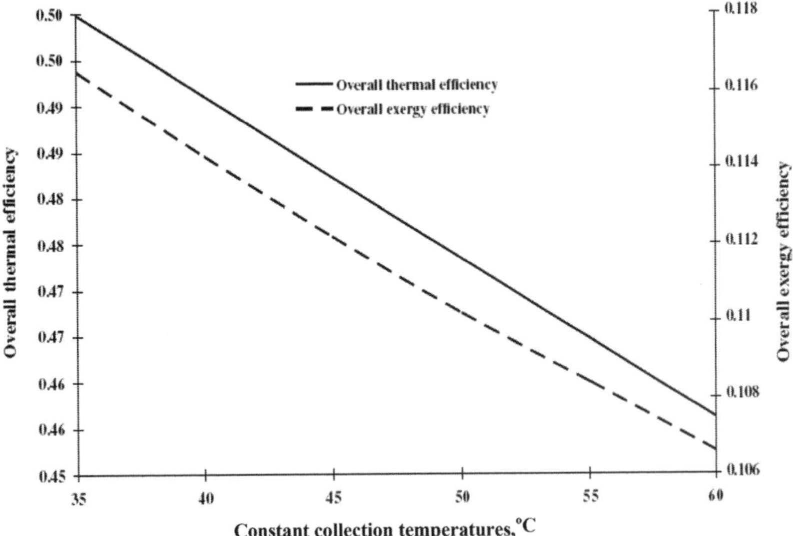

Figure 3.22a Variation of overall thermal and exergy efficiency with constant collection temperature.

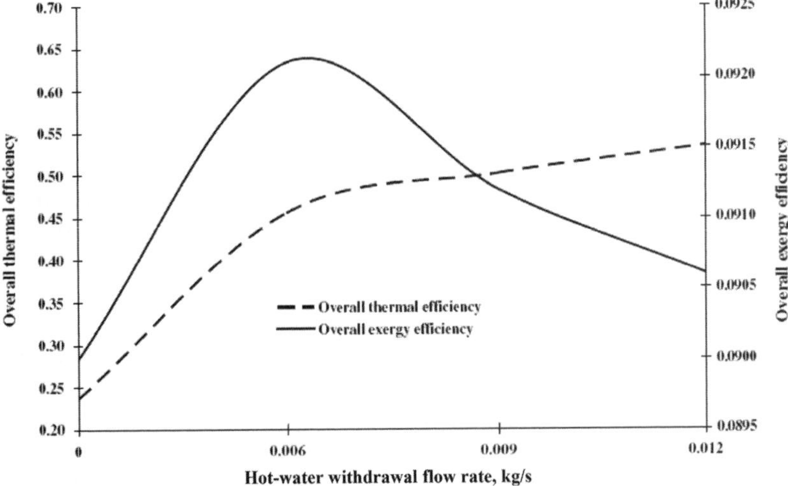

Figure 3.22b Variation of overalthermal and exergy efficiency with different hot water withdrawal flow rate.

systems. It is the ratio of embodied energy and the annual energy output from the system, which can be expressed as

$$\text{EPBT} = \frac{\text{Embodied Energy}\,(E_{\text{in}})}{\text{Annual Energy Output}\;(E_{\text{out}})} \tag{3.80}$$

(ii) Energy Production Factor (EPF)

This is used to predict the overall performance of the system. It is defined as the ratio of the output energy and the input energy or it can also be expressed as the inverse of EPBT. The energy production factor is defined by two types

(i) On an annual basis, and

$$\chi_a = \frac{E_{out}}{E_{in}} \tag{3.81a}$$

or,

$$\chi_a = \frac{1}{T_{epb}}$$

If $\chi_a \rightarrow 1$, for $T_{epb} = 1$ the system is worthwhile, otherwise it is not worthwhile from the energy point of view.

(v) On a lifetime basis.

$$\chi_L = \frac{E_{out} \times T}{E_{in}} \tag{3.81b}$$

(iii) Lifecycle Conversion Efficiency (LCCE)

LCCE is the net energy productivity of the system with respect to the solar input (radiation) over the lifetime of the system, (T years) given by

$$\phi(t) = \frac{E_{out} \times T - E_{in}}{E_{sol} \times T} \tag{3.82}$$

The energy payback time for a crystalline-silicon (c-Si) solar cell module under Indian climatic condition for annual peak load duration is about four years. The energy payback time (EPBT) for an amorphous silicon (a-Si) solar cell module with efficiency of 5% as 7.4 years for the climatic conditions of Detroit, USA; the EPBT gets reduced to 4.1 years with the increase in the efficiency of the module to 9%. The energy payback time for an amorphous silicon (a-Si) solar cell module reduces to 2.6 years after considering the gross energy requirement (GER) and the hidden energy. The energy payback time for a conventional multicrystalline building integrated system, retrofitted on a tilted roof, located in Rome (Italy); with the yearly global insolation on a horizontal plane was taken as 1530 kWh/m^2 yr. They concluded that the energy payback time reduces from 3.3 year to 2.8 yr.

FPBT, EPF and LCCE for a life span of 60 years of building using different types of solar cell materials are given in Table 3.3.

The table shows that the EPBT for the CIGS BIPVT system is a minimum (19.31 years) while that for the amorphous silicon BIPVT is a maximum (29.13), taking the life span of the BIPVT as

Table 3.3 Energy payback time, energy production factor on lifetime basis and life-cycle conversion efficiency for life time 60 years of BIPVT system.

PV Technology	*Energy payback period (EPBT) in years*	*Energy production factor (EPF) on life time basis*	*Life-cycle conversion efficiency (LCCE)*
m-si	20.81	2.884	0.349
p-si	20.93	2.867	0.308
r-si	20.94	2.866	0.269
a-si	29.13	2.060	0.132
CdTe	24.67	2.432	0.185
CIGS	19.31	3.107	0.247

60 years. Also, for the CIGS BIPVT system EPF is a maximum (3.107), while that for the amorphous silicon BIPVT is a minimum (2.06). It has also been found that for the monocrystalline silicon BIPVT system LCCE is a maximum (0.349), while that of amorphous silicon BIPVT is a minimum (0.132). Hence, from the EPBT and EPF point of view, the CIGS solar cells in the BIPVT system is the most suitable, while from the LCCE point of view the use of monocrystalline solar cells in the BIPVT system is the most suitable.

(iv) Annualised Uniform Cost (Unacost)

Annualised uniform cost is defined as a product of present value of the system and capital recovery factor (CRF)

$$\text{Unacost} = \text{Net present value} \times \text{Capital recovery factor} \tag{3.83}$$

$\text{CRF} = \frac{i(i+1)^n}{(i+1)^n - 1}$, Here, n = no. of years and i = interest rate per year.

Let P be the present value and R_1, R_2,..... R_n are operational, maintenance and pump replacement costs per year and $R_{3,1}$, $R_{6,2}$,.....$R_{n,n}$ are black painting, cleaning and glass replacement costs occurred in every three year. Then the net present value is evaluated as,

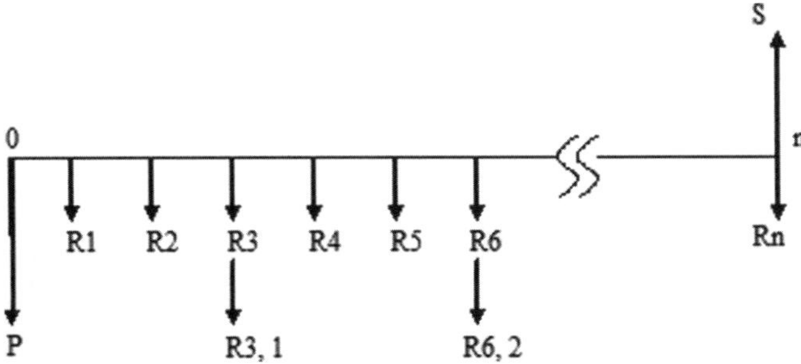

$$\text{Net Present Value (NPV)} = P + R_1 \times \left[\frac{(i+1)^n - 1}{i(i+1)^n}\right]_{i,n} + R_{3,1} \times \left[\frac{1}{(i+1)^n}\right]_{i,3} + R_{6,2} \times \left[\frac{1}{(i+1)^n}\right]_{i,6} + R_{9,3} \times \left[\frac{1}{(i+1)^n}\right]_{i,9} + \dots\dots\dots - S \times \left[\frac{1}{(i+1)^n}\right]_{i,n} \tag{3.84}$$

The unit cost produced by a PV system can be evaluated as

$$\text{Unit power generationcost} = \frac{\text{NPV}}{\text{annual power genertaion}} \tag{3.85}$$

The annualised cost in the case of monocrystalline silicon (m-Si), polycrystalline silicon (p-Si), ribbon silicon (r-Si), amorphous silicon (a-Si), cadmium telluride (CdTe) and copper indium gallium selenide (CIGS) BIPVT systems are US \$ 1931, US \$ 1641, US \$ 1487, US \$ 786, US \$ 1129 and US \$ 2929, respectively, as shown in Figure 3.23a. The figure also shows that the annualised cost of BIPVT systems are 2–7% higher than the similar BIPV systems. It is further to be noted that the use of BIPVT system reduces the unit power generation cost by 12–25% below that of the similar BIPV systems, as shown in Figure 3.23b. The figure also shows that the power generation cost for unit kilowatt (kW) of the monocrystalline silicon (m-Si), polycrystalline silicon (p-Si), ribbon silicon (r-Si), amorphous silicon (a-Si), cadmium telluride (CdTe) and copper indium

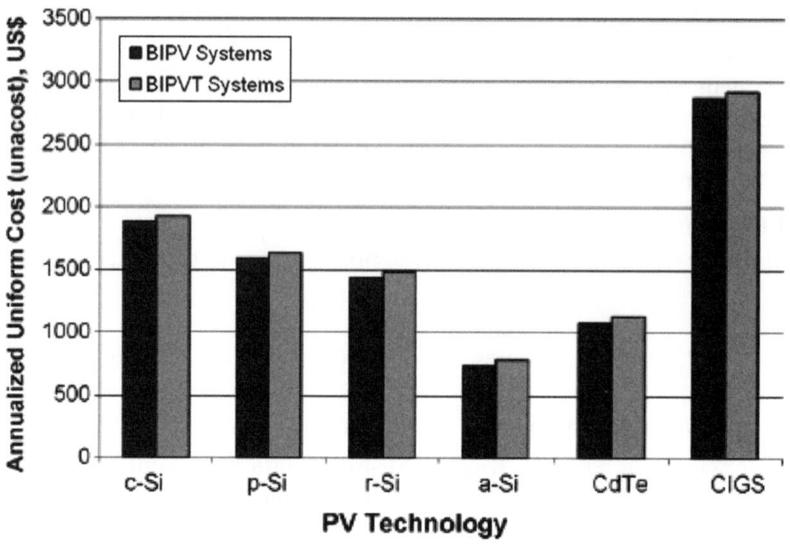

Figure 3.23a Annualised uniform cost of BIPV and BIPVT systems using different solar cells.

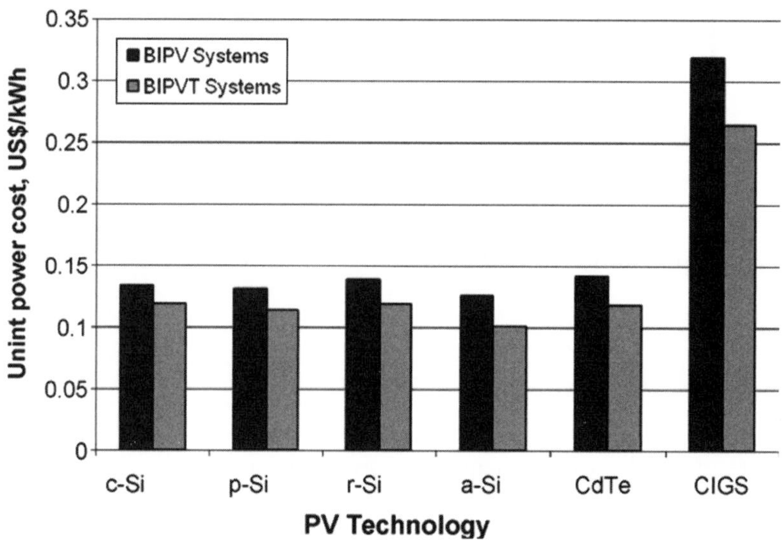

Figure 3.23b Unit power generation cost of the BIPV and BIPVT systems using different solar cells.

gallium selenide (CIGS) BIPVT systems are US $ 12.44, US $ 14.51, US $ 17.35, US $ 24.97, US $ 19.77 and US $ 20.41, respectively. Thus, from the economic point of view the a-Si BIPVT are more suitable for rooftops. Also, the cost of unit power generation from the a-Si BIPVT system is quite close to the cost of unit power generation through the conventional grid. Therefore, the application of such systems in residential and commercial buildings will help in reducing greenhouse gas emission, which is necessary for sustainable development.

For details and cross-references see the review by Tiwari *et al.*[24]

ADDITIONAL EXERCISES

Exercise 3.1: Determine the Fermi energy level for a silicon crystal doped with an acceptor impurity of concentration 10^{17} cm^3. Given the effective density of states in the valence band, at room temperature, is 1.04×10^{-19} cm^3.

Solution: We have

$$E_F = E_V - kT \ln \frac{N_A}{N_V}$$

If the conduction band is taken at the reference level, $E_V = 1.1$ eV. Substitution of values gives,

$$E_F = 1.1 - \left[\left(\frac{1.38 \times 10^{-23}}{1.6 \times 10^{-19}} \right) \times 310 \ln \frac{10^{17}}{1.04 \times 10^{-19}} \right]$$

$$E_F = -1.16 \text{ eV}$$

The shift is –1.16 –0.55 = –1.71 eV.
Answer: The shift in Fermi energy level = –1.71 eV.

Exercise 3.2: Determine the bandgap in a GaAs crystal at 38 °C.

Solution: The variation of bandgap with temperature is given by the relation:

$$E_g(T) = E_g(O) - \frac{aT^2}{T+b}$$

Substituting values of a and b, we get

$$E_g(38) = 1.52 - \left[\frac{(5.84 \times 10^{-4}) \times (311)^2}{311 + 300} \right]$$

$$= 1.52 - 0.092$$

$$E_g(38) = 1.428 \text{ eV}$$

Answer: Bandgap at 38 °C = 1.428 eV.

Exercise 3.3: Determine the value of saturation current for silicon at 32 °C.

Solution: The dependence of saturation current for silicon at 32 °C is given by the relation

$$I_o = AT^3 \exp(E_g/kT)$$

Here, A is the nonideality factor and its value is taken as 1.
Substituting the known values in the above relation, we get

$$I_o = 305^3 \exp\left[\frac{-1.776 \times 10^{-19}}{4.209 \times 10^{-21}} \right]$$

$$I_o = +1.34 \times 10^{-11} \text{A/m}^2$$

Answer: The value of saturation current at 32 °C = 1.34×10^{-11} A/m^2.

Exercise 3.4: Calculate the fill factor if a solar cell of area 4 cm^2 is irradiated with an intensity of 100 W/m^2. Given $V_{oc}=0.24$ V, $I_{sc}=-10$ mA, $V_{max}=0.14$ V, $I_{max}=-6.5$ mA. Also calculate R_{OP}.

Solution: We have the expression for fill factor as

$$\text{Fill factor} = \frac{V_{max} \times I_{max}}{V_{oc} \times I_{sc}}$$

Substituting the appropriate values in the above equation we get,

$$\text{Fill factor} = \frac{V_{max} \times I_{max}}{V_{oc} \times I_{sc}} = \frac{0.14 \times -6.5}{0.24 \times -10} = 0.37$$

and

$$R_{op} = \frac{V_m}{I_m} = \frac{0.14}{-6.5 \times 10^{-3}} = -21.53 \, \text{ohm (Volt/Ampere)}$$

Answer: Fill factor = 0.37.

Exercise 3.5: What will be solar cell current if dark and light-induced current are equal?

Solution: The overall cell current is determined by subtracting the light induced current I_L from the diode dark current I_D

$$I = I_D - I_L$$

if $$I_D = I_L$$

then $$I = 0$$

Thus, under this condition, cell will not generate any current.

Answer: $I=0$.

Exercise 3.6: Calculate the dark current for a solar cell for reverse and forward bias modes.

Solution: We have the expression for dark current as

$$\text{Dark current } I_D = I_o\left[\exp\left(\frac{eV}{kT}\right) - 1\right]$$

For good solar cells, $I_o = 10^{-8}$ A/m^2 or 10^{-5} mA/m^2

$$e = 1.6 \times 10^{-19} \text{ J}, \; k = 1.38 \times 10^{-23} \text{ J/k}, \; T = 300 \text{ K}$$

Then

$$I_D = 10^{-5}\left[\exp\left(\frac{1.6 \times 10^{-19} V}{1.38 \times 10^{-23} \times 300}\right) - 1\right]$$

$$I_D = 10^{-5}[\exp(38.6V) - 1] mA$$

Now, the above equation has been used to calculate I_D under forward and reverse bias

Variation of dark current with voltage in forward bias mode

Voltage (V)	*Dark current* I_D *(mA)*
0.00	0.00
0.01	0.47×10^{-5}
0.02	1.16×10^{-5}
0.03	2.18×10^{-5}
0.04	3.68×10^{-5}
0.05	5.88×10^{-5}
0.06	9.10×10^{-5}
0.07	13.90×10^{-5}
0.08	20.90×10^{-5}
0.10	46.40×10^{-5}
0.50	2.40×10^{3}

Voltage is taken negative in reverse bias mode Variation of dark current with voltage in reverse bias mode

Voltage (V)	*Dark current* I_D *(mA)*
0.00	0
0.01	-0.32×10^{-5}
0.02	-0.53×10^{-5}
0.03	-0.68×10^{-5}
0.04	-0.78×10^{-5}
0.05	-0.85×10^{-5}
0.06	-0.89×10^{-5}
0.07	-0.93×10^{-5}
0.08	-0.97×10^{-5}
0.09	-0.97×10^{-5}
0.10	-0.97×10^{-5}
0.50	-0.99×10^{-5}

OBJECTIVE QUESTIONS

3.1 In what form can solar energy be used?
- (a) Thermal energy
- (b) Electrical energy
- (c) Mechanical energy
- (d) All of them

3.2 What is the most common material used in making solar cells?
- (a) Silver
- (b) Iron
- (c) Aluminium
- (d) Silicon

3.3 The electrical output of a solar cell depends on
 (a) Intensity of solar radiation
 (b) Heat component of solar radiation
 (c) Ultraviolet radiation
 (d) Infrared radiation

3.4 Solar photovoltaic cells convert solar energy directly into
 (a) Mechanical energy
 (b) Electricity
 (c) Heat energy
 (d) Transportation

3.5 What does SPVT stand for with respect to solar energy?
 (a) Solar photovoltaic thermal
 (b) Solar platevoltaic thermal
 (c) Solar platevoids thermal
 (d) None of the above

3.6 Which of the following appliances use solar photovoltaic technology?
 (a) Solar lantern
 (b) Biogas plant
 (c) Solar water heater
 (d) Solar air heater

3.7 Which material has the highest reported solar cell efficiency?
 (a) Amorphous silicon
 (b) Thin-film silicon
 (c) Polycrystalline silicon
 (d) Single-crystal silicon

3.8 At present, the maximum efficiency of a commercial solar cell is?
 (a) 3%
 (b) 12–30%
 (c) 50–65%
 (d) 65–70%

3.9 Where is the world's largest solar power plant located?
 (a) Germany
 (b) USA
 (c) India
 (d) UK

3.10 Which of the following materials has the lowest reported solar cell efficiency?
 (a) Amorphous silicon
 (b) Gallium arsenide
 (c) Polycrystalline silicon
 (d) Single-crystal silicon

3.11 Exergy efficincy of PVT system
 (a) is less than thermal efficiency
 (b) is more than thermal efficiency
 (c) is equal to thermal efficiency
 (d) None

3.12 Energy production factor (EPF) on lifetime basis
 (a) is more than one
 (b) is equal to one
 (c) is less than one
 (d) None

3.13 Lifecycle conversion efficiency (LCCE)
   (a) is more than one
   (b) is equal to one
   (c) is less than one
   (d) None
3.14 Energy payback time (EPBT) should be
   (a) more than life of PV system
   (b) equal to life of PV system
   (c) no relation
   (d) less than life of PV system
3.15 A PVT system can be used for
   (a) air heating
   (b) water heating
   (c) air/water heating
   (d) all of them
3.16 A PVT system can also be used for
   (a) lighting
   (b) underground water pumping
   (c) building
   (d) all of them
3.17 A PV system is
   (a) nonrenewable source of energy
   (b) renewable source of energy
   (c) finite source
   (d) all of them
3.18 A PV system provides
   (a) clean power
   (b) good environment
   (c) sustainable climate
   (d) all of them
3.19 A PVT system is more economical
   (a) for building integration
   (b) stand alone
   (c) for both (a) and (b)
   (d) all of them
3.20 Electrical efficiency of semitransparent PV module is
   (a) more than opaque PV module
   (b) equal to opaque PV module
   (c) less than opaque PV module
   (d) None of them

ANSWERS

3.1 **(d)**; 3.2 **(d)**; 3.3 **(a)**; 3.4 **(b)**; 3.5 **(a)**; 3.6 **(a)**; 3.7 **(d)**; 3.8 **(b)**; 3.9 **(b)**; 3.10 **(b)**; 3.11 **(a)**; 3.12 **(c)**; 3.13 **(c)**; 3.14 **(d)**; 3.15 **(d)**; 3.16 **(d)**; 3.17 **(b)**; 3.18 **(d)**; 3.19 **(a)**; 3.20 **(c)**.

REFERENCES

1. Pierce, B. 2008. Very high efficient solar cells, http://www.arpa.mil/sto/smallunitops/vhesc.html, accessed 25 July 2008.

2. Hance, J. 2008. Breakthrough in solar energy, http://news.mongabay.com/2008/0710-hance_solar.html, accessed 18 August 2008.
3. De Soto, W., Klein, S. A. and Beckman W. A., 2006, *Solar Energy* **80**, 78–88.
4. Mondol, J. D., Yohanis, Y. G. and Norton, B. 2006. *Solar Energy* **80**, 1517–1539.
5. Decker, B., Jahn, U., Rindelhardt, U. and Vaaben, W. 1992. In *11th European Photovoltaic Solar Energy Conference*, Montreux, Switzerland, pp. 1497–1500.
6. Macagnan, M. H. and Lorenzo, E. 1992. In *11th European Photovoltaic Solar Energy Conference*, Montreux, Switzerland, pp. 1167–1170.
7. Jantsch, M., Schmidt, H. and Schmid, J. 1992. In *11th Photovoltaic Solar Energy Conference*, Montreux, Switzerland, pp. 1589–1593.
8. Louche, A., Notton, G. Poggi, P. and Peri, G. 1994. In *12th European Photovoltaic Solar Energy Conference*, Amsterdam, The Netherlands, pp. 1638–1641.
9. Gautam, N. K. and Kaushik, N. D. 2002. *Energy* **27**, 347–361.
10. Tsalides, P. and Thanailakis, A. 1985. *Sol. Cells* **14**, 83–94.
11. Kern, J. and Harris, I. 1975. *Solar Energy* **17**, 97–102.
12. Fortman, M., Zhou, T., Malone, C., Gunes, M., and Wronski, R. 1990. Deposition conditions, hydrogen content and the Staebler–Wronski effect in amorphous silicon. In: *Conference Record of the 21st Photovoltaic Specialist Conference*; pp. 1648–1652.
13. Green, A. M. 1998. *Solar Cells Operating Principles Technology And System Application*, 1st edn, University of New South Wales Press, New South Wales.
14. Roedern, B. and Ullal, H. 2008. The role of polycrystalline thin film PV technologies in competitive PV module markets. In: *33rd IEEE Photovoltaic Specialists Conference Proceedings*.
15. Schock, H. 2007. Chalcopyrite (CIGS) based solar cells and production in Europe. In: *Technical digest 17th International Photovoltaic Science And Engineering Conference (PVSEC-17)*; pp. 40–43.
16. Fraunhofer ISE. 2009. World Record: 41.1% efficiency reached for multi-junction solar cells at Fraunhofer ISE, press release 2009.
17. O'Regan, B. and Gratzel, M. 1991. *Nature* **353**, 737–739.
18. Tiwari, G. N. and Dubey, S. 2010. *Fundamentals of Photovoltaic Modules and their Applications*, RSC Publishing, Cambridge, UK.
19. Norton, M. 2006. Investigation of a novel, building-integrated photovoltaic concentrator. In: Department of Construction Management and Engineering. PhD in Engineering. Reading, University of Reading.
20. Jones, A. D. and Underwood, C. P. 2001. *Solar Energy* **70**(4), 349–359.
21. Tiwari, A. and Sodha, M. S. 2006. *Renew. Energy* **31**(15), 2460–2474.
22. Evans, D. L. 1981. *Solar Energy* **27**, 555–560.
23. Duffie, J. A. and Beckman, W. 1991. *Solar Engineering of Thermal Processes*, John Wiley and Sons, New York.
24. Tiwari, G. N., Mishra, R. K., and Solanki, S. C. 2011. *Rev. Thermal Modell.* **88**, 2287–2304.

Biofuels that are Produced from Biomass are Renewable Energy Sources and are Sustainable with the Environment and Human Beings to Meet their Energy Demand.

CHAPTER 4

Biofuels

4.1 INTRODUCTION

Any material that can be burned to release thermal energy is called a **conventional fuel**, which consists primarily of hydrogen (H) and carbon (C). They are referred to as hydrocarbon fuels. They are denoted by the general formula C_nH_m. Hydrocarbon fuels exist in all phases, namely coal, petrol and natural gas. The main constituent of coal is carbon. Coal also contains oxygen, hydrogen, nitrogen, sulfur, moisture and ash. It is difficult to give an exact mass analysis of coal since its composition varies considerably from one geographical area to the next. Even within the same geographical location the composition of coal varies. Most liquid hydrocarbon fuels are a mixture of various hydrocarbons namely petrol, kerosene, diesel fuel, fuel oil, *etc.* These are obtained from crude oil by distillation. The most volatile hydrocarbon vaporises first, which is known as petrol. The less volatile fuels obtained during distillation are kerosene, diesel fuels and fuel oil.

The liquid hydrocarbon fuels are mixtures of different hydrocarbons, which are usually considered to be a single hydrocarbon. For example, petrol and diesel fuels are treated as octane (C_8H_{18}) and dodecane ($C_{12}H_{26}$), respectively. The gaseous hydrocarbon fuel, which is a natural gas, is a mixture of methane (CH_4) and smaller amounts of other gases. It is produced from gas/oil wells. On vehicles, it is either in gas phase (CNG, compressed natural gas) at pressures of 150 to 250 atm or in the liquid phase (LNG, liquefied natural gas) at –162 °C. Liquefied petroleum gas (LPG) is a byproduct of natural gas processing or crude oil refining. LPG consists of mainly propane (over 90 percent). Thus, it is usually referred to as propane. However, it also contains various amounts of butane, propylene and butylenes.

All petroleum-based fuels are the major sources of air pollutants such as nitric oxides (NO), carbon monoxide (CO) and the greenhouse gas (GHG), carbon dioxide (CO_2), when they are used in vehicles. Hence, there is currently a shift in the transportation industry from commercial petroleum-based conventional fuels (petrol and diesel) to the cleaner burning alternative fuels (natural gas, alcohols, ethanol and methanol) which are friendly to the environment. Due to the limited sources of natural gas, researchers have to explore the appropriate renewable source of energy as the fuel for sustainability and pollution-free characteristics with the natural ecology. One such natural resource is biomass. **Biofuels** are produced through the chemical or biological processes from **biomass**. The term biomass generally refers to renewable organic matter generated by plants through photosynthesis. The solar energy combines the carbon dioxide and water to form

Advanced Renewable Energy Sources
G. N. Tiwari and R. K. Mishra

Published by the Royal Society of Chemistry, www.rsc.org

carbohydrate and oxygen in photosynthesis. Materials having combustible organic matter are also referred to as biomass. It contains C, H and O. It is the oxygenated hydrocarbon. Biomass includes forest and mill residues, agricultural crops and wastes, wood and wood wastes, animal wastes, livestock operation residues, aquatic plants, municipal and industrial wastes. Fossil fuels can also be referred to as biomass, since they are the fossilised remains of plants.

Wood was initially the primary biomass fuel for humans. The goal of many biomass conversion processes is to convert solid fuel into more useful forms, *i.e.* gaseous or liquid fuels. Examples are (i) conversion of biomass to gaseous fuels include anaerobic digestion of wet biomass to produce methane gas (ii) high-temperature gasification of dry biomass to produce a flammable gas mixture of hydrogen, carbon monoxide and methane (iii) fermentation of sugars to ethanol, thermo-chemical conversion of biomass to pyrolysis oils (iv) processing of vegetable oils to biodiesel.

The resulting liquid and gaseous fuels can then be used in machinery to produce heat and electrical power. Biomass can also be converted to heat and electrical power by burning it, as occurs in boilers and steam power plants.

Biomass is a versatile resource of energy that is expressed in what is termed mixed **"*f*" fields, namely food, feed, fuel, feedstock, fibre and fertiliser**. These form the essential ingredients for our survival. It is obvious that the demand for biomass can be defined in a complex field of competitive parameters. This chapter deals with the properties of various biomass feed stocks for producing useful energy of fuels.

The broad spectrum of natural vegetation and residual deposits from human and animal activities constitute the major sources of biomass. The term **"biomass"** is therefore related to the quantity of all living matter from the five kingdoms in biology namely plants, animals, fungi, bacteria and algae. All the five kingdoms of life are in a renewable manner. Photosynthesis can occur in algae and bacteria as well as in plants. The term "Bio" is derived from *bios* (Greek word) meaning life. Biomass is a highly diverse and complex resource. It has to be studied in a wholly holistic context with recognition of the interdependencies in the overall system of land, water, nutrients.

4.2 BIOMASS RENEWABILITY

Biomass is a solid fuel that is a renewable resource of energy. The term **"renewable"** is defined as a material that can be restored when its initial stock is exhausted. In nature, biomass is formed by the process of inorganic material molecules (mainly chlorophyll) splitting water in organic cells (photolysis) in the presence of solar energy. The originating hydrogen along with carbon dioxide in the air forms the biomass (component of the carbon dioxide loop) as shown in Figure 4.1. Hence, broadly speaking, the term "renewable resource" is used as a synonym for "biomass" from a resource of geological origin. The energy source that causes the renewal of biomass is the Sun, which is renewable in nature.

Further, biomass is also considered as a form of stored solar energy. Solar energy is the only renewable energy source that can be converted into solids, liquids or gaseous fuels. Unlike fossil fuels, it is fairly well distributed over the world. Also, unlike fossil fuels, the use of biomass as fuel is carbon neutral. If the biomass is not burned, it still releases carbon into the environment by decaying.

4.3 BIOMASS SUSTAINABILITY

Biomass is the single most important source of energy for most of the world's people. About two-thirds of the Earth's population is virtually dependent on biomass for their cooking and heating. Most of these people live in rural areas. Biomass energy is also an important fuel for agroindustry, urban industries, institutions, restaurants, hotels, *etc.* for urban populations. Therefore, it is a fuel used in domestic sector in nearly every corner of the developing world as a source of heat. Biomass energy includes energy from all plant matter (trees, shrubs, and crops) and animal dung. It is also

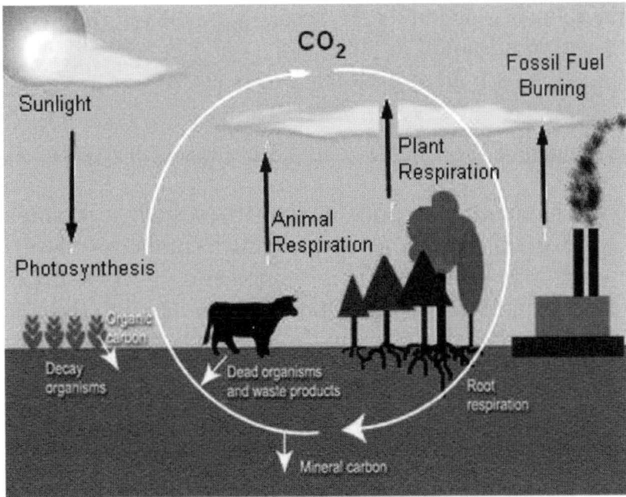

Figure 4.1 Natural carbon dioxide (CO_2) cycle.

characterised by a low efficiency of use and a low quality of life of human being associated with its gathering and use. It balances the environment, the economy and the social equality. Developing countries as a whole derive 35% of their energy from biomass.

To make the biomass more sustainable, it must at least keep pace with use. The dry mass of biological material cycling is about 250×10^9 t y^{-1} in the biosphere, which includes about 100×10^9 t y^{-1} of carbon. The associated energy bound in photosynthesis is 2×10^{21} J Y^{-1} (0.7×10^{14} W). Of this, about 0.5% by weight is biomass that is used for human food. The remaining biomass needs to be used efficiently and effectively for sustainable production of energy in the form of biofuels to meet the global energy requirements. If biomass is to play a sustainable part in the world's energy economy, attention must be given to developing strategies for sustaining high yields over large areas for long period of time. It needs to be efficiently and cost effectively converted into modern energy carriers to play a major part in the global energy economy.

Biomass crops have a lifecycle of several years. In order to compare returns from biomass production with alternative investments the stream of biomass revenues and costs has to be discounted. With a fixed amount of biomass production, the competitive price is set by the units of sources coming into the production expenses to make the system economically viable. The estimation of supply price for biomass per acre of land can be done by calculating the present value of estimated revenue and costs over the production of heat and electricity. There are particularly attractive for decentralised applications for producing gaseous fuels or electricity. Unlike solar, wind or microhydroelectric systems, modern biomass energy systems can be set up in virtually any location where plants can be grown or domestic animals can be reared. Renewable sources of energy such as solar, wind and microhydroelectric systems that are intermittent require spare or additional capacity to produce adequate energy when conditions are right such as water flow or wind speed. Such renewable energy sources necessitate electricity storage facilities, especially with small and local systems.

Bioenergy sources such as producer-gas system do not require electricity storage and offer an opportunity for sustainable (as biomass can be grown sustainably), self reliant and equitable development. The word "sustain" is derived from the Latin *sustinere* (*sus* – from below and *tenere* – to hold). It defines to keep in existence or to maintain and implies long-term support or permanence. As the word "sustainability" pertains to biomass, it describes biomass or bioenergy systems that are capable of maintaining their productivity and usefulness to the society indefinitely. It should be environmentally sound. The word **"sustainability"** is also the characteristic of a process or state that can be maintained indefinitely. Hence, biomass is referred as the sustainable energy that can be replenished within a human lifetime.

The discounted revenue for each acre of biomass is given by

$$\text{Discount revenue} = \Sigma_t[(p \times y_t)/(1+r)^t] = \Sigma_t[(y_t)/(1+r)^2] \tag{4.1}$$

where p = average annual biomass crop price per ton harvested, y_t = biomass yield per acre (in period "t"), C_t = total production cost per acre (including a competitive return to the land), r = the discount rate over the crop cycle period and p is defined as an average price over the crop cycle.

This discounted revenue from biomass production must be equal to the discounted cost given by

$$\text{Discount cost} = \Sigma_t[(y_t)/(1+r)^t] \tag{4.2}$$

By solving Eqs. (4.1) and (4.2) for p, one gets

$$p = \frac{\Sigma_t\left[\dfrac{C_t}{(1+r)^t}\right]}{\Sigma_t\left[\dfrac{Y_t}{(1+r)^t}\right]} \quad \text{where} \quad y_t = \frac{C_t\left[\dfrac{(1+r)^t-1}{r(1+r)^t}\right]}{y_t\left[\dfrac{(1+r)^t-1}{r(1+r)^t}\right]} = \frac{C_t}{y_t} \tag{4.3}$$

The average biomass supply price depends on the ratio of discounted production costs to discounted yields (resource productivity).

Therefore, to achieve the sustainability of biomass production, the supplies should be on an equitable basis and at a fair price to maintain the system on a long-term basis.

EXAMPLE 4.1

Calculate an average price for the following given parameters namely $t = 10$ years, $r = 0.10$, $C_t = \$10\,000$ and $y_t = 100$ kg

Solution:

From Eq. (4.3), one has

$$p = \frac{10000}{100} = \$100/\text{kg}$$

4.4 ORIGIN OF BIOMASS: THE PHOTOSYNTHETIC PROCESS

Photosynthesis refers to a chemical reaction occurring on the Earth between sunlight and green plants within the plants in the form of chemical energy. Photo means light and synthesis means the making. In the photosynthesis process, solar energy is absorbed by green plants and some micro-organisms to synthesise organic compounds from low energy carbon dioxide (CO_2) and water (H_2O). The organic compounds formed are simply (i) the biomass, which is a renewable energy source due to its natural and repeated occurrence in the environment in the presence of sunlight (ii) It is this same photosynthesis that converted the Sun's energy into living organisms millions of years ago to provide the fossil fuels that we are using today.

The Sun is the primary source of energy for all living organisms on the Earth. This energy reaching the Earth from the Sun is electromagnetic radiation with a spectrum from about 0.3 to 3 μm in wavelength. This corresponds to the radiation from the near ultraviolet (UV) through the visible and into the infrared (IR). The energy association with a ***quantum*** of light (**photon**) is hv where h is Planck's constant and v is the frequency of the light. The photons driving the process are mostly from the red end of the visible spectrum where the energy per photon is 1.7 eV.

The energy of a photon (hv) in the ultraviolet portion of the solar spectrum is sufficient to break a chemical bond. A photon in the visible portion of the spectrum has sufficient energy to raise an atom to an excited state. The excited atomic state may make it possible for bonding to take place with a neighbouring atom. This is the way that photochemical reactions precede. In photosynthetic process sunlight interacts with the molecules of water and carbon dioxide to form carbohydrates as shown in the following equations

$$CO_2 + 2H_2\dot{O} \xrightarrow{light} CH_2O + H_2O + CO_2 + \dot{O}_2 \tag{4.4}$$

In the above equation, 112 kcal of light energy is needed per mole of CH_2O formed. A mole (1 gram molecular weight) is the weight in grams equal to the molecular weight. For example, the mass number of oxygen (O) is 16, so the mass number of the oxygen molecule (O_2) is 32 (16 times 2). Hence, 1 mole of O_2 is 32 g. It also contains Avogadro's number (6.023×10^{23}) of molecules. The oxygen atoms initially in CO_2 and H_2O are to be distinguished.

The general formulae of the organic material produced during photosynthesis process is $(CH_2O)_n$ which is mainly carbohydrate. Some of the simple carbohydrates involved in this process are glucose ($C_6H_{12}O_6$) and sucrose ($C_{12}H_{22}O_{11}$). The sugars, glucose and fructose are simple carbohydrates. The photosynthetic reaction leading to glucose is

$$6CO_2 + 6H_2\dot{O} \xrightarrow{light} C_6H_{12}O_6 + 6CO_2 + 6\dot{O}_2 \tag{4.5}$$

In the above equation, 672 kcal of light energy is needed to form a mole of glucose. This reaction does not occur in a single step but requires several steps in which the various components of the sugar molecule enter into the reaction.

Photosynthesis is a two-stage process:

(i) **Light reactions:** The first step in the light dependent process requires the direct energy of light to make energy-carrier molecules that are used in the second phase (Eqs. (4.4) and (4.5)). In light reactions, light strikes chlorophyll in such a way as to excite electrons to a higher energy state. In a series of reactions, the energy is converted (along an electron transport process) into energy rich adenosine triphosphate (ATP) and a strong reducing agent nicotinamide adenine dinucleotide phosphate (NADPH). Water is split in the process, releasing oxygen as a byproduct of the reaction.

(ii) **Dark reactions:** The light independent process occurs when the products of the light reactions are used to form C–C covalent bonds of carbohydrates. This usually occurs in the dark if the energy carriers from the light reaction process are present. These reactions take place in the stroma of the chloroplasts. The ATP and NADPH are used to make C–C bonds in dark reactions. In dark reactions, carbon dioxide from the atmosphere is captured and modified by the addition of hydrogen to form carbohydrate. The incorporation of inorganic carbon dioxide into organic compounds is known as carbon fixation. The energy for this comes from the first phase of the photosynthesis process.

Living organisms cannot utilise light energy directly. The unusable sunlight energy is converted into usable chemical energy for the origin of biomass through a complicated series of reactions.

Human beings depend on the chemical energy (food) formed by the plants (biomass) during the photosynthesis process. When we eat food, our body oxidises or burns the carbohydrate with oxygen from the air. One of the carbohydrates resulting from photosynthesis is cellulose, which makes up the bulk of wood and other plant materials. When wood is burnt, the cellulose is converted into carbon dioxide and stored energy is released as heat. Thus, biomass is referred to as biofuels. Burning fuel is basically the same oxidation process that occurs in our body, Hence, the stored solar energy is released in the form of chemical energy in a useful form and releases carbon dioxide to the atmosphere.

The amount of biomass that can be grown certainly depends on the availability of sunlight to drive the conversion of CO_2 and H_2O into carbohydrates ((CH_2O), Eqs. (4.4) and (4.5)). In addition to limitations of sunlight, there is a limit placed by the availability of appropriate land, temperature, climate and nutrients namely nitrogen, phosphorus and trace minerals in the soil. Plant diseases and insects also affect photosynthesis.

Let us examine the overall efficiency with which biomass can be produced from sunlight. The average solar energy per unit horizontal area and time in extraterrestrial region is about 0.5 cal/min cm^2 (Example 4.2). For the purposes of evaluating plant production, it is easier to determine the number of calories in a day per square centimeter. It is assumed that 47% of the solar energy incident on the atmosphere (Figure 4.2) reaches the ground.

The energy available for food production (biomass)

$$= 0.5 \frac{\text{cal}}{\text{min} \cdot \text{cm}^2} \times (0.47) \times \frac{60\,\text{min}}{1\,\text{hr}} \times \frac{24\,\text{hr}}{1\,\text{day}} = 338 \frac{\text{cal}}{\text{cm}^2\text{day}} (1420\ \text{J/cm}^2\text{day}) \tag{4.6}$$

For a typical summer day, the value of solar energy = 500 to 700 cal/cm^2 day. The overall efficiency of photosynthesis is as follows:

(i) Plants use visible light (0.4–0.7 μm); 50% of the solar spectrum.
(ii) Chloroplasts absorb; 80%
(iii) Theoretical photon to glucose efficiency for absorption process; 35% (8 hf for each CO_2)
(iv) Dark respiration (metabolism) uses 50%; 50% (remaining)

By taking all of these factors into account,

$$\text{The overall efficiency} = 0.5 \times 0.8 \times 0.35 \times 0.5 \times 100 = 7\% \tag{4.7}$$

The amount of energy stored per carbohydrate unit synthesised is about 5 eV.

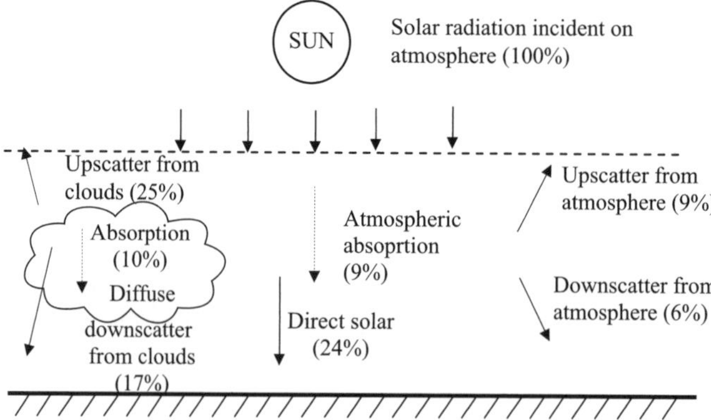

Figure 4.2 Propagation of solar radiation in the atmosphere.

If an energy content of 3744 cal/g with the dry biomass production resulting from photosynthesis is assumed then

$$\text{The biomass production per year} = 120 \times 10^{12}\frac{\text{J}}{\text{s}} \times \frac{1\,\text{cal}}{4.18\,\text{J}} \times \frac{3.15 \times 10^{7}\,\text{s}}{\text{year}} \times \frac{1\,\text{g}}{3744\,\text{cal}}$$

$$= 24 \times \frac{10^{16}\ \text{g}}{\text{year}} \approx 250 \times 10^{19}\ ty^{-1} \tag{4.8}$$

EXAMPLE 4.2

Calculate the average solar energy available per unit area and time (cal/min cm^2) in the extra-terrestrial region.

Solution

The solar constant that is measured in the extraterrestrial region and its value is now adopted to be 1367 W/m^2. Hence,

The energy associated with solar constant per minute

$$= 1367\frac{\text{W} \times 60\,\text{s}}{\text{m}^2 \times 60\,\text{s}} = 1367 \times 60\frac{\text{J}}{\text{m}^2\,\text{min}}$$

$$= 1367 \times 60\frac{\text{J}}{\text{m}^2 \cdot \text{min}} \times \frac{1\,\text{cal}}{4.187\,\text{J}} \times \frac{1\,\text{m}^2}{10^4\,\text{cm}^2}$$

$$= 1.96\frac{\text{cal}}{\text{min cm}^2} \approx 2\frac{\text{cal}}{\text{min cm}^2}$$

The Earth appears to be a disk with a surface area of $\pi r_e{}^2$. But the Earth's total surface area is $4\pi r_e{}^2$. Hence, the average solar energy per unit horizontal area at the top of the atmosphere is the total incident energy divided by the total area, which is given by

$$\textit{The average solar energy per unit area per minute} = \frac{2\,\text{cal/min cm}^2 \times \pi r_e^2}{4\pi r_e^2} = 0.5\,\text{cal/ min cm}^2$$

This average value is constant for day and night for all latitudes.

EXAMPLE 4.3

Calculate the energy associated with photons that have a wavelength λ of 700 nm (red end of visible solar spectrum).

Solution

The energy associated with the photon (E) is given by

$$E = hv$$

$-\dfrac{hc}{\lambda}$, where c is the velocity of light.

For $h = 6.63 \times 10^{-34}$ J s and $c = 3 \times 10^{8}$ m/s, we have

$$E = \frac{6.63 \times 10^{-34} \mathrm{J \cdot s} \times 3 \times 10^{8}\,\mathrm{m/s}}{700 \times 10^{-9}\,\mathrm{m}}$$

$$= 2.84 \times 10^{-19}\,\mathrm{J}$$

The value of E in terms of J is given by

$$E = \frac{2.84 \times 10^{-19}\,\mathrm{J}}{1.60 \times 10^{-19}\,\mathrm{J/eV}} = 1.7\,\mathrm{eV}$$

as 1 eV = 1.6 $\times 10^{-19}$ J.

EXAMPLE 4.4

Calculate the production of biomass (carbohydrate or glucose) per unit area per day by using solar energy in the photosynthesis process for the following parameters:

Glucose ($C_6H_{12}O_6$) produced = 674 kcal per mole. Solar radiation incident on ground = 500 cal/cm^2 g. The overall efficiency of photosynthesis = 7%.

Solution

The carbon, hydrogen and oxygen have atomic masses of 12, 1 and 16 respectively.

A mole of glucose = a mass of 12(6) + 1(12) + 16(6) = 180 g/mole. Since 674 kcal per mole of energy are stored, so

$$\textit{The energy stored per gram} = 674 \frac{\mathrm{Kcal}}{\mathrm{mole}} \times \frac{1\,\mathrm{mole}}{180\,\mathrm{g}} \times \frac{10^{3}\,\mathrm{cal}}{1\,\mathrm{Kcal}} = 3744 \frac{\mathrm{cal}}{\mathrm{g}}$$

Now, net biomass production = incident solar radiation × overall efficiency of photosynthesis Hence,

$$\text{Net biomass production} = 500 \frac{\mathrm{cal}}{\mathrm{cm^2 \cdot day}} \times \frac{10^{4}\,\mathrm{cm^2}}{1\,\mathrm{m^2}} \times \frac{1\,\mathrm{g}}{3744\,\mathrm{cal}} \times 0.07 = 93 \frac{\mathrm{g}}{\mathrm{m^2 \cdot day}}$$

The gross production of biomass becomes

$$\frac{93}{0.5} = 186 \frac{\mathrm{g}}{\mathrm{m^2 \cdot day}}$$

The gross production of biomass reduces due to the energy lost during the respiration process, as, in overall efficiency of photosynthesis, dark respiration (metabolism) uses 50% of the energy after glucose efficiency.

Hence, the quantity of biomass utilised for the respiration process to occur = 186−93 = 93 g/m^2 · day.

The main biomass sources are described as:

(i) wood and sawdust;
(ii) agricultural residues namely rice husk, bagasse, groundnut shells, coffee husk, straws, coconut shells, coconut husk, arhar stalks, jute sticks *etc.*;

(iii) aquatic and marine biomass namely algae, water hyacinth, aquatic weeds and plants, sea grass beds, kelp, coral reep, *etc.*;
(iv) wastes namely municipal solid waste, municipal sewage sludge animal waste, industrial waste, *etc.*

4.5 BIOMASS CHARACTERISTICS

The main components of biomass material are as follows (i) lignin (ii) hemicelluloses (iii) cellulose (iv) mineral matter and (v) ash.

Wood is a solid lignocellulosic material naturally produced in trees and some shrubs. It is made up of 40–50% cellulose, 20–30% hemicelluloses and 20–30% lignin. The percentage of the above-mentioned components of biomass varies from species to species. Evaluation of biomass resources requires information about their composition, *i.e.* heating value, production yields (in the case of energy crops) and bulk density. Compositional information can be reported in terms of (i) biochemical analysis, (ii) proximate analysis and (iii) ultimate analysis.

4.5.1 Biochemical Analysis

Biochemical analysis describes the kinds and amounts of plant chemicals as proteins, oils, sugar, starches and lignocelluloses (fibre). Engineers are particularly interested in the lignocellulosic component and how it is partitioned among cellulose, hemicellulose and lignin in the case of energy crops. Tables 4.1a and b give information about the biochemical composition of important categories of energy crops. This information is particularly useful in designing biological processes that convert plant chemicals into liquid fuels. Energy crops are defined as plants grown specifically as an energy resource.

Cellulose is the carbohydrate and it is the principal constituent of wood. It forms the structural framework of the wood cells. It is a polymer of glucose with a repeating unit of $C_6H_{10}O_5$. Hemicellulose consists of short, highly branched chains of sugars. In contrast to cellulose, a hemicellulose is a polymer of five different sugars. Lignin is the major noncarbohydrate constituent of wood. Lignocellulose refers to plant materials made up primarily of lignin, cellulose and hemicelluloses. Herbaceous plants are generally nonwoody species of vegetation, usually of low lignin content such as grasses.

Table 4.1a Biochemical composition of cellulosic biomass (dry basis).[1]

Feedstock	*Cellulose*	*Hemicellulose*	*Lignin*	*Others**
Bagasse	35	25	20	20
Com cobs	32	44	13	11
Wheat straw	38	36	16	10
Short-rotation woody crops	50	23	22	5
Herbaceous energy crops	45	30	15	10
Waste paper	76	13	11	0

Table 4.1b Biochemical composition of starch and sugar biomass (dry basis).[2]

Feedstock	*Protein*	*Oil*	*Starch*	*Sugar*	*Fibre*
Corn grain	10	5	20	<1	13
Wheat grain	14	<1	13	<1	5
Sugar cane	<1	<1	<1	50	50
Sweet sorghum	<1	<1	<1	50	50

4.5.2 Proximate Analysis

Proximate analysis is important in developing thermochemical conversion processes for biomass. Proximate analysis reports the yields (% mass basis) of various products obtained upon heating the material under controlled conditions in presence of air. These products consist of moisture, volatile matter, fixed carbon and ash. The proximate analysis of biomass is commonly reported on a dry basis. Volatile matter is the fraction of biomass. It decomposes and escapes as gases upon heating a sample at moderate temperatures (about 400 °C) in an inert environment. Knowledge of volatile material is important in designing burners and gasifiers for biomass. The remaining fraction is a mixture of fixed carbon and ash. It can be distinguished by further heating the sample in the presence of oxygen. The carbon is converted to carbon dioxide, leaving only the ash. Table 4.2 gives the proximate analysis on a dry basis of some biomass materials as well as fossil fuels. It is to be noted that the relatively high volatile content of biomass, (50–75%) compared to coal (typically less than 25%) makes biomass very suitable for gasification.

4.5.3 Ultimate Analysis

Table 4.3 gives the ultimate analysis of some biomass. It is seen from the table that biomass on average consists of 40–50% carbon, 4–7% hydrogen and 30–45% oxygen on moisture and ash-free basis. Biomass also contains negligible amounts of nitrogen and sulfur.

The heating value is the net energy released upon reaction of a particular fuel with oxygen under isothermal conditions. If the water vapours, formed during reaction, condense at the end of the process, the latent heat of condensation is known as the higher heating value (HHV). If latent heat does not contribute, then the lower heating value (LHV) is considered. These measurements are typically performed in a bomb calorimeter and then higher heating values (HHV) of the biofuels are recorded. Heating values of biomass/biofuels are important in performing energy balances on a biomass conversion process.

Table 4.4 gives information on alkali in ash for selected biomass materials that is useful in designing biomass combustion systems.

Bulk density is determined by weighing a known volume of biomass that is packed or baled in the form anticipated for its transportation or use. Clearly, solid logs have higher bulk density than the same wood chipped. Bulk density is an important determinant of transportation costs and the size of fuel storage and handling equipments. Volumetric energy content is also important in transportation and storage issues. Table 4.5 compares the bulk densities and volumetric energy contents

Table 4.2 Proximate analysis of some biomass.[3]

Proximate analysis (%)	*Coal*	*Sawdust*	*Groundnut*	*Rice Husk*
Moisture	5	10–20	–	10
Volatile matter	25	50–70	73.3	55
Fixed carbon	30	20–25	22.9	15
Ash	40	1–3	3.8	20

Table 4.3 Ultimate analysis of some biomass.[3]

Feedstock	*C(%)*	*H(%)*	*N(%)*	*O(%)*	*Ash(%)*
Wood	44–52	5–7	0.5–0.9	40–48	1–3
Rice husk	37	5.5	0.5	37	20
Bagasse	47	6.5	0.0	42.5	4
Groundnut shell	34–45	2–4.6	1.1–1.4	43–60	3–5

Table 4.4 Alkali content of biomass.[4]

Feedstock	*Heating values (MJ/kg)*	*Ash in Fuel (%)*	*Alkali in ash (%)*
Hybrid poplar	19.0	1.9	19.8
Pine chips	19.9	0.7	3.0
Tree trimming	18.9	3.6	16.5
Urban wood waste	19.0	6.0	6.2
White oak	19.0	0.4	31.8
Almond shell	17.6	3.5	21.1
Bagasse (washed)	19.1	1.7	12.3
Rice straw	15.1	18.7	13.3
Switch grass	18.0	10.1	15.1
Wheat straw	18.5	5.1	31.5

Table 4.5 Density and volumetric energy content of various solid and liquid fuels.[5]

Fuel	*Density (kg/m^3)*	*Volumetric energy content (GJ/m^3)*
Ethanol	790	23.5
Methanol	790	17.6
Biodiesel	900	35.6
Pyrolysis oil	1280	10.6
Petrol	740	35.7
Diesel fuel	850	39.1
Agricultural residues	50–200	0.8–3.6
Wood	160–400	3–9
Coal	600–900	11–33

of various liquids and solid fuels. The cost of collecting large quantities of biomass is significant. Wood or other biomass resources must generally be produced within no more than a 50-mile radius of the power plant for economical viability due to the high transportation costs and low densities of biomass.

4.6 ENERGY FARMING

Energy farming is defined as the production of crops to be used as fuels or energy as a main or subsidiary product of (i) agriculture (fields), (ii) silviculture (forests) and (iii) aquaculture (fresh and sea water). The crops grown are called energy crops in the energy-farming system. It is to be noted that wood obtained from an old-growth forest does not constitute an energy crop. An energy crop is planted and harvested periodically. Harvesting may occur on an annual basis. The cycle of planting and harvesting over a relatively short time assures that the resource is used in a sustainable fashion.

Energy crops can fulfil one or more market values. The grown plant may also be used as feedstock for production of electricity or liquid fuels, or both. Such is the case when trees are grown and harvested specifically as boiler fuel for steam power plants. The energy as well as food, feed and fibre coproducts are available from a single crop grown. For example, alfalfa is being evaluated for its potential to yield both energy and feed from a single crop. The high protein leaves are removed after harvesting and it is processed into animal feed, while the fibrous stems are used as fuel for gasification of power plants. The least desirable and most wasteful scenario for energy crops is the extraction of the highest valued portion of the crop for conversion into an energy product and discarding of the rest of the plant as waste. Milling sugarcane extracts sugar for fermentation to ethanol and discards the rest of the plant material (known as **bagasse**). It is an example of a conversion process that is wasteful of biomass resource. A better strategy for utilising this resource is to extract the sugar as food and to utilise **bagasse** for electricity generation.

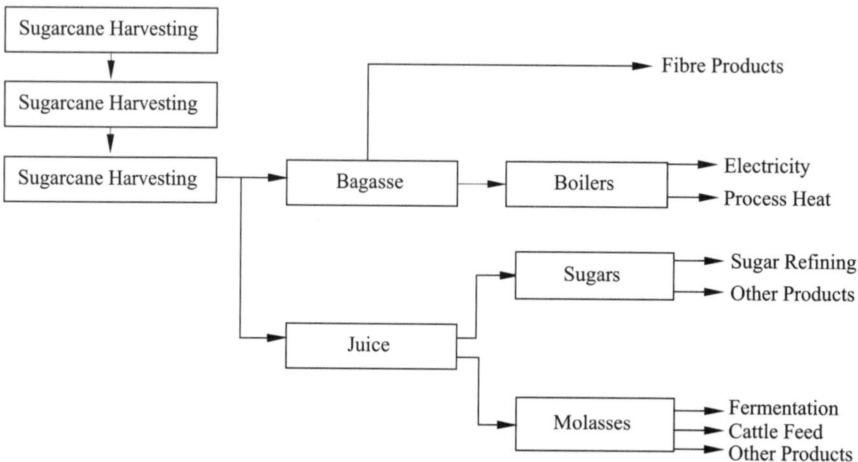

Figure 4.3 Flow diagram of sugar cane agroindustry.

Energy crops contain significant quantities of one or more of four important energy rich components: oils, sugars, starches and lignocelluloses (fibre). Oils arise from soybeans and nuts; sugars from sugar beets, sorghum and sugarcane and starches from corn and cereal crops. An outstanding and established example of energy farming is the sugarcane industry, as shown in Figure 4.3.

Energy crops are conveniently divided into woody energy crops and herbaceous energy crops. Woody crops grown on a sustainable basis and these are harvested on a rotation of 5–7 years. Thus, they are often referred to as short rotation woody crops (SRWC). Hardwoods are more promising than soft woods due to their higher productivity potentials, lower costs and the ability to resprout from stumps. Promising hardwoods include hybrid poplar, willow and black locust. Research in recent years has increased SRWC yields from 4.5 ton/ha to 11–16 ton/ha. Herbaceous energy crops (HEC) include both annual crops and perennial crops. The examples of HEC annual crops are corn, sweet sorghum, *etc*. Example of HEC perennial crops are switchgrass, Indian grass, *etc*.

Due to the importance of producing biomass for energy, the ministry of New Renewable Energy Sources (MNRE) set up biomass research centres in 11 agroclimatic regions in India. One of the main goals was to identify productive tree species and to develop a package of practices for high yields from short rotation tree crops. A summary of experimental results on productivity for various biomass are given in Table 4.6. For the species like Leucaena leucocephala and Acacia auriculiformis the yield of species were recorded as 37 t/ha/yr and 11 t/ha/yr.

4.6.1 Factors Affecting Yields from Energy Farming

The yield of a plant is directly related to the soil composition, inputs (fertiliser) and irrigation (water transmission efficiency). Other parameters that affect the yield are (i) tree species (ii) range of soil (iii) spacing and tree density (iv) management practices (v) weeding and fertiliser timings. Another important factor is the utilisation of the land sources. This involves the maximum usage of the farm land available to cover it with green plants, using methods like multiple cropping, intercropping, *etc*. Water supply, nutrient supply and temperature control are the other fundamental aspects to be given importance. Fast growing and high yielding varieties should also be introduced. Thus, for large-scale renewable energy farming, technological and managerial skills are needed to maximise energy capture. Analytical evaluations of the inputs and output of the energy plantations provide valuable insights into their performances, reveal defects in the system and suggest ways of improving productivity (Table 4.7).

Table 4.6 Biomass productivity of selected energy plant species in biomass research centres.[6]

Location	*Name of Species*	*Age (yr)*	*Total Yield (ton/ha)*	*Yield (ton/ha/yr)*
Bhubaneswar	Cassia siamea	3.5	59.2	16.9
	Leucaena Leucocephala	3.5	39.5	11.3
	Casuarina equisetifolia	3.5	22.1	6.3
	Acacia auriculiformis	3.5	37.7	10.8
	Gliricidia sepium	3.5	25.1	7.2
Srinagar (UP)	Prunus cerasoides	7.0	126.0	18.0
	Alnus nepalensis	5.0	70.0	14.0
	Albizia stipulate	6.0	120.0	20.0
	Leucaena Leucocephala	3.0	110.4	36.8
	Albizia lebbek	3.0	60.0	20.0
	Hardwickia binnata	3.0	100.5	33.5
NBRI	Samania Saman	5.0	80.1	26.7
Lucknow	Cassia Siamea	3.0	62.1	20.7
	Erythrina indica	3.0	79.8	26.7
	Prosopisj utiflora	8.0	96.5	12.1
NARI Phaltan	Terminalia arjuna	6.0	42.0	7.0
	Acacia nilotica	8.0	59.5	7.4
	Acacia auriculiformis	8.0	54.5	6.8
	Prosopisju liflora	8.5	187.0	22.0

Table 4.7 Analysis of typical firewood energy plantations.[7]

	Species			
Particulars	*Casurina*	*Eucalyptus*	*Kubabul*	*Prosopis juliflora*
Location	Coastal	Inland	Inland	Inland
Logging efficiency (%)	65	75	70	50
Plantation area (ha)	3	1.0	1.0	4.0
Plantation time	October	September	September	–
Previous crop	Casurina	Ground nut	Crass land	–
Regeneration	Seed	Coppice	Coppice	Coppice
Rotation cycle (yr)	5	5	4	5
Soil	Sandy	Loamy	Loamy	Loamy
Solar energy efficiency (%)	0.005	0.005	0.005	0.005
Spacing *(m × m)*	1 × 1	1 × 1	0.9 × 0.9	Seed generation
Tree survival (%)	58	65	70	80
Trees/ha	10 000	10 000	12 000	3000
Type of plantation	Commercial	Commercial	Commercial	Noncommercial
Water transmission efficiency (%)	70	45	65	–
Water/cycle (cm)	450–550	500–650	400–550	260–350
Watering	Pot watering	Pump	Pot/pump	No water

4.6.2 Advantages and Disadvantages of Energy Farming

The following are advantages and disadvantages of energy farming:

Advantages

(i) storage of solar energy in chemical energy form;
(ii) large potential supply (for transport fuel and electricity generation);
(iii) linked with established agriculture and forestry;
(iv) encourages integrated farming practice to farmers;

(v) efficient uses of byproducts, residues and wastes for power generation through gasification required;
(vi) environmental improvement by reducing the level of carbon dioxide;
(vii) establishes agroindustry that may include full range of technical tasks and processes, for skilled and trained personnel.

Disadvantages

(i) may lead to soil infertility and erosion;
(ii) may compete with food production;
(iii) due to low energy density, large land requirements, transport and storage problems become uneconomical;
(iv) large-scale agroindustry may be too complex for efficient operation;
(v) poorly designed and incompletely integrated systems may produce pollution of water and air.

4.7 BIOFUEL PRODUCTION PROCESSES

Biomass is the primary energy source to produce **biofuels**. Adequate and specific technologies are required to convert biomass into energy efficient forms (**biofuels**). This is particularly desirable as the energy content of many biomass feedstocks are very low and converting them into high-energy content ones requires technological advancement. Biomass materials are either burnt directly for heat (thermal energy) or upgraded physically or chemically to produce better fuels, giving a higher calorific value. There is a wide range of production routes. End products are generally obtained through the use of wood as an energy input or feedstock. The flow chart for the biofuel production processes is shown in Figure 4.4.

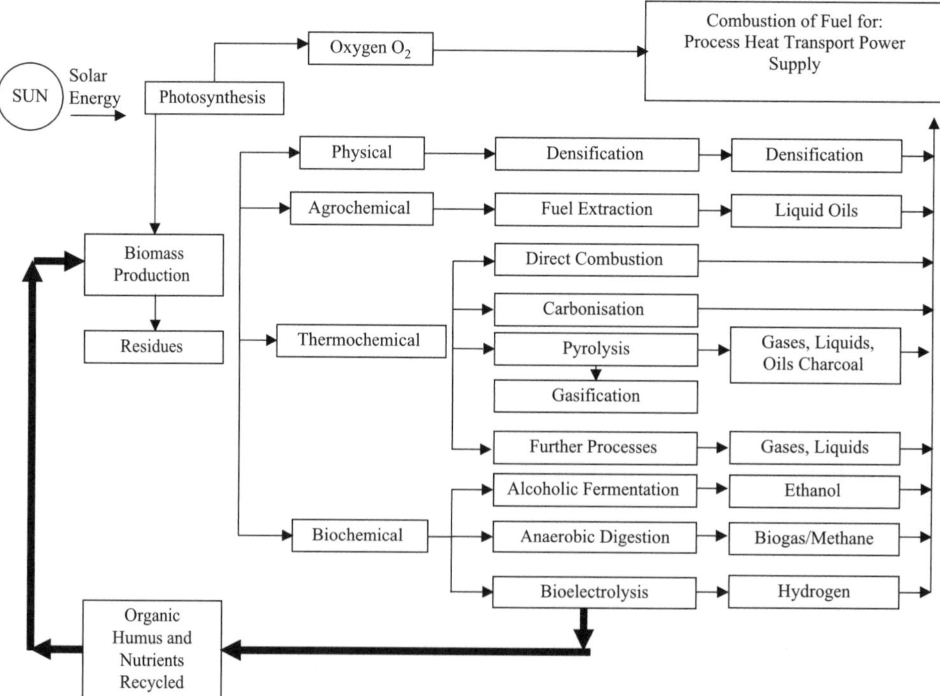

Figure 4.4 Flow chart for biofuel production processes.

4.7.1 Biomass Conversion to Biofuel

Fresh biomass in comparison with the conventional fossil fuels has the following relative inferior characteristics:

(i) they have only a modest thermal content;
(ii) they have a high moisture content that causes significant energy loss on combustion;
(iii) they usually have a low bulk density and hence become difficult for handling, storage and burning;
(iv) the physical form is often not homogeneous and hence becomes difficult for transportation and feeding to end use equipments.

The conversion processes of biomass usually involve the following:

(i) the reduction of the moisture content of the material, for simultaneous increase in its thermal value and ensure its preservation;
(ii) improving the handling characteristics of the material (converting them into fluid either gas or liquid).

4.7.2 Biomass Conversion

There are many biomass conversion routes to prepare energy efficient biofuels. The conversion routes are broadly divided in 4 categories. These are outlined as follows:

(i) Physical → Densification of biomass into solid briquettes

(ii) Agrochemical → Fuel (solid or liquid) extraction from freshly cut plant.

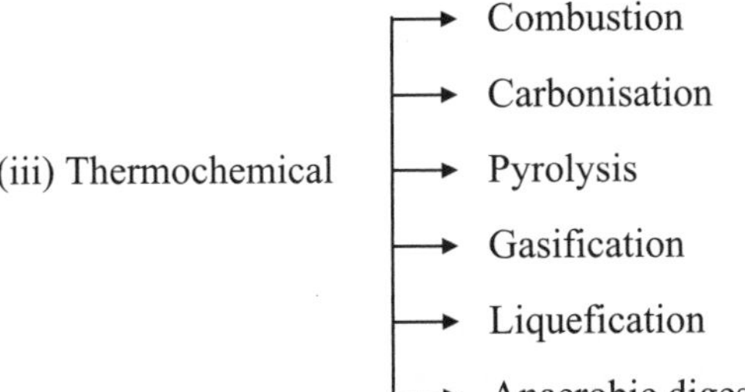

(iii) Thermochemical
→ Combustion
→ Carbonisation
→ Pyrolysis
→ Gasification
→ Liquefication
→ Anaerobic digestion to methane

(iv) Biochemical → Ethanol fermentation
→ Hydrogen formation for fuel cell

4.7.3 Physical Method of Bioconversion

The simplest method of physical conversion of biomass is through the compression of combustible material. It is densified by compression through the processes called briquetting and pelletisation.

4.7.3.1 Briquetting

Briquetting is a well known technique. This is brought about by compression baling. If elasticity is not sufficiently removed, the compressed wood tends to regain its predensifying volume. Densification is carried out by compression under a die at high temperature for moisture removal and pressure. Sometimes, waxes from external sources are used to bind the wood. Briquettes (66 mm dia and 96 mm thick) made from paddy husk or sawdust are a cheap and effective fuel for the tobacco-curing industry.

Commercially useful briquettes are produced by two types of machines. These are (i) reciprocating ram-type and (ii) screw presses. The screw-press briquettes are more homogeneous. It has better crushing strength and better storage properties with extraordinary combustion properties due to large surface area per unit weight.

4.7.3.2 Pelletisation

Pelletisation is a process in which wood is compressed and extracted in the form of rods (5–12 mm diameter and 12 mm long). It has applications in steam power plants and gasification systems. The purpose of pelletisation is to reduce the moisture contents and increase the energy density of wood for longer transportation haulage. Pelletising reduces the moisture content and increases the heat values of the biomass. Pallets are more uniform fuel and have better and more efficient combustion characteristic than directly burned chips. The palletisation process involves several stages as follows.

(i) Wood chips are reclaimed from storage. It is pulverised to reduce the chips to a specific particle size.
(ii) Sized particles are passed through a drying process, for moisture reduction up to a level 7 to 10 percent.
(iii) Material is forced under pressure though an extrusion device. The final pellet is formed but is still "hot".
(iv) Hot pellets then pass through an air cooling device to make them, they are then stored and then transported.

A nominal size wood pelletisation plant is about 250 to 300 tons per day. Since most moisture is removed from the wood waste, approximately 2 tons of green wood chips are required to produce approximately 1 ton of pellets.

4.7.4 Agrochemical Fuel Extraction

Agrochemical fuel extraction describes the production of fuels from plants. The plant usually remains alive and unharmed. Generally, liquid or solid fuels may be obtained directly from living or freshly cut plants. The materials are called exudates. They are obtained by cutting into the stems and trunks of the living plant or by crushing freshly harvested material. A well known similar process is the production of natural rubber latex.

Some of the plants form not only partly oxidised C–H bonds (cellulose or lignin), but also form completely oxygen-free hydrocarbons. The oil of the plant itself can directly be used as an energy source. The oils are essentially used for the production of food products and for the manufacturing of paint, colours, soap and cosmetic articles. Categories of suitable materials are as follows:

(i) seeds (sunflower with 50% oil);
(ii) nuts(oil palm; coconut copra to 50% by mass of oil);
(iii) fruits (olive);
(iv) leaves (eucalyptus with 25% oil);
(v) tapped exudates (rubber latex);
(vi) harvested plants (oils and solvents to 15% of the tree dry mass, *e.g.* turpentine from pine trees; oil from Euphorbia).

The advantages and disadvantages of oil extracting plants are as follows:

4.7.4.1 Advantages

(i) good chemical feedstock for high quality fuels;
(ii) ecologically sound (the plants remain in the field and heavy machines are not used for oil extraction;
(iii) intercropping or cattle farming may be integrated.

4.7.4.2 Disadvantages

(i) small yields (average 2 t $ha^{-1}y^{-1}$ to a maximum of 10 t $ha^{-1}y^{-1}$);
(ii) displaces food production unless on marginal land;
(iii) much labour required for oil extraction.

4.7.5 Thermochemical

4.7.5.1 Direct Combustion for Heat

The direct combustion of biomass is the simplest method of utilisation of this material to produce heat. This method is known as the thermochemical method of bioconversion. The main biomass for combustion is wood, which has been used over the years. It is burnt to provide heat for cooking, comfort heat (space heat), crop drying, factory processes and forming steam for electricity production and transport. In most developing countries, biomass combustion provides the largest component of total national fuel use. This is due to the extensive use of firewood for cooking and occasional heavy industrial use of biomass for sugarcane milling, tea drying, oil-palm processing and paper making. But in industrialised countries, there is an entirely dominant use of fossil fuels and nuclear energy. Table 4.8 gives the heat of combustion for a range of energy crops, residues, derivative fuels and organic products, for dry material. The most common uses of biomass for direct combustion are employed in domestic cooking and process heat in many developing countries and are described below.

4.7.5.2 Domestic Cooking and Heating

About half of the world's population depends on biomass for cooking and domestic uses. Average daily consumption of fuel is about 0.5 to 1 kg of dry biomass per person (10–20 MJ per day) which is equivalent to nearly 150 W. An average consumption of 150 W solely for cooking is definitely high. This happens due to the use of inefficient cooking methods (thermal efficiency of only about 5%), due to an open fire. Only about 5% of the heat released by complete combustion reaches the interior of the cooking pot. The rest is mostly lost by incomplete combustion of the wood, light breezes carrying heat away and by radiation and convective losses, *etc*. This results from the mismatch between fire and pot size. Therefore, cooking efficiency needs to be improved. It can be achieved by

(i) introducing alternative foods and cooking methods (steam cookers);
(ii) decreasing heat losses using enclosed burners or stoves;
(iii) introducing stove controls that are robust and easy to use (improving the cooking efficiency). The use of firewood can also be diverted to other purposes particularly for process heat and electricity generation.

Table 4.8 Calorific values of various fuels.

	Gross Calorific Value		
	MJ/kg	*MJ/l*	*Remarks*
Crops			
Wood			
Green	~8	~6	Varies more with moisture content than species of wood
Seasonal	~13	~10	
Oven dry	~16	~12	
Vegetation: dry	~15		Examples: grasses, hay
Crop residues			For dry material
Rice husk	12 to 15		In practice residues may be very wet
Bagasse (sugarcane solids)			
Cow dung			
Peat			
Secondary biofuels			
Ethanol	30	25	C_2H_5OH: 789 kg m^{-3}
Methanol	23	18	CH_3OH
Biogas	28	20×10^{-3}	50% methane + 50% CO_2
Producer gas	5 to 10	$(4–8) \times 10^{-3}$	Depends on composition
Charcoal			
Solid pieces	32	11	
Powder	32	20	
Coconut oil	39	36	
'Cocohol'	39	33	Ethyl esters of coconut oil
Fossil fuels			
Methane	55	38×10^{-3}	Natural gas
Petrol	47	34	Motor spirit, petrol
Kerosene	46	37	
Diesoline	46	38	Automobile distillate, derv
Crude oil	44	35	
Coal	27		Black, coking grade

4.7.5.3 Process Heat and Electricity

Process heat is the heat used in an industrial process. Steam process heat is commonly obtained for industries by burning wood and other biomass residues in boilers. It is physically sensible to use higher temperatures to generate electricity. Thus, electricity is produced and the process heat is retained for future use, particularly in the industrial process.

4.7.5.4 Carbonisation

Wood is heated with a restricted air flow to form a high carbon product by removing volatile materials from it is termed carbonisation. The final product is known as charcoal. It is extensively used as a domestic fuel. Charcoal contains 20–25% volatiles and 75–80% fixed carbon on a dry basis. It burns smokeless and can be preserved for longer periods. Charcoal stoves have a higher overall burning efficiency than wood stoves.

The carbonisation process takes place in four main stages determined by the temperature attained in each stage. The first stage: This is endothermic and involves the initial drying of the wood to be carbonised. This stage occurs at temperatures up to 200 °C.

The second stage: This is the precarbonisation stage, which is also endothermic. It occurs in the temperature range of 170–300 °C, producing some pyroligneous liquids as well as small quantities of noncondensable gases that are mainly CO and CO_2.

The third stage: This is exothermic and takes place in the 250–300 °C temperature range. At this stage, the greater proportion of the light tars and pyroligneous acids produced in the second stage are released from the wood steadily to produce the carbonised residue of the wood, which is charcoal.

The fourth stage: This follows at temperatures above 300 °C during which the bulk of the remaining volatile components of the charcoal are driven off, thus increasing the carbon content of the charcoal. Following the carbonisation, the charcoal product is allowed to cool, which may take a few hours to many days depending on the type of kiln used for the production.

4.7.5.5 Pyrolysis

Pyrolysis is similar to carbonisation. In this process, the temperatures are higher and unlike carbonisation, the energy rich gaseous products of the process are restored in addition to the charcoal. The products of pyrolysis of wood are mainly charcoal (25%), wood gas (20%′) pyroligneous acid (40%) and tar or wood oil (15%), excluding the moisture content. The last two liquid products being obtained by condensation of the volatiles from the wood. Both the liquid and gaseous products of pyrolysis are combustible and are potential fuel feedstocks.

"Pyrolysis" is a general term for all processes whereby organic material is heated or partially combusted with restricted supply of air to produce secondary fuels and chemical products. The input may be wood, biomass residues, municipal wastes or indeed coal. The products are gases, condensed vapours as liquids, tars and oils and solid residue as char (charcoal) and ash. Charcoal making or charbonisation is the pyrolysis with the vapours and gases not collected. Similarly another term **"gasification"** is also the pyrolysis adopted to produce maximum amount of secondary fuel gases (wood gases). Partial-combustion devices, which are designed to maximise the amount of combustible gases rather than char and volatiles are usually called gasifiers. The process is essentially pyrolysis, but may not be described as such. Hence, pyrolysis is defined as the destructive distillation of organic matter resulting in the production of charcoal, pyrolytic oil and fuel gas. Destructive distillation means the continued distillation of a substance in absence of air until all the volatile materials are driven off, resulting in the decomposition of the substance as well.

The rate of **pyrolysis** depends on factors like

(a) composition of the material;
(b) heating rate;
(c) residence time;
(d) temperature level.

The nature of the pyrolysis products varies according to the above factors. Usually, slow pyrolysis is adopted to maximise solid char. The fast pyrolysis is used for getting more of liquid and gaseous products. So a pyrolysis process can be classified based on heating rate as very slow, slow, or fast. The various process parameters and typical products are summarised in Table 4.9.

Table 4.9 Pyrolysis process parameters.

Process	*Residence Time*	*Heating Rate (°C/min)*	*Temp Range Max(°C)*	*Product (max wt. %) G–L–S*
Very slow (Carbonisation)	h–days	$<<1$	(300–400) 400	Char
Slow	Up to 30 min	5–100	(400–600) 600	Char, Bio-oil gas
Flash	0.5–50 s	100	(450–600) 650	Bio-oil chemicals, fuel gas
Ultra	<0.5 s	1000–10 000	(700 900) 1000	Chemicals, fuel gas

4.7.5.6 Pyrolysis Reactions - The Summative Analysis

Biomass is a complex mixture of three main constituents; cellulose, hemicelluloses and lignin. The composition of these constituents is given by chemical summative analysis. The pyrolysis of biomass proceeds through a series of complex reactions. The operating parameters have a great influence on the proportion composition of the three major pyrolysis products, *i.e.* char, tar and gases.

(A) Cellulose

The cellulose component is the same in all types of biomass except for the degree of polymerisation. It is a glucose polymer with a repeating unit of $C_6H_{10}O_5$ strung together by β-glycosidic linkages.

The β-linkages in cellulose for linear chains are highly stable and resistant to chemical attack due to the high degree of hydrogen bonding. It can occur between chains of cellulose. Aggregation of the linear chains makes cellulose crystalline. Cellulose component normally constitutes 40–45% of the dry biomass. Shafizadeh (1982)[11] has studied the pyrolysis of cellulose as follows:

(a) At temperatures $<300\,°C$: The dominant processes are the reduction in the degree of polymerisation; the appearance of free radicals; the elimination of water; and the formation of carbonyl and carboxyl groups that are assumed to give rise to the evolution of carbon dioxide and carbon monoxide with the formation of some char.
(b) At temperatures $>300\,°C$: There is formation of char, tar and gaseous products. The major components of tar is levoglucosan (38–50%) that vaporises and then decomposes with increasing temperatures.

(B) Hemicellulose

Hemicellulose is a mixture of polysaccharides with much lower molecular weight than cellulose. The branched nature of hemicelluloses renders it amorphous as compared to crystalline cellulose. It contains 20–40% of dry biomass. Hemicellulose is thermally most sensitive. It decomposes in the temperature range of 200 °C to 260 °C. As compared to cellulose, hemicellulose gives rise to more gas, less tar and char. The components of tar are organic acids (acetic acid, formic acid).

(C) Lignin

Lignin is amorphous in nature. It is a random polymer of substituted phenyl propane units that can be processed to yield aromatics. It is considered as the main binder for agglomeration of fibrous components. The lignin contents in biomass vary from 17–30%. According to Soltes and Elder (1981), lignin decomposes with heat between 280 °C and 500 °C. The char is a more abundant constituent product than for cellulose pyrolysis, with a yield of 55%. Liquid product known as pyroligneous acid consists of 20% aqueous components and 15% far residue on a dry lignin basis. The aqueous portion consists of methanol, acetic acid, acetone and water. The tar residue consists of mainly homologous phenolic compounds such as phenol, gualacol, 2, 6-dimethoxy phenol, *etc.* The gaseous products consist of methane, ethane and carbon monoxide, represent 10% of the lignin.

(D) Wood or Biomass

A summary of the wide range of conditions and products of biomass pyrolysis are given as follows:[8]

(i) At around 160 °C:The removal of moisture occurs, the rate peaks at about 130 °C and drops to negligible levels by 200°C.
(ii) The temperature range between 200 °C and 280 °C:All the hemicelluloses decompose, yielding predominantly volatile products such as CO, CO_2 and condensable vapours.
(iii) In the temperature range 280 °C to 500 °C:Cellulose decomposes at an increasing rate that reaches a maximum at around 320 °C. The products are again predominantly volatiles.

Lignin, which has already experienced some decomposition, also begins to give off significant quantities of volatiles as temperatures exceed 320 °C.

4.8 GASIFICATION

Gasification of biomass is thermal decomposition in the presence of controlled air. It is the conversion process of solid, carbonaceous fuels into combustible gas mixtures, known as producer gas. It is also referred to as wood gas, water gas and synthesis gas. This gas can be burned directly in a furnace to generate process heat for electricity generation. It can also fuel internal combustion engines and gas turbines of fuel cells. It can serve as feedstock for production of liquid fuels.

All internal combustion engines actually run on vapour. The liquid fuels (used in petrol engines) are vaporised before they enter the combustion chamber above the pistons. In diesel engines, the fuel is sprayed into the combustion chamber as fine droplets. It burns as they vaporise. Alternative fuel from biomass is prepared through a gasification process in gaseous form, which substitute for fuel in internal combustion engine.

Now, the question is "Where does the combustible gas come from?"

Answer: Light a wooden match; hold it in a horizontal position and notice that while the wood becomes charcoal, it is not actually burning but is releasing a gas that begins to burn brightly a short distance away from the matchstick.

Notice the gap between the matchstick and luminous flame. This gap contains the wood gas. This starts burning only when properly mixed with oxygen in air. The same chemical laws of combustion processes also apply to gasification. By weight, wood gas or producer gas obtained from the charring wood contains approximately 10–20% hydrogen (H_2), 15–30% carbon monoxide (CO) and small amount of methane. All are combustible. It also contains 45–60% non-combustible nitrogen (N_2). However it does occupy volume and dilutes the producer gas as it enters and burns in an engine. Gasification is the partial combustion of biomass that occurs when the air supply is less than adequate for combustion of biomass to be completed. Biomass contains carbon, hydrogen and oxygen molecules for complete combustion to produce carbon dioxide (CO_2) and water vapour (H_2O). Partial combustion produces carbon monoxide (CO) as well as hydrogen (H_2), which are both combustible gases.

Abundant quantities of agricultural wastes like rice husk, bagasse (sugarcane waste) coconut husk, cereal straw, *etc.*, are produced worldwide every year. They are now underutilised. The major portion of these wastes undergoes natural decomposition resulting in the production of various greenhouse gases, posing environmental problems. As mentioned earlier, solid biomass fuels are usually inconvenient, have low efficiency of utilisation and can only be used for certain limited applications. Conversion of the same biomass to a combustible mixture like producer gas solves most of these problems with use of solid biomass fuels.

Though the conversion to gas results in loss of energy up to 25%, use of producer gas can be highly efficient and hence overall efficiency becomes very high. Furthermore, the producer gas can be fed directly into internal combustion engines (I.C. engines) where the costly and precious polluting petroleum fuels are saved. Also, it can be employed at any scale for decentralised applications like shaft power, electricity or thermal energy. Due to the flexibility of application of producer gas, gasification has been proposed as the basis for energy refineries. This would provide a variety of energy and chemical products including electricity and transportation fuels.

The solid biomass fuels consist primarily of carbon with varying amounts of hydrogen, oxygen and impurities such as sulfur, ash and moisture. The aim of the gasification is the almost complete transformation of these constituents into gaseous form leaving ashes and inert materials. Further, gasification is a form of incomplete combustion. Heat from the burning solid fuel creates gases that are unable to burn completely due to insufficient amounts of oxygen from the available supply of air.

4.8.1 Chemical Reactions in Gasification

The gasifier is essentially a chemical reactor where various complex and physical as well as chemical processes take place. Biomass gets dried, heated, pyrolysed, and partially oxidised. The essence of gasification process is the conversion of solid carbon fuels into carbon monoxide and hydrogen by thermochemical process. A general layout of gasification process of biomass has been shown in Figure 4.5.

Gasification is a complex thermochemical process. Splitting of a gasifier into strictly separate zones is not realistic. Four distinct processes take place in a gasifier (i) drying of the fuel, (ii) pyrolysis, (iii) combustion and (iv) reduction. Heat is supplied during drying, pyrolysis and reduction processes with restricted amounts of air. Although there is the considerable overlap in the above processes. Each can be considered as occupying a separate zone in which fundamentally different chemical and thermal reactions take place.

The combustion zone is generally situated near the base of the gasifier as shown in Figure 4.6. It is also called the oxidation or hearth zone, where air is fed into the gasifier allowing combustion of the fuel to take place. The necessary air draught may be created by the suction of an engine or by an appropriate arrangement of fans. The key feature is that the air supply is restricted to avoid the burning from spreading to the whole full load. In such a situation, the gasifier would simply become a stove producing heat and incombustible gases. The fuel column is ignited at one point exposed to the air blast. The producer gas is drawn off at another location in the reactor. The following processes begin with the addition of heat to raise the temperature of the fuel particles during combustion.

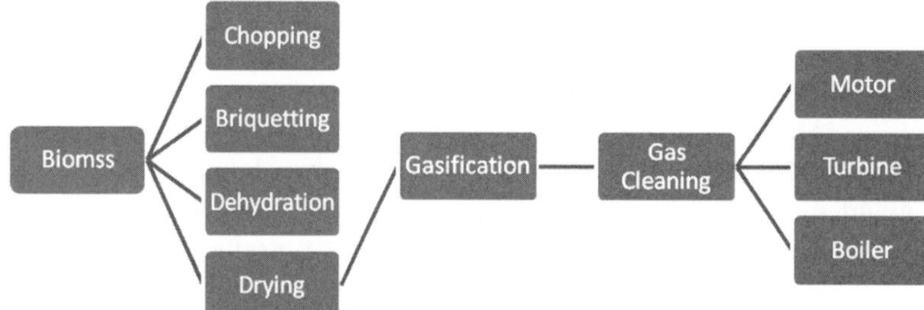

Figure 4.5 Complete gasification process.

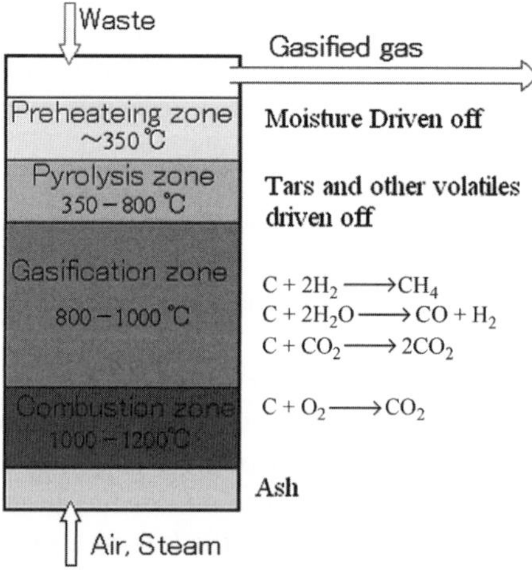

Figure 4.6 Schematic diagram of the reaction zones in an updraught gasifier.

4.8.1.1 Drying

Biomass fuels consist of moisture (water) ranging from 5 to 35%. The moisture is removed at temperatures above 120 °C and converted to steam. In the drying, fuels do not experience any kind of decomposition. The drying may require several minutes to accomplish or may occur almost instantaneously depending upon the kind of reactor, the fuel composition and the size of fuel.

4.8.1.2 Pyrolysis

The complex structure of biomass begins to breakdown with the release of gases, vapours and liquids at about 400 °C. Many of these released components are combustible. It contributes significantly to the heating value of the product gas from the gasifier. The products are influenced by the chemical composition of biomass fuels and their operating conditions. Pyrolysis reaction times and product yields also depend on fuel properties and reaction conditions. Reaction times range from milliseconds to minutes. Reaction yields range from mostly liquids to exclusively low molecular weight gases. It is noted that no matter how a gasifier is built, there is always a low-temperature zone where pyrolysis takes place generating condensable hydrocarbons.

4.8.1.3 Oxidation

Introduced air in the oxidation zone contains inert gases (nitrogen and argon besides oxygen and water vapours). The oxidation takes place at the temperature range of 700–1300 °C. Heterogeneous reaction takes place between oxygen in the air and solid carbonised (char) fuel, producing carbon dioxide (CO_2) and water vapour as:

$$C + O_2 = CO_2 + 393\,800\,\text{kJ/kg mol}\ (\textit{Exothermic Reaction}) \tag{4.8a}$$

Hydrogen in the fuel reacts with oxygen in the air blast, producing steam as

$$H_2 + \frac{1}{2}O_2 = H_2O + 242\,000\ \text{kJ/kg mol}\ (\textit{Exothermic Reaction}) \tag{4.8b}$$

4.8.1.4 Reduction

In the reduction zone, a number of high temperature chemical reactions take place in the absence of oxygen. Most of the reactions are endothermic. The heat released during exothermic reactions in oxidation is also utilised in the reaction zone. Hence, the temperature of the gas goes down in this zone. The temperature in this zone ranges from 800–1000 °C. The principal reactions that take place in reduction are as follows:

$$C + CO_2 = 2CO - 172\,600\,\text{kJ/kg mol}\ (\textit{Boundouard reaction}) \tag{4.9a}$$

$$C + H_2O = H_2 + 2CO - 131\,000\,\text{kJ/kg mol}\ (\textit{Water gas reaction}) \tag{4.9b}$$

$$CO + H_2O = CO_2 + H_2 + 420\,000\ \text{kJ/kgmol}\ (\textit{Water shift reaction}) \tag{4.9c}$$

$$C + 2H_2 = 2CH_4 + 75\,000\,\text{kJ/kgmol}\ (\textit{Hydro generation reaction}) \tag{4.9d}$$

Hence, final gas produced in the gasifier is composed of mainly CO and H_2.

4.8.2 Producer Gas and its Constituents

Producer gas is the mixture of combustible and noncombustible gases. The quantity of the constituents of producer gases depends upon the type of fuel for given operating conditions. Figures 4.7a and b show the constituents and heating value of different gases of producer gas, respectively.

The heating values of producer gas vary from 4.5 to 6 MJ/m^3. Carbon monoxide (CO_2) is produced from the reduction of carbon dioxide. Its quantity varies from 15 to 30 N% by volume. Carbon monoxide possesses a higher octane number of 106, its ignition speed is low. This gas is toxic in nature. Hence, human operators need to be careful during handling of producer gas.

Hydrogen is also a product of the reduction process in the gasifier. Hydrogen possesses the octane number of 60–66. It increases the ignition ability of producer gas. Methane and hydrogen are responsible for the higher heating value of producer gas. The amount of methane present in producer gas is less than 4%. Carbon dioxide (CO_2) and nitrogen (N_2) are noncombustible gases present in the producer gas. Producer gas contains the highest amount (45–60%) of nitrogen compared to other gases. The amount of carbon dioxide varies from 5 to 15%. A higher percentage of carbon dioxide indicates incomplete reduction. Water vapour in the producer gas occurs due to the moisture content of the air introduced during the oxidation process, injection of steam in the gasifier or the moisture content of biomass fuels.

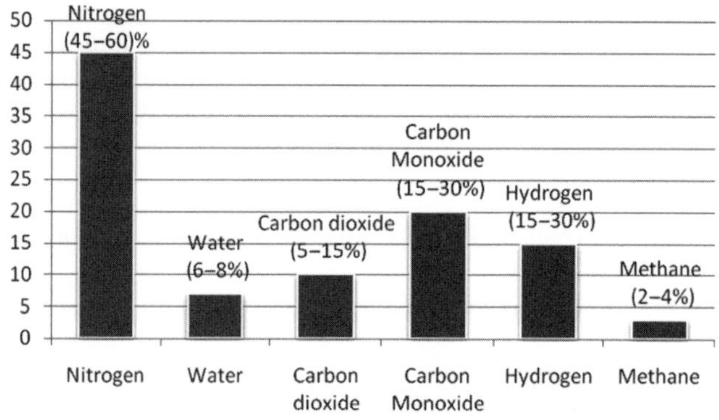

Figure 4.7a Constituent of producer gas.

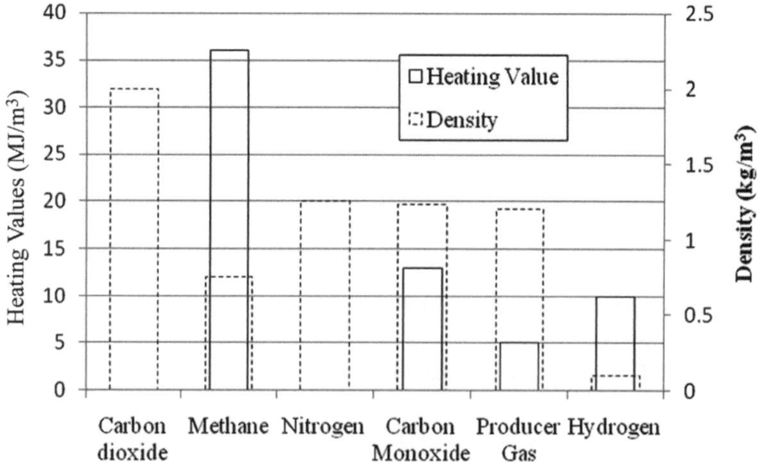

Figure 4.7b Density and heating values of different gases present in producer gas.

4.8.2.1 Applications of Producer Gas

Producer gas can be used as:

(i) **thermal applications:** (cooking, water boiling, steam generation, drying *etc.*);
(ii) **motive power applications:** (fuel in IC engines and water pumping); and
(iii) **electricity generation:** (dual-fuel mode in diesel engine, spark ignition engine, gas turbine *etc.*).

4.8.3 Classification of Gasifier

Gasifiers are classified according to the method of contacting fuel, direction of air/gas movement, types of bed and types of fuel used. The most important classification of gasifier depends on the types of bed, *i.e.* fixed bed or fluidised bed.

(A) Fixed-Bed Gasifier

There are three main designs of fixed bed gasifiers namely (i) updraught (ii) downdraught (iii) crossdraught.

(i) Updraught-Type Gasifier

An updraught type gasifier is known as a counterflow gasifier. It is the simplest as well as the first type of gasifier developed. It was the natural evolution from charcoal kilns. In this type of gasifier, the air enters at the bottom. The producer gas is drawn off at the top. The updraught gasifier achieves the highest efficiency as the hot gas passes through the fuel bed and leave the gasifier at low temperature. The producer gas is produced in the reduction zone that leaves the gasifier together with the pyrolysis products and steam from the drying zone. The resulting gas is rich in hydrocarbons (tars) and is suitable only for direct heating purposes in industrial furnaces. If it is to be used for electricity generation by I.C. engines, it has to be cleaned thoroughly. It is most unsuitable for highly volatile fuels. Figure 4.8a shows the schematic diagram of an updraught-type gasifier.

(ii) Downdraught-Type Gasifier

In downdraught or cocurrent gasifiers air enters at the combustion zone as sown in the Figure 4.8b. The producer gas leaves near the bottom of the gasifier. The purpose of this type of gasifier is to convert the tar (produced in the pyrolysis) to gaseous products by complete thermal cracking. This is not possible in an updraught-type gasifier. The essential characteristic of this type of gasifier is to draw tars (given off in the pyrolysis zone) through the combustion zone. They are broken down or burned in the combustion zone. As a result, the energy they contain is usefully released. The mixture of gases in the exit stream is relatively clean. The arrangement of the combustion or hearth zone is thus a critical element in a downdraught gasifier. In most downdraught gasifiers, the internal diameter is reduced in the combustion zone to create a throat. This is frequently made of replaceable ceramic material. Air inlet nozzles are commonly set in a lining round the throat to distribute air as uniformly as possible. In some designs, the nozzles protrude into the fuel bed itself. In that case they are usually called tuyeres. Because of the heat, these may have to be cooled with a water jacket or made with specifically heat resistant steel. Another variation is to use a vertical air inlet tube, running down the centre of the gasifier.

This type of gasifier is most commonly used for engine applications due to its ability to produce a relatively clean gas. A disadvantage of this type of gasifier is that slagging or sintering of ash may occur due to the concentrated oxidation (combustion) zone. Rotating ash grates or similar mechanisms can solve this problem. The gasifier efficiency is less than in an updraught gasifier due

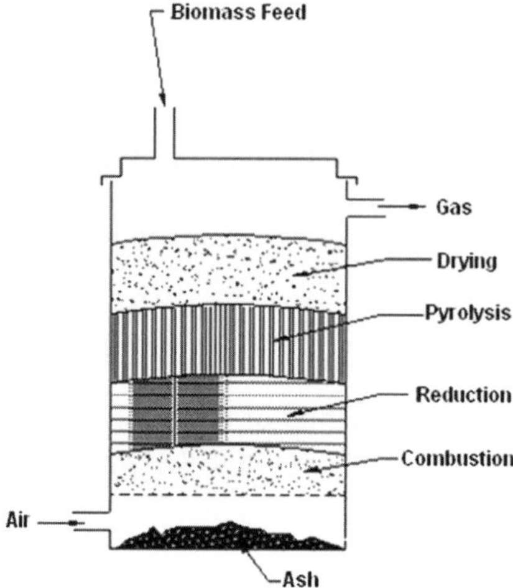

Figure 4.8a Schematic diagram of an updraught gasifier.

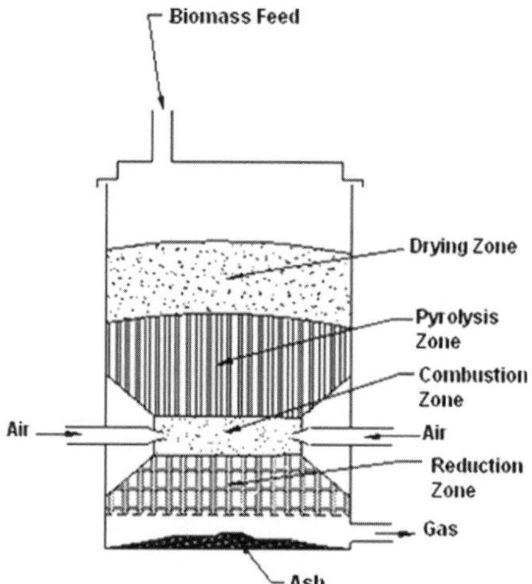

Figure 4.8b Schematic diagram of a downdraught gasifier.

to the higher temperature. This gasifier is not suitable for high ash fuels, high moisture and low ash fusion temperature fuels.

(iii) Crossdraught-Type Gasifier

The flow of air and gas is across the gasifier in a crossdraught gasifier as shown in Figure 4.8c. It operates at very high temperatures. It confines its combustion and reduction zones by using a small-diameter air inlet nozzle. Water cooling of the cast iron or steel tuyere is essential due to the high temperature. This type of gasifier responds most rapidly to changes in gas production due to the short pathlength for the gasification reactions. Ash formed due to the high temperature falls to the bottom but does not hinder operation. The high exit temperature of the gases and low CO_2 reduction results in poor quality of the gas with low efficiency. The fuel in the hopper behaves as a heat shield against the radiant heat. When operated with charcoal, the gasifier does not need to be refractory lined. Crossdraught gasifiers have very few applications due to their poor efficiency.

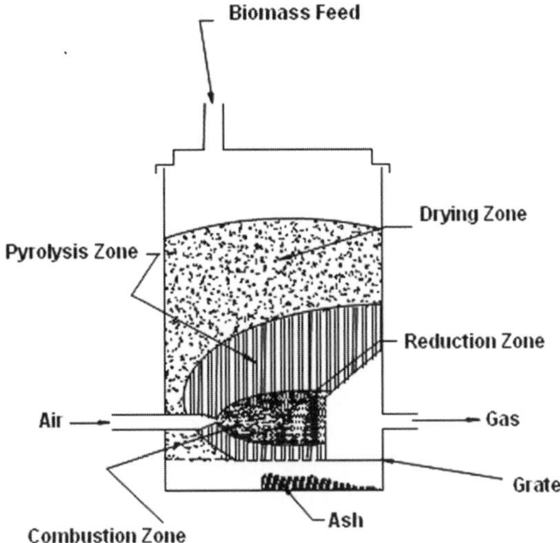

Figure 4.8c Schematic diagram of a cross-draught gasifier.

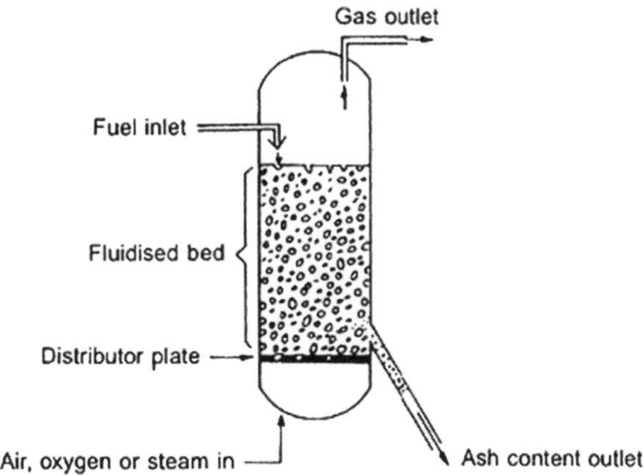

Figure 4.9 Fluidised-bed gasifier.

(B) Fluidised-Bed Gasifier

A fluidised-bed gasifier takes advantage of the excellent mixing characteristics and high reaction rates of gas–solid mixtures. A simple fluidised bed gasifier as shown in Figure 4.9 consists of a chamber containing a bed of inert particles such as sand or limestone supported by distributor plates. Pressurised air is passed through the distributor plate. The velocity of air is progressively increased so as to support the entire weight of the bed by the fluid drag on the bed particles due to upward flowing air. The bed is then said to be incipiently fluidised. It exhibits fluid-like properties above this particular velocity, called the minimum fluidisation velocity. This moving mass of solid particles is called a fluidised bed. The turbulence of the bed increases with velocity greater than the minimum fluidisation velocity.

In combustion/gasification process, the fluidised bed is first heated externally close to the operating temperature. The bed material, usually sand, absorbs and stores the heat, while the turbulence and mixing of the bed keeps the temperature very uniform throughout the bed. When biomass fuel is introduced into the fluidised bed, the high heat and mass transfer characteristics of the bed permits the rapid energy conversion under practically isothermal conditions. The high surface area available in fluidised beds and the constantly moving area per unit volume on which reactions occur, result in good conversion efficiency and lower operating temperature when compared to fixed beds. Uniform temperatures and high heating capacities of sand media permit a wide range of low grade fuels of even nonuniform size and varying moisture content to be converted to the desired products. The processes of pyrolysis and char conversion occur throughout the bed. Although a fluidised bed gasifier can handle a wide range of biomass fuels, the fuel particles must be less than 10 cm in length and must have no more than 65 percent moisture content. The fluidised bed design produces a gas with low tar content but a higher value of particulate compared with fixed-bed designs.

Fluidised beds are gaining favour as the gasifier for biomass gasification. The high thermal mass of the bed imparts a high degree of flexibility in the kinds of fuels with high moisture content that can be gasified. Disadvantages include relatively high exit temperature relatively high particulate burdens in the gas, which complicates efficient energy recovery due to abrasive forces acting within the fluidised bed.

4.8.4 Liquefaction

Liquefaction of biomass can be achieved through two processes:

(i) liquefaction through pyrolysis without any gasification medium; and
(ii) liquefaction through methanol synthesis with gasification medium.

(i) Liquefaction Through Pyrolysis

Pyrolysis is the thermal decomposition of organic compounds in the absence of air, as mentioned earlier. The resulting product streams depend on the rate and duration of heating. Liquid yields exceeding 70 percent are produced under fast pyrolysis conditions. It is characterised as having short residence time (<0.5 s) at moderate temperatures of 450–600 °C, and rapid quenching at the end of the process. Rapid quenching is essential if high molecular weight liquids are to be condensed rather than further decomposed to low molecular weight gases.

Pyrolysis liquid from flash pyrolysis is a low viscosity, dark brown fluid up to 15 to 20% water. It contrasts with the black, tarry liquid resulting from slow pyrolysis. Fast pyrolysis liquid is a complicated mixture of hydrocarbons arising from the uncontrolled degradation of lignin in lignocelluloses biomass. The liquid is highly oxygenated, which makes it highly unstable. It contains many different compounds, namely phenols, sugars and both aliphatic and aromatic hydrocarbons. The liquid, despite its high water content, shows no appreciable phase separation. The low pH of pyrolysis liquids makes the liquids highly corrosive. The liquid also contains particulate char. The higher heating values of pyrolysis liquids range from 17 MJ/kg to 20 MJ/kg with liquid densities of about 1280 kg/m^3. For a conversion efficiency of 72% of the biomass feedstock to liquid on a weight basis, yields of pyrolysis oil are about 135 gal/ton.

Pyrolysis liquids can be used directly as a substitute for heating oil. In some cases, they are also suitable as fuel for combustion turbines or modified diesel engines. Receiving high-value chemicals is another possibility. This suggests an integrated approach to production of both chemicals and fuel.

Lignocellulosic feedstocks (wood or agricultural residues) are milled to a fine powder to promote rapid reaction. The particles that transport the material to the pyrolysis reactor are entrained in steam. The particles are rapidly heated and converted into condensable vapours, noncondensable gases and solid charcoal within the reactor. These products are transported out of the reactor into a cyclone operating above the condensation point of pyrolysis vapours where the charcoal is removed. Vapours and gases are transported to a quench vessel where a spray of pyrolysis liquid cools vapours sufficiently. The noncondensable gases are burned in air to provide heat for the pyrolysis reactor, which include combustible carbon monoxide, hydrogen and methane.

There are several problems with pyrolysis liquids. The storage of these liquids becomes difficult due to phase separation and polymerisation of the liquids and corrosion of containers. The high oxygen and moisture content of pyrolysis liquids makes them incompatible with conventional hydrocarbon fuels. Upgrading pyrolysis liquids to more conventional hydrocarbon fuels such as petrol is highly desirable.

(ii) Liquefaction Through Methanol Synthesis

Methanol is a liquid fuel that can be burned in a modified internal combustion engines. It is also known as methyl alcohol. It can be blended with petrol (85% methanol and 15% petrol – a mixture known as M85). Methanol is also used as a feedstock for production of methyl tertiary butyl ether (MTBE) (an octane-enhancing petrol additive). Octane number is a measure of the resistance of a fuel to preignition (knock) when burned in an internal combustion engine. The higher the number, the higher the antiknock quality.

Methanol (CH_3OH) is produced by the reaction of CO and H_2 at 330 °C and 150 atmospheric pressure as

$$CO + 2H_2 \overset{yields}{\longrightarrow} CH_3OH; \Delta H = -90.77 \text{ kJ/mol} \tag{4.10a}$$

The CO and H_2 are required for this process. It can be produced by gasifying biomass fuels. Gasification often produces less hydrogen than the 2:1 ratio of H_2 to CO indicated above for methanol synthesis. Thus, the gas mixture (producer gas/synthesis gas/syn gas) is often reacted with steam in the presence of a catalyst to promote a shift to higher hydrogen content as:

$$CO + H_2O \overset{yields}{\longrightarrow} CO + H_2; \Delta H = -40.5 \text{ kJ/mol} \tag{4.10b}$$

CO_2 and H_2S in the syn gas are removed prior to the methanol reactor. The gas reacts with a catalyst at elevated temperatures and pressures to produce methanol in a highly exothermic reaction.

The yields of methanol from woody biomass are expected to be in the range of 480–568 l/ton. Methanol can be used as a liquid fuel in petroleum engines. It has an energy density of 23 MJ/kg. Variations in methanol yields from different biomass feedstocks are expected to be relatively small. The heating values of biomass materials are remarkably similar. The process of thermal decomposition by gasification is essentially the same for all biomass feedstocks.

4.9 BIOGAS

4.9.1 Anaerobic Digestion

Anaerobic digestion is the decomposition of organic waste by bacteria in an oxygen free environment to gaseous fuel. The process breaks down the organic matter into simpler organic compounds. The final product is a mixture of methane (CH_4), carbon dioxide (CO_2) and some trace gases known as biogas. The process is called anaerobic fermentation. It has been known to exist for quite a long time. Biogas is also known as swamp gas, sewer gas, fuel gas, marsh gas, wet gas and in India more commonly as "gobar" gas. The main fuel component of biogas is methane gas. Biogas is produced in a digester by anaerobic fermentation. A digester is a sealed tank or container in which the biological requirements of anaerobic digestion are controlled to achieve fermentation and to produce biogas. Anaerobic digestion is a simple and low cost process that can be economically carried out is rural areas where organic wastes are generated in plenty. Biogas is a renewable and nonfossil fuel that is created as a byproduct of plant and animal materials. Wastes in large quantities on a renewable basis are also available from agricultural crops and residues, fruit and vegetable plants and municipal refuse. The potential of generating gaseous fuel and enriched fertiliser through anaerobic digestion is enormous which can bring economic and environmental gain to the vast population.[9]

4.9.2 Basics of Anaerobic Digestion

For production of biogas the three basic steps have been explained as:

Step I: Cellulose + Nutrients Extracellular Enzymes

$$(C_6H_{12}O_5)_n + nH_2O \xrightarrow{Enzyme} n(C_6H_{12}O_6)$$

Soluble Glucose + Nutrients

Step II: Acid-Producing Bacteria

$$C_6H_{12}O_6 = 2CH_3 \cdot CHOH \cdot COOH \quad \text{(Lactic Acid)}$$

$$C_6H_{12}O_6 = CH_3 \cdot CH_2 \cdot CH_2 \cdot COOH + 2CO_2 + 2H_2 \quad \text{(Butyric Acid)}$$

$$C_6H_{12}O_6 = 2CH_3 \cdot CH_2 \cdot OH + 2CO_2 \quad \text{(Ethyl Alcohol)}$$

$$+\begin{bmatrix}\text{Volatile} \\ \text{FattyAcid}\end{bmatrix} + CO_2 + H_2 + \begin{bmatrix}\text{Alcohol} \\ \text{Lactate, etc.}\end{bmatrix} \tag{4.11}$$

Step III: Methane-Producing Bacteria

$$4H_2 + CO_2 = CH_4 + 2H_2O$$

$$2CH_3 \cdot CH_2 \cdot OH + CO_2 = CH_3 \cdot COOH + CH_4$$

$$CH_3 \cdot COOH = CH_4 + CO_2$$

$$CH_3 \cdot CH_2 \cdot CH_2 \cdot COOH + 2H_2 + CO_2 = CH_3 \cdot COOH + CH_4$$

$$CH_4 + CO_2$$

Anaerobic digestion basically involves three phases namely (i) hydrolysis, (ii) an acid phase and (iii) the methane phase.

The hydrolysis phase covers breakdown of large molecules into smaller ones by enzymes that are decomposable by bacteria. During the acid phase, complicated molecules such as proteins, fats and carbohydrates are decomposed by acid-forming bacteria into organic acids, carbon dioxide, hydrogen, ammonia and some impurities. Organic acids are mainly short-chain fatty acids. During the methane phase, the methane-forming bacteria convert fatty acids into methane. Complete details of the biochemistry of anaerobic digestion process are given below.

(A) Hydrolysis Phase

During enzymatic hydrolysis, polymers are converted into soluble monomers. It acts as substrates for micro-organisms during the acid phase in which soluble organic compounds are converted into organic acids. In organic wastes, carbohydrates are mostly in the form of cellulose and other components of plant fibre. Thus, for breaking these components during anaerobic digestion, bacteria need to possess what is known as cellulolytic, lipolytic and proteolytic activity.

(B) Acid Phase

The monomeric components released during the hydrolysis phase act as substrate for acid-forming phase. The acid phase can be viewed as a period during which simple organic materials are converted into simpler acids such as volatile fatty acids. Acetic acid is one common byproduct of digestion of fats, starch and proteins. Methane bacteria are strictly anaerobic. It can produce methane either (i) by fermenting acetic acid to form methane and carbon dioxide, or (ii) by reducing carbon dioxide to methane using hydrogen gas. Acetic acid accounts for 70 percent of the methane produced.

(C) Methane-Forming Phase

Methanogenic bacteria are selective in their reaction with substrate components. They react with acetic acid, methanol, carbon dioxide and hydrogen to produce methane. The presence of oxygen retards the activity of methanogenic bacteria. Methanogenic bacteria are sensitive to the pH of the digester contents. The optimum pH lies between 7 and 7.2.

4.9.3 Factors Influencing Biogas Yield

The biogas production rate depends on the organic material being digested, the digester loading rate and environmental conditions. The loading rate is defined as the amount of raw material fed to the digester per day per unit volume. Under ideal conditions, the temperature of the slurry is about 35 °C and proper pH as given in Table 4.10. It is possible to produce about 1.9 m^3 of biogas at atmospheric pressure from one day of faeces (54 kg) of a cow. The cow dung is mixed usually in the proportion of 1:1 (by weight) in order to bring the total solid content to 8–10%. The raw cowdung contains 80–85% moisture. The balance 15–20% is termed total solids.

The period of retention of the material slurry for biogas generation inside the digester is known as the retention period. It represents the average time the slurry remains in the digester. The volume of slurry in the digester remains constant. The incoming slurry displaces an equal amount of processed slurry from the digester each time it is loaded. Since the volume is constant, the fraction of digester liquid volume replaced each day determines the retention time. For example, if slurry equaling one-tenth of the digester's liquid volume is added daily, digested slurry has an average retention time of 10 days. This period depends on the type of feed stocks and the temperature.

The optimum conditions for biogas production are given in Table 4.10.

Table 4.10 Optimum conditions for biogas production.

Parameters	*Optimum Value*
Temperature (°C)	30–35
pH	6.8–7.5
Carbon/nitrogen ratio (C/N)	20–30
Solid content (%)	8–10
Retention time (days)	20–50

(i) Feed Materials for Biogas Generation.

Common feed materials are described below:	
(a) Animal wastes:	Cattle dung, poultry droppings, slaughterhouse waste, fish wastes, elephant dung and piggery wastes, *etc.*
(b) Human wastes:	Faeces, urine and night soils.
(c) Agricultural wastes:	Aquatic and terrestrial weeds crop residues, stubbles of crops, sugarcane trash, spoiled fodder, bagasse vegetable processing waste, cotton and textile waste
(d) Waste of aquatic origin:	Marine plants, algae, water hyacinth and waterweeds, *etc.*
(e) Industrial wastes:	Sugar factory, tannery and paper *etc.*

The generation of biogas from different animal dungs and human faeces are given in Table 4.11.

4.9.4 Application and Advantages of Anaerobic Digestion

Biogas is used in many ways given as follows:

(A) Cooking fuel: The most common use of biogas is for cooking due to being cheap and extremely convenient. In terms of cost and use, biogas is cheaper as well as beneficial in comparison to direct use of animal dung as shown below.

Direct Burning	*25 kg fresh dung*	*Biogas*
5 kg dry dung	Product	1 m^3 bio gas
10 460 kcal	Gross Energy	4713 kcal
10%	Device Efficiency	55%
1046 kcal	Useful Energy	2592 kcal
None	Manure	10 kg air-dried manure

Based on effective heat produced, a 2 m^3 biogas plant can replace, in a month, fuel equivalent of 26 kg of LPG or 37 l of kerosene or 88 kg of charcoal, or 210 kg of fuelwood or 740 kg of animal dung. Replacement values of biogas (gobar gas) with different fuels are given in Table 4.12. Biogas can therefore be used as an energy source to produce steam or hot water.

The comparison of calorific value of biogas with different fuels are given in Table 4.13.

Table 4.11 Gas production from different feedstocks.

Type of Feedstock (dung)	*Gas Yield per kg (m^3)*	*Normal Manure availability per animal per day*	*Gas Yield per day (m^3)*
Cattle	0.036	10.00	0.36
Buffalo	0.036	15.00	0.54
Pig	0.078	2.25	0.18
Chicken	0.062	0.18	0.01
Human (adult)	0.070	0.40	0.028

Table 4.12 Replacement values of different fuels.

Name of Fuel	*Unit*	*Gobar gas 1 m³*	*Kerosene 1l*	*Fire wood 1 kg*	*Cow dung cakes 1 kg*	*Charcoal 1 kg*	*Soft coke 1 kg*	*Butane 1 kg*	*Furnace Oil 1l*	*Coal gas 1 m³*	*Electricity 1 kWh*
Gobar gas	m^3	1.0	1.613	0.288	0.081	0.686	0.623	2.309	2.398	0.849	0.213
Kerosene	l	0.62	1.0	0.178	0.05	0.425	0.386	1.431	1.487	0.527	0.132
Fire wood	kg	3.474	5.603	1.0	0.283	2.383	2.165	8.21	8.33	2.951	0.74
Cow dung cakes	kg	12.296	19.83	3.539	1.0	8.435	7.64	28.387	29.483	10.443	2.617
Charcoal	kg	1.458	2.351	0.42	0.119	1.0	0.908	3.365	3.495	1.238	0.31
Soft coke	kg	1.605	2.589	0.462	0.13	1.101	1.0	3.705	3.848	1.363	0.342
Butane	kg	0.433	0.699	0.125	0.035	0.297	0.27	1.0	1.039	0.368	0.092
Furnace Oil	l	0.417	0.673	0.12	0.034	0.286	0.26	0.963	1.0	0.354	0.089
Coal gas	m^3	1.117	1.899	0.339	0.096	0.808	0.734	2.788	2.823	1.0	0.251
Electricity	kWh	4.698	7.576	1.352	0.382	3.223	2.927	10.846	11.264	3.99	1.0

Table 4.13 Comparison of various fuels.

Name of Fuel	*Calorific Value*	*Mode of burning (kcal)*	*Thermal efficiency (%)*	*Effective heat (kcal)*
Gobar gas (m^3)	4713	In standard burner	60	2828
Kerosene (l)	9122	Pressure stove	50	4561
Fire wood (kg)	4708	In open chulha	17.3	814
Cow dung cakes (kg)	2092	In open chulha	11	230
Charcoal (kg)	6930	In open chulha	28	1940
Soft coke (kg)	6292	In open chulha	28	1762
Butane (kg)	10 882	In standard burners	60	6529
Furnace Oil (l)	9041	In water-tube boiler	75	6181
Coal gas (m^3)	4004	In standard burners	60	2402
Electricity (kWh)	860	Hot plate	70	602

*Butane: Bottled domestic cooking gas sold as Indane, Burshane, ESSO, *etc.*

4.9.4.1 Reducing Health Hazards

Biogas is easy to use and saves time in the kitchen. Cooking on biogas is free from smoke and soot and can substantially reduce the health problems to the human life.

(B) Power Generation

The use of biogas is not confined to cooking alone. It can be used for lighting too. Further, biogas can partially replace diesel to run IC (internal combustion) engines for water pumping. This would not only reduce the dependence on diesel, but also help in reducing carbon pollutants. Biogas can similarly be used to produce electricity. Dual-fuel engines (80% biogas and 20% diesel), are nowadays gaining importance for using alternate fuels in power generation. The basic need of biogas for power generation purpose are given in Table 4.14.

4.9.4.2 Aid for Environment Improvement

Biogas provides a nontraditional way for disposal of sewage, animal and human faeces in urban and rural areas. In biogas plants, the menace of mosquitoes, flies and hookworms is sufficiently

Table 4.14 Biogas requirements for power generation.

Purpose	*Specification*	*Gas required (m^3/h)*
Lighting	200-candle power	0.1
	40-watt bulb	0.13
	2-mantle	0.14
Petrol engine	Per HP	0.43
Diesel engine	Per HP	0.45
Incubator	Per m^3	0.6
Space heater	30-cm diameter	0.16

Table 4.15 Relative manurial efficiency of anaerobically treated digested effluents and aerobically treated farmyard manure.

Crop	*Crop Yield in jin/mu with Fertiliser as*		*Increase in Crop Yield*		
	Digester Effluent	*Aerobically Treated Manure*	*Actual (jin/mu)*	*Percent*	*Total Tests for each Crop*
Rice	636.4	597.5	38.9	6.5	18
Maize	555.9	510.4	45.5	8.9	9
Wheat	450.0	390.5	59.5	15.2	29
Cotton	154.5	133.5	21.5	15.7	2
Rape	258.4	233.6	24.8	10.6	15

Source: SPOBD, Biogas Technology and Utilisation, Chengdu Seminar, Sichuan Provincial Office of Biogas Development, Sichuan, China, 1979.
Note: 1 jin (Chinese) = 0.550 kg.
1 mu (Chinese) = 660 m^2 = 0.66 hectare.

controlled. Anaerobic digestion also helps to control diseases like schistosomiasis, hepatitis and intestinal infections. It helps to bring improvements in ecology and the environment.

(C) Manurial Aspects

Animal manure has been used for centuries as a source of plant nutrients for fertilising the soil. In a biogas plant, nutrient recovery represents an economic return to the farmer using this technology. The liquid effluent from anaerobic digestion is a relatively stable product that contains almost all of the nitrogen present in raw feed material prior to digestion in comparison to the influent. The effluent is an excellent fertiliser because the nitrogen in the effluent is more readily absorbed by plants. Furthermore, the effluent serves as a source of energy.

Effluent can also be used as a fertiliser for fish ponds. An ideal feed for Singi, an air-breathing cat fish contains equal quantities of mustard oil seed cake, bran and digested slurry. This supplementary diet should be provided at a rate of 3–5% of the body weight of the fish. Similarly, the nitrogen containing animal and agricultural wastes is transformed into ammonia that crops can easily absorb. Relative manurial efficiencies of anaerobically treated digester effluents and aerobically treated farm yard manure are given in Table 4.15.

4.9.5 Classification of Biogas Digester

There are basically two designs of biogas digester namely (i) floating-gas-holder-type digester and (ii) the fixed-dome-type digester.

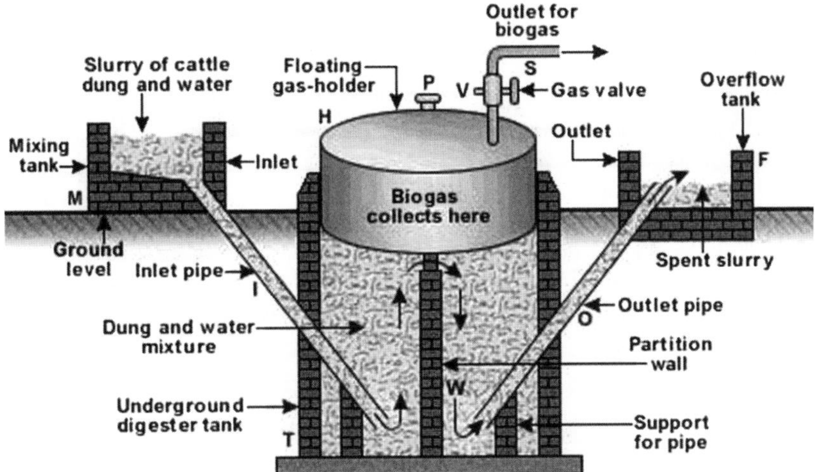

Figure 4.10 Schematic diagram of floating-gas-holder-type biogas plant.

(i) Floating-Gasholder-Type Plants

This type of biogas plant was designed by Khadi and Village Industries Commission (KVIC) in 1961 and is now being adopted and popularised by various agencies in many countries. It is a floating drum digester type biogas plant. The design consists of a deep well-shaped underground digester connected with inlet and outlet pipes at its bottom. A schematic view of a KVIC biogas plant is shown in Figure 4.10. The inlet and outlet pipes are separated by a partition wall dividing the 3/4s of the total height into two parts. A mild-steel gas storage drum that is inverted over the slurry and goes up and down around a guide pipe with the accumulation and withdrawal of gas. Now FRP (fibre-reinforced plastic) and ferrocement gas holders are also being used in this type of plant. The gas holder is separated from the digester. A partition is provided in the digester to encourage circulation. The floating gas holder provided at the top of the digester helps to keep the pressure constant. The gasholder rises when the pressure is increased due to production of gas and allows the generated gas to be let out through the gas supply pipe. It lowers when the pressure is decreased to stop the supply of the biogas.

(ii) Fixed Dome

In a fixed-dome digester, the gas holder and the digester are combined. Gas is stored in the upper part of the digester. The upper portion of the digester pit itself acts as gasholder. The displaced level of slurry provides the requisite pressure for release of gas for its subsequent use. The pressure inside the digester varies as the gas is collected. A fixed-dome digester is usually built below the ground level and is suitable for cold regions. As the plant does not involve any steel parts, it can be built with local materials and hence its construction costs are low. A schematic view of a fixed-dome-type biogas plant is shown in Figure 4.11.

A comparison of these two designs of biogas digester is presented in Table 4.16.

There are other popular Indian designs of biogas plants namely: Janata and Deenbandhu biogas plants.

(A) Janata Design

This is a fixed-dome-type biogas plant. It was developed by the Planning, Research and Action Division (PRAD), Lucknow India, in 1978. In this case, the digester and gas holder are the parts of

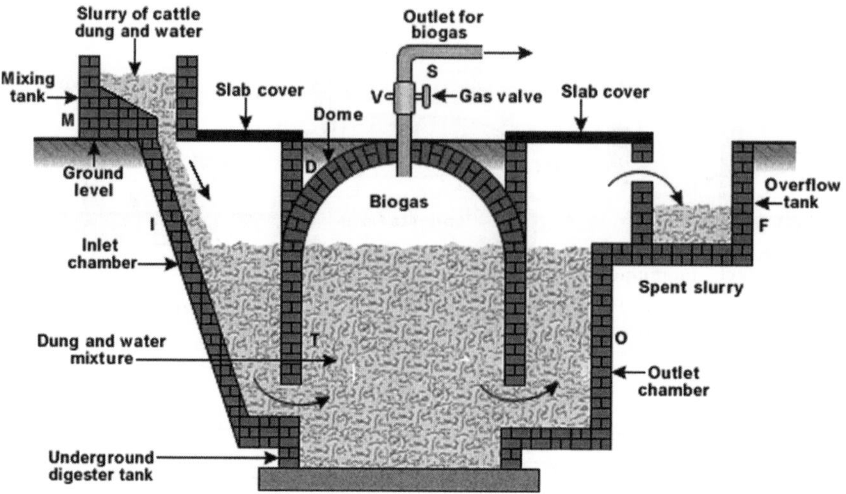

Figure 4.11 Schematic diagram of fixed-dome-type biogas plant.

Table 4.16 Comparison of two types of biogas digesters.

	Floating-gas-holder type	*Fixed-dome type*
(i)	Gas is released at constant pressure.	Gas is released at variable pressure.
(ii)	Identifying the defects in gas holder easy.	Identifying defects is difficult.
(iii)	Cost of maintenance is high.	Cost of maintenance is low.
(iv)	Capital cost is high.	Capital cost is low (for same capacity).
(v)	Floating drum does not allow the use of space for other purposes.	Space above the drum can be used.
(vi)	Temperature is low during winter.	Temperature is high during winter.
(vii)	Life span is short.	Life span is comparatively long.
(viii)	Requires relatively less excavation.	Requires more excavation work.

a composite unit made of bricks and cement masonry. It has a cylindrical digester with a dome-shaped roof and large inlet and outlet tanks on its two sides. Produced biogas is stored under a fixed dome of reinforced concrete cement (RCC); the requirement for skilled and trained masons is a bottleneck in the whole implementation programme. For the lack of necessary skill and precision, leakage of biogas is observed from the gas portion of many fixed-dome plants installed so far. RCC or brick masonry construction is not completely gas leak proof. Therefore, at least three layers of extra plasters are made to prevent any leakage. The upper part of the digester above the liquid surface provides storage space for biogas. When gas is produced, the level of the digester liquid drops, whereas that in outlet rises with the height difference between the two varying according to gas pressure. This difference in height helps to regulate gas pressure within the digester over a wide range. This self-regulating mechanism for gas pressure is the unique feature of fixed-dome-type plant. To obtain higher gas yield per unit digester volume, it is always desirable and necessary to keep the free surface slurry area of the digester larger in relation to the height of the digester.

(B) Deenbandhu Design

In order to reduce the cost of biogas plants, AFPRO (Action for Food Production), an NGO in the year 1984 developed a low cost fixed-dome model plant called the Deenbandhu Model, meaning "friend of the poor". The cost of a Deenbandhu biogas plant has been reduced without adversely

Table 4.17 Comparison among different models biogas plants.

	KVIC	*Janata*	*Deenbandhu*
1.	The digester is a deep well-shaped masonry structure. There is a partition in the middle of the digester.	Digester is a shallow well-shaped masonry structure. No partition wall is provided.	Digester is made of segments of two spheres one each for the top and bottom.
2.	Gas holder is generally made of mild steel. It is inverted into the digester and goes up and down with formation and utilisation of gas.	Gas holder is integral part of the masonry structure of the plant. Slurry from the gas storage portion is displaced out of the digester with the formation of gas and comes back when it is used.	Biogas is stored in the same way as in the case of Janta plants.
3.	The biogas is available at a constant pressure of about 10 cm of water.	Gas pressure varies from 0 to 90 cm of water.	Gas pressure varies from 0 to 75 cm of water.
4.	Inlet and outlet connections are provided through PVC pipes.	Inlet and outlet tanks are large masonry structures.	Inlet connection is through PVC pipe. Outlet tank is large masonry designed to store displaced slurry.
5.	The volume of the biogas holder governs the gas storage capacity of the plant.	It is the combined volumes of the inlet and outlet chambers.	It is the volume of the outlet displacement chamber.
6.	The floating mild steel gas holder needs regular maintenance to prevent corrosion.	There is no moving part and hence no recurring expenditure.	There is no moving part and hence no recurring expenditure.
7.	Gas holder has a short life.	It has a long working life	It also has a long working life.
8.	Installation cost is very high.	It is cheaper than the KVIC type plants.	It is much cheaper than KVIC- and Janta-type plants.
9.	Digester can be constructed locally. The gas holder needs sophisticated workshop facilities.	A trained mason using locally available materials can build the entire plant.	Entire plant can be built by a trained mason using locally available materials.

affecting the efficiency of biogas plants. After intensive trials and testing under controlled and field conditions, designs of such plant have been standardised for family size use. It has a curved bottom and a hemispherical top that are joined at their bases with no cylindrical portion in between. The slurry after digestion moves to the outlet chamber as there is no displacement space on the inlet side. An inlet pipe connects the mixing tank with the digester. This model is 30 percent cheaper than the Janata model. It is also 45 percent cheaper than the KVIC plants of comparable capacities.

Brief descriptions of these three models are given below and their comparisons are presented in Table 4.17.

4.9.6 Digester Sizing

The energy (E) available in MJ from a biogas digester is given by

$$E = \eta H_b V_b \tag{4.12a}$$

where η = the combustion efficiency of burner or boiler. H_b = the heat of combustion per unit volume biogas (20 MJ/m^3) at 10 cm water gauge pressure (0.01 atmos) (Appendix VI) and V_b = the volume of biogas (m^3).

It is important to note that some of the heat of combustion of methane is used to heat the CO_2 of the biogas and hence is unavailable for other applications.

Equation (4.12) a can also be written in terms of the fraction (f_m) of methane in biogas ($\sim$0.7) as

$$E = \eta H_m f_m V_b \tag{4.12b}$$

where H_m is the heat of combustion of methane (56 MJ/kg, 28 MJ/m^3 at STP).

If c is the biogas yield per unit dry mass of whole input (0.2 to 0.4 m^3/kg) and m_0 is the mass of dry input (*e.g.* 9.2 kg per day per cow), then the volume of biogas is given by

$$V_b = cm_0 \tag{4.13}$$

The volume of fluid per day in the digester can be written as

$$\dot{V}_f = m_0/\rho_m \tag{4.14}$$

where ρ_m is the density of dry matter in the fluid ($\sim$50 kg/m^3).

If $\dot{V}_f$ is the flow rate of the digester fluid and t_r is the retention period in the digester for anaerobic fermentation. The value of t_r ranges approximately from 10 to 50 days, the volume of digester is given by volume of digester is given by

$$V_d = \dot{V}_f t_r \tag{4.15}$$

EXAMPLE 4.5

Evaluate the following parameters of a biogas system.

(i) the volume of biogas digester;
(ii) the power available.

Given: number of cows = 8, retention period = 20 days, temperature for fermentation = 30 °C, dry matter consumed per cow per day = 2 kg, burner efficiency = 0.7 and methane proportion = 0.7.

Solution

The total mass of dry input (m_0) = 2 × 8 = 16 kg/day

From Eq. (4.14), the volume of fluid per day is evaluated as

$$\dot{V}_f = 16/50 = 0.32\,\text{m}^3/\text{day}$$

From Eq. (4.15), the volume of the digester is evaluated as

$$V_d = \dot{V}_f t_r = 0.32 \times 20 = 6.4\,\text{m}^3$$

From Eq. (4.13), the volume of biogas produced is

$$V_b = 0.2 \times 16 = 3.2\,\text{m}^3 \text{ per day}$$

Hence, the energy available from the digester can be obtained as:

$$E = 0.7 \times 28 \times 0.7 \times 3.2 = 43.90\,\text{MJ per day} = 510\,\text{W}$$

4.9.7 Temperature Control for Anaerobic Digestion in Biogas Plant

The temperature of slurry in digester contents has a major effect on biogas yield. There are different temperature ranges during which mesophilic and thermophilic bacteria are most active causing maximum yield of biogas. It is generally found that the former are most active between the range 35 ° to 40 °C and the latter between 50 to 60 °C. The choice between mesophilic and thermophilic fermentation is governed by the natural climatic conditions in which the plant is located. However, it is possible to create conditions for thermophilic fermentation by an external heat source. The length of the fermentation period is linked with digester temperature. For the same yield, when the digester temperature is high, the fermentation period (retention period) is usually kept shorter. On the other hand, when the digester temperature is low, the fermentation period is generally enhanced following reduced bacterial activity. It has been observed that (i) If the digester operates at a temperature of 15 °C, it takes nearly a year for digestion cycle to complete, (ii) if the temperature is around 35 °C, a cycle can be easily completed in less than a month's time and (iii) if the digester temperature is around 25 °C, it takes about 50 days for digestion of cattle wastes to complete, (iv) if the temperature ranges between 32 and 38 °C, digestion is generally completed within 25 days. Hence, the optimum temperature for complete digestion in biogas plant is 35 °C.

The possible methods to maintain the required temperature for desired gas production can be through either passive or active methods. These have been described as follows:

4.9.7.1 Passive Methods

(a) use of movable insulation during off-sunshine hours to reduce top heat loss;
(b) use of water/solar still over the dome to provide additional heat inside the digester and reduce top loss too;
(c) erection of greenhouse over the dome with a provision of movable insulation;
(d) construction of digester with insulating material to reduce the bottom and side losses.

4.9.7.2 Active Methods

(a) Hot charging of the slurry: This method is most suitable for places where large heating is required due to the low mean ambient air temperature.
(b) Integration of collectors with digester through heat exchanger as shown in Figure 4.12.

Biogas plant is integrated with a panel of collectors through a heat exchanger placed inside the digester (Figure 4.12). The heated water from the collector is passed through the heat exchanger, Heat is transferred to the slurry by conduction and convection, thus, raising its temperature, although an excessive rise in the slurry temperature may lead to the death of anaerobic bacteria. Hence, the slurry should be heated slowly so that anaerobic bacteria exist for the microbiological process.

EXAMPLE 4.6

Evaluate the overall heat transfer coefficient from 37 °C slurry to 15 °C ambient air temperature in the case of a floating-type gas holder.

Solution

The air within the dome may be considered to be at 35 °C. Thus, the heat transfer coefficient from the slurry to the air within the dome would be the sum of radiative, convective and evaporative heat transfer coefficients.

The radiative heat transfer coefficient (Eq. (1.24c)) is evaluated as,

$$h_{rs} = 0.8 \times 5.67 \times 10^{-8}\left[\frac{(37+273)^4 - (35+273)^4}{(37-35)}\right] = 5.35\,\mathrm{W/m^{2\circ}C}$$

The convective loss coefficient is evaluated as

$$h_{cs} = 0.884 \times \left[(37-35) + \frac{(6145.35 - 0.9 \times 5517.62)(310)}{268 \times 10^3 - 6145.45}\right]^{\frac{1}{3}} = 1.33\,\mathrm{W/m^{2\circ}C}$$

The evaporative heat-transfer coefficient, from Eq. (1.27c) is evaluated as:

$$h_{es} = 0.016 \times 1.33 \times \frac{(6145.35 - 0.9 \times 5517.62)}{37-35} = 12.55\ \mathrm{W/m^{2\circ}C}$$

Thus, the total heat transfer coefficient from the slurry to the enclosed air is = 5.35 + 1.33 + 12.55 = 19.23 W/m^2 °C.

If the thickness of the metal dome is 0.005 m and its conductivity is 62.403 W/m K, and the heat transfer coefficient from the top of the dome to ambient is given by Eq. (1.17a), and is 17.1 W/m^2 °C for V = 3 m/s.

Thus, the overall heat transfer coefficient is given as:

$$U = \left[\frac{1}{19.2.} + \frac{0.005}{62.403} + \frac{1}{17.1}\right]^{-1} = 9.04\ \mathrm{W/m^{2\circ}C}$$

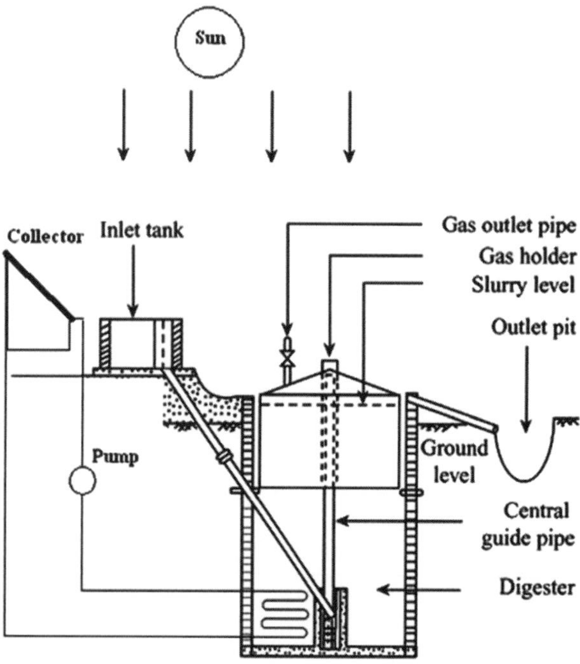

Figure 4.12 Vertical floating-type biogas plant integrated with panel of collectors Ester Layer.

4.9.8 Thermal Analysis

In order to determine the useful energy collected by different components of the biogas-cum-greenhouse system, the following assumptions have been made

(i) The dome and gas has negligible heat capacity.
(ii) The slurry has no temperature gradient.
(iii) The trapped air within the greenhouse has no temperature gradient.
(iv) The Earth surrounding the digester wall is in isothermal condition.
(v) The system is in a steady-state condition.

Hence, simple energy balance equation can be written as follows:
For the dome

$$\dot{Q} = \alpha\tau I_{eff}(t) - h_{pa}(T_p - T_a)A_i \tag{4.16}$$

where, $I_{eff}(t) = \sum_{i=1}^{n/2} I_i \Delta A_i$ and n is the number of vertical strip of green house circular wall/roof and ΔA_i is the area of the ith strip.

$$\text{Forbiogas}: \quad \dot{Q} = h_{pg}(T_p - T_g)A_p \tag{4.17}$$

$$\text{Forslurry}, \quad \dot{Q} = h_{pg}(T_p - T_g)A_p \tag{4.18}$$

Equations (4.16)–(4.18) can be simplified to get an expression for net energy gain by the slurry, as follows:

$$\dot{Q} = U_L\left[\frac{\alpha\tau}{h_{pa}}\frac{I_{eff}(t)}{A_i} + T_a - T_s\right] \tag{4.19}$$

where, $$h_{pa} = \left[\frac{1}{5.7} + \frac{1}{(5.7 + 3.8v)}\right]^{-1} \text{ and}$$

$$U_L = \left[\frac{1}{A_i h_{pa}} + \frac{1}{A_p h_{pg}} + \frac{1}{A_p h_{gs}}\right]^{-1} \tag{4.20}$$

The convective heat transfer coefficient from gas to slurry (h_{gs}) can be calculated from the following expression,

$$h_{gs} = 0.884\left[T_g - T_s + \frac{(P_g - P_s)(T_g + 273)}{268.9 \times 10^3 - P_g}\right]^{\frac{1}{3}} \tag{4.21}$$

EXAMPLE 4.7

Calculate the total heat transfer coefficient from dome to slurry and *vice versa* during the day and off-sunshine hours through gas for the following parameters.

For day $T_s = 25\ ^\circ\text{C}$, and $T_p = 60\ ^\circ\text{C}$, height of exposed surface of dome from slurry (characteristic length) = 5 m. For off sunshine hours $T_s = 40\ ^\circ\text{C}$, and $T_p = 25\ ^\circ\text{C}$.

Solution

(i) During the day time: Evaporation heat losses are assumed to be zero because of the large heat capacity of the slurry. In this case,

The radiative heat transfer coefficient from dome to slurry will be calculated as:

$$h_{rps} = 0.8 \times 5.67 \times 10^{-8}\left[\frac{(60+273)^4 - (25+273)^4}{(60-25)}\right] = 5.74\,\mathrm{W/m^2 - {}^\circ C}$$

In the present case, the hot plate is facing downward, hence $Nu = 0.27(GrPr)^{0.25}$ where Gr and Pr are calculated at an average temperature of dome and slurry. The values of different parameters at average temperatures are $\rho = 1.12$ kg/m^3, $\beta = 1/\Delta T = 3.169 \times 10^{-3}$ K^{-1}, $\mu = 1.98 \times 10^{-5}$ N s/m^2, $K = 0.0027$ W/m °C and Pr = 0.7 (Appendix V). The value of Pr and Gr can be evaluated using expression mentioned in Eqs. (1.12a–d)

$$Gr = \frac{(9.8)(3.169 \times 10^{-3})(5^3)(1.12^2)(35)}{(1.98 \times 10^{-5})} = 4.34 \times 10^{11}$$

Now, calculating the convective heat transfer coefficient (h_{cps}), we get,

$$h_{cps} = \frac{Nu \cdot K}{L} = \frac{0.27(\mathrm{Gr}\cdot\mathrm{Pr})^{0.25}\cdot K}{L}$$

$$= \frac{0.27(434 \times 10^{11} \times 0.7)^{0.25} \cdot 0.027}{5} = 1.08\,\mathrm{W/m^2 - {}^\circ C}$$

The total heat transfer coefficient from dome to slurry during day time

$$= 5.74 + 1.08 = 6.82\,\mathrm{W/m^2\,{}^\circ C}$$

(ii) During the night time: The process will be reversed and heat transfer will take place from the slurry to the dome and can be calculated as,

The radiative heat transfer coefficient from the slurry to the dome will be,

$$h_{rsp} = 0.8 \times 5.7 \times 10^{-8}\left[\frac{(40+273)^4 - (25+273)^4}{(40-25)}\right] = 5.20\,\mathrm{W/m^2 - {}^\circ C}$$

The convective heat transfer can be determined by using vapour pressure at the slurry and the dome temperatures as 7335 and 3166 N/m^2, we get,

$$h_{csp} = 0.884\left[40 - 25 + \frac{(7335-3166)(40+273)}{268.9 \times 10^3 - 7335}\right]^{\frac{1}{3}} = 2.39\,\mathrm{W/m^{2\circ}C}$$

The evaporative heat transfer is determined as,

$$h_{esp} = 0.016 \times h_{csp}\left[\frac{P_s - P_p}{T_s - T_p}\right]$$

$$h_{esp} = 0.016 \times 2.39\left[\frac{(7335 - 3166)}{40 - 25}\right] = 10.62\ \mathrm{W/m^{2\circ}C}$$

Hence, the total heat transfer from the slurry to the dome during the night $= h_{rsp} + h_{csp} + h_{esp} = 5.20 + 2.39 + 10.62 = 78.2\ \mathrm{W/m^{2\circ}C}$

EXAMPLE 4.8

Write down the energy balance for the slurry of a biogas cum greenhouse system of capacity M_s in a quasisteady-state condition.

Solution

In a quasisteady-state condition, the net energy gain, $\dot{Q}$, will be equal to the energy stored by the mass of slurry. It can be expressed as,

$$M_s C_s \frac{dT_s}{dt} = U_L\left[\frac{\alpha\tau}{h_{pa}}\frac{I_{eff}(t)}{A_i} + T_a - T_s\right]$$

4.9.9 Present Status of Family-Type Biogas Plants in India

Against an estimated potential of 12 million family-type biogas plants, over 2.9 million family-type biogas plants have been installed in the country up to December 1999. Table 4.18 shows the State-wise achievements up to December 1999.

EXAMPLE 4.9

Derive an expression for the time interval to raise the temperature of slurry from T_{s0} to T_s for a given $I(t)$.

Solution

The rate of net thermal energy available to slurry will be responsible to raise its temperature; hence from Eq. (4.7a), we get from Example 4.8 as

$$\dot{Q} = U_L\left[\frac{\alpha\tau}{h_{pa}}\frac{I_{eff}(t)}{A_i} + T_a - T_s\right] = M_s C_s \frac{dT_s}{dt}$$

or,

$$\frac{dT_s}{\frac{\alpha\tau}{h_{pa}}\frac{I_{eff}(t)}{A_i} + T_a - T_s} = \frac{U_L}{M_s C_s}dt$$

The solution of the above equation after integration will be:

$$\ln\left[T_a - T_s - \frac{\alpha\tau}{h_{pa}}\frac{I_{eff}(t)}{A_i}\right] = \Delta t + C$$

At the initial condition $t = 0$, $T_s = T_{s0}$, then $C = \ln\left(T_{s0} - T_a - \frac{\alpha\tau}{h_{pa}}\frac{I_{eff}(t)}{A_i}\right)$, thus,

$$-\frac{U_L}{M_S C_S}\Delta t = \ln\left[\frac{T_s - T_a - \frac{\alpha\tau}{h_{pa}}\frac{I_{eff}(t)}{A_i}}{T_{s0} - T_a - \frac{\alpha\tau}{h_{pa}}\frac{I_{eff}(t)}{A_i}}\right]$$

or,

$$\Delta t = \frac{M_S C_S}{U_L}\ln\left[\frac{T_s - T_a - \frac{\alpha\tau}{h_{pa}}\frac{I_{eff}(t)}{A_i}}{T_{s0} - T_a - \frac{\alpha\tau}{h_{pa}}\frac{I_{eff}(t)}{A_i}}\right]$$

Table 4.18 State-wise achievements up to December, 1999.

State/UTs	*Total estimated potential (number)*	*Number of biogas plants installed*
Andhra Pradesh	1 065 600	266 336
Arunachal Pradesh	7500	541
Assam	307 700	35 800
Bihar	939 900	111 343
Goa	8000	3089
Gujarat	554 000	328 395
Haryana	300 000	39 072
Himachal Pradesh	125 600	42 112
Jammu & Kashmir	128 500	1759*
Kamataka	680 000	258 878
Kerala	150 500	56 867
Madhya Pradesh	1 491 200	167 668
Maharashtra	897 000	636 218
Manipur	38 700	1706
Meghalaya	24 000	989*
Mizoran	2200	1861
Nagaland	6700	1032*
Orissa	605 500	154 365
Punjab	411 600	53 146
Rajasthan	915 300	64 430*
Sikkim	7300	2345
Tamilnadu	615 800	195 333
Tripura	28 500	999
Uttar Pradesh	2 021 000	327 722
West Bengal	695 000	155 956
A & N Islands	2200	138*
Chandigarh	1400	97
Dadra & Nagar Haveli	2000	168
Delhi	12 900	671
Pondicherry	4300	539
Total	**12 049 900**	**290 957**

*Provisional

Source: Annual report 1999–2000, Ministry of Non-Conventional Energy Sources, Government of India. New Delhi.

4.10 ALCOHOLIC FERMENTATION

The history of alcoholic fermentation is as follows:

- Alcohol came into wide use long before refined petroleum fuels were developed.
- In the mid-1800s: alcohol replaced whale oil as a lamp fuel due to being odorless and fairly widely available.
- In 1876: Nikolaus Otto, the recognised inventor of the four-stroke-cycle internal combustion engine, advocated alcohol as the fuel for his new engine. Henry Ford was an early champion of alcohol for fuel. His earliest automobile, the quadricycle, was designed to run on alcohol.
- The model T was built with an adjustable carburetor that could be set to burn pure alcohol in the engine.
- In 1922: Alexander Graham Bell suggested alcohol as an alternative to petroleum-based fuel, as a "beautifully clean and efficient fuel. It can be produced from vegetable matter of almost any kind, even garbage of our city".
- After World War I: France and Germany mixed petrol and alcohol in ratios as high as 50:50.
- From the 1950s, the research towards alcohol as an alternative fuel for petrol started throughout the world.

Ethyl alcohol, known as alcohol and technically termed ethanol is a colourless and flammable liquid with a chemical formula C_2H_5OH. It is the alcoholic product of fermenting the sugars in natural raw materials with yeast. The natural raw materials include vegetable matter, growing crops, farm waste, waste organic products (straw and saw dust), molasses, wastes of paper and pulp industries *etc.* Its use may be regarded as obtaining energy by a direct method from the Sun without the intermediate storage. Plants will perform their synthesis of starch from the abundant carbon dioxide and water as long as the Sun shines. From this annually renewed store of raw materials, ethanol can be readily produced in quantities sufficient to meet the world demand. It thus has attributes of renewal in nature.

4.10.1 Alcohol as Automobile Fuel

Ethanol offers certain advantages if used as an automobile engine fuel. It is an antiknock fuel and has the ability to stand very high compression ratios. It is a high octane, water-free alcohol. Octane rating is defined as the fuel's ability to resist engine knock. Ethanol is both a fuel and an octane enhancer. It is an ideal replacement for lead. It has a high latent heat of vaporisation. This property can be utilised to achieve lower charge temperature during induction for higher charge density. Hence, higher volumetric efficiency is obtained. It tends to produce less carbon deposits than normal petrol and the deposits are softer and easier to remove.

Its lower calorific value, higher viscosity, greater surface tension and hygroscopic nature are some of the difficulties in its use as a complete fuel in present day combustion engines. It can, however, be mixed with other fuel or fuel mixtures so as to impart to the resulting blend some of its important properties namely higher compression operation without knock, cleaner combustion and cooler engine operation.

Ethanol has the potential as the alternative fuel to solve the problems of limited source of fossil fuels. Ethanol is most commonly used as a 10 percent blend with petrol known as gasohol that can be burned in unmodified automotive engines.

4.10.2 Unsuitability of Ethanol in Diesel Engines

Diesel engines operate on an entirely different combustion principle in comparison to spark ignition engines. It relies on the fuel to self ignite from an increased pressure and temperature of the compression cycle. The ability of a fuel to self ignite is measured by its octane rating. Diesel fuel has a high cetane rating. It also has a low octane rating. The converse is true for ethanol, which is the major problem in the use of ethanol in diesel engines. Ethanol can be burned in diesel engines, but because it resists self ignition, it can have accompanying knock and engine stresses that may be unacceptable due to noise and lower engine life. Cetane-enhancer additives are available but expensive and of questionable performance. Also, phase separation of diesel and ethanol is more pronounced than with petrol blends, which limits its performance during cold weather. Hence, ethanol is not suitable to use as a fuel in diesel engine systems.

4.10.3 Feedstocks for Ethanol Production

Ethanol can be produced by the fermentation of any feedstock that contains sugar or starches. It can be produced from cellulose materials that can be converted into fermentable sugar. These three groups of biomass are as follows:

(i) sugars (sugar beets, sugarcane, sweet sorghum, fruits);
(ii) starches (small grains such as corn and wheat, potatoes, cassava); and
(iii) cellulose (wood, solid waste, agricultural residues).

Sugar crops contain monosaccharide forms of sugar such as glucose. This can readily be fermented into alcohol by yeast with no intermediate processing. Starch crops contain sugar units that are tied together in long chains. Yeast cannot use these disaccharide forms of sugar until the starch chains are converted into individual six-carbon groups such as glucose or fructose. This conversion process can be done fairly simply by the use of cooking in a dilute acid solution or reacting the starch with thermophilic (heat tolerant) enzymes. Cellulose crops contain chemicals called polysaccharides. These chains must be broken down to release the sugar. Breaking the chemical bonds of cellulose is more complicated than breaking down starch to simple sugar. The conversion is typically done physically (milling or heat treatment) or chemically (basic or acid reactions). Ethanol production from various crops is presented in Table 4.19.

Table 4.19 Ethanol yields of various crops, based on average yields in Brazil. Two crops a year are possible in some areas, Cassava yields could be boosted to 3600 l ha^{-1} y^{-1} by improved cultivation technologies.

	Liters of ethanol per tonne of crop	*Liters of ethanol per hectare per year*
Sugarcane	70	3500
Cassava	180	2160
Sweet sorghum	86	3010
Sweet potato	125	1875
Corn (maize)	370	2220
Wood	160	3200

4.10.4 Ethanol Production Process

Ethanol production proceeds as follows:

(i) formation of a fermentable sugar solution;
(ii) fermentation of the sugar solution to ethanol; and
(iii) separation of ethanol from other process ingredients.

The ethanol production process is outlined as follows:

1. **Milling:** Feedstock is first passed through hammer mills and then it is ground into a fine powder.
2. **Sterilisation:** If distressed or damaged crop are used, sterilisation facilities are included to avoid biological contamination.
3. **Cooking:** Enzymes (alpha-amylase) and water are added to form a mash that is cooked for about 30 min. Heat is applied at this stage to enable liquefaction.
4. **Cooling:** The mash from the cooker is cooled. The secondary enzyme (glucoamylase) is added to convert the liquefied mash to fermentable sugars (dextrose).
5. **Fermentation:** Yeast is added to the mash to ferment the sugars to ethanol and carbon dioxide. Mash ferments for about 2 days, during which the sugar is converted to alcohol.
6. **Distillation:** The fermented mash, now called "beer" contains about 10% alcohol as well as the nonfermentable solids. Initially the mass is introduced into the first-stage distillation column that is regulated to maintain a temperature just a little higher than the boiling point of ethanol. Alcohol vaporises and ascends out of the column. Liquid water mixed with residual grain and yeast (distilliers grain) flows down and out. The vaporised alcohol (now about 50 percent) is passed through a second distillation column. It concentrates the alcohol content to 95 percent (*e.g.* 190 proof). Proof is a measure of ethanol content. One percent ethanol content equals 2 proofs.
7. **Dehydration:** Since the alcohol from the second-stage column contains 5% water, which is unusual for commercial gasohol, it must be totally dehydrated. Benzene and petrol are common additives in a further process to separate the residual water and produce anhydrous ethanol (100 percent or 200 proof alcohol).
8. **Denaturing:** Ethanol that will be used for fuel is then denatured with a small amount (2–5%) of some product, like petrol, to make it unfit for human consumption.
9. **Fuel:** The denatured anhydrous alcohol is transported to the gasohol market. It is blended with nonleaded petrol at the ratio of 10 percent ethanol to 90 percent petrol.
10. **Coproducts:** Two primary coproducts of ethanol production are distillers grain and carbon dioxide. Both of which have market value. Distiller grain consists of the protein value of corn, yeast, fibre and water discharge from the first-stage distillation column. The distillers grain has a high protein content of about 27 percent. Therefore, it makes an excellent animal feedstock supplement. It is competitive with soybean meal, which is about 44 percent protein. Carbon dioxide (CO_2) that evolves from the fermentation process can be used in grain storage silos to retard spoilage.

4.11 VEGETABLE OILS/BIODIESEL

A variety of grains and seeds contain relatively high concentrations of vegetable oils that are of high caloric value. It can be used as liquid fuels. However, the high viscosity of these raw vegetable oils (20 times that of diesel), would lead to serious lubrication oil contamination. On the other hand, chemical modification of vegetable oils to methyl or ethyl esters yields excellent diesel engine fuel without the viscosity problems. Biodiesel is the generic name given to these vegetable oil esters. Hence, biodiesel is defined as the biofuel produced through transesterification. This is a process in which organically derived oils are combined with alcohol (ethanol or methanol) in the presence of a catalyst to form ethyl or methyl ester. The biomass-derived ethyl or methyl esters can be blended with conventional diesel fuel. It is used as a neat fuel (100% biodiesel). Biodiesel can be made from soybean, sunflower, cottonseed, corn, groundnut (peanut), sunflower, rapeseed, waste vegetable oils and animal fats, *etc*. Therefore, biodiesel is made from renewable biological sources such as vegetable oils and animal fats and as an alternative diesel fuel.

One hundred years ago, Rudolf Diesel tested vegetable oil as a fuel for his engine.[10] With the advent of cheap petroleum, appropriate crude oil fractions were refined to serve as fuel. In the 1930s and 1940s, vegetable oils were used as diesel fuels from time to time. It was usually only in emergency situations. Recently, there has been a renewed focus on vegetable oils and animal fats to make biodiesel fuels due to the increase in crude oil prices, limited resources of fossil oil and environmental concerns. Intensification of local air pollution and magnification of global warming problems caused by CO_2, through continued and increasing use of petroleum can be reduced by the use of biodiesel fuels.

The advantages of vegetable oils as diesel fuel

(i) liquid-nature portability;
(ii) heat content (80% of diesel fuel);
(iii) ready availability and
(iv) renewability.

The disadvantages are

(i) higher viscosity; and
(ii) lower volatility.

It is the action of a fat or oil with an alcohol to form esters and glycerol. In addition to lowering the viscosity of vegetable oils, transesterification improves other characteristics such as (i) total removal of the glycerine (ii) increase in boiling point and (iii) lowering of flash point.

4.11.1 Production of Biodiesel

The production of biodiesel from oil seed is shown in Figure 4.13. Oil is squeezed from the seed in an oil press. Byproduct meal can be sold as a feed additive. One triglyceride molecule ($GL(FA)_3$) reacts with three methanol molecules (MeOH) to produce three ester molecules (MeFA) and one glycerol molecule ($GL(OH)_3$). The chemical reaction has been given as

$$GL(FA)_3 + 3MeOH \xrightarrow{yields} 3.MeFA + GL(OH)_3$$

where FA represents the fatty acid component of the ester molecule. The reaction can be catalysed by alkalis, acids or enzymes. The alkalis include NaOH or KOH. A mixture is formed by adding 400 ml of methyl alcohol to 1000 ml vegetable oil along with 10 ml of catalyst. The reaction takes place at room temperature in about an hour. Small amounts of soap are also produced by the reaction of alkali with fatty acids. Upon completion, the glycerol and soap are removed in a phase separator. Finally, methyl ester layer is collected as biodiesel. The heat of combustion of biodiesel is within 95 percent by weight of conventional diesel. They burn more efficiently, they have essentially the same fuel value as diesel. Biodiesel can be used in unmodified diesel engines with no excess wear or operational problems.

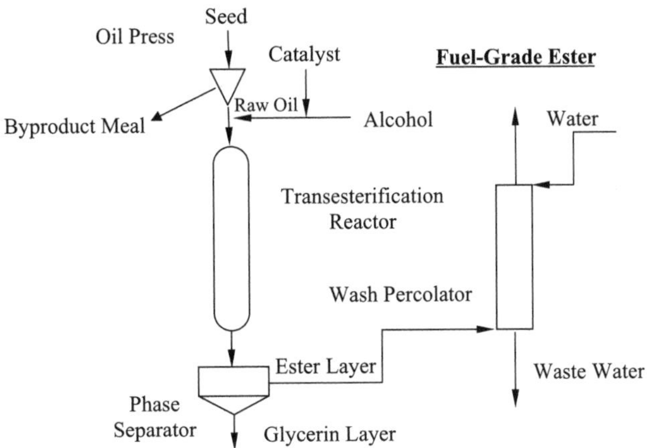

Figure 4.13 Production of fuel-grade biodiesel from Seed.

4.12 FUEL CELLS

Fuel cells are electrochemical devices that convert a fuel's chemical energy directly to electrical energy without an intermediate combustion or thermal cycle. With no internal moving parts, fuel cells operate similar to batteries. An important difference is that batteries store energy, while fuel cells produce electricity continuously as long as fuel (usually natural gas or hydrogen) and air are supplied. Fuel cells electrochemically combine a fuel (typically hydrogen) and oxygen from air without burning, thereby dispensing with the inefficiencies and pollution of traditional energy-conversion systems. Fuel cells emit virtually no pollution as the waste "exhaust" is simply water vapour and heat. In many applications, the waste heat can be used, making a fuel cell system much more efficient than conventional power supplies. In some applications, fuel cell systems can convert 80% of the energy available in the fuel into electrical and heat energy.

Although the fuel cell is not a renewable energy technology, it is certainly a core element in a renewable energy system, particularly if the hydrogen comes from a renewable fuel or process, such as a biofuel or electrolysis *via* solar generated electricity. Fossil fuels like natural gas can also be reformed for use in fuel cells. Gases from coal and diesel fuel are poor choices of fuel since they cause an overall increase in CO_2 emission.

Fuel cells forego the traditional fuel to electricity production route common in modern power production that consists of heat extraction from fuel, conversion of heat to mechanical energy and, finally, transformation of mechanical energy into electrical energy. In a sense, our bodies operate like fuel cells because we oxidise hydrocarbon compounds in our food and release chemical energy without combustion.

Though the idea of a fuel cell was suggested 200 years before in 1802 by Sir Humphrey Davy, it was demonstrated first in the laboratory in 1839 by Sir William Grove and used for the first time for its practical applications in the Gemini and Apollo spacecrafts in 1960. There are numerous applications of fuel cells in military, space, central power station or even power for off-the-road vehicles. Despite their benefits, high profile use and the fact that they have been around for over 200 years, the widespread adoption of fuel cells has yet to take off. The primary reason is cost. During the past three decades, significant efforts have been made to develop more practical and affordable designs for stationary power applications. But the progress has been slow. Today, the most widely marketed fuel cells cost about \$4500 per kilowatt; by contrast, a diesel generator costs \$800 to \$1500 per kilowatt and a natural gas turbine can be even less. Recent technological advances in material research, however, have significantly improved the economic outlook for fuel cells. The state-of-the-art fuel cells now being tested are likely to cost around $1200 per kilowatt comparable to a large-scale coal-fired power plant. The goal is to cut costs to as low as $400 per kilowatt by the end of this decade. Also, it is forecasted that the fuel cells are now on the verge of being introduced commercially, revolutionising the way to produce power.

4.12.1 Working of Fuel Cell

The operation of a fuel cell is illustrated schematically in Figure 4.14. The fuel cell consists of two gas-permeable electrodes separated by an electrolyte, which is a transport medium for electrically charged ions. Hydrogen gas, the ultimate fuel in all current designs of fuel cells, enters the fuel cell through the anode, while oxygen is admitted through the cathode. Hydrogen is generated from biofuels and oxygen is supplied from the air. Depending on the fuel cell design, either positively charged hydrogen ions form at the anode or negatively charged ions containing oxygen form at the cathode. In either case, the resulting ions migrate through the electrolyte to the opposite electrode from which they are formed. Hydrogen ions migrate to the cathode where they react with oxygen to form water. Oxygen-bearing ions migrate to the anode where they react with hydrogen to form water. Both ionic processes release chemical energy in the form of electrons at the anode, which flow to the cathode through an external electric circuit. The flow of electrons from anode to cathode represents the direct generation of electric power from flameless oxidation of fuel.

4.12.2 Material Characteristic Requirements for Fuel Cells

Though the fuel cell appears to be a simple device, the requirements of electrodes (anode and cathode) and electrolyte are quite specific. These are listed as follows:

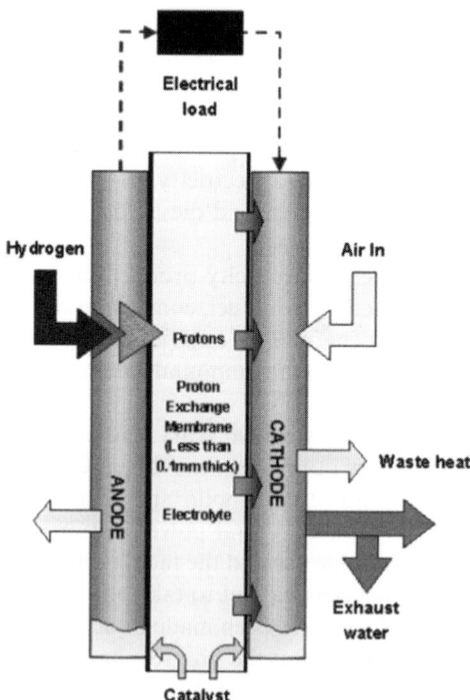

Figure 4.14 Schematic diagram of a fuel cell.

(a) Requirements of an Electrode

(i) It must be a good conductor, to transport electronic charges.
(ii) It must serve as a catalyst to convert hydrogen and oxygen molecules into their ions.
(iii) It must be chemically inert so that there should be no corrosion when it is in contact with an electrolyte.
(iv) It must be moldable into any shape and size as the stacking of fuel cells need to be done to generate the desired fuel cell power generation system.
(v) It must withstand high temperatures, as some fuel cells need to be operated at higher temperature.

(b) Requirements of an Electrolyte

(i) It must be ionically conducting.
(ii) The two electrodes must be prevented from coming into electrical contact.
(iii) It must allow passage of ions from one electrode to the other.

4.12.3 Types of Fuel Cells

There are varieties of fuel cells developed working from room temperature to 1000 °C. These are classified mainly into two groups, low temperature fuel cells and high temperature fuel cells. The fuel cells can be classified further depending on the electrolyte used. These are of five types:

(i) alkaline fuel cells;
(ii) polymer membrane or proton-exchange membrane fuel cells;
(iii) phosphoric acid fuel cells;
(iv) molten carbonate fuel cells;
(v) solid oxide fuel cells.

4.12.4 Benefits from Fuel Cells

(i) The efficiency of fuel cells is as high as 50 to 70%.
(ii) Their nonpolluting nature facilitates the use in cities where power is needed, rather than far away.
(iii) A variety of fuel can be used including natural gas, hydrogen, methanol and biogas.
(iv) Fuel cells are finding many uses in stationary, mobile and portable applications, *i.e.* from large buildings to cellphones.
(v) Fuel cells can virtually eliminate emissions of nitrous oxide, carbon monoxide, hydrocarbon and particulate matter.

OBJECTIVE QUESTIONS

4.1 Biomass is formed
  (a) By the process of inorganic materials
  (b) By the process of organic materials
  (c) By originating hydrogen along with CO_2 in the air
  (d) All of them

4.2 The basic source of biomass is
   (a) Wind
   (b) Earth
   (c) Sun
   (d) Air
4.3 The biomass is a
   (a) Liquid fuel
   (b) Gas fuel
   (c) Solid fuel
   (d) None
4.4 The biomass can be converted into
   (a) Liquid fuel
   (b) Gas fuel
   (c) Solid fuel
   (d) None
4.5 The biomass contains
   (a) C, N and O
   (b) C, S and O
   (c) C, H and O
   (d) None
4.6 The biomass can be used as the storage media of
   (a) Wind energy
   (b) Solar energy
   (c) Wave energy
   (d) All of them
4.7 The biomass is
   (a) Nonsustainable
   (b) Sustainable
   (c) Fixed
   (d) None
4.8 The term "bio" means
   (a) Life
   (b) Neutral
   (c) Dead
   (d) All of them
4.9 The sustainability means
   (a) Finite resources
   (b) Maintaining productivity
   (c) Infinite resources
   (d) All of them
4.10 For photosynthesis process, the basic need is
   (a) Presence of air
   (b) Presence of sunlight
   (c) Presence of green plant
   (d) All
4.11 Finite product of photosynthesis is
   (a) Glucose, $C_6H_{12}O_6$
   (b) Sucrose, $C_{12}H_{22}O_{11}$
   (c) Both (a) & (b)
   (d) All of them

4.12 For biomass production solar energy (light energy) is used
   (a) Indirectly (chemical process)
   (b) Directly
   (c) Both (a) & (b)
   (d) None
4.13 The percentage of biomass produced on the Earth used for human food is
   (a) 5%
   (b) 15%
   (c) 0.5%
   (d) 0.05%
4.14 The plants use the following wavelength of light
   (a) 0–∞ μm
   (b) 0.7–∞ μm
   (c) 0.4–0.7 μm
   (d) 0–0.7 μm
4.15 The conversion efficiency from photon (light energy) to glucose is
   (a) 3.5%
   (b) 35%
   (c) 0.35%
   (d) 7%
4.16 The photosynthesis in the plants takes place by
   (a) Light reaction
   (b) Dark reaction
   (c) Both light and dark reaction
   (d) All of them
4.17 The percentage of solar radiation (light intensity) available for photosynthesis is
   (a) 6.8%
   (b) 0.68%
   (c) 0.068%
   (d) 68%
4.18 The overall conversion efficiency of photosynthesis process for visible light (0.4–0.7 μm)
   (a) 0.7%
   (b) 7%
   (c) 70%
   (d) 0.07%
4.19 The energy content in one photon for 700 nm (7×10^{-7}m) wavelength is
   (a) 1.7 eV
   (b) 17 eV
   (c) 0.17 eV
   (d) none
4.20 The maximum atomic mass number of one element out of C, H and O of glucose is
   (a) 12
   (b) 1
   (c) 16
   (d) none
4.21 The conventional fuel is generally denoted by the formula
   (a) H_mC_c
   (b) C_nH_m
   (c) $(CH_2O)_n$
   (d) All of them

4.22 The primary source to produce biofuel is
(a) Biomass
(b) Biogas
(c) Solar energy
(d) None

4.23 During carbonisation of biomass, the carbon contents remains in biomass (charcoal) is
(a) 50%
(b) 25%
(c) 75–80%
(d) 100%

4.24 The temperature range of the carbonisation process is
(a) 170–300 °C
(b) 200 °C
(c) 400–1000 °C
(d) All

4.25 The air flow during carbonisation should be
(a) Not restricted
(b) Restricted
(c) Open air
(d) All

4.26 The temperature range of pyrolysis process is
(a) 170–300 °C
(b) 200 °C
(c) 400–1000 °C
(d) All

4.27 The final product of carbonisation is
(a) Solid
(b) Liquid
(c) Gas
(d) All of them

4.28 The final product of pyrolysis is
(a) Solid
(b) Liquid
(c) Gas
(d) All of them

4.29 During the pyrolysis process, the carbon contents remains in bio fuel is
(a) 50%
(b) 25%
(c) 75–80%
(d) 100%

4.30 Which fuel has the maximum gross calorific value or heat of combustion?
(a) Coal
(b) Kerosene
(c) Charcoal
(d) Methane

4.31 The grass calorific value or heat of combustion of methane is
(a) 15 MJ/kg
(b) 5–10 MJ/kg
(c) 55 MJ/kg
(d) None

4.32 Energy farming can be done from
(a) Agriculture (field)
(b) Silviculture (forest)
(c) Aquaculture (fresh and sea water)
(d) All of them

4.33 Energy crops can be harvested in the form of
(a) Solid
(b) Liquid
(c) Gas
(d) All of them

4.34 The maximum element in grown biomass is
(a) "C" and "N"
(b) "H" and "N"
(c) "N" and ash
(d) "C" and "O"

4.35 Producer gas obtained from gasification of biomass is
(a) Combustion gas mixture
(b) Wood gas
(c) Water gas
(d) Synthesis gas

4.36 The combustible element in producer gas is
(a) H_2
(b) CO and CH_4
(c) N_2
(d) All of them

4.37 The noncombustible element in producer gas is
(a) H_2
(b) CO and CH_4
(c) N_2
(d) All of them

4.38 The combustible and nonconventional element is producer gas is about
(a) 30% and 70%
(b) 50% each
(c) 70% and 30%
(d) None

4.39 The gasifier process is a
(a) Physical reaction
(b) Chemical reaction
(c) Biological reaction
(d) All of them

4.40 The temperature of the combustion zone of a gasifier is
(a) 1300–1400 °C
(b) 1000 °C
(c) 500 °C
(d) None

4.41 Which component of producer gas has the maximum calorific value
(a) N_2
(b) CO_2

(c) CH_4
(d) H_2

4.42 Producer gas can be used for
(a) Thermal applications
(b) As fuel in IC engines
(c) Gas turbines
(d) All of them

4.43 Liquefaction of biomass can be done by
(a) Pyrolysis without any gasification medium
(b) Pyrolysis with any gasification medium
(c) Methanol synthesis with gasification medium
(d) Methanol synthesis without gasification medium

4.44 Pyrolysis reaction (decomposition of organic compounds) takes place
(a) In presence of O_2
(b) In presence of H_2O
(c) In presence of CO_2
(d) In absence of O_2

4.45 The operating temperature range of liquefaction of biomass through pyrolysis is
(a) 450–600 °C
(b) 100–450 °C
(c) 600–1000 °C
(d) None

4.46 Methanol (CH_3OH) is produced by the reaction between
(a) CO_2 and H_2
(b) CO and H_2
(c) CO and H_2O
(d) None

4.47 Methanol synthesis takes place
(a) 330 °C at 1 atmospheric pressure
(b) 330 °C at 15 atmospheric pressure
(c) 330 °C at 150 atmospheric pressure
(d) None of them

4.48 The composition of biogas is
(a) Methane (CH_4) and carbon monoxide (CO)
(b) Methane (CH_4) and carbon dioxide (CO_2)
(c) Methane (CH_4) and sulfur dioxide (SO_2)
(d) None of them

4.49 Biogas is produced by
(a) Anaerobic digestion
(b) Gasification
(c) Liquefaction
(d) Pyrolysis

4.50 The retention period of biogas lies between
(a) 0–20 days
(b) 20–50 days
(c) 50–100 days
(d) None

ANSWERS

4.1 **(a) & (c)**; 4.2 **(c)**; 4.3 **(c)**; 4.4 **(a), (b) & (c)**; 4.5 **(c)**; 4.6 **(b)**; 4.7 **(b)**; 4.8 **(a)**; 4.9 **(b)**; 4.10 **(c)**; 4.11 **(d)**; 4.12 **(b)**; 4.13 **(c)**; 4.14 **(c)**; 4.15 **(b)**; 4.16 **(b)**; 4.17 **(c)**; 4.18 **(b)**; 4.19 **(a)**; 4.20 **(c)**; 4.21 **(b)**; 4.22 **(b)**; 4.23 **(c)**; 4.24 **(a)**; 4.25 **(b)**; 4.26 **(c)**; 4.27 **(a)**; 4.28 **(d)**; 4.29 **(b)**; 4.30 **(d)**; 4.31 **(c)**; 4.32 **(d)**; 4.33 **(a)**; 4.34 **(d)**; 4.35 **(a), (b), (c) & (d)**; 4.36 **(a) & (b)**; 4.37 **(c)**; 4.38 **(b)**; 4.39 **(b)**; 4.40 **(a)**; 4.41 **(c)**; 4.42 **(d)**; 4.43 **(a) & (c)**; 4.44 **(d)**; 4.45 **(a)**; 4.46 **(b)**; 4.47 **(c)**; 4.48 **(b)**; 4.49 **(a)**; 4.50 **(b)**

REFERENCES

1. Bull, S. R. 1991. The U.S. Department of Energy Biofuels Research Program, Energy Sources, **13**, 433–442.
2. Wayman, M. and Parekh, S. 1990. *Biotechnology of Biomass Conversion*. Open University Press, Philadelphia, PA.
3. Iyer, P. V. R., Rao, T. R. and Grover, P. D. 2002. Biomass, Thermo-Chemical Characterisation. Published under MNES sponsored Gasifier Action Research project, Indian Institute of Technology, Delhi, India.
4. Miles, T. R., Baxter, Tr. L., Jenkins, B. and Oden, L. 1995. Alkali deposits found in biomass power plants. Summary Report, National Renewable Energy Laboratory, NREL, April 15.
5. Grohmann, K., Wyman, C. E. and Himmel, M. E. 1992. Potential for fuels from biomass and wastes. ACS Symposium series 476. Washington, D.C: American Chemical Society.
6. Chaturvedi P. 1993. Bioenergy production and utilisation in India, expert consultation on biofuels for sustainable development; their potential as suitable to fossil fuels and CO_2 emission reduction. FAO, Rome.
7. Reddy, B. Sudhakara 1994. *Energy Conver. Manag.* **35**, No. 4 341–361.
8. Connors, M. A. and Salazar, C. M. 1985. Proceedings Symposium on Forest Products Research Achievement and the Future, vol 5, 7–12, April, 85, Pretoria, S. Africa.
9. Mittal, K. M. 1996. *Biogas Systems; Principles and Applications*. New Age International (p) Limited Publishers, New Delhi, India.
10. Shay, E. G. 1993. *Biomass Bioenergy* **4**, 227–242.
11. Shafizadeh, F. 1982. *J. Appl. Anal. Appl. Pyrol.* **3**, 283–305.

Biopower (Electricity), Derived from Biomass, is Clean and Friendly with the Ecosystem, Unlike Fossil Fuels. It Helps to Sustain the Climate.

CHAPTER 5

Biopower

5.1 INTRODUCTION

Biopower is the electricity produced from biomass fuels. Biopower is also called biomass power. Biopower technologies convert biomass into heat as well as electricity using modern boilers, gasifiers, turbines, generators, fuel cells and other method, which is good for the environment. Biomass fuels are renewable help in reducing the greenhouse gas emissions from fossil fuels. This is good for the environment. Biopower also makes productive use of (i) crop residues (ii) wood-manufacturing wastes, (iii) residues of wood and paper products industries, (iv) wastes from food production and processing activities, (v) animal wastes, *etc.*

The "useless" wastes are generally open-burned, left to rot in fields, or buried in landfills. Landfills are the waste that rots in the field, often producing methane, a greenhouse gas even more potent than carbon dioxide. Burying energy-rich wastes in a landfill is like burying petroleum instead of using it. But today's biopower plants have generation costs more than for production of fossil fuels. Biomass fuels contain less concentrated energy. Therefore, these are less economic to transport over long distances. It requires more preparation as well as handling than fossil fuels. These factors contribute to higher costs. The further improvement in biopower generation can be achieved through these bioresources. There is greater flexibility in providing a much wider spectrum of final energy forms, *i.e.* thermal energy, fuels (gaseous, liquid or solid) as well as electrical power. A number of technological choices are available for conversion of bioresources to the energy form of interest here, *i.e.* electricity and mechanical power.

5.2 BIOPOWER GENERATION ROUTES

The six major types of **biopower** generation are as follows:

5.2.1 Direct Combustion

This involves the direct burning of biomass with excess air to produce hot flue gases. The gases are used to produce steam in the boilers. Most of the biopower plants in the world use direct-fired systems. They burn bioenergy feed stocks directly to produce steam to run a turbine and then generate electricity by a generator. The steam from the power plant is also used for manufacturing processes or for space heating.

Advanced Renewable Energy Sources
G. N. Tiwari and R. K. Mishra

Published by the Royal Society of Chemistry, www.rsc.org

5.2.2 Cofiring

This refers to the practice of introducing biomass in high efficiency coal-fired boilers. It involves replacing a portion of the coal with biomass at an existing power plant boiler. This is achieved either (i) by mixing biomass with coal before fuel is introduced into the boiler or (ii) by using separate fuel feeds for coal and biomass. Depending on the boiler design and fuel feed system employed, biomass can replace up to 20% of coal in a cofiring operation. The purpose of using biomass with coal is that biomass contains less sulfur, les heavy metals and ash resulting in the reduction of greenhouse gas (GHG) emissions. On a dry ash-free basis, biomass has chemical composition of about 50% carbon, 6% hydrogen and 44% oxygen. It also produces more than 80% of volatiles, in gaseous and vapour form when biomass is heated. Coal by comparison produces usually less than 50% volatiles. Simply by heating, biomass drives off these volatiles rapidly within a few seconds. Biomass is therefore more reactive than coal. It makes its energy content easier to mobilise.

5.2.3 Biomass Gasification

This involves heating biomass in an oxygen-starved environment to produce a medium or low calorific gas (a mixture of hydrogen, carbon monoxide and methane) generally called producer gas. It is then used as fuel in a combined cycle power-generation plant that includes a gas turbine topping cycle and a steam turbine bottoming cycle. The topping cycle is a cogeneration system in which electric power is generated first. The reject heat from power generation is then used to produce useful process heat. Similarly in bottoming cycle, steam is used first for process heat and then for electric power production.

5.2.4 Biomass Pyrolysis

This refers to a process where biomass is exposed to high temperatures in the absence of air. It causes the biomass to break down by producing a high proportion of vapour products that exist in the liquid state at room temperature, called "**biocrude**". Like petroleum, biocrude can be refined to produce a high quality road transport fuel. It can, however, be burned directly as fuel for a slow-speed stationary diesel engine. The advantage of taking this liquid fuel route to biomass power is that such liquid fuel can be produced at any time, stored and transported like other flue-gas hydrocarbon fuel. Its energy density may be high enough to make it worth transporting over longer distances than raw biomass.

5.2.5 Anaerobic Digestion

This is a process by which organic matter is decomposed by bacteria in the absence of oxygen to produce methane and other byproducts. The primary energy product is a low to medium calorific gas called "**biogas**" normally consisting of 50–60% methane, which is a combustible gas.

5.2.6 Fuel Cell

This is a device that converts the chemical energy of a fuel directly into electricity without combustion. The advantage of a fuel cell is that it generates electricity with very little pollution as the final byproduct is the water. With no internal moving parts, a fuel cell operates similar to batteries. An important difference is that batteries store energy while fuel cell can produce electricity continuously, as long as fuel and air are supplied. A fuel cell electrochemically combines a fuel (typically hydrogen) and an oxidant without burning, thereby dispensing with the inefficiencies and pollution of traditional energy conversion systems.

5.3 BASIC THERMODYNAMIC CYCLES IN BIOPOWER GENERATION

Biomass is converted into electricity through several processes, as mentioned earlier. The majority of biopower is generated by using steam and a gas turbine cycle. The commonly used option for biopower generations is shown in Figure 5.1.

The devices or systems used to produce a power output are often called **engines**. Heat engines are categorised as internal combustion and external combustion engines. This depends on how the heat is supplied to the working fluid. In external combustion engines (such as steam power plants), energy is supplied to the working fluid from an external source such as a furnace. In internal combustion engines (such as automobile engines), this is done by burning the fuel within the system boundaries.

The thermodynamic cycles they operate on them are called **power cycles**. The power cycles can also be categorised as gas cycles or **vapour cycles**, depending on the phase of the working fluid. In gas cycles, the working fluid remains in the gaseous phase throughout the entire cycle. In the vapour cycles, the working fluid exists in the vapour phase during one part of the cycle and liquid phase during another part. In vapour power cycles, the working fluids are alternately vaporised and then condensed. Steam is the most common working fluid used in vapour power cycles due to its many desirable characteristics, such as low cost, availability and high enthalpy of vaporisation. Steam power plants are commonly referred to as coal plants, nuclear plants, natural gas plants or biopower plant, depending on the type of fuel used to supply heat to the steam.

Thermodynamic cycles can be categorised, in yet another way: closed and open cycles. In **closed cycles**, the working fluid is returned to the initial state at the end of the cycle and it is recirculated. In **open cycles**, the working fluid is renewed at the end of each cycle instead of being recirculated. In automobile engines, the combustion gases are exhausted and replaced by a fresh air–fuel mixture at the end of each cycle. The engine operates on a mechanical cycle. The working fluid does not go through a complete thermodynamic cycle.

The model cycle of gas power cycles is the Otto cycle for spark-ignition engine, the Diesel cycle for compression ignition engines, the Stirling and Ericsson cycles for gas engines, and the Brayton

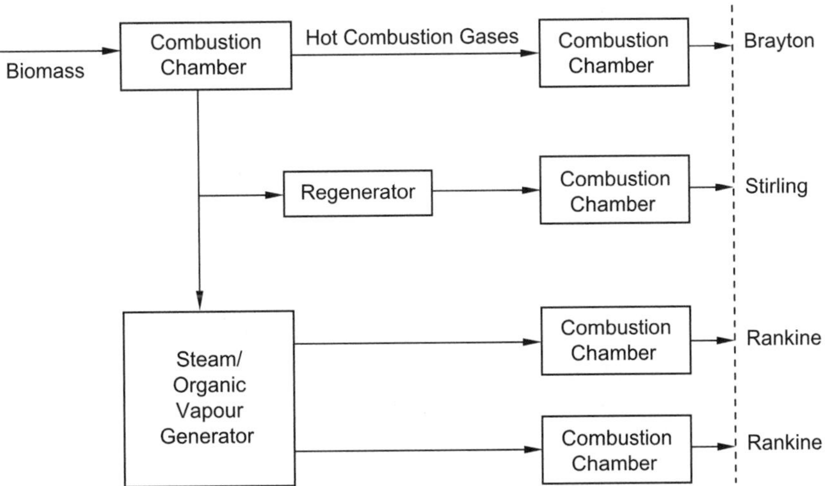

Figure 5.1 Flow diagram of biopower generation.

cycle for gas-turbine engines. Similarly, the model cycle for vapour power cycle is the Rankine cycle. The Carnot cycle is not a suitable model for gas power and vapour power cycles because it cannot be approximated in practice.

The most efficient cycle operating between a heat source at temperature T_{H} and a sink at temperature T_{L} is the **Carnot cycle**. The thermal efficiency of the Carnot cycle is given by

$$\eta_{\mathrm{th,carnot}} = 1 - \frac{T_{\mathrm{L}}}{T_{\mathrm{H}}} \tag{5.1}$$

The T_{L} and T_{H} are in Kelvin. The actual power cycles are rather complex. The approximations used to simplify the analysis are as follows:

(i) air is assumed to have constant specific heats at room temperature,
(ii) air behaves as an ideal gas,
(iii) the combustion and exhaust processes are replaced by heat addition and heat rejection processes, respectively,
(iv) the processes are assumed to be internally reversible
(v) heat transfer through finite temperature difference. These approximations are known as **the standard assumptions** or **cold-air standard assumptions**.

The property diagrams such as *P–v* and *T–s* diagrams have served as valuable aids in the analysis of thermodynamic processes. On both the *P–v* (pressure–volume) and *T–s* (temperature–entropy) diagrams, the area enclosed by the process curves of a cycle represents the net work produced during the cycle. It is also equivalent to the net heat transfer for that cycle. On a *T–s* diagram, a heat addition process proceeds in the direction of increasing entropy, a heat-rejection process proceeds in the direction of decreasing entropy and an isentropic (internally irreversible, adiabatic) process proceeds at constant entropy.

A process is said to be **reversible** if both the system and the surroundings can be restored to their original conditions. The effects such as friction, nonquasiequilibrium expansion or compression and heat transfer through a finite temperature difference are referred to as irreversible processes which are called **irreversibilities**.

An **adiabatic process is a process during which** there is no heat transfer taking place. The process during which the temperature remains constant is called an **isothermal process**. An adiabatic process should not be confused with an isothermal process. Even though there is no heat transfer during an adiabatic process, the energy content and thus the temperature of the system can still be changed by other means such as work.

The power generation coupled with process heating is referred to as **cogeneration**. Similarly, power cycles that consist of two separate cycles are known as **binary cycles** and **combined cycles**. The heat rejected by one fluid is used as the heat input to another fluid operating at a lower temperature.

5.3.1 Brayton Cycle: The Ideal Cycle for Gas-Turbine Engines

The **Brayton cycle** was first proposed by George Brayton in around 1870 for use in the reciprocating oil-burning engine. Today, it is used for gas turbines only where both compression and expansion processes take place in rotating machinery. Gas turbines operate on an open cycle, as shown in Figure 5.2. Fresh air is drawn under ambient conditions into the compressor, where its temperature and pressure are raised. The high-pressure air proceeds into the combustion chamber for burning of the fuel at constant pressure. The resulting high temperature gases then enter the turbine, for

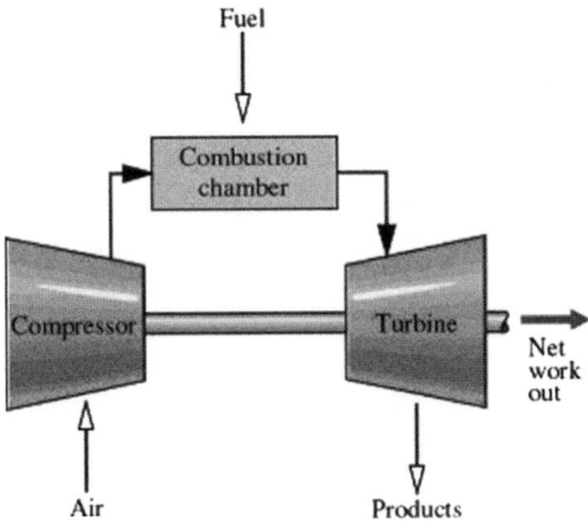

Figure 5.2 Flow diagram for open-cycle gas-turbine engine.

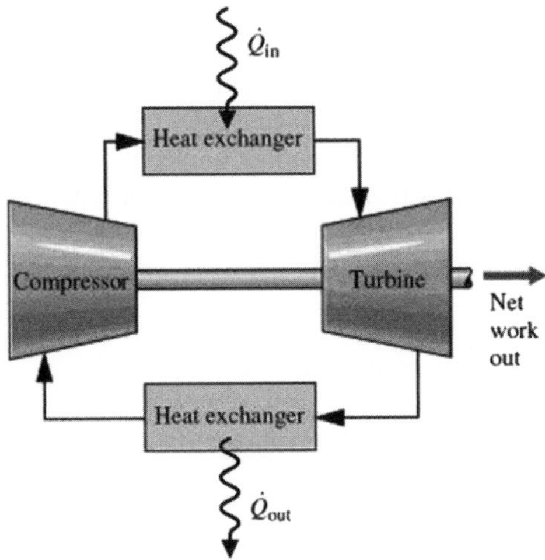

Figure 5.3 Flow diagram for closed-cycle gas-turbine engine.

expansion to the atmosphere pressure, thus producing power. The exhaust gases leaving the turbine are thrown out. This causes the cycle to be referred to as an **open cycle**.

The open gas-turbine cycle described above can be modelled as a closed cycle, as shown in Figure 5.3. Here, the compression and expansion processes remain the same. The combustion process is replaced by a constant-pressure heat addition process from an external source. The exhaust process is replaced by a constant-pressure heat rejection process to the ambient air. The ideal cycle that the working fluid undergoes in this closed loop is the Brayton cycle (Figure 5.4), which is made up of four internally reversible processes:

1 – 2 isentropic compressor (in a compressor);
2 – 3 constant-pressure heat addition;
3 – 4 isentropic expansion (in a turbine);
4 – 1 constant-pressure heat rejection.

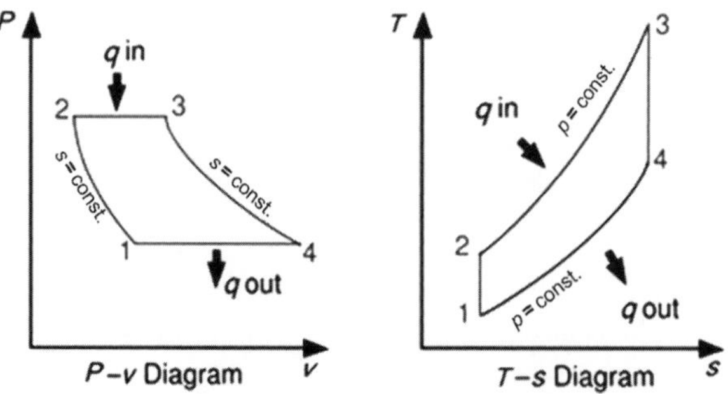

Figure 5.4 *P–v* and *T–s* diagram for the ideal Brayton cycle.

All four processes of the Brayton cycle are executed in steady-flow devices; thus they should be analysed as steady-flow processes. When the changes in kinetic and potential energies are neglected, the energy balance for a steady-flow process can be expressed on a unit-mass basis, as

$$(q_{\text{in}} - q_{\text{out}}) + (w - w_{\text{out}}) = h_{\text{exit}} - h_{\text{inlet}} \tag{5.2}$$

where q is the heat transfer per unit mass of a system in kJ/kg
w is the work done per unit mass of a system in Joule and
h is the enthalpy (heat contents or total heat) per unit mass of a system in kJ/kg.
Therefore, the heat transfers to and from the working fluid are

$$q_{\text{in}} = h_3 - h_2 = C_p(T_3 - T_2) \tag{5.3a}$$

and

$$q_{\text{out}} = h_4 - h_1 = C_p(T_4 - T_1) \tag{5.3b}$$

Then, the thermal efficiency of the ideal Brayton cycle under the cold-air standard assumptions becomes

$$\eta_{\text{th,Brayton}} = \frac{w_{\text{net}}}{q_{\text{in}}} = 1 - \frac{q_{\text{out}}}{q_{\text{in}}} = 1 - \frac{C_p(T_4 - T_1)}{C_p(T_3 - T_2)} = 1 - \frac{T_1(T_4/T_1 - 1)}{T_2(T_3/T_2 - 1)} \tag{5.4}$$

Applying isentropic processes for ideal gases, the processes 1 – 2 and 3 – 4 are isentropic and if $P_2 = P_3$ and $P_4 = P_1$, then

$$\frac{T_2}{T_1} = \left(\frac{P_2}{P_1}\right)^{(k-1)/k} = \left(\frac{P_3}{P_4}\right)^{(k-1)/k} = \frac{T_3}{T_4} \tag{5.5}$$

where $k = C_p/C_v$; C_p and C_v are the specific heat at constant pressure and at constant volume, respectively. Substituting these equations into the thermal efficiency relation and simplifying, the expression for thermal efficiency becomes

$$\eta_{\text{th,Brayton}} = 1 - \frac{1}{r_p^{(k-1)/k}} \tag{5.6}$$

where
$r_p = P_2/P_1$
r_p and k are the **pressure ratio** and the **specific heat ratio** respectively. It is clear from Eq. (5.6) that the thermal efficiency of the ideal Brayton cycle increases with r_p and k, which is also the case for actual gas turbines.

5.3.2 Stirling Cycle

A thermodynamic cycle involves an isothermal processes at T_H and an isothermal heat-rejection process at T_L. It differs from the Carnot cycle. Two isentropic processes are replaced by two constant-volume regeneration processes in the Stirling cycle. The basic principle of the Stirling engine was first proposed by Robert Stirling in 1816.

This cycle utilises regeneration. This is a process during which heat is transferred to a thermal energy storage device (called a regenerator) during one part of the cycle. It is transferred back to the working fluid during another part of the cycle as shown in Figure 5.5. Figure 5.6 shows the *T–s* and *P–v* diagrams of the Carnot and Stirling cycles. It is made up of four totally reversible processes:

1 – 2 (temperature constant) expansion (heat addition from external source);

2 –3 (constant volume) regeneration (internal heat transfer from the working fluid to the regenerator);

3 – 4 (temperature constant) compression (heat rejection to the external sink);

4 – 1 (constant volume) regeneration (internal heat transfer from the regenerator to the working fluid).

The execution of the Stirling cycle requires rather innovative hardware. The execution of the Stirling cycle in a closed system is explained with the help of the hypothetical engine shown in Figure 5.7. This system consists of a cylinder with two pistons on each side and a regenerator in the middle. The regenerator can be a wire or a ceramic mesh or any kind of porous plug with a high thermal mass (mass times specific heat). It is used for the temporary storage of thermal energy. The mass of the working fluid contained within the regenerator at any instant is considered negligible.

Initially, the left chamber houses the entire working fluid (a gas). It is at a high temperature and pressure. During process 1 – 2, heat is transferred to the gas at T_H from a source at T_H. The left piston moves outward as the gas expands isothermally, doing work and the gas pressure drops. Both pistons are moved to the right at the same rate (to keep the volume constant) during process 2 – 3, until the entire gas is forced into the right chamber. Heat is transferred to the regenerator as the gas passes through the regenerator. The gas temperature drops from T_H to T_L. For a heat-transfer

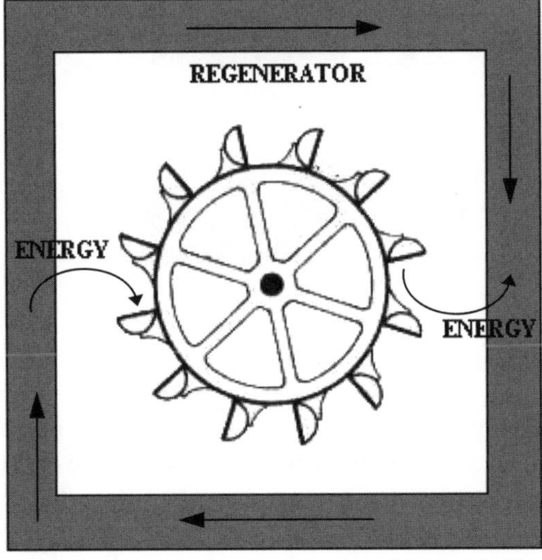

Figure 5.5 The working principle of regeneration Process.

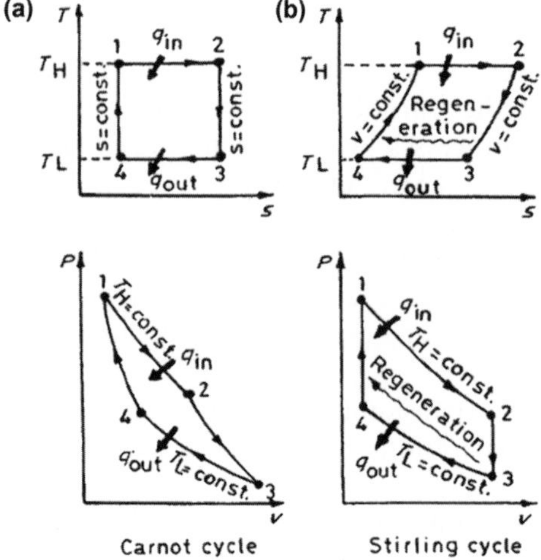

Figure 5.6 Flow chart of T–s and P–v diagrams of Carnot and Stirling cycles.

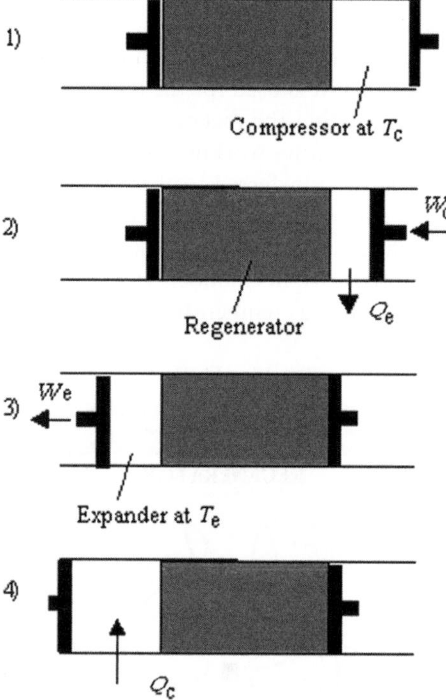

Figure 5.7 The working principle of the Stirling cycle.

process to be reversible, the temperature difference between the gas and the regenerator should not exceed a differential amount dT at any point. Thus, the temperature of the regenerator will be T_H at the left end and T_L at the right end of the regenerator when state 3 is reached. The right piston is moved inward during process 3–1, compressing the gas. Heat is transferred from the gas to a sink at temperature T_L. So, the gas temperature remains constant at T_L, while the pressure rises. Finally, during process 4–1, both pistons are moved to the left at the same rate (to keep the volume constant) forcing the entire gas into the left chamber. The gas temperature rises from T_L to T_H as it passes through the regenerator and picks up the thermal energy stored there during process 2 – 3. This completes the cycle.

It is observed that the second constant-volume process takes place at a smaller volume than the first one. The net heat transfer to the regenerator during a cycle is zero. That is, the amount of energy stored in the regenerator during process 2–3 is equal to the amount picked up by the gas during process 4–1.

The thermal efficiency of the Stirling cycle is identical to the efficiency of a Carnot cycle operating between the same temperature limits for an ideal gas as the working fluid. Heat is transferred to the working fluid isothermally from an external source at temperature T_H during process 1–2. It is rejected again isothermally to an external sink at temperature T_L during process 3–4. For a reversible isothermal process, heat transfer is related to the entropy change by

$$q = T\Delta s \tag{5.7}$$

where s is the entropy (kJ/kg K) per unit mass of the system.

The entropy change of an ideal gas during an isothermal process is given by

$$\Delta s = C_p \ln\frac{T_{exit}}{T_{in}} - R\ln\frac{P_{exit}}{P_{in}} \tag{5.8}$$

For an isothermal process, then Eq. (5.8) reduces to

$$\Delta s = -R\ln\frac{P_{exit}}{P_{in}} = s_2 - s_1 \tag{5.9}$$

Then, the heat input and heat output can be expressed as

$$q_{in} = T_H(s_2 - s_1) = T_H\left(-R\ln\frac{P_2}{P_1}\right) = RT_H\ln\frac{P_1}{P_2} \tag{5.10}$$

$$q_{out} = -T_L(s_4 - s_3) = -T_L\left(-R\ln\frac{P_4}{P_3}\right) = RT_L\ln\frac{P_4}{P_3} \tag{5.11}$$

where R is the gas constant from the ideal gas equation of state (kJ/kg K).

Using the ideal gas equation for the process 1–2 and 3–4.

$$\frac{P_1}{P_2} = \frac{v_2}{v_1} \text{ and } \frac{P_4}{P_3} = \frac{v_3}{v_4} \tag{5.12}$$

But $v_2 = v_3$ and $v_4 = v_1$.

Hence, $P_1/P_2 = P_4/P_3$

Now,

$$\eta_{th,\,Stirling} = 1 - \frac{q_{out}}{q_{in}} = 1 - \frac{RT_L\ln\left(\frac{P_4}{P_3}\right)}{RT_H\ln\left(\frac{P_1}{P_2}\right)} = 1 - \frac{T_L}{T_H}$$

Hence,

$$\eta_{\text{th,Stirling}} = 1 - \frac{T_L}{T_H} \tag{5.13}$$

Equation (5.13) is identical with the efficiency of the Carnot cycle.

The Stirling cycle is difficult to achieve in practice because it involves heat transfer through a differential temperature difference in all components including the regenerator. This requires either infinitely large surface areas for heat transfer or allowing an infinitely long time for the process. In reality, all heat transfer processes will take place through a finite temperature difference. The regenerator will not have an efficiency of 100 percent. The pressure losses in the regenerator will be considerable. Because of these limitations, the Stirling cycle has long been of only theoretical interest.

The Stirling cycle is an external combustion engine. The biomass is burned outside the system in the furnace. External combustion offers several advantages. First, a variety of fuels can be used as a source of thermal energy. Secondly, there is more time for combustion. Thus, the combustion process is more complete, which means less air pollution and more energy extraction from the fuel. Thirdly, if the engine operates on a closed cycle, then a working fluid has the most desirable characteristics namely stable, chemically inert, high thermal conductivity.

5.3.3 Rankine Cycle: The Ideal Cycle for Vapour Power Cycle

The Carnot cycle is not a suitable model for vapour power cycle. Many of the impracticalities associated with the Carnot cycle can be eliminated by superheating the steam in the boiler and condensing it completely in the condenser, as shown in Figure 5.8. The cycle is the Rankine cycle that is the ideal cycle for vapour power plants. The ideal Rankine cycle does not involve any internal irreversibility. It consists of the following four processes:

1–2 isentropic compression in a pump;
2–3 constant pressure heat addition in a boiler;
3–4 isentropic expansion in a turbine;
4–1 constant pressure heat rejection in a condenser.

Water enters the pump at state 1 as saturated liquid. It is compressed isentropically to the operating pressure of the boiler. The water temperature increases during this isentropric compression process due to a slight decrease in the specific volume of the water. The vertical distance between states 1 and 2 on the *T*–*s* diagram is raised for clarity.

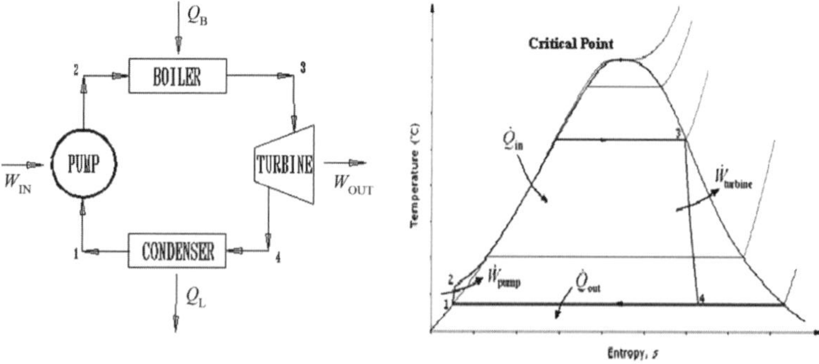

Figure 5.8 The simple ideal Rankine cycle.

Water enters the boiler as a compressed liquid at state 2. It leaves as a superheated vapour at state 3. The boiler is basically a large heat exchanger where heat originating from combustion gases (biomass) is transformed to the water essentially at constant pressure. The boiler is often called the steam generator (the superheater).

The superheated vapour at state 3 enters the turbine. It expands isentropically in the turbine and produces work by rotating the shaft connected to an electric generator. The pressure and temperature of the steam drop during this process to the values at state 4. The steam enters the condenser with low values of pressure and temperature. At this state, steam is usually a saturated liquid–vapour mixture with a high quality. The steam is condensed at constant pressure in the condenser, through a heat exchanger by rejecting heat to a cooling medium like the atmosphere. Steam leaves the condenser as saturated liquid. It enters the pump, completing the cycle.

The area under the process curve on *T–s* diagram represents the heat transfer for internally reversible processes. The area under process curve 2–3 represents the heat transfer to the water in the boiler. The area under the process curve 4–1 represents the heat rejected in the condenser. The difference between these two (the area enclosed by the cycle curve) is the net work produced during the cycle.

(A) Energy Analysis of the Ideal Rankine Cycle

All four components in the Rankine cycle (the pump, boiler, turbine and condenser) are steady-flow devices and thus all four processes that make up the Rankine cycle can be analysed as steady-flow processes. The kinetic- and potential-energy changes of the steam are usually small relative to the work and heat transfer terms. These are therefore usually neglected. Then, the steady-flow energy equation per unit mass of steam reduces to

$$(q_{\text{in}} - q_{\text{out}}) + (w_{\text{in}} - w_{\text{out}}) = h_{\text{exit}} - h_{\text{inlet}} \tag{5.14}$$

The boiler and the condenser do not involve any work and the pump and the turbine are assumed to be isentropic. Then, the conservation of energy relation for each device can be expressed as follows:

Pump (q = 0):

$$w_{\text{pump,in}} = h_2 - h_1 \tag{5.15}$$

or,

$$w = v(P_2 - P_1)$$

where

$$h_1 = h_{\text{f@}P_1} \quad \text{and} \quad v \approx v_1 = v_{\text{f@}P_1}$$

Boiler (w = 0):

$$q_{\text{in}} = h_3 - h_2 \tag{5.16}$$

Turbine (q = 0):

$$w_{\text{turbine,out}} = h_3 - h_4 \tag{5.17}$$

Condenser (w = 0):

$$q_{\text{out}} = h_4 - h_1 \tag{5.18}$$

The thermal efficiency of the Rankine cycle is determined from

$$\eta_{th} = \frac{w_{\text{net}}}{q_{\text{in}}} = 1 - \frac{q_{\text{out}}}{q_{\text{in}}} \tag{5.19}$$

where

$$w_{\text{net}} = q_{\text{in}} - q_{\text{out}} = w_{\text{turbine,out}} - w_{\text{pump,in}}$$

The thermal efficiency can also be interpreted as the ratio of the area enclosed by the cycle on a *T–s* diagram to the area under the heat-addition process.

(B) Deviation of Actual Vapour Power Cycle from Idealised Rankine Cycle

The actual vapour power cycle differs from the ideal Rankine cycle due to irreversibilities in various components. Fluid friction and heat loss to the surroundings are the two common sources of irreversibillties.

Fluid friction causes a pressure drop in the boiler, the condenser and the piping between various components. As a result, steam leaves the boiler at a somewhat lower pressure. The pressure at the turbine inlet is somewhat lower than that at the boiler exit due to the pressure drop in the connecting pipes. The pressure drop in the condenser is usually very small. To compensate for the pressure drops, the water must be pumped to a sufficiently higher pressure. This requires a larger pump and larger work input to the pump.

The other major source of irreversibility is the heat loss from the steam to the surroundings as the steam flows through various components. To maintain the same level of net work output, more heat needs to be transferred to use steam in the boiler to compensate for these undesired heat losses. As a result, cycle efficiency decreases.

The irreversibilities also occur within the pump and the turbine. A pump requires a greater work input. A turbine produces a smaller work output as a result of irreversibilities. Under ideal conditions, the flow through these devices is isentropic. The deviation of actual pumps and turbines from the isentropic ones can be accurately accounted for in both adiabatic and reversible processes. However, isentropic efficiencies defined as

$$\eta_{\text{Pump}} = \frac{w_{\text{s}}}{w_{\text{a}}} = \frac{h_{2\text{s}} - h_1}{h_{2\text{a}} - h_1}$$

and

$$\eta_{\text{Turbine}} = \frac{w_{\text{a}}}{w_{\text{s}}} = \frac{h_3 - h_{4\text{a}}}{h_3 - h_{4\text{s}}}$$

where states 2a and 4a are the actual exit states of the pump and the turbine, respectively. The 2s and 4s are the corresponding states for the isentropic case.

5.3.4 Combined Gas–Vapour Power Cycles

The continued effort for higher thermal efficiencies has resulted in innovative modifications to conventional power plants. A more popular modification involves a gas power cycle topping a vapour power cycle. It is known as a **combined gas–vapour cycle**, or just the **combined cycle**. The combined cycle is the gas-turbine (Brayton) cycle topping a steam-turbine (Rankine) cycle. This has a higher thermal efficiency than either of the cycles executed individually.

Gas-turbine cycles typically operate at considerably higher temperatures. Steam cycles operate at low temperature. The maximum fluid temperature at the turbine inlet is about 620 °C for modern steam power plants. Fluid temperature over 1150 °C is used in gas-turbine power plants. The use of such higher temperatures in gas turbines is made possible by recent developments in cooling the turbine blades and coating the blades with ceramics, which are high-temperature-resistant materials. Heat is supplied at higher average temperature, gas-turbine cycles have a greater potential for higher thermal efficiencies. The gas leaves the gas-turbine cycle at very high temperatures (usually above 500 °C), which reduce any potential gains in the thermal efficiency. The situation can be improved by using regeneration.

It is required to take advantage of the very desirable characteristics of the gas-turbine cycle at high temperatures. One can use high temperature exhaust gases as the energy source for the bottoming cycle such as a steam power cycle. In this case, energy is recovered from the exhaust gases by transferring it to the steam in a heat exchanger that serves as the boiler. In general, more than one gas turbine cycle is needed to supply sufficient heat to the steam. Also, the steam cycle may involve regeneration as well as reheating. Energy for the reheating process can be supplied by burning some additional fuel in the oxygen-rich exhaust gases.

Such a system is economical economically very attractive without increasing the initial cost greatly. Thermal efficiencies of combined cycle well over 40 percent are reported.

5.3.5 Cogeneration Cycle

As discussed above, the sole purpose was to convert a portion of the heat to work due to the most valuable form of energy. The remaining part of heat is rejected to the atmosphere as waste heat, because of its quality (grade) being too low to be of any practical use. Many systems or devices, however, require energy input in the form of heat known as process heat. Some industries that rely heavily on process heat are chemical, pulp and paper, steel making, food processing and textile industries. Process heat in these industries is usually supplied by steam at 5 to 7 atmosphere pressure and at 150 to 200 °C. Energy is usually transferred to the steam by burning biomass, coal, natural gas or another fuel in a furnace.

Cogeneration is the production of more than one useful form of energy, namely process heat and electric power from the same energy source. Industries that use large amounts of process heat also consume a large amount of electric power. Therefore, it makes it economical to use the already-existing work potential to produce power instead of letting it go to waste.

Either a steam-turbine (Rankine) cycle or a gas-turbine (Brayton) cycle or even a combined cycle can be used as the power cycle in a cogeneration plant. The schematic diagram of an ideal steam-turbine cogeneration plant is shown in Figure 5.9. Let us say that this plant is to supply process heat Q_p at 500 kPa at a rate of 100 kW. To meet this requirement, steam is expanded in the turbine to a pressure of 500 kPa and producing electrical power at a rate of, say 20 kW. The flow rate of the steam can be adjusted such that steam leaves the process heating section as a saturated liquid at 500 kPa. Steam is then pumped to the boiler pressure. It is heated in the boiler to state 3. The pump work is usually very small and can be neglected. Disregarding any heat losses, the rate of heat input in the boiler is determined from an energy balance to be 120 kW.

It is seen that we would need to supply heat to the steam in the boiler at a rate of 100 kW instead of at 120 kW without the turbine. The additional 20 kW of heat supplied is converted to work. Therefore, a cogeneration power plant is equivalent to a process-heating plant combined with a power plant. It has a thermal efficiency of 100 percent.

The ideal steam-turbine cogeneration plant is not practical due to the variations in power and process-heat loads. The schematic diagram of a more practical cogeneration plant is shown in Figure 5.10. Under normal operation, some steam is extracted from the turbine at some predetermined intermediate pressure P6. The rest of the steam expands to the condenser pressure P7. It is then cooled at constant pressure. The heat rejected from the condenser represents the waste heat for the cogeneration cycle.

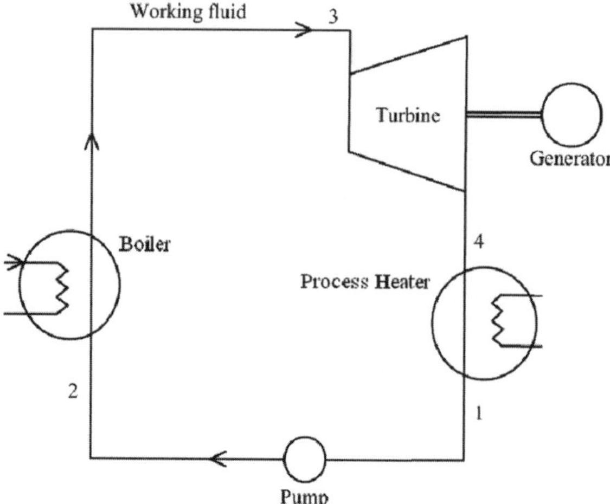

Figure 5.9 View of an ideal cogeneration plant.

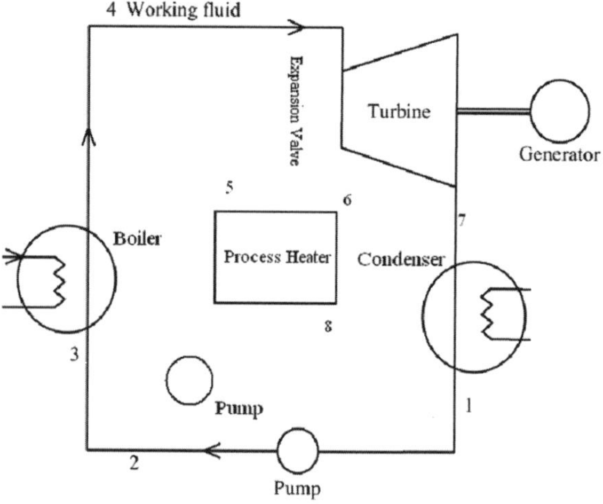

Figure 5.10 View of a cogeneration plant with adjustable loads.

At times of high demand for process heat, all the steam is routed to the process-heating units and none to the condenser ($\dot{m}_7 = 0$). The waste heat is zero in this mode. If this is not sufficient, some steam leaving the boiler is throttled by an expansion or pressure-reducing value (PRV) to the extraction pressure P_6. It is directed to the process-heating unit. Maximum process heating is realised when all the steam leaving the boiler passes through the PRV ($\dot{m}_5 = \dot{m}_4$). No power is produced in this mode. When there is no demand for process heat, all the steam passes through the turbine and the condenser $\dot{m}_5 = \dot{m}_6 = 0$. The cogeneration plant operates as an ordinary steam power plant.

The rates of heat input, heat rejected and process heat supply as well as the power produced for this cogeneration plant can be expressed as follows:

$$\dot{Q}_{\text{in}} = \dot{m}_3(h_4 - h_3)$$

$$\dot{Q}_{\text{out}} = \dot{m}_7(h_7 - h_1)$$

$$\dot{Q}_{in} = \dot{m}_3(h_4 - h_3)$$

$$\dot{Q}_{in} = \dot{m}_3(h_4 - h_3)$$

Under optimum conditions, a cogeneration plant simulates the ideal cogeneration plant. All the steam expands in the turbine to the extraction pressure and continues to the process-heating unit. No steam passes through the PRV or the condenser. Thus, no waste heat is rejected ($\dot{m}_4 = \dot{m}_6$ *and* $\dot{m}_5 = \dot{m}_7 = 0$). This condition may be difficult to achieve in practice due to the constant variations in the process-heat and power loads. But the plant should be designed for the optimum operating conditions at most of the time.

Cogeneration plants have proved to be economically very attractive.

EXAMPLE 5.1

A stationary power plant operating on an ideal Brayton cycle has a pressure ratio of 8. The gas temperature is 300 K at the *T* (K) compressor inlet and 1300 K at the turbine inlet. Utilising the air-standard assumptions namely (i) steady operating conditions exist, (ii) the air-standard assumptions are applicable, (iii) kinetic and potential energy s changes are negligible and (iv) constant specific *T–s* diagram for the Brayton cycle heats are considered. Determine (a) the gas temperature at the exits of the compressor and the turbine (b) the back work ratio and (c) the thermal efficiency

Solution

The *T–s* diagram of the ideal Brayton cycle is shown in the figure below.

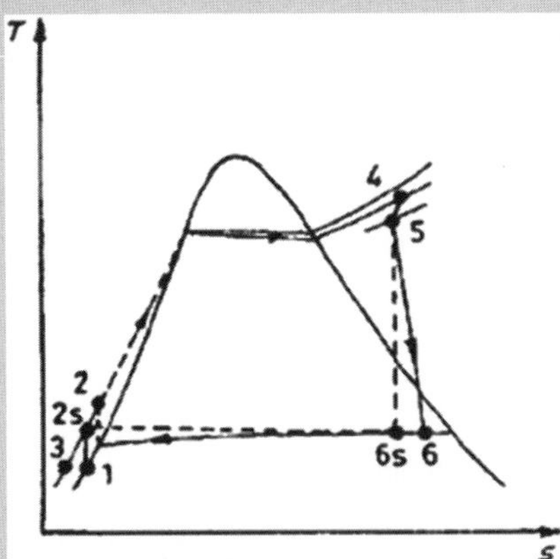

(a) The air temperatures at the compressor and turbine exits are determined from isentropic relations.
Process 1–2 (isentropic compression of an ideal gas)
At $T_1 = 300$ K, $h_1 = 300.19$ kJ/kg. (Appendix VIIIc)
Applying the isentropic process for an ideal gas from Eq. (5.5)

$$T_2/T_1 = (P_2/P_1)^{(k-1)/k}$$

Since $P_2/P_1 = 8$ (pressure ratio) and $k = 1.4$ (Eq. (5.6))
Hence,

$$T_2 = T_1\left(\frac{P_2}{P_1}\right)^{\frac{k-1}{k}} = 300(8)^{0.286} = 544K$$

Similarly at $T_2 = 544$ K, the corresponding value of enthalpy.
$h_2 = 544$ kJ/kg
and at $T_3 = 1300$ K, $h_3 = 1395.97$ kJ/kg
Since $P_2 = P_3$ and $P_1 = P_4$
Hence, from Eq. (5.5)

$$\left(\frac{P_3}{P_4}\right)^{\frac{k-1}{k}} = \frac{T_3}{T_4} = \left(\frac{P_2}{P_1}\right)^{\frac{k-1}{k}}$$

Also, we can write the above equation as

$$T_4 = T_3\left(\frac{P_1}{P_2}\right)^{\frac{k-1}{k}} = 1300\left(\frac{1}{8}\right)^{0.286} = 717K$$

Hence, the gas temperatures at the exits of compressor and turbine are 544 K and 717 K, respectively.

(b) To find the work input to the compressor and the work output of the turbine, we have at $h_4 = 731.58$ kJ/kg (Appendix VIII(c))

$$\omega_{\text{compressor,in}} = (h_2 - h_1) = 549 - 300.19 = 249.19kJ/kg \qquad \text{(Eq.5.15)}$$

and,

$$\omega_{\text{turbine,out}} = (h_3 - h_4) = 1395.97 - 731.58 = 664.39kJ/kg \qquad \text{(Eq.5.19)}$$

Thus, the back work ratio is given by

$$\frac{\omega_{\text{compressor,in}}}{\omega_{\text{turbine,out}}} = \frac{249.19}{664.39} = 0.375$$

(c) As per Eq. (5.6), the thermal efficiency of the cycle is

$$\eta_{\text{th,Brayton}} = 1 - \frac{1}{r_p^{(k-1)/k}}$$

where $r_p = (P_2/P_1) = 8$ (given)
Hence, the thermal efficiency can be obtained as

$$\eta_{\text{th,Brayton}} = 0.448$$

EXAMPLE 5.2

Compare the efficiency of the Stirling cycle with the Brayton cycle when both are working in the same temperature limits of 300 K and 1300 K, as mentioned in Example 5.1.

Solution

Both *P–v* and *T–s* diagrams for Stirling cycle are shown in the figure below

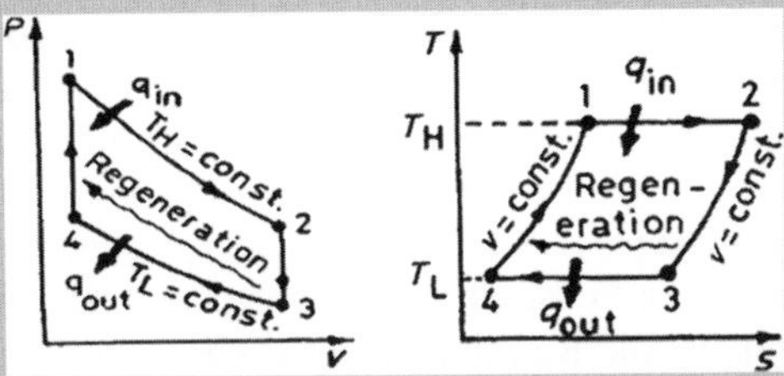

As per Eq. (5.13), the efficiency of the Stirling cycle is

$$\eta_{\text{Stirling}} = 1 - \frac{T_L}{T_H}$$

Given,
T_L = Lower operating temperature = 300 K
T_H = Upper operating temperature = 1300 K
Hence, thermal efficiency is given by

$$\eta_{\text{Stirling}} = 1 - \frac{300}{1300} = 0.769$$

Therefore, comparing Example 5.1, the Stirling cycle is more efficient than the Brayton cycle.

5.4 BIOMASS-BASED STEAM POWER PLANT

Combustion is the rapid oxidation of biomass fuel to obtain energy in the form of heat. Biomass fuels are composed primarily of carbon, hydrogen and oxygen and hence the main oxidation products are carbon dioxide and water. Depending on the heating value and moisture content of the biomass fuel, the amount of air used to burn the fuel and the construction of the furnace, flames can exceed 1500 °C.

Solid-fuel combustors can be categorised as

(a) Grate-Fired Systems

This was the first burner system to be developed. The most common system is the spreader-stoker, consisting of a fuel feeder that mechanically or pneumatically flings fuel onto a moving grate where the fuel burns.

(b) Suspension Burners

This suspends the fuel as a fine powder in a steam of vertically rising air. The fuel burns in a fireball. It radiates heat to tubes that contain water to be converted into steam. Suspension burners have dominated the power industry in many developed countries since World War II. This is because of high volumetric heat release rate and their ability to achieve combustion efficiencies often exceeding 90 percent. It is also known as a pulverised biomass boiler.

(c) Fluidised Bed Combustors

This is a recent innovation in boiler design. Air injected into the bottom of the boiler suspends a bed of sand or other granular refractory material. The turbulent mixture of air and sand is heated to high temperature. This allows a variety of fuels to be efficiently burned. The large thermal mass of the bed allows the unit to be operated at relatively low temperatures around 800 °C. This lowers the emission of nitrogen compound. Biomass is fed from a bunker through pulverisers designed to reduce fuel particle size enough to burn in suspension. The fuel particles are suspended in the primary air flow. It is fed to the furnace section of the boiler through burner ports where it burns as a rising fire ball. Secondary air injected into the boiler helps in completing the combustion process. Heat is absorbed by steam tubes arrayed in banks of heat exchangers (superheaters and economiser) before exiting through a bag house designed to capture ash released from the fuel. Steam produced in the boiler is a part of the Rankine power cycle.

The **mixtures of biomass and coals can be burned together in a process known as cofiring**. Cofiring offers several advantages for industrial boilers. Industries can use cofiring as an alternative to costly land filling of wastes that generate large quantities of biomass wastes such as pulp and paper industries, sulfur emission can be reduced from the boilers through cofiring with biomass. This contains much less sulfur than coal, as discussed earlier. Cofiring capability also provides fuel flexibility. It is important during times of unstable fuel pricing.

Alkali present in biomass fuel presents a difficult problem for direct combustion systems. Compounds of alkali metals (potassium and sodium salts) are common in rapidly growing plant tissues. Annual biomass crops contain large quantities of alkali. The old-growth parts of perennial biomass contain relatively small quantities of alkali. These alkali compounds appear as oxides in the residue left after combustion of volatiles and char. They would not be troublesome except that they act as a fluxing agent after allowing other minerals in the fuel to melt. These sticky compounds bind ash particles to fuel grates and heat exchanger surface. Boiler performance degrades as air flow and heat transfer are restricted by ash deposits. One solution to ash fouling is cofiring; the process by which biomass and coal are burned together. Some studies suggest that limiting biomass to 5–15 percent of the fuel requirement of a boiler can prevent ash fouling.

The best wood-fired power plants have heat rates exceeding 12 500 Btu/kWh. These are typically 20–100 MW in capacity. In contrast, large, coal-fired power plants have heat rates of only 10 250 Btu/kWh (1 Btu = 1.04 kJ).

5.5 BIOMASS-BASED COMBINED CYCLE POWER PLANT

A combined power cycle can be followed for generating electricity with the use of biomass fuel. Gas turbines generate mechanical power by expanding a stream of hot, high-pressure gas through an array of vanes attached to a rotating hub. In a conventional gas-turbine cycle, mixing compressing air with fuel and burning the mixture at high pressure produces the energetic gas stream. This approach works very well with fuels such as natural gas and kerosene. These are clean burning and do not generate particulate matter or corrosive vapours that can damage turbine vanes. Gas-turbine cycles are increasingly attractive for electric power generation. This is due to the relative ease of plant construction, cost effectiveness in a wide range of sizes (from tens of kilowatts to hundreds of megawatts). It has potential for very high thermodynamic efficiencies when employed in advanced cycles.

Gas turbines cannot be directly fired with biomass due to the ash particles and alkali. These are released from the burning biomass and would damage turbine blades. The biomass must first be converted to clean-burning fuel of producer gas before it can be used in gas turbines. Economics currently favour producer gas as a biomass-derived fuel for gas turbines.

This is an integrated gasification/combined cycle (IGCC). Air is compressed and enters an oxygen plant. It separates oxygen from the air. The oxygen is used to gasify biomass in a pressurised gasifier to produce medium-heating-value producer gas. The producer gas passes through cyclones and a gas clean-up system to remove particulate matter, tar and other contaminants. It may adversely affect gas-turbine performance (alkali and chloride most prominent among these). These clean-up operations are best performed at high temperature and pressure to achieve high cycle efficiency. The clean gas is then burned in air and it is expanded through a gas turbine operating in a topping cycle. The gas exits the turbine at temperatures ranging between 400 °C and 600 °C. A heat-recovery steam generator produces steam for a bottoming cycle that employs a steam turbine. Electric power is produced at two locations in this plant. This gives thermodynamic efficiencies approaching around 47 percent.

Integrated gasifier/combined cycle (IGCC) systems based on gas turbines are attractive for several reasons. These include (i) their relative commercial readiness, (ii) the ability to construct small generating capacity units and (iii) the expectation that they can generate electricity at the lowest cost of all possible biomass power options. An alternative to IGCC is to generate steam for injection into the gas turbine combustor to increase mass flow and power output from the turbine. This variation is less capital intensive than IGCC. It is referred to as a steam-injected gas-turbine (STIG) cycle, which employs a steam turbine. The STIG cycle is commercially developed for natural gas. It has lower flammability limits for producer gas. It makes the steam injection more problematic for biomass-derived producer gas.

5.6 BIOMASS-BASED COGENERATION PLANT

Cogeneration is the simultaneous production of process heat and electric power using a single fuel. Biomass fuel can also be used in cogeneration plants for enhancing their efficiency. Biomass combustion facilities produce electricity from steam driven turbine generators. It has a conversion efficiency of nearly 17 to 25 percent. Using a boiler to produce both heat and electricity (cogeneration) improves overall system efficiency to 85 percent.

Agro industries where cogeneration potential is possible are sugarcane, rice mills, paper and pulp, plywood industries, dairies, *etc*. Cogeneration is most popular in sugarcane industries. It is an environmentally friendly way of producing electricity using sugarcane bagasse. Similarly, rice husk in rice mills is the most promising and convenient method of generating electricity. In sugarcane industries, reasonably large quantities of steam at low pressure and temperature are required. This is apart from the electrical and mechanical energy required to drive various equipments. In such cases, it is advantageous to raise steam at higher pressure. It uses the extra energy (in the form of pressure) to drive generators for production of electricity. Such energy produced is identical to the systems. It requires only marginal quantities (about 20%) of additional fuel. Cogeneration facilities increase the economic viability and profitability of an industry.

Husk producing in rice mills is the outer cover of rice. It accounts for 20% of the paddy produced on weight basis. The main characteristics of rice husk are; 16.3 MJ/kg heating value, a content of 74% volatile matter and 12.8% ash. These characteristics indicate that rice husk is a good fuel for generating either producer gas or steam by combustion. The steam generated by a husk-fired boiler can be partly utilised for steam turbines in generating electricity. The rest can be used as process heat for drying of paddy. Drying of paddy is one of the most important postharvest operations in rice mills for safe preservation and quality retention of the paddy. A rice-husk-fired cogeneration plant is therefore of great importance in rice-mill industries.

Cogeneration projects based on agro waste, like rice husk, bagasse, *etc*., as fuel result in (i) lowering the cost of energy generation, (ii) low capital investment, (iii) higher profitability of plant due to substantial reduction in the cost of production and (iv) enhanced productivity and less consumption of costly and scarce fuels like diesel oil.

Traditionally, cogeneration is adopted as a means of utilising bagasse in a useful way by generating steam in boilers for process heating and electricity generation. This is regarded as the clever way of converting waste into useful energy. In the sugarcane industry, it is required to produce both the steam and the electricity for driving the sugar processes. Bagasse is the waste of the industry and can be the best fuel for utilising in steam as well as electricity generation in a cogeneration cycle. The topping cycle is used to generate high-pressure and high temperature steam. Electricity has found wide usage in the sugarcane industry. When the sugar has been extracted, the remaining fibrous residue after dewatering (bagasse) is conveyed to the boilers for burning as fuel or storage purposes. Its moisture content is around 45–55%. The sugar processes use low-pressure steam from the turbine used in cogenerations plant. The schematic flow diagram of a bagasse-based cogeneration plant is shown in Figure 5.11.

Bagasse power development has advantages: it is environmentally friendly, uses renewable energy and it can also encourage the use of sugar trash. Leaving sugar trash to rot and to provide compost manure has some environmental problems due to release of methane gas, which is a greenhouse gas that contributes to global warming. For meeting the power requirement as well as process heat, bagasse is the right alternative in cogeneration plants. Similarly for process heat in rice mills, rice husk is also an essential byproduct of the industry concerned. Hence, for agro industries' power requirement, cogeneration is a suitable, feasible and economical alternative.

5.7 FUEL CELL CYCLE

Fuel cells convert chemical energy directly into electricity without combustion. Actually, fuel cells operate like batteries. But unlike batteries, fuel cells do not run down or require recharging. They produce electricity as long as the fuel is supplied. They are basically devices that convert the chemical energy of a fuel directly and very efficiently into electricity, without combustion. The advantages include: no transmission of fuel cells, they produce no emissions and distribution losses, they are made up of very compact systems, and refuelling in the system is very easy. However, fuel cells are very costly, and there are no facilities for hydrogen storage in them. Fuel cells function on the principle of electrolytic charge exchange between a positively charged anode plate and a negatively charged cathode plate. When hydrogen is used as the basic fuel, reverse hydrolysis occurs, yielding only water and heat as byproducts. The operating principle of a fuel cell is shown in Figure 5.12.

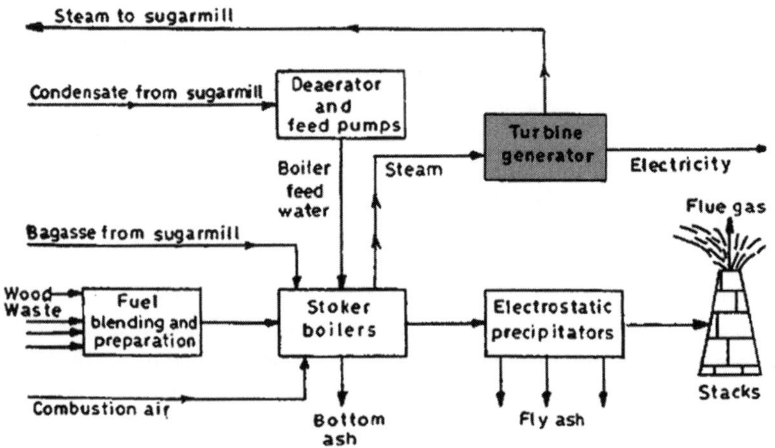

Figure 5.11 Schematic flow diagram of bagasse-based cogeneration plant.

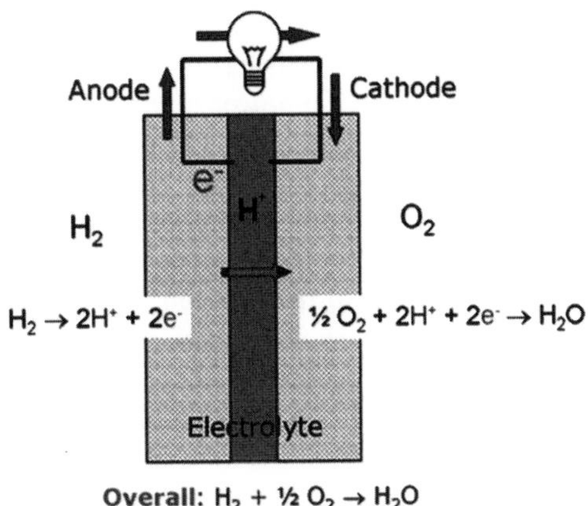

Figure 5.12 View of operating principle of fuel cell.

Several types of fuel cells have been developed or are under development. These fuel cells are classified according to the kind of electrolyte employed: phosphoric acid, polymeric, molten carbonate and solid oxide. Despite differences in materials and operating conditions, all these fuel cells are based on the electrochemical reaction of hydrogen and oxygen for biomass power applications. The molten carbonate and solid-oxide systems are of particular interest. These types of fuel cells operate at elevated temperatures. They present opportunities for heat recovery and integration into combined cycles.

The molten-carbonate fuel cells operate at about 650 °C. Although hydrogen is the ultimate energy carrier in the electrochemical reactions of this fuel cell, they have been designed to operate on a variety of hydrogen-rich fuels, including methane, diesel fuel, ethanol and producer gas. The fuel cell is a reformer that converts these fuels into mixtures of hydrogen, carbon monoxide, carbon dioxide and water along with varying amounts of unreformed fuel. A reformer is a vessel within which fuel and other gaseous recycle streams are reacted with water vapours and heat, in the presence of a catalyst to produce hydrogen-rich gas for use within the fuel cell. These decomposition reactions are strongly endothermic. A reformer also removes waste heat generated during the electrochemical oxidation of hydrogen in the fuel cell. The reformed fuel flows past the anode of the fuel cell. This absorbs hydrogen and releases water vapour and carbon dioxide. The exhaust gas stream, contains methane and hydrogen. This is mixed with air and burned external to the fuel cell to produce a hot mixture of oxygen, carbon dioxide and water vapour. This mixture passes over the cathode of the fuel cell to supply carbon dioxide and oxygen required for the proper operation of the system. The overall efficiency of converting methane into electricity is around 50 percent or a heat rate of 6850 Btu/kWh. Unlike many other power-conversion devices, fuel cells can maintain high efficiencies even when operated at half their rated capacities. Molten-carbonate fuel cells have completed demonstration trials and are close to market entry.

The solid-oxide system is the next-generation fuel cell. The electrolyte is a solid. It is a nonporous metal oxide usually based upon yttrium and zirconium. The fuel cell operates at 650–1000 °C, where ionic conduction by oxygen ions takes place. The solid electrolyte provides for a simpler, less expensive design and longer expected life than current fuel cell systems. The higher operating temperature compared to the molten-carbonate system also enhances its attractiveness for combined-cycle operation. However, solid-oxide systems are still in the early stages of development.

Obviously, fuel cells designed to oxidise hydrogen electrochemically require some modification before they can convert the chemical energy stored in solid biomass into electricity. Several options are possible for this. Anaerobic digestion produces a mixture of methane and carbon dioxide. This is ideal for carbonate fuel cells. Ethanol can also be reformed in carbonate fuel cells to a suitable gaseous fuel. However, biogas and ethanol are relatively expensive fuels compared to producer gas. Much of the interest in using biomass with fuel cells has focused on integrating them with gasifiers.

An integrated gasifier/combined cycle (IGCC) power plant is based on a molten-carbonate fuel cell. Biomass is gasified in oxygen to yield producer gas. Gasification occurs at elevated pressures to improve the yield of methane. This is important for proper thermal balance of the fuel cell. Hot gas clean-up to remove particulate matter, tar and other contaminants is followed by expansion through a gas turbine as part of a topping cycle. The pressure and temperature of the producer gas is sufficiently reduced after this to admit it into the fuel cell. High temperature exhaust gas exiting the cathode of a fuel cell enters a heat-recovery steam generator. This is also part of a bottoming cycle in the integrated plant. Thus, electricity is generated at three locations in the plant for an overall thermodynamic efficiency reaching 60 percent or more.

5.8 ENGINE POWER USING BIODIESEL

Increased environmental concerns and depletion of petroleum feedstocks have directed scientists to work on alternative fuel sources that burn more cleanly and are renewable in nature. Hence, the studies of vegetable oils are becoming more important among other investigations. The idea of using vegetable oils for diesel engines as fuel is not new. Rudolph Diesel used peanut oil to fuel his engine at the Paris Exposition of 1900. The short-term tests are almost always positive.[1] Long-term use of neat vegetable oils leads to severe engine problems, such as deposition, ring sticking and injector coking. High viscosity and a tendency for polymerisation within the cylinder appear to be at the root of many problems associated with direct use of these oils as fuels.[2]

Fatty esters are known as biodiesel. Biodiesels have shown suitability as alternative diesel fuels as a result of improved viscosity and volatility relative to the triglyceride (vegetable oil). Therefore, efficient transesterification plays a vital role for conversion of raw vegetable oils to biodiesel. The esterification process involves the transformation of the large, branched, triglyceride molecules of bio-oils and fats into smaller, straight-chain molecules, similar in size to components of diesel fuel. In the conventional transesterification of vegetable oils or animal fats, the reaction reactor is initially charged with vegetable oil or animal fat, methyl alcohol and a small amount of sodium hydroxide catalyst. The reaction mixture is then heated to the boiling temperature of methyl alcohol and refluxed for about an hour under agitation. The reaction is at atmospheric pressure. Methyl alcohol vapours condense at about 65–70 °C. A conversion of 90–95% is usually obtained to form a blend of bioiesel. With the stoppage of agitation, the reaction mixture separates into an upper layer of methyl esters and lower layer of glycerol diluted with unreacted methyl alcohol. The biodiesel product in the upper layer should be further neutralised and vacuum distilled for removal of excess alcohol before use as a fuel.

There is a resemblance between chemical and physical properties of biodiesel and diesel fuel. This has been documented by many workers.[3] Biodiesel's cetane number, energy content, viscosity and phase changes are similar to those of petroleum based diesel fuel (Table 5.1). Moreover, biodiesel is essentially sulfur free. Engines fueled by biodiesel emit significantly fewer particulates, hydrocarbons and less carbon monoxide. Emissions of NO_xs are also less in biodiesel than diesel fuel.

Biodiesel also offers enhanced safety characteristics in comparison to diesel fuel. It has a higher flash point. It does not produce explosive air/fuel vapours. Moreover, it is biodegradable and less toxic in comparison to petroleum fuel. Biodiesel emissions from an engine operating on these fuels are less toxic due to the absence of aromatic hydrocarbons. The esterified vegetable oils have a higher cloud point than diesel fuels. This requires engine and fuel heaters as a pure fuel in climates for temperature drops below 0 °C. However, for a 20–25% blend of the esterified oil and petroleum based diesel, no heater is required from about –10 to about –15 °C. Cloud-point suppressants may also be used to eliminate this problem.

Table 5.1 Physical and chemical specifications of the vegetable oil fuels used.

Fuel Type	*Calorific Value (kJ/kg)*	*Density (g/dm³)*	*Viscosity (27 °C)*	*mm²/s (75 °C)*	*Cetane number*	*Flame point (°C)*	*Chemical Formula*
Diesel fuel	43 350	815	4.3	1.5	47	58	$C_{16}H_{34}$
Raw sunflower (rsf) oil	39 525	918	58	15	37.1	220	$C_{57}H_{103}O_6$
Sun flower methyl ester	40 579	878	10	7.5	45–52	85	$C_{55}H_{105}O_6$
Raw cottonseed oil	39 648	912	50	16	48.1	210	$C_{55}H_{102}O_6$
Cottonseed methyl ester	40 580	874	11	7.2	45–52	70	$C_{54}H_{101}O_6$
Raw soybean oil	39 623	914	65	9	37.9	230	$C_{56}H_{102}O_6$
Soybean methyl ester	39 160	872	11	4.2	37	69	$C_{53}H_{101}O_6$
Corn oil	37 825	915	46	10.5	37.6	270–295	$C_{56}H_{103}O_6$
Opium poppy oil	38 920	921	56	13	—	—	$C_{57}H_{103}O_6$
Rapeseed oil	37 620	914	39.5	10.5	37.6	275–290	$C_{57}H_{105}O_6$

Even though the popularisation of biodiesel fuels has been implemented since World War II (WW II), their high cost is a major obstacle for commercialisation. Their current role in the market place is largely tied to environmental concerns. This may change if the crude glycerol coproduct from the transesterification process is utilised for production of derivatives of glycerol such as mono- and diglycerides. Glycerolysis of vegetable oils and animal fats produces a mixture of mono- and diglycerides. This is a valuable chemical intermediate. The production cost of biodiesel depends on the regional prices of biofuel, labour, land and processing plant cost, *etc*. The coproducts from the transesterification of vegetable oils may be the vital factor for lowering the economics of biodiesel industry. Of course, biodiesel oil is sold commercially in some European countries.[4] However, in other countries biodiesel's commercialisation is gaining popularity slowly due to the cost factor. Some of US demonstration programs are using biodiesel in more than 200 vehicles including buses, trucks, construction/maintenance equipments and motor boats, *etc*. In France, there are 2000 vehicles including buses and trucks running on rape seed methyl ester to see the effect of rape seed methyl ester on lubricating oil performance. Some countries have started reservation of biodiesel for emergency uses to reduce the dependency on imported fuels. Malaysia is producing 3000 Mt/year of palm-oil diesel. This is used in transit fleet, bus and cars.[5] Therefore, the demand of biodiesel movement to the market is fully dependent on the availability of diesel oil and its official price. After 5–10 years, the demand for biodiesel will be increased both in Europe and the US. Research and development programmes are also in full swing to produce more efficient biodiesel.

5.8.1 Blending of Biodiesel

Research has been concentrated on mixing (blending) of biodiesels with conventional diesel fuel at different proportions to make them energy efficient, environmentally friendly and cost effective. The Institut Francais du Petrole has developed and coordinated a test programme aiming at introduction of 95% diesel and 5% rapeseed methyl ester (RME) blends at the pumps without further mention of its composition in France.[6]

As a consequence, RME is now sold to the general public in filling stations blended with Diesel fuel in proportions up to 5%. The results of the testing procedures with RME have generally shown a slight reduction of power, a slight increase of fuel consumption, and a decrease of visible pollutant and particulate emissions. There is also a decrease of SO_2 emission and a slight increase of NO_x emissions. Similarly, the results of sunflower methyl ester (30%) blended with 70% diesel fuel have shown that (i) the blended fuel contributes to the reduction of CO specific emissions, (ii) to a slight increase of NO_x-specific emissions and (iii) to a reduction of the opacity of the exhaust gases.

Blumberg and Fort[7] have discussed the short and long term (200 h) engine performance. They have also done tests using different fuel samples, namely 2D diesel fuel; 30% cotton seed oil, 50%, 2D diesel fuel; 50% cotton seed oil, 35% 2D diesel fuel; 65% cotton seed oil, 50% 2D diesel fuel; 50% transesterified cotton seed oil and 100% cotton seed methyl ester. They observed that short-term results were more desirable than long term results. Long term tests showed carbon deposits, ash and wear in combustion chamber and sticky gum content in fuel line elements.

Schinstock *et al.*[8] have made 200-h engine performance tests using 25/75 blends (volumetric) of soybean and sunflower methyl ester with diesel fuel. They observed that the values of engine torque with the mixtures were greater than with pure diesel operation.

In general, it has been observed by most researchers that if raw vegetable oils are used directly or blended with diesel fuel, engine performance decreases and CO and HC emissions increase. NO_x emissions also decrease accordingly. But the use of biodiesel directly or blended with diesel fuel offers similar alternative fuel as compared to diesel fuel. The most important advantage of biodiesel is that they are produced from vegetable oils that are renewable energy sources. Vegetable oils show promise of providing all the liquid fuel needed on a typical farm by diverting 10% or less of the total acreage to fuel production. The extraction and processing of vegetable oils are simple low-energy processes. They make use of equipment not unlike that with which farmers are already familiar. Of the several methods available for producing biodiesel, transesterification of natural oils and fats is currently the method of choice. The purpose of the process is to lower the viscosity of the oil or fat. Although blending of oils and other solvents and microemulsions of vegetable oils lowers the viscosity, there are engine performance problems (carbon deposits and lubricating oil contamination). This minor problem may be subsequently solved with more research and choices of various biodiesels.

5.8.2 Engine Performance

A test bench that consisted of an electric dynamometer, a cooling tower, an exhaust gas aspiration system and engine-mounting elements is shown in Figure 5.13.[9] Diesel fuel and nine different vegetable oil fuels (raw sunflower oil, sunflower methyl ester, raw cottonseed oil, cottonseed methyl ester, raw soybean oil, soybean methyl ester, opium poppy oil, corn oil and rapeseed oil) are considered as fuels for the test. The physical and chemical specifications of the vegetable oil fuels are given in Table 5.1. They are usually heated before the fuel pump and before the injectors to minimise their resistance to flow due to the higher viscosity of raw vegetable oils. Two

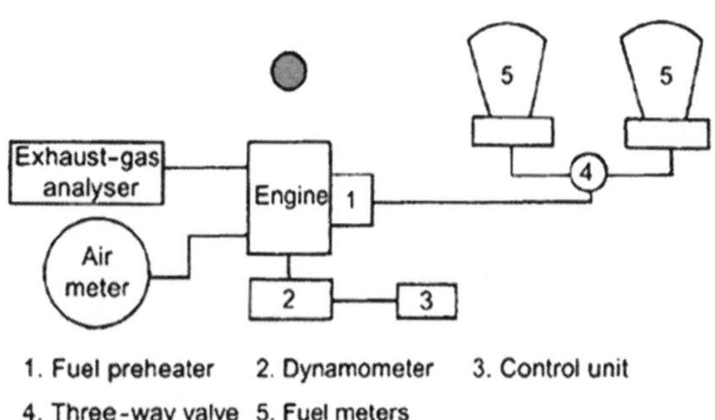

Figure 5.13 Schematic layout of the test system.

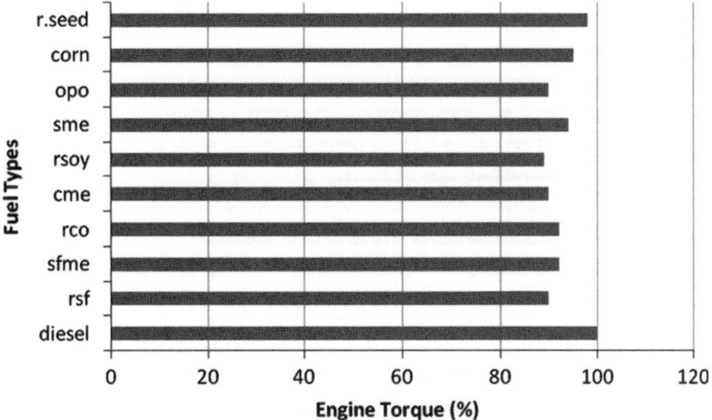

Figure 5.14 The variation of engine torque in relation with the fuel types (max engine torque at 1300 rpm).

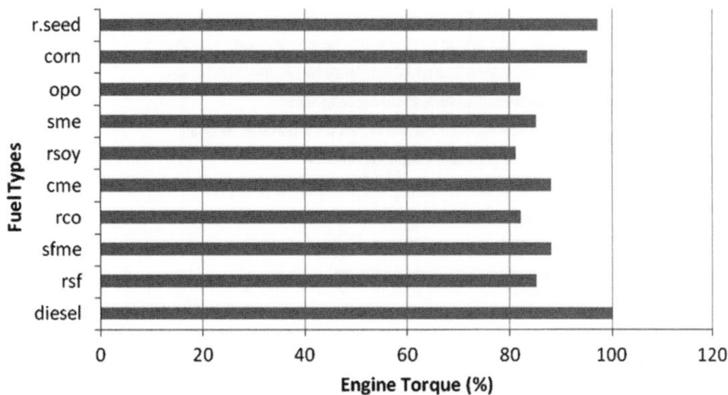

Figure 5.15 The variation of engine torque in relation with the fuel types (max engine torque at 1700 rpm).

thermostatically controlled electrical heaters are used for this purpose. The temperatures are maintained at about 80 °C. The engine was operated on diesel fuel first and then on vegetable oil and methyl esters. The engine was run for about 20 min at idle speed to replace the test fuel in the system with diesel fuel after the tests of raw vegetable oils.

The variations of maximum engine torque values for different fuels are shown in Figure 5.14.[9] The maximum torque with diesel fuel operation for the above test is 43.1 N m at 1300 rpm. For comparison, this torque is assumed to be 100% as the reference point. They also observed maximum torque values of the alternate fuels to be about 1300 rpm. This is less than the diesel fuel value for each fuel. The maximum torque differences between the reference value and the peak values of the vegetable oil fuels are about 10%. This is obtained with raw sunflower oil, raw soybean oil and opium poppy oil fuels. The minimum torque difference is about 3% between the reference value and the peak values of corn oil and rapeseed oil fuels. These results are due to the higher viscosity and lower heating values of vegetable oils.

The variations of maximum engine power values with different fuel types are given in Figure 5.15. The maximum power with diesel fuel operation is 7.45 kW at 1700 rpm. This power is also assumed to be 100% as the reference point. Observed maximum power values of the vegetable oil fuels are also at about 1700 rpm. But this is less than the diesel fuel value for each fuel. The maximum power differences between the reference value and the peak values of the vegetable oil fuels are about 18%. It is obtained with raw cottonseed oil and raw soybean oil fuels. The minimum power difference is

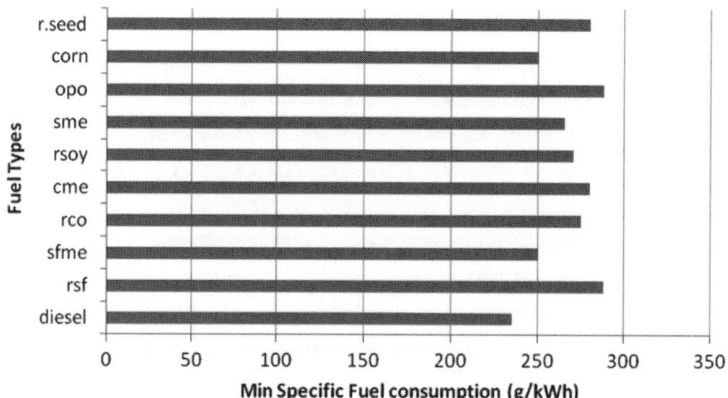

Figure 5.16 The variation of minimum specific fuel consumption in relation with the fuel types (at 1300 rpm).

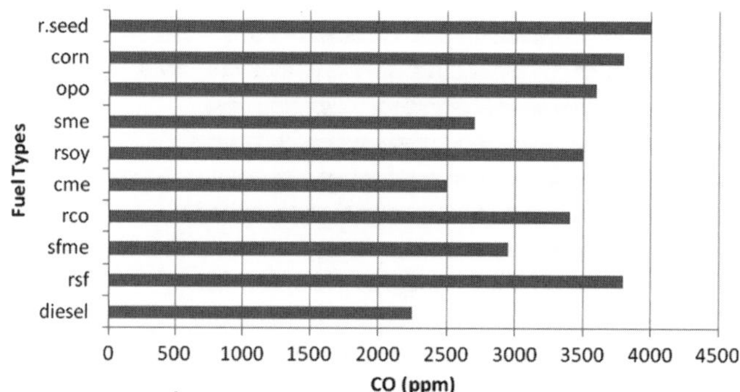

Figure 5.17 The variation of CO emissions in relation with the fuel types (at 1300 rpm 35 N m constant engine torque).

about 3% between the reference value and the maximum value of rapeseed oil fuels. These results are also due to the higher viscosity and lower heating values of vegetable oils.

Specific fuel consumption is defined as the consumption per unit of power in a unit of time. The obtained minimum specific fuel consumption values are in the vicinity of the maximum torque area. The minimum specific fuel consumptions values are 245 g/kWh with diesel fuel, 290 g/kWh with raw sunflower oil and 289 g/kWh with opium poppy oil at 1300 rpm (Figure 5.16). Specific fuel-consumption values of the methyl esters are generally less than those of the raw oil fuels.

The minimum CO emission is about 2225 ppm with diesel fuel (Figure 5.17). The maximum CO emissions are about 4000 ppm with rapeseed oil fuel, 3850 ppm with raw sunflower oil and 3800 ppm with corn oil fuel. Comparing raw vegetable oils, relatively lower CO emissions are obtained with the esters. This is due to the better spraying qualities and more uniform mixture preparation of methyl ester fuels. The maximum CO_2 emissions are about 10.5% with diesel fuel, 10.25% with raw sunflower oil, 10.4% with sunflower methyl ester and 10.2% with raw soybean oil fuel (Figure 5.18). This is again due to the better spraying qualities and more uniform mixture preparation of these vegetable oil fuels.

NO_x emissions are usually a result of the higher combustion temperatures (Figure 5.19). Maximum NO_2 emission for the above test is 2100 mg/m^3 with diesel fuel. Further, NO_2 emissions with vegetable oil fuels are lower than those with diesel fuel. NO_2 values of the methyl esters are higher than those of the raw oil fuels. The most significant factor that causes NO_2 formation is the maximum combustion temperature. The combustion efficiency and maximum combustion

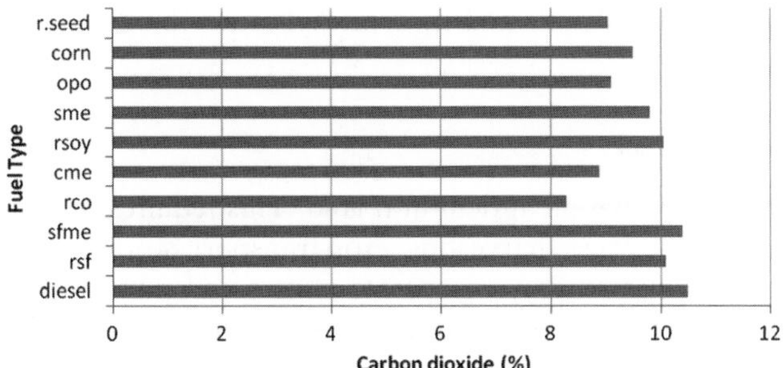

Figure 5.18 The variation of CO_2 emissions in relation with the engine speed (at 1300 rpm 35 N m constant engine torque).

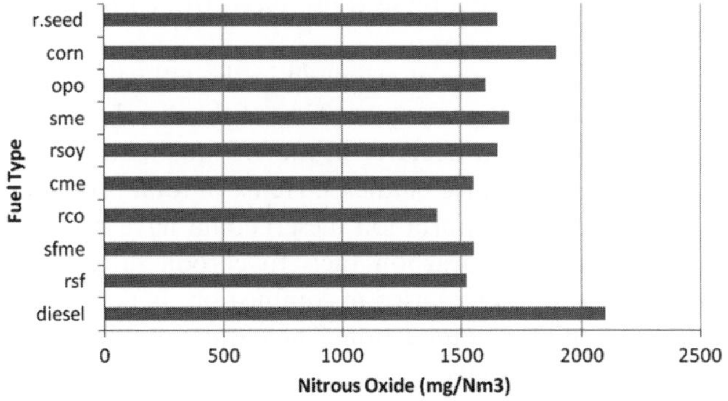

Figure 5.19 The variation of NO_2 emissions in relation with the fuel types (at 1300 rpm 35 N m constant engine torque).

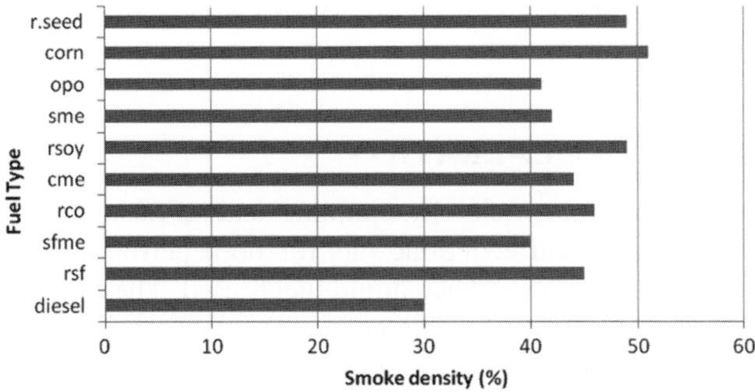

Figure 5.20 The variation of particulate emissions in relation with the fuel types (at 1300 rpm 35 N m constant engine torque).

temperatures with each of the vegetable oils are lower and hence NO_2 emissions are less. This happens due to the injection particles of the vegetable oils are greater than those of diesel fuel.

Smoke opacity percentages during each of the vegetable oil operations are greater than that of diesel fuel (Figure 5.20). The minimum smoke opacity is 29.3% with diesel fuel. The maximum smoke opacity values are 49% with soybean oil fuel and rapeseed oil fuel and 51% with corn oil fuel due to the heavier molecules of hydrocarbons. The capacity values of the methyl esters are between those of diesel fuel and raw oil fuels.

Hence, the uses of raw vegetable oils and their methyl esters (biodiesel) in the diesel engine are comparable with diesel fuel.

5.8.3 Production of Biodiesel

The vegetable oil plants are grown on agricultural land. This requires inputs of energy for tillage, chemical applications, machinery use, harvesting, crop transport, storage and labour. Processing of vegetable plants for oils requires energy to operate mixers and pumps, presses and labour. The vegetable plants for oils also require transportation to the point of use.[10] It is estimated that 4.2 units of energy are produced by rapeseed oil for every unit of energy input. The 3 to 4 units of energy output for each unit of energy input for biodiesel has been estimated by other researchers.

It is observed that the crop of rapeseed from 1 hectare of land produces 2700 kg seeds. From 1000 kg seeds, 930 kg moisture-free seeds are obtained. This contains 372 kg oil and 558 kg solid matter. Similarly, 1000 kg oil provides (adding 110 kg methanol) 1000 kg rape oil methyl ester and 110 kg glycerine after transesterification. Hence, rapeseed produced from 1 hectare of land provides 1005 kg oil or 1105 l of biodiesel.[11]

It is estimated that 27.5 million hectares of cropland in the US was idle.[11] Land could produce about 25.7 billion litres of vegetable oil per year or 27% of the diesel used in transportation. They reported that an additional 25.1 million hectares in the US was used for producing exports. The same land could have been used to produce biodiesel. Hence, a total production in the order of 45–54 billion liters (about 50% of the transportation diesel) could have been replaced by biodiesel.

Complete combustion converts hydrocarbon fuels to carbon dioxide and water. Diesel fuel releases 3.11 kg of CO_2 per kilogram of fuel used in combustion. In contrast, biodiesel releases 2.86 kg of CO_2 per kilogram of fuel used in combustion. The reduction of carbon dioxide due to use of this renewable source, biodiesel would have been in the range of 113–136 billion kilogram per year. This has been computed assuming a reduction in CO_2 of 3.11 kg for each kilogram of diesel fuel replaced by biodiesel. It also assumes that it requires 1.1 liters of biodiesel for each litre of diesel's replacement. Moreover, the carbon dioxide released by petroleum diesel fuel is deposited in the atmosphere, which causes global warning. The carbon dioxide (CO_2) released by biodiesel is fixed by plants in the recent or next cultivation. It is recycled by the next generation of crops mainly used for biodiesel productions.

5.9 BIOGAS FOR POWER GENERATION

Biogas can be used to produce electricity by coupling a dual-fuel engine to an asynchronous generator. Biogas is an excellent and economic fuel for both petrol and diesel engines.[12,13] The power obtained is less than that obtained when liquid fuel is used. The engines using biogas become hotter than those running on liquid fuels. The enhanced cooling is required in this case. Petrol engines can be run 100% on biogas except that a little petrol is consumed for starting up. However, diesel engines need to be modified to dual-fuel engines. They use both biogas and diesel oil. Biogas is introduced in the inlet pipe after it passes through the air filter. Gas-inlet devices are designed to suit different engine designs and inlet pipes in order to give a proper biogas/air mixture. Injection of a little diesel oil to ignite the gas mixture in each stroke is essential for normal running of the engine. This is due to the fact that in diesel engines the temperature at the end of the compression stroke is usually not above 700 °C. The ignition temperature of methane/air mixtures is 814 °C.

The 1 kWh of electricity can be generated from 0.75 m^3 of gas. It can light 25 electric bulbs of 40 W rating whereas 0.75 m^3 of gas if directly burnt can light only 7 biogas lamps for one hour. Hence, it is advantageous to first generate electricity to light a large number of electric bulbs. In China, asynchronous generators of 3-, 3.5- and 7.5-kW rating coupled to biogas-based IC engines are commonly used. However, electricity generation from biogas is economical only when gas is supplied by large community plants. It requires high initial capital investment. Decentralised power systems can be cost effective to minimise transmission and distribution costs.

Biogas is used to generate electric power as a decentralised source of energy in several countries. Biogas-fuelled engines can be used for water pumping, crop processing and power generation. The economics of a biogas-based decentralised energy system is dependent on capital costs of the system. Biogas-based power systems and biomass gasifiers are ideal to supply conventional power to remote villages, but this may still take many more years.[13]

A 5-kW biogas power generator set has been successfully tested in the Research and Development division of Bharat Heavy Electricals Ltd. (BHEL) Hyderabad, India. The estimated cost of biogas-based power was as Rs. 0.70 as against Rs. 170 per unit with a diesel generating set. This shows the cost effectiveness of biogas-based power.

Following are some of the major technical specifications of the BHEL-designed biogas power system.

Type: Single-cylinder 4-stroke, vertical, air-cooled engine directly coupled to a 6-kVA generator
Speed: 1500 rpm; speed control; by centrifugal governor.
Generator output: 5 kWh; 415 V; 3 phase
Biogas consumption: 5 m^3 h^{-1}
Starting: Self-starting DC motor and DC Generator with cutouts for starting and charging battery.

The Tulsi Shyam Temple at Gujrat, India has been using a 85 m^3 gas unit based on 300 cattle to run an engine. It drives a water pump and flour mill, in addition to generating 7.5 kVA of power for 4 h at night. The plant has been working since 1996. It was installed by KVIC (Khadi Village Industries Commission). The desirability of utilising biogas for power on the farm is an apparent choice in view of the high usage of energy for this purpose.[14]

A large sized plant in Italy produces biogas and electricity from animal wastes on a farm with 35 000 hens and 1000 pigs. Before fermentation, wastes are purified by aerated lagooning. Biogas produced by the plant is used without further treatment as fuel for boilers. It produces both electric power and hot water using a cogeneration unit, while during winters the energy needs of the farm match the energy supply. During summers, the gaseous energy from biogas becomes surplus, which is compressed and used as fuel for tractors and farm vehicles.[15] Tentcher[16] provided the technical details of the construction of a full-scale plug-flow pilot biogas plant and design of an associated engine-generator set for power generation. Wastes of about 800 pigs were used as feed material. The power generation was estimated to be 1.4 kWh m^{-3} of biogas.

5.9.1 Engine Power using Biogas

Biogas is a clean fuel to run internal combustion engines. Biogas can act as a promising alternative fuel. It can substitute for a considerable amount of diesel fuels. Diesel engines have higher thermal efficiency compared to other types of engines and hence they are more extensively used in rural areas.

The two most common sources of biogas are landfill gas and digester gas. Both are the byproducts of anaerobic decomposition of organic matter. They are primarily composed of methane (CH_4) and carbon dioxide (CO_2). Landfill gas is produced during the decomposition of organic matter in sanitary landfills. Digester gas is manufactured at sewage treatment plants during the treatment of municipal and industrial sewage. More general proportions are typically 60% methane and 40% (CO_2) for commercial operation with natural gas can also be used for biogas. The presence of carbon dioxide in the biogas reduces the burning velocity that ultimately affects the performance of the engine. The peak pressure inside the cylinder as well as the maximum power decreases due to the presence of carbon dioxide in the biogas.[17] Biogas containing more than 45% carbon dioxide causes harsh and irregular running of the engine.

The carbon dioxide (CO_2) content of biogas can, however, be reduced by methanation processes. These processes are usually expensive. This complicates the development of technology implementations. Decreasing carbon dioxide (CO_2) content will certainly improve the quality of biogas. This research needs essential to study how the engine performance varies with the carbon dioxide content in biogas.

Nearly 575 million tonnes of cattle dung is generated per annum in a country like India. Assuming 75 percent of it is available for biomethanation, an estimated 27 billion m^3 of gas can be available for use. If two-thirds of this gas is available for fuelling IC engines. 1 m^3 of gas can replace 1/2 l of diesel oil. An estimated 13 billion l of diesel can be saved annually. This is much more than the total diesel needed for operating irrigation pump-sets and multifarious agricultural activities at large.

The normal consumption rate of biogas for running IC engines is 0.45 to 0.54 m^3/hour per hp or 0.60 to 0.75 m^3/h per kW for power generation. Biogas pressure is found to vary from 25 to 100 mm water gauge. If an engine consumes 0.50 m^3 of biogas per hp per hour, then the quantity of gas needed for running a 10-hp engine for 10-h operation per day becomes 50 m^3. Similarly, if an engine consumes at a rate of 0.65 m^3/kW/h, then quantity of gas needed for running a 10-kW generator for 10-h operation each day becomes 65 m^3.

Dual-fuel engines offer the flexibility of easy switchover from dual fuel to pure diesel operation and *vice versa*. Important elements of a dual-fuel engine are air-filter, dual-fuel intake pipe, biogas choke and exhaust pipe. It is necessary to consider the relative costs of biogas and diesel for the economics of biogas use in dual-fuel operation. The extent of diesel replacement with biogas and consequent higher maintenance, lubricant needs and comparative thermal efficiency of the two systems should also be considered for economics. Thermal efficiency is governed by dual-fuel mix, engine output and speed. The dual-fuel engine is first started with diesel fuel only. After it has attained normal running for some time, a biogas choke is opened to admit gas into combustion chamber. Biogas admission can be controlled by adjusting the choke. To stop the dual-fuel engine, the biogas choke should be stopped first followed by the throttle. It should be endeavored to keep the engine speed uniform. The engine should be allowed to run idle for some time if the engine does not pick up speed either due to engine overload or due to a higher intake of biogas. This can be again put on load after its running becomes normal. The biogas choke should be adjusted during operation, if occasional sound or knocking is noticed. If it is on account of small loads on the engine, it can be corrected by suitably increasing the load on it. In a steady engine operation, dual-fuel use achieves nearly 80 percent saving in diesel consumption.

Dual-fuel operation is smoother if the compression ratio is kept low. Otherwise, it leads to low power output and gives rise to start-up problems. Diesel engines are generally set with an advanced injection angle. This makes them appropriate for running on dual fuel as well as diesel alone. Furthermore, biogas possesses reasonably good antidetonation properties. The original compression ratio and advance injection angle set for diesel operation should not be disturbed during running on dual-fuel. This also facilitates quick reversal to diesel fuel if some difficulties in running with dual fuel are encountered.

As the consumption of biogas in diesel engines is about 0.5 m^3 per hour per hp (0.65 m^3 gas per hour per kW electricity), large-size biogas units are required to run the engines. Kirloskar Oil Engines Ltd (KOEL) at Khadki, Pune, India took the lead in the past to manufacture dual-fuel engines utilising a mixed fuel of 80 percent biogas and 20 percent diesel. The company began with a 5-hp, 1500-rpm engine. It came to the market first in June 1977. Subsequently, engines of higher ratings up to 75 hp were developed with facilities of both water-cooled and air-cooled systems. The engine can quickly alternate between dual fuel and diesel according to the need and convenience. For operating a 5-hp engine, 2.25 m^3 of biogas is needed per hour. The KOEL dual-fuel engine consumes 0.45 m^3 of biogas per hp per hour. The engine can be easily coupled to an irrigation pump set. It is also used for operating farm machinery like a crusher, thresher, flour mill, power generating sets, *etc*. The density of exhaust smoke is found to be less with dual fuel than when diesel was used alone. The temperature of the exhaust gas is, however, found to be the same with no significant difference. As for engine deposits, it was found that the engine run on dual fuel was cleaner as compared to diesel.

The gaseous mixture of natural gas and carbon dioxide of different composition can be used in diesel engines to study the performance of engines with biogas containing varying percentages of carbon dioxide. Natural gas contains about 92 percent of methane as shown in Table 5.2. It plays the same role as methane in the biogas. Carbon dioxide is an incombustible gas. Its presence in biogas lowers the combustion enthalpy and the combustion rate of mixtures in the engine cylinder.

The engine was water cooled, naturally aspirated with a double-swirl combustion chamber. The flow charts of the technological processes used in the experiments are shown in Figure 5.21. The engine was operated under dual-fuel principles. Accordingly, the air-inlet system was modified in order to introduce the gaseous mixture with the air before being inducted inside the engine cylinder. At a certain percentage of diesel substitution by natural gas (15, 30, 50 and 75%), the pure carbon dioxide was introduced with the natural gas. The results are summarised as follows:

(i) The biogas supply to produce the same power increases with increased percentage of carbon dioxide in the biogas. The rate of increase of biogas at higher substitution is higher.[18]

Table 5.2 Natural gas composition.

Name	*Content (%)*
Methane	92.0408
Ethane	3.2406
Propane	0.654
Butane	0.0157
Pentane	0.0999
C_6^+	0.111
Nitrogen	3.0079
Carbon dioxide	0.587

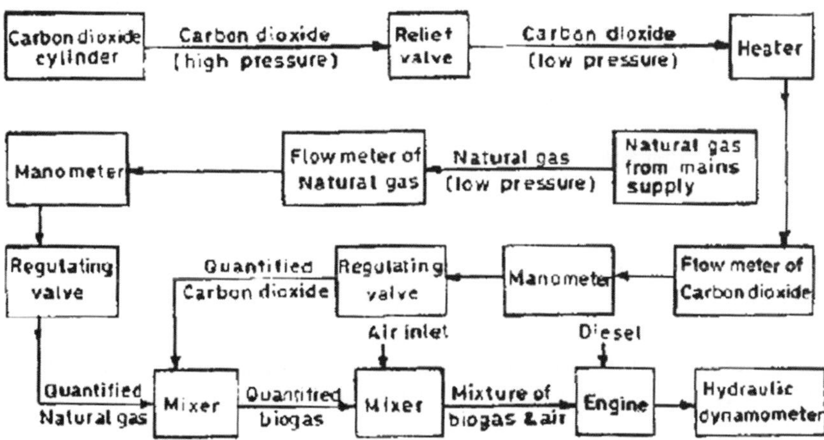

Figure 5.21 Flow charts of the dual-fuel diesel engine operated by biogas (at 1300 rpm 35 N m constant engine torque).

(ii) For the same power output the diesel supply rate also increases with the increase in the percentage of carbon dioxide in biogas. The rate of increase of diesel at lower substitution is higher.
(iii) Brake specific fuel consumption (bsfc) is higher at higher substitution of diesel.

It is found that the presence of as high as 40% carbon dioxide in the biogas does not deteriorate the engine performance in dual-fuel mode. Biogas containing more than 40% carbon dioxide needs scrubbing. The engine normally runs harshly with biogas containing high carbon dioxide (>40%). Also, biogas from different feedstock, causing different percentages of carbon dioxide, can be used as diesel substitute in diesel engine. This not only saves diesel fuels but also biogas is renewable in nature, thereby it is not a net contributors to the greenhouse gases.

Therefore, a diesel engine can be easily converted to a dual-fuel engine. It is the most practical and efficient method of utilising alternate fuels like biogas having higher spontaneous ignition temperature than diesel fuel. In the case of a shortfall in biogas supply during an important operation, the engine switches over smoothly without interruption to conventional diesel operation.

ADDITIONAL EXERCISE

Exercise 5.1: Consider a steam power plant operating on the simple ideal Rankine cycle by the use of biomass fuel. The steam enters the turbine at 3 MPa and 350 °C and it is condensed at a pressure of 75 kPa. Determine the thermal efficiency of this cycle for (i) steady operating conditions exist and (ii) kinetic- and potential-energy changes are negligible.

Solution

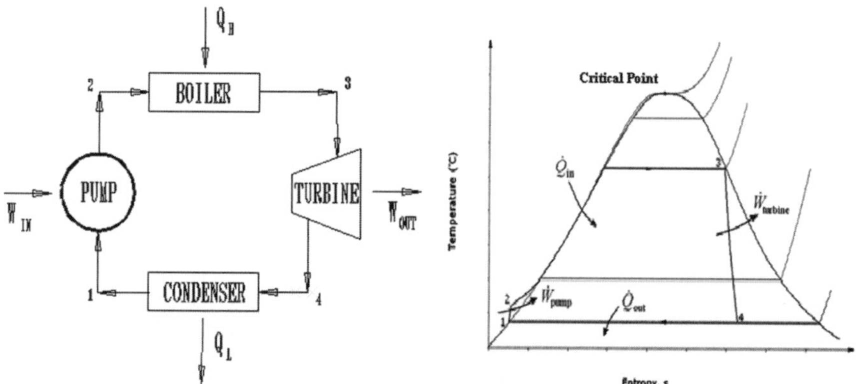

First, the enthalpies at various points in the cycle are determined using data from steam tables (saturated water temperature, saturated water pressure table and superheated water table Appendix V).

Stage 1: At saturated water $P_1 = 75$ kPa, $h_1 = h_f$@75 kPa = 384.39 kJ/kg and $V_1 = V_f$ h_f@75 kPa = 0.001037 m^3/kg (Appendix VIII(b))

Stage 2: Isentropic process $P_2 = 3$ MPa $= 3000$ kPa and $S_2 = s_1$

$$\omega_{\text{pump,in}} = v_1(P_2 - P_1) = \left(0.001037 \frac{\text{m}^3}{\text{kg}}\right)[(3000 - 75)\text{kPa}]\left(\frac{1\,\text{kJ}}{1\,\text{kPa} \cdot \text{m}^2}\right)$$

$$= 3.03\ \text{kJ/kg}$$

Stage 3: 2 – 3 process is at constant pressure condition. At $P_3 = 3$ MPa $= 3000$ kPa and $T_3 = 350\,°\text{C}$

$h_3 = 3115.3$ kJ/kg; $s_3 = 6.7428$ kJ/kg · K (Appendix VIII(c)).

Stage 4: Process 4 – 1 is at constant pressure condition (saturated mixture). At $P_4 = 75$ kPa $s_4 = s_3 = 6.7428$ (3 – 4 is at isentropic process), $s_f = 1.213$, $s_{fg} = 6.2464$ (Appendix VIII (c)). Now,

$$x_4(\textit{quality of mixture}) = \frac{s_4 - s_f}{s_{fg}} = \frac{6.7428 - 1.213}{6.2464} = 0.8857$$

At

$$x_4 = 0.8857,\ h_4 = h_f + x_4 h_{fg} = 384.39 + 0.8857(2278.6) = 2402.6 \frac{\text{kJ}}{\text{kg}}$$

$$q_{in} = h_3 - h_2 = (3115.3 - 387.42)\frac{\text{kJ}}{\text{kg}} = 2727.9 \frac{\text{kJ}}{\text{kg}} \qquad \text{(From Eq. 5.16)}$$

$$q_{out} = h_4 - h_1 = (2402.6 - 384.39)\frac{\text{kJ}}{\text{kg}} = 2018.2 \frac{\text{kJ}}{\text{kg}} \qquad \text{(From Eq. 5.18)}$$

$$\eta_{\text{th,Rankine}} = 1 - \frac{q_{out}}{q_{in}} = 1 - \frac{2018.2}{2727.9} = 0.260 - 26\% \qquad \text{(From Eq. 5.19)}$$

This indicates that the power plant converts 26 percent of the heat to net work. An actual power plant, operating between the same temperature and pressure limits will have a lower efficiency because of the irreversibilities such as friction.

OBJECTIVE QUESTIONS

5.1 A fuel cell converts
 (a) Chemical energy directly into electrical energy
 (b) Chemical energy indirectly into electrical energy
 (c) Mechanical energy directly into electrical energy
 (d) None of them

5.2 Biomass fuels contain
 (a) Concentrated energy (higher energy density)
 (b) Less energy density (low energy density)

(c) Both (a) and (b)
(d) None of them

5.3 Pyrolysis of biomass produces electricity
(a) Directly from liquid fuel (biocrude)
(b) Indirectly from liquid fuel (biocrude)
(c) Directly from diesel engine
(d) None of them

5.4 Biopower (biomass) is the electricity produced from
(a) Biogas
(b) Biodiesel
(c) Biomass
(d) All of them

5.5 Biopower generation is based on
(a) Gas turbine (Brayton) cycle
(b) Vapour turbine cycle
(c) Steam turbine (Rankine) cycle
(d) All of them

5.6 Biopower generation cycle is a
(a) Reversible cycle
(b) Irreversible cycle
(c) Both (a) & (b)
(d) None of them

5.7 An open-cycle and closed-cycle gas turbine is a
(a) Reversible cycle
(b) Irreversible cycle
(c) Both (a) & (b)
(d) None of them

5.8 The value of specific heat ratio $\left(k = \frac{C_p}{C_v}\right)$ is always
(a) <1
(b) >1
(c) $=1$
(d) None

5.9 The pressure depends on temperature in all cycles
(a) Linearly
(b) Nonlinearly
(c) Exponentially
(d) All of them

5.10 The overall efficiency of bioengine working on power cycle is
(a) <1
(b) >1
(c) $=1$
(d) None

5.11 Pressure ratio $\left(r_p = \frac{P_2}{P_1}\right)$ of a bioengine operating on the Brayton cycle is
(a) <1
(b) >1
(c) $=1$
(d) None

5.12 The efficiency of carbine cycle is
(a) $>$ gas turbine
(b) $>$ steam turbine

(c) Both (a) & (b)
(d) None of them

5.13 In an adiabatic process, there is
(a) No heat transfer
(b) No temperature variation (constant temperature)
(c) Both (a) and (b)
(d) None of them

5.14 In an isothermal process, there is
(a) No heat transfer
(b) No temperature variation (constant temperature)
(c) Both (a) and (b)
(d) None of them

5.15 The efficiency of a binary/combined cycle is higher than
(a) Brayton cycle
(b) Rankine cycle
(c) Carnot cycle
(d) All of them

5.16 During an isothermal process, there is a change in
(a) Entropy
(b) Enthalpy
(c) Both (a) and (b)
(d) None of them

5.17 During an adiabatic process, there is a change in
(a) Entropy
(b) Enthalpy
(c) Both (a) and (b)
(d) None of them

5.18 For a change in entropy (s), there is
(a) Variation in temperature
(b) Constant temperature
(c) Constant energy
(d) None of them

5.19 The working of Brayton and Stirling cycles depends on
(a) T–s diagram
(b) P–v diagram
(c) Both (a) and (b)
(d) None of them

5.20 The working of the Rankine and combined gas–vapour power cycle depends on
(a) T–s diagram
(b) P–v diagram
(c) Both (a) and (b)
(d) None of them

5.21 The Stirling cycle is
(a) $>$ Brayton cycle
(b) $<$ Brayton cycle
(c) $=$ Brayton
(d) None of them

5.22 The pressure ratio ($r = P_2/P_1$) of a bioengine operating on the Stirling cycle is
(a) < 1
(b) > 1

(c) $=1$
(d) None of them

5.23 The operation of pump and turbine on an ideal Rankine cycle is
(a) Reversible
(b) Irreversible
(c) Both (a) and (b)
(d) None of them

5.24 Isentropic flow is
(a) Adiabatic
(b) Reversible
(c) Both (a) and (b)
(d) None of them

5.25 The torque and power of an engine based on most biodiesel fuels is about
(a) 50%
(b) 80–100%
(c) 25%
(d) None of them

5.26 The variation of CO_2 emission from most biodiesel engines varies between
(a) 8–10%
(b) 50%
(c) 25%
(d) None of them

5.27 Which of the following process densifies biomass?
(a) Anaerobic digestion
(b) Aerobic digestion
(c) Incineration
(d) Briquetting

5.28 Jatropha, jojoba and karanj bear oil seeds that can be used to produce
(a) Petrol
(b) Biodiesel
(c) Natural gas
(d) Aviation fuel

5.29 Ethanol is produced from
(a) Jatropha
(b) Sugarcane
(c) Karanj
(d) None of them

5.30 Which of the following processes is not used for conversion of biomass?
(a) Direct combustion
(b) Thermochemical conversion
(c) Biochemical conversion
(d) Electromagnetic conversion

5.31 Biomass can be used as fuel through
(a) Combustion
(b) Fermentation
(c) Digestion
(d) All of them

5.32 The biomass can be converted into
(a) Solid fuel
(b) Liquid fuel

(c) Gaseous fuel
(d) All of them

5.33 Which of the following is not a biomass?
(a) Plants and trees
(b) Wood
(c) Cow dung
(d) Water

5.34 Which of the following is a biogas plant model of low cost?
(a) Floating-dome-type biogas plant
(b) Fixed-dome-type plant
(c) Ferrocement-type biogas plant
(d) All of them

ANSWERS

5.1 **(a)**; 5.2 **(b)**; 5.3 **(b)**; 5.4 **(d)**; 5.5 **(d)**; 5.6 **(c)**; 5.7 **(a)**; 5.8 **(b)**; 5.9 **(c)**; 5.10 **(a)**; 5.11 **(b)**; 5.12 **(c)**; 5.13 **(a)**; 5.14 **(b)**; 5.15 **(d)**; 5.16 **(a)**; 5.17 **(b)**; 5.18 **(a)**; 5.19 **(c)**; 5.20 **(a)**; 5.21 **(b)**; 5.22 **(c)**; 5.23 **(b)**; 5.24 **(c)**; 5.25 **(b)**; 5.26 **(a)**; 5.27 **(d)**; 5.28 **(b)**; 5.29 **(b)**; 5.30 **(d)**; 5.31 **(d)**; 5.32 **(d)**; 5.33 **(d)**; 5.34 **(b)**.

REFERENCES

1. Mazel, M. A., Summers, J. D. and Batchelder, D. G. 1985. *Trans. ASAE* **28**(s), 1375–1384.
2. Peterson, G. L. 1986. *Trans. ASAE* **29**(5), 1413–1422.
3. Mittelbach M. and Tritthart P. 1988. *JAOCS* **65**(7), 1185–1187.
4. Kalam, M. A. and Masjuki, H. H. 2002. *Biomass Bioenergy* **23**, 47 1479.
5. Abdullah, A. and Basri, M. N. H. 2002. *Selected Readings on Palm Oil and its Uses.* PORIM, Malaysia, p. 25–176.
6. Silva Fernando Neto da, Prata Antonio Salgado and Teixeira Jorge Rocha 2003. *Energy Conver Manag.* **44**, 2857–2878.
7. Blumberg, P. N. and Fort, E. F. 1982. Performance and durability of a turbocharged diesel fueled with cottonseed oil blends. Vegetable oil fuels proceedings of the International Conference on plant and Vegetable Oils as Fuels, ASAE.
8. Schinstock J. L., Hanna M. A. and Schliek M. L., 1988. *Biomass.* **31**, 5–15.
9. Altin Recep, Cetinkaya Selim and Yucesu Huseyin Serdar 2001. *Energy Conver. Manag.* **42**, 529–538.
10. Auld, D. and Peterson, C. 1989. Biodiesel as a means of reducing U.S. dependence on imported oil. Brochure, College of Agriculture, Moscow, Idaho 83844.
11. Peterson, G. L., Wagner, G. L. and Auld, D. L. 1983. *Trans. ASAE* **26**(2), 322–327.
12. Tiwari, G. N. and Chandra, A. 1986. *Energy Conver. Manag.* **26**, 117–150.
13. Usmani, J. A., Tiwari, G. N. and Chandra, A. 1996. *Energy Conver. Manag.* **9**, 1423–1433.
14. Singhal, O. P. 1987. *Urja* **21**(2), 125–128.
15. La Torre, C. 1987. *Condiz. dell Aria* **31**(5), 603–60g [In Italian].
16. Tentcher, W. 1987. *RERIC News* **10**(2), 1–3.
17. Jawurek, H. H., Lane, N. W. and Rallis, C. J. 1990. *Biomass.* **13**, 87–103.
18. WREC, 1996. *World Renewable Energy Congress Bulletin*, p. 153.

The Harnessing of Hydropower from Stored Potential Energy in Water in Mountain/Hills is Renewable, Clean, Sustainable and Environmentally Friendly.

CHAPTER 6

Hydropower

6.1 BACKGROUND

Hydropower is one of the most established renewable sources for electricity generation from stored water at a given height. It is also known as water power. Kinetic energy in falling water from a height is converted into mechanical energy by a turbine and then electrical energy by a generator to meet the energy needs for a variety of tasks. Thus, the power is known as hydroelectric power. The other sources of renewable energy that are utilised for rotation of turbine are wind, geothermal and bioenergy.

The idea of utilising hydraulic energy to develop mechanical energy has prevailed for more than 2000 years. The amount of electrical energy generated from a water source depends on two aspects, namely (i) the water to fall from a height and (ii) the quantity of water flowing. In water turbine, blades are attached to the shaft and when flowing water passes against the blades of a turbine, the shaft rotates. After giving up its energy to the turbine, water is discharged through the drainage pipes or channel called the tailrace of the hydropower plant for irrigation or water-use purposes. The coupling of a generator with a turbine shaft finally produces electrical energy. A general layout of a hydroelectric plant is shown in Figure 6.1. Therefore, two main factors, *i.e.* the amount of water flow per unit time and the vertical fall of water, are very important for determination of the generating potential of any hydroelectric power station. Vertical fall (or head) of water may be natural due to the topographical situation or may be created artificially by means of dams. Once developed, it remains fairly constant. Water flow on the other hand is a direct result of the intensity, distribution and duration of rainfall. Also, hydropower potential can be estimated with the help of river flows around the world. Hence, one of the essential components of the hydraulic power generation is the availability of a continuous source of water with a large amount of hydraulic energy. If available, water energy from river, stream, canal system or reservoir is utilised properly, then the efficiency of hydropower plant would be more than or even double the conventional thermal power plant. This is due to the fact that a volume of water that is allowed to flow through a vertical distance represents the kinetic energy that can be easily converted to mechanical energy needed for generating electricity as compared to the caloric energy produced in a thermal power plant. Equipments associated with hydropower are well developed, relatively simple and very reliable. As no heat is involved during hydroelectric power generation, the equipments work for a longer period of about 40 years or more. In general, the main component of a hydropower system is the water turbine that consists of a wheel called runner (or rotor) having a number of specially

Advanced Renewable Energy Sources
G. N. Tiwari and R. K. Mishra

Published by the Royal Society of Chemistry, www.rsc.org

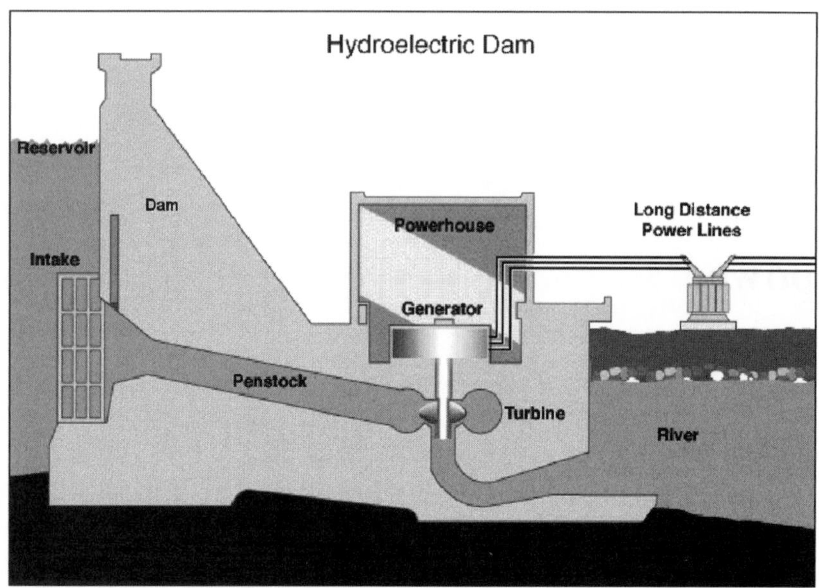

Figure 6.1 View of hydroelectric power plant.

designed vanes or blades. This possesses a large amount of hydraulic energy when it strikes the runner, it does work on the runner and causes it to rotate. The mechanical energy so developed is supplied to the generator coupled to the runner for generating electrical energy. Thus, the fuel for hydropower is water which is itself renewable and is not consumed in the electricity generating process. But the efficiency of the system depends on the collection and the management of water sources.

Falling water as a source of kinetic energy was known in ancient times. The Greeks used hydropower to turn waterwheels for grinding of wheat in a flourmill. The hydraulic energy was first produced in Asia (China and India) in the form of mechanical energy by passing water through a water wheel. It was reported that the water wheel was used in Europe 600 years after its origin in India. The actual design of a water wheel was first developed by Leonardo Da Vinci (1452 to 1519 AD), the great Italian artist. The practical experiments for the use of water wheels were carried out by Smearrn and Bossut in the year 1759. A French scientist Jean Victor Poncelet (1788 to 1876) first designed the water wheel on the basis of the theory given by Bolda in 1766. It was mostly manufactured in England in 1828. A Swiss scientist, Daniel Bernoulli, wrote the theory for the conversion of water power into other forms of energy in his book "Hydrodynamica". Then, Leonard Euler from Switzerland wrote the theory of hydraulic machines in 1750. It is used as a fundamental book in the subject of hydraulic machine. In 1824, a French scientist named Burdin designed a radial water wheel named a water turbine. This turbine was further developed by his students Fourneyron in 1827, which was the first water turbine. Towards the end of that century, many mills were replaced by water wheels with turbines for mechanical power. To produce electrical energy with the help of water turbines coupling with the generator became easier and more widespread with the invention of the water turbine in the end of the 18th century.

Dams and hydropower stations were built in Europe and North America at a rapid rate, exploiting up to 50% of the technically available potential. The first hydropower station in India was a small hydropower station of 130 kW commissioned in 1897 at Darjeeling in West Bengal. With the advancement in technologies and increasing requirement of electricity, attention was shifted to large-sized hydropower stations. In 1963, hydropower had attained a share of 50% in total installed capacity of power generation in India. But in recent years with the depletion of fossil fuel, the situation is different. Governments, policy makers, funding agencies are again taking a growing interest in the area of hydropower development. There is now more than 105 000 MW of hydrocapacity under construction in the world.[1] It is a part of multipurpose developments such as irrigation water, industrial and drinking supply, flood control, improved navigation and so on.

Water turbines can be broadly classified as shown below:

(i) On the Basis of the Head and the Quantity of Water Required

(a) **High-head turbine:** High-head turbines are those that are capable of working under very high heads ranging from several hundred metres to a few thousand metres. These turbines thus require relatively less water. Impulse turbines are generally high-head turbines. In **impulse turbines**, the following water hits the turbine as a jet in an open environment with the power deriving from the kinetic energy of flow.

(b) **Medium-head turbine:** Medium-head turbines are those that are capable of working under medium heads ranging from about 60 m to 250 m. These turbines require a relatively large quantity of water. A Francis turbine may be classified as a medium-head turbine. In **reaction turbines**, the turbine is totally embedded in the fluid and is powered from the pressure drop across the device.

(c) **Low-head turbine:** Low-head turbines are those that are capable of working under heads of less than 60 m. These turbines thus require a large quantity of water. Kaplan and propeller turbines may be classified as low-head turbines, also a reaction turbine.

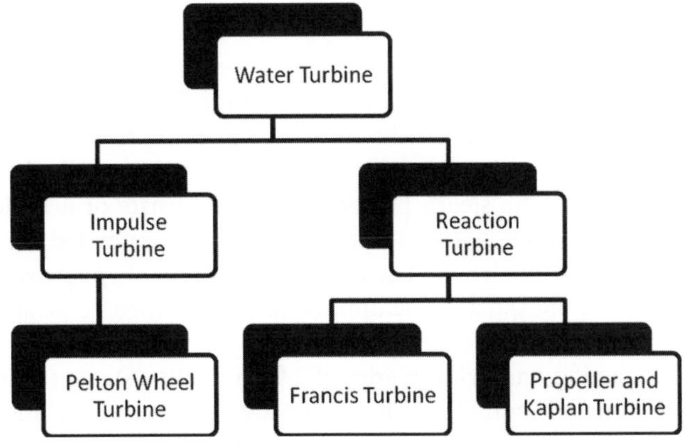

(ii) On the Basis of the Direction of Water in Turbine Blades

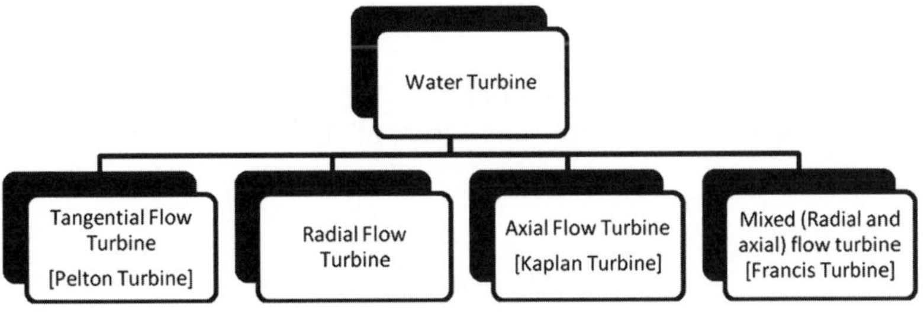

(a) **Tangential flow turbine:** Water flows along the tangent to the path of the rotation of turbine. A Pelton wheel (turbine) is a tangential flow turbine.
(b) **Radial flow turbine:** Water flows along the radial direction and remains wholly in the plane normal to the axis of rotation as it passes through the blades of turbine. Again, a radial flow turbine may be either an inward radial flow type or an outward radial flow type. In an inward radial flow turbine, water enters at the outer circumference and flows radially inwards towards the centres of the runner (the Francis turbine – named after James Bichens Francis (1815–1892) of England, is a reaction type of turbine). In an outward radial flow turbine, water enters at the centre and flows radially outwards towards the periphery of the runner (Fourneyron turbine).
(c) **Axial flow turbine:** The flow of water through the runner is mainly along the direction parallel to the axis of rotation of the runner. (Propeller turbines and Kaplan turbines, the latter was named in the honour of Dr. Victor Kaplan (1816–1934) of Germany. It is also a reaction type of turbine.)
(d) **Mixed-flow turbine:** Water enters radially and emerges out axially so the discharge is parallel to the axis of the shaft (modern Francis turbine).

(iii) According to Disposition of Turbine Shaft

The turbine shaft may be either vertical or horizontal. In modern turbine practice, a Pelton turbine has a horizontal shaft, whereas other turbines have vertical shafts.

6.2 WATER TURBINE THEORY

A turbine is a device to transfer energy from a continuously flowing fluid associated with steam, gas, air or water by the dynamic action of the moving blades on the runner. The word turbo or turbines is of Latin origin, and implies spins or whirls around. Hero built the first steam powered engine around 2000 years ago. The steam powered engine consists of a boiler, two hollow bent tubes mounted to a sphere. It is known as an aeolipile. The steam exited through the bent tubes (nozzles) caused the sphere to rotate freely. The resulting spinning of sphere was due to the reaction force exerted by the steam on the sphere. The application of the turbine principle has been used in many different types of machines such as windmills and water wheels. The power is produced from the wind or flowing water acting on a set of blades of a turbine. In a steam turbine, steam served the same purpose as the wind in a windmill and flowing water in a water turbine.

Let "m" be the mass of fluid moving with velocity v_j and let the change of velocity dv_j in time dt, then

$$\text{Change of momentum} = m \cdot dv_j$$

$$\text{and, rate of change of momentum} = m \cdot \frac{dv_j}{dt}$$

According to Newton's second law of motion

The dynamic force applied in the x-direction = Rate of change of momentum in the x-direction.

or,

$$F_x = m \cdot \frac{dv_j}{dt} \tag{6.1}$$

where suffix "x" denotes the components in the x-direction. The above equation is known as the **linear momentum equation**. It can also be written as

$$F_x \cdot dt = m \cdot dv_j \tag{6.2}$$

The left-hand term of this equation is the product of the force and the time increment during which it acts. This is known as the impulse of the applied force. Hence, Eq. (6.2) is known as impulse-momentum equation that states

"Impulse of dynamic force = resulting change in momentum of body"

6.2.1 Dynamic Force Exerted by Water Jet on Stationary Flat Plate

The following assumptions have been made:

(a) The plate is quite smooth and hence the friction between the water jet and the vertical flat plate is neglected.
(b) There is no energy loss in the flow due to the impact of the water jet.
(c) The difference in elevation between the incoming and outgoing water jet is neglected.

(i) Stationary Vertical Flat Plate Normal to Water Jet

Let a water jet with diameter "d" and velocity u_j from a nozzle strikes a vertical stationary flat plate as shown in Figure 6.3a.

Let Q be the volume of water striking the vertical plate per second (m^3/s) and ρ be the mass density of water (kg/m^3), u_j is the velocity of the jet of water (m/s). Then,

$$\text{mass of fluid per second (kg/s)} = au_j\rho = \rho Q \tag{6.3a}$$

Also,

$$Q = au_j$$

where "a" is the cross-sectional area of the water jet (m^2). The initial velocity of water jet in the x-direction before striking the vertical plate is u_j; and the final velocity of the water jet in the x-direction after striking the vertical plate is zero. By applying the impulse-momentum equation

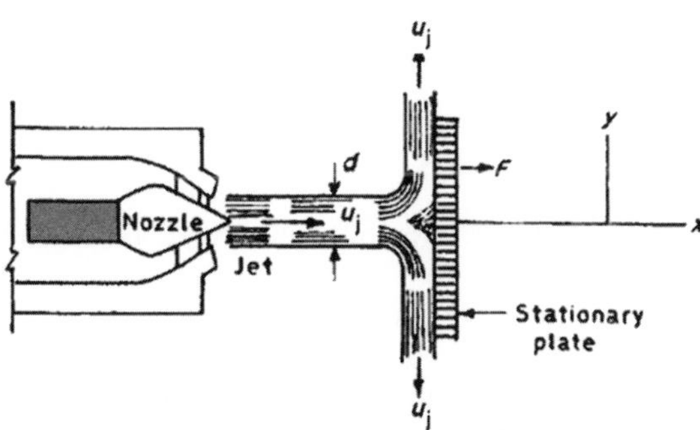

Figure 6.2a View of water jet striking a stationary vertical plate.

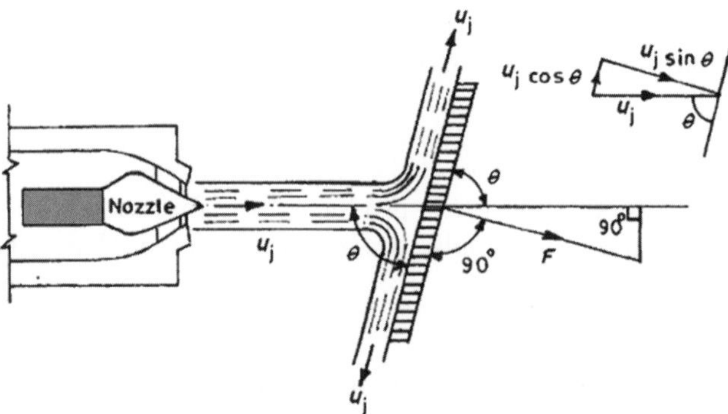

Figure 6.2b View of water jet striking on a stationary inclined plate.

Eq. (6.1), the force F exerted by the stationary plate on the water jet of fluid in the direction normal to the plate is given as

$$0 - F = \rho Q(0 - u_j)$$

or,

$$F = \rho Q u_j \tag{6.3b}$$

The sign for the force F has been considered negative because the force exerted by the vertical plate on the water jet is in the negative x-direction. But the force that the water jet exerts on the plate is equal and opposite to the force exerted by the vertical plate on the water jet. Hence, the force exerted by fluid on the plate is "F" acting in the positive x-direction and is given by Eq. (6.3b).

(ii) Stationary Flat Plate Inclined to Water Jet

As shown in Figure 6.2b, the stationary flat plate is held inclined at an angle θ to the direction of flow of the water jet. The dynamic force acting normal to the inclined plate is given by

F = mass of water striking the flat plate per second × change of velocity normal to inclined plate.

The component of the velocity (u_j) of the water jet before impact in the direction normal to the inclined plate $= u_j \sin\theta$.

The water jet, after impact, leaves the plate tangentially, as shown in Figure 6.2b, then the component of the velocity of the leaving water jet in the direction normal to the flat plate becomes zero. By applying the impulse-momentum equation, the thrust exerted by the water jet in the direction normal to the flat plate may be written as

$$F = \rho Q(u_j \sin\theta - 0) = \rho Q u_j \sin\theta \tag{6.4}$$

The component of this force F in the direction of the water jet, as shown in Figure 6.3b, is given by

$$F_x = F \sin\theta = \rho Q u_j \sin^2\theta \tag{6.4a}$$

The component F_y normal to the direction of water jet is given by

$$F_y = F \cos\theta = \rho Q u_j \sin\theta \cos\theta \tag{6.4b}$$

In both the cases, the work done by the force exerted by the water jet on the inclined flat plate is zero as the plate is held stationary.

6.2.2 Dynamic Force Exerted by Water Jet on Moving Flat Plate

(i) Striking Moving Vertical Flat Plate Normal to Water Jet

Let the vertical flat plate move with a velocity u_t= (m/s) in the same direction as the water jet as shown in Figure 6.3. In this case, the absolute velocity of water jet u_j will not be the effective velocity with which the jet strikes the plate. This is due to the fact that the water jet is about to strike the plate, the vertical flat plate has also moved away from the water jet with a velocity u_t. Only that mass of fluid that really overtakes the vertical flat plate will be striking it to cause the impinging action. In this situation, the problem is solved with the help of the principle of relative motion for the whole system. The effective velocity with which the jet strikes the vertical flat plate will be $(u_j - u_t)$. It is the velocity of the water jet relative to the plate. The quantity of fluid striking the vertical flat plate per second is

$$Q = \left(\frac{\pi d^2}{4}\right)(u_j - u_t) = a(u_j - u_t)(m^3/s)$$

The mass of water striking the vertical flat plate per unit time is given by

$$\dot{m} = \rho Q = \rho a(u_j - u_t)$$

If the assumptions taken in Section 6.4.1 are also applicable to this case, then the water jet leaving the moving vertical flat plate will have a velocity equal to $(u_j - u_t)$. Further, since the water jet after impact deflects through 90°, the component of velocity leaving the water jet in the direction normal to the plate is equal to zero.

By applying the impulse-momentum Eq. (6.1), the dynamic thrust exerted by the water jet on the vertical flat plate in the direction normal to the moving plate may be expressed as

$$F = \rho a(u_j - u_t)[(u_j - u_t) - 0] = \rho a(u_j - u_t)^2 \tag{6.5}$$

The work done by the water jet on the moving vertical flat plate per second is given by

$$P(u_t) = (F \times u_t) = \rho a(u_j - u_t)^2 u_t \tag{6.6}$$

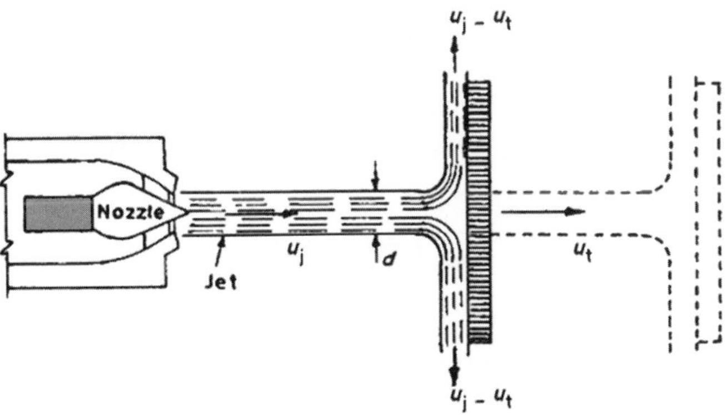

Figure 6.3 View of water jet striking moving vertical flat plate.

Since the distance between the moving plate and the nozzle constantly increases at the rate of u_t per second, which in turn requires a continuous lengthening of the water jet at this rate in order to strike the plate. This problem may, however, be solved if instead of a single plate, there is a continuous series of plates arranged at a fixed distance apart so that each plate moving with a velocity u_t may appear successively before the water jet in the same position and all moving in the same direction as the jet. This condition may be achieved if a number of flat smooth plates are mounted radially at equal spacing on the periphery of a large wheel that is capable of rotating in a vertical plane. Such arrangements or system is known as water turbines.

In this case, since the vertical plates face the water jet normally one after the other the vertical position intercepts the water jet. After one vertical flat plate leaves its position due to the force applied by water jet, the other preceding vertical flat plate of wheel occupies its position. This process continues as long as the water jet exists. Hence, the entire fluid emerging from the nozzle is utilised to strike the vertical flat plates when all plates are considered. Therefore, the mass of water striking the vertical plates per second is

$$\dot{m} = \rho\left(\frac{\pi d^2}{4}\right)u_j = \rho a u_j \tag{6.6a}$$

The relative velocity between moving vertical plates and water jet is $(u_j - u_t)$. After impact the water jet leaves the vertical plate tangentially, hence the component of the velocity of water jet leaving the vertical plate in the direction normal to the plate is zero. By applying the impulse-momentum equation (Eq. (6.1)), the force exerted by the water jet in the direction normal to the vertical plate is expressed as

$$F = \rho a u_j[(u_j - u_t) - 0] = \rho a u_j(u_j - u_t) \tag{6.7}$$

In this case, the work done by the water jet per second is the output of the water jet and the expression for power derived by vertical flat plate is given by

The output power $P(u_t)$ of jet is given by

$$P(u_t) = F \times u_t = \rho Q(u_j - u_t)u_t \tag{6.8a}$$

The kinetic energy per second associated with water jet is referred to as the input energy to the water turbine. The expression for the same can be obtained as

$$Input\ Energy = \frac{\frac{1}{2}mv^2}{t} = \frac{1}{2}\dot{m}u_j^2 = \frac{1}{2}\rho a\, u_j^3 = \frac{1}{2}\rho^2 Q\, u_j^2 \tag{6.8b}$$

The efficiency of the water turbine (system) can be obtained as

$$\eta = \frac{Work\ obtained}{Energy\ input} = \frac{Work\ output}{Kinetic\ energy\ of\ the\ jet}$$

From Eqs. (6.8a) and (6.8b) one gets,

$$\eta = \frac{\rho a u_j(u_j - u_t)u_t}{\frac{1}{2}\rho a u_j(u_j)^2}$$

or,

$$\eta = \frac{2u_t(u_j - u_t)}{(u_j)^2} \tag{6.9}$$

For η_{max}, differentiating the above equation w.r.t. "u_j" and making it equal zero, one gets

$$\frac{d\eta}{dt} = 0 \quad \text{for a given } u_j.$$

$$\text{or,} \quad \frac{d}{d\,u_t}(u_t u_j - u_t^2) \quad \text{or,} \quad u_j - 2u_t = 0 \quad \text{or,} \quad u_j = \frac{u_t}{2}$$

Substituting the value of u_t, in Eq. (6.9), one gets

$$\eta_{max} = \frac{1}{2} = 0.5 = 50\%$$

From the above formulation it is clear that the plate needs to move to produce work output. Instead of a single plate, a series of plates are to be arranged radially around a wheel in order to utilise the whole kinetic energy of incoming water jet. As a result, the kinetic energy of the incoming water jet can be used to produce rotating-shaft power for mechanical power output. In this arrangement, the efficiency of the water turbine (system) can be achieved up to a maximum value of 50%.

EXAMPLE 6.1a

Calculate the condition for the maximum power for a water jet striking tangentially a number of plates attached to the periphery of a large wheel.

Solution

The expression for power can be written from Eq. (6.8) as

$$P(u_t) = \rho Q(u_j - u_t) \cdot u_t$$

After differentiating the above equation w.r.t. to u_t and making it equal to zero one gets:

$$\frac{dP(u_t)}{dt} = \rho Q[(u_j - u_t) \cdot 1 + u_t \cdot (-1)] = 0$$

or,

$$u_t = \frac{u_j}{2}$$

This is the same condition obtained earlier by using the expression for mechanical efficiency of a water turbine. Now,

$$P_j(max) = \rho Q\left(u_j - \frac{u_j}{2}\right) \cdot \frac{u_j}{2} = \frac{1}{4}\rho Q u_j^2$$

(ii) Moving Inclined Flat Plate at an Angle θ to the Water Jet (Figure 6.4)

As shown in Figure 6.4, the inclined flat plate is held at an angle θ to the direction of the flow of the water jet. The effective velocity with which the jet strikes the moving plate is $(u_j - u_t)$.

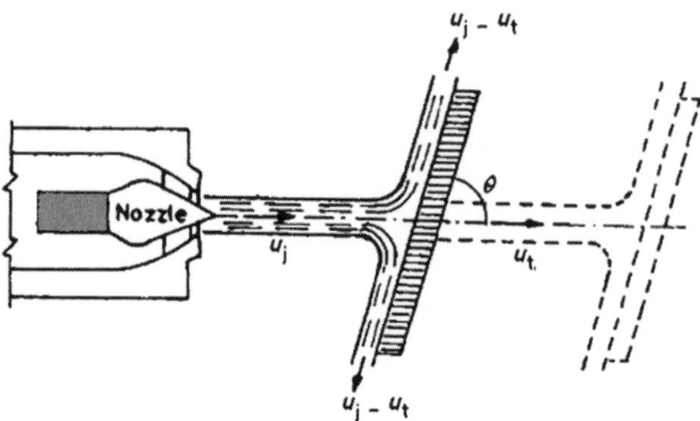

Figure 6.4 View of water jet striking a moving inclined flat plate.

Other conditions are the same as discussed in the previous Section (6.4.1(i)). The component of relative velocity $(u_j - u_t)$ in the direction normal to the inclined flat plate per second is given by

$$\dot{m} = \rho\left(\frac{\pi d^2}{4}\right)(u_j - u_t) - \rho a(u_j - u_t)$$

After impact, the water jet leaves the inclined plate tangentially. Hence, the component of this relative velocity in the direction normal to the inclined plate becomes zero. Applying the same impulse-momentum equation (Eq. 6.1), the force exerted on the inclined plate in the direction normal to the inclined plate is expressed as

$$F = \rho a(u_j - u_t)[(u_j - u_t)\sin\theta - 0] = \rho a(u_j - u_t)^2 \sin\theta \qquad (6.10)$$

Here, $(u_j - u_t)\sin\theta$ is the component of relative velocity normal to the inclined moving flat plate.

The normal thrust F can be resolved into two components (i) F_x is parallel to the direction of the water jet and F_y is normal to the direction of the water jet.

Hence, its expression can be written as

$$F_x = F\sin\theta = \rho a(u_j - u_t)^2 \sin^2\theta \qquad (6.11)$$

and

$$F_y = F\cos\theta = \rho a(u_j - u_t)^2 \sin\theta \cdot \cos\theta \qquad (6.12)$$

The work done by the water jet per second, *i.e.* the power delivered by the water jet is given by

$$P_j(t) = F_x \times u_t = F\sin\theta \times u_t = \rho a(u_j - u_t)^2 u_t \cdot \sin^2\theta \qquad (6.13)$$

Substituting the value of θ as 90° for the vertical flat plate [Section 6.4.2(i)] in Eq. (6.13), the work done obtained will be same as Eq. (6.5), for the work done of the moving vertical flat plate normal to the water jet.

EXAMPLE 6.1b

Evaluate the condition for maximum power for a moving inclined flat plate at an angle θ (Figure 6.4).

Solution

For a moving inclined flat plate, an expression for power $P_j(t)$ is given by Eq. (6.13) as

$$P_j = \rho a \sin^2\theta \cdot (u_j - u_t)^2 u_t$$

Differentiate the above equation w.r.t. u_t and make it to zero

$$\frac{dP_j}{du_t} = 0$$

or,

$$\frac{dP_j}{du_t} = \rho a \sin^2\theta \cdot [(u_j - u_t)^2 + u_t \cdot 2(u_j - u_t)(-1)] = 0$$

which gives

$$u_j = 3u_t \quad \text{or} \quad u_t = \frac{u_j}{3}$$

Substituting $u_t = \frac{u_j}{3}$ in expression for $P_j(t)$, one gets

$$P_j(t) = \rho a \sin^2\theta \cdot \left(u_j - \frac{u_j}{3}\right)^2 \cdot \frac{u_j}{3} - \frac{4}{27}\rho a u_j^{\,3} \sin^2\theta$$

$$\text{For } \theta = 90°, \qquad P_j(t) = \frac{4}{27}\rho a u_j^{\,3}$$

6.2.3 Water Jet Striking Stationary Curved Vane

(i) Water Jet Striking Symmetrically a Curved Plate (or Vane)

On the basis of the direction of water jet striking the curved vane, two distinct cases, namely symmetrical and unsymmetrical, of striking are discussed in Figure 6.5a. As the vane is assumed to be smooth, the water jet after striking the vane glides over it with the same velocity as shown in Figure 6.5a. Let 2θ be the angle between the two tangents drawn to the vane at its outlet ends. Thus, the water jet after striking the vane deflects on each side at an angle $(180° - \theta)$. As a result, the component of the velocity of the leaving water jet in the direction of incoming water jet becomes $u_j \cos(180 - \theta) = -u_j \cos\theta$.

By applying the impulse-momentum equation (Eq. (6.1)), the force exerted by water jet on the stationary curved vane in the direction of flow of water jet can be expressed as

$$F = \rho a u_j \cdot \{u_j - (-u_j \cos\theta)\}$$

or,

$$F = \rho a u_j \cdot \{u_j + u_j \cos\theta\} = \rho a u_j^{\,2} \cdot (1 + \cos\theta) \tag{6.14}$$

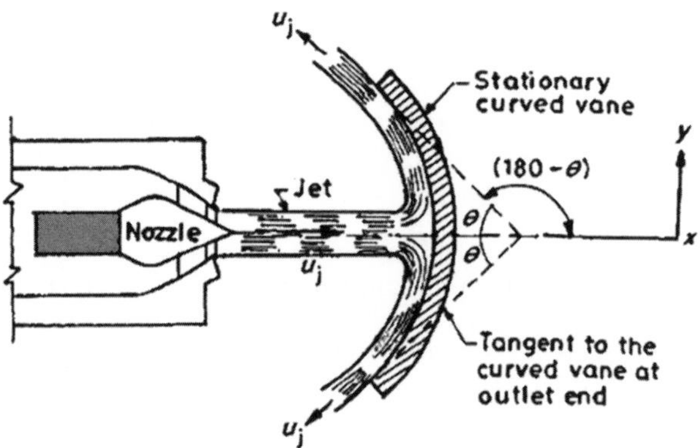

Figure 6.5a View of water jet striking symmetrically a curved vane at its centre.

For $\theta = 90°$ (or $180° - \theta = 90°$), then the curved plate reduces to a vertical flat plate [Section 6.4.1(i)] becomes

$$F = \rho a u_j^2 = \rho Q u_j$$

which is the same as Eq. (6.3).

If $\theta = 0°$ (or $180° - \theta = 180°$), then the plate reduces to a semicircular plate and the water jet leaves the outlet ends parallel to the incoming water jet. The directions of outlet water jet are opposite to the incoming water jet. The expression of F when $\theta = 0$ becomes

$$F = 2\rho a u_j^2 = 2\rho Q u_j \tag{6.15}$$

From Eq. (6.14), it is inferred that the force exerted by a water jet in its direction of flow on a curved vane is always greater than that exerted on a flat plate.

This is due to

$$\rho a u_j^2 \cdot (1 + \cos\theta) > \rho a u_j^2$$

Similarly one can see that as the angle of deflection of water jet increases from 90° (in the case of the flat plate) to 180° (in the case of a semicircular plate), the value of the exerted force F varies from $\rho a u_j^2$ to $2\rho a u_j^2 = (\rho Q u_j \text{ to } 2\rho Q u_j)$.

One can conclude that (i) If the angle of deflection (θ) of the water jet at the outlet end is obtuse ($\theta > 90°$), then the exerted force becomes more than $\rho a u_j^2$ (ii) If the deflection angle is acute ($\theta < 90°$), then the exerted force becomes less than $\rho a u_j^2$.

Hence, in order to get more force, the curvature of the plate should be such that the outlet angle (deflection angle) with respect to the direction of fluid jet should be obtuse or equal to 180°.

(ii) Water Jet Striking Asymmetrically Stationary Curved Vane Tangentially at One of the Ends

As shown in Figure 6.5b, the jet of water strikes the stationary curved vane tangentially at one of its ends asymmetrically. This means the angles made by the tangents at both of its ends with the horizontal axis are not equal. The angles made by the tangents at the inlet end and outlet end are, respectively, θ and ϕ ($\theta \neq \phi$). Therefore, after striking, the water jet gets deflected through an angle

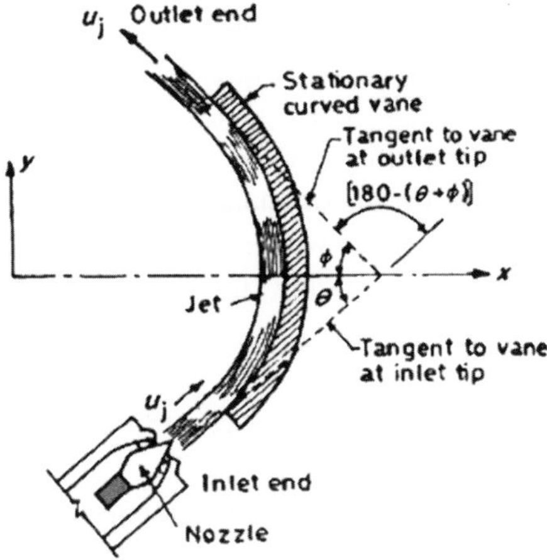

Figure 6.5b View of water jet striking tangentially at one of its ends of a stationary curved vane.

(Figure 6.5b) $[180 - (\theta + \phi)]$. Thus, the components of velocity of water jet at the inlet and the outlet ends of the vane in the x-direction become, $u_j \cos\theta$ and $u_j \cos(180 - \theta) = -u_j \cos\theta$, respectively. Therefore, by applying the impulse-momentum equation (Eq. (6.1)), the force exerted by the jet on the vane in the x-direction can be written as,

$$F_x = \rho a u_j \cdot \{u_j \cos\theta - (-u_j \cos\phi)\}$$

or,

$$F_x = \rho a {u_j}^2 \cdot \{\cos\theta + \cos\phi\} = \rho Q u_j \cdot \{\cos\theta + \cos\phi)\} \tag{6.16}$$

Similarly, the components of the velocity of the water jet at the inlet end and outlet ends in the y-direction will be $u_j \sin\theta$ and $u_j \cos\theta$, respectively. In this case, the force exerted by the water jet in the y-direction becomes

$$F_y = \rho a {u_j}^2 \cdot \{\sin\theta - \sin\phi\} = \rho Q u_j \cdot \{\sin\theta - \sin\phi\} \tag{6.17}$$

The resultant force (thrust) exerted by the water jet on the vane will be $F = \sqrt{{F_x}^2 + {F_y}^2}$ and its inclination with the x-axis is given by

$$\alpha = \tan^{-1}\left(\frac{F_y}{F_x}\right)$$

If $\theta = \phi = 0$ [or $180° - (\theta = \phi) = 180°$], the vane becomes semicircular and therefore incoming and outgoing jets become parallel to the horizontal axis of symmetry. But the two water jets are in opposite directions. The expression of F_x in this case becomes

$$F_x = 2\,\rho Q u_j \tag{6.18}$$

The above equation is the same as Eq. (6.15).

As the vane is stationary, the work done is equal to zero.

6.2.4 Force Exerted by a Water Jet on Moving Curved Vane

The curved vane is moving with a velocity u_t, as shown in Figure 6.6. The velocity of the water jet impinging the vane is u_j. The effective velocity of the water jet w.r.t the vane is the relative velocity between the water jet and the velocity of the moving vane. Hence, the effective velocity is $(u_j - u_t)$. If the vane is assumed to be smooth for no frictional resistance between the jet and the vane, the velocity of the water jet after striking the vane will be the same as the effective velocity, *i.e.* $(u_j - u_t)$. The component of the velocity of the jet leaving the vane in the direction of the incoming jet is given by

$$\text{component of the velocity of jet} = (u_j - u_t)\cos(180^\circ - \theta) \quad \text{or} \quad -(u_j - u_t)\cos\theta$$

The force exerted by the water jet on the plate in the direction of flow of the incoming water jet is given as

$$F = \rho a(u_j - u_t)[(u_j - u_t) - \{(u_j - u_t)\cos\theta\}]$$

or,

$$F = \rho a(u_j - u_t)^2[1 + \cos\theta] \tag{6.19}$$

If $\theta = 90^\circ$ (or $180^\circ - \theta = 90^\circ$), the above equation is reduced to

$$F = \rho a(u_j - u_t)^2 \tag{6.20}$$

which is the same expression as derived in the case of the force exerted by a jet on a moving flat plate, Eq. (6.5). From Eqs. (6.19) and (6.20) it is observed that the force exerted by a jet on a moving curved vane is greater than that on a moving flat plate. Again if $\theta = 0$ (or $180^\circ - \theta = 180^\circ$), the vane becomes semicircular and the expression for the force exerted by the water jet in its direction of flow becomes

$$F = 2\,\rho a(u_j - u_t)^2 \tag{6.21a}$$

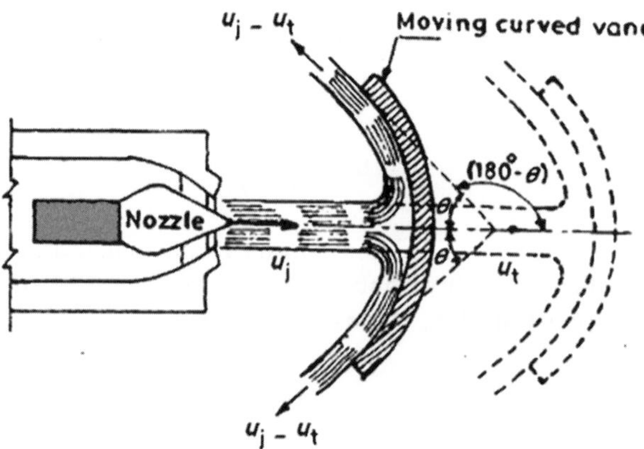

Figure 6.6 View of water jet striking symmetrically a curved vane at its centre direction.

The work done by the water jet on the vane per second is given by

$$F \times u_t = \rho a(u_j - u_t)^2(1 + \cos\theta)u_t \tag{6.21b}$$

The above equation gives the output of the water jet.

As the distance between the vane and the water jet will be constantly increasing at a rate of u_t per second, a continuous lengthening of the jet is required. This will not be practically feasible. Hence, this problem will be solved if instead of a single curved vane, there is a series of such curved vanes arranged at equal spacing on the periphery of a large wheel. This is capable of rotating in a vertical plane. As the wheel rotates, each vane will become normal to the water jet in turn. The entire volume of water of the incoming jet issued from the nozzle will be utilised in striking the vanes.

Then, the mass of water striking the wheel per second $(\dot{m}) = 2\,\rho a u_j$

The force exerted by the water jet on the vanes in the direction of the water jet is calculated by

$$F = \rho a u_j[(u_j - u_t) - (u_j - u_t)\cos(180° - \theta)]$$

or,

$$F = \rho a u_j[(u_j - u_t) - \{-(u_j - u_t)\cos\theta\}]$$

or,

$$F = \rho a u_j(u_j - u_t)(1 + \cos\theta) = \rho Q(u_j - u_t)(1 + \cos\theta) \tag{6.22}$$

The work done on the wheel per second (power) is given by

$$P(u_t) = F \times u_t = \rho a u_j(u_j - u_t)(1 + \cos\theta)u_t \tag{6.22a}$$

The mechanical efficiency of the water turbine can be obtained from Eqs. (6.6), (6.6a) and (6.22a), respectively. It is given by

$$\eta = \frac{\textit{work output}}{\textit{kinetic energy of jet}} = \frac{\rho a u_j(u_j - u_t)(1 + \cos\theta)u_t}{\frac{1}{2}\rho a u_j \cdot {u_j}^2}$$

or,

$$\eta = \frac{2u_t(u_j - u_t)(1 + \cos\theta)}{{u_j}^2}$$

For a water jet of constant velocity u_j, striking a given wheel, the efficiency will be maximum when $d\eta/dt = 0$ *i.e.*

$$\frac{2(u_j - u_t)(1 + \cos\theta)}{{u_j}^2} = 0$$

Since, $\frac{2(1 + \cos\theta)}{{u_j}^2} \neq 0$, being constant for a given system, $(u_j - 2u_t) = 0$ or

$$u_j = 2u_t \Rightarrow u_t = \frac{u_j}{2}$$

Hence, for maximum efficiency, $u_j = 2u_t$ and by substitution, the value of maximum efficiency becomes

$$\eta_{max} = \frac{1}{2}(1 + \cos\theta) \tag{6.22b}$$

EXAMPLE 6.2

Prove that the maximum power for symmetrically attached to a rotating wheel curved vanes occurs at $u_j = 2u_t$.

Solution

From Eq. (6.22a), we get

$$P_j = \rho a(1 + \cos\theta)u_j(u_j - u_t)u_t$$

$$= \rho Q(1 + \cos\theta)(u_j - u_t)u_t$$

as

$$\rho Q = \rho a u_j$$

After differentiation of the above equation, one gets

$$\frac{dP_j}{du_t} = \rho Q(1 + \cos\theta)[(u_j - u_t) + u_t(-1)]$$

For maximum P_j,

$$\frac{dP_j}{du_t} = 0 \Rightarrow [(u_j - u_t) + u_t(-1)] = 0 \Rightarrow u_j = 2u_t, \text{ since } \rho Q(1 + cos\,\theta) \neq 0$$

or,

$$u_j = 2u_t$$

The maximum efficiency of a wheel in this case is 100%. This is a theoretical approach. The same is employed in the case of the bucket of a Pelton wheel of an impulse turbine. The water jet impinges at the centre of the bucket and it deflects through 160° to 170° depending on the particular design. Now, it is inferred that the efficiency of a moving semicircular vane is higher than the moving flat plate with the same velocity of jet.

6.2.5 Working Principle of Different Water Turbines

(a) Impulse turbine

In this case, the water is brought in through the penstock ending in a single nozzle as shown in Figure 6.7. The potential energy of water is converted into kinetic energy or velocity head. The water coming out of the nozzle in the form of a free water jet is made to strike a series of buckets mounted on the periphery of a wheel (runner). The resulting deflection of water constitutes a change in momentum of the fluid. The bucket exerts a force on the water and therefore water likewise exerts a force on the bucket. The tangential force applied to the bucket causes it to rotate. There occurs no pressure difference between the inlet water and the outlet water. The wheel revolves freely in air and hence the water is at atmospheric pressure. A casing is, however, provided around the wheel to prevent splashing and to lead the water to the tailrace. This is also referred to as a Pelton wheel turbine (Figure 6.7).

The runner of the Pelton wheel turbine consists of a circular disc with a number of buckets (cups) evenly spaced round its periphery. The bucket or cup is of double semiellipsoidal shape. Each cup is divided into two symmetrical parts by a sharp-edged ridge known as a splitter. The water jet strikes

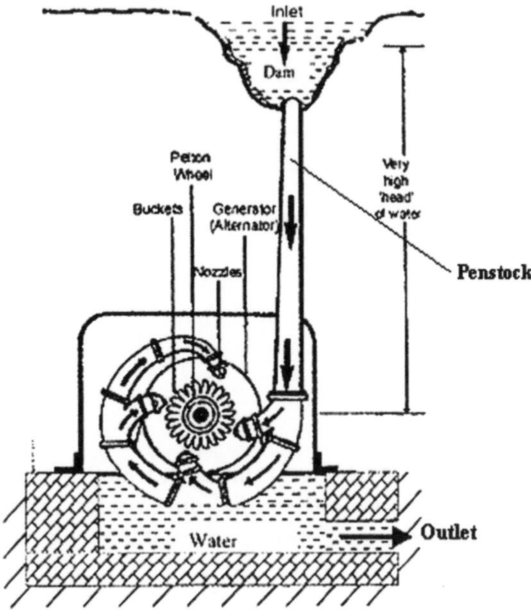

Figure 6.7 A schematic diagram of an impulse turbine.

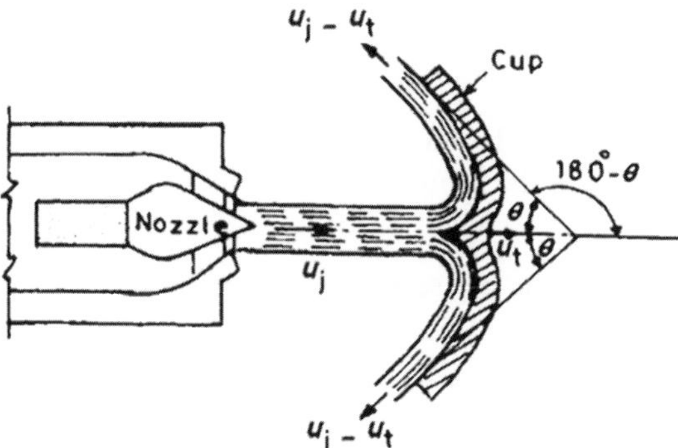

Figure 6.8 View of velocity of cup and water jet in Pelton wheel turbine.

the splitter that divides the jet into two equal portions. Each of them after flowing round the smooth inner surface of the cup leaves at its outer edge. The cups are so shaped that the angle at the outlet tip varies from 10° to 20°, so that the jet of water gets deflected through 170° to 160°. But theoretically, it is assumed that the jet of water gets deflected through 180°. In this case, water at the outlet end gets deflected opposite to the direction of the water at the inlet end. If the inner surface of the cup is polished, then losses due to frictional resistance can be neglected. As a result, the initial relative velocity between the jet and cup at inlet end may be assumed to be the same as the relative velocity at the outlet end, as shown in Figure 6.8.

The work done per second by the water jet on the vane (power developed by the wheel) is given by

$$P = \rho a u_j (u_j - u_t) u_t (1 + \cos\theta) \tag{6.22a}$$

$$= \rho Q (u_j - u_t) u_t (1 + \cos\theta) \tag{6.23}$$

The expression for efficiency,

$$\eta_{max} = \frac{1}{2}(1 + \cos\theta) \tag{6.22b}$$

(b) Reaction Turbine

In this case, a portion of the potential energy of the water is converted into kinetic energy before the water reaches the runner, unlike in the impulse turbine. The kinetic energy is then taken from the jet by suitable flow through moving vanes. The vanes of the impulse turbine are partly filled with the jet and open to the atmosphere throughout their travel through the runner. However, water completely fills all the passage in the runner in a reaction turbine. The water leaving the reaction turbine has still some of the pressure as well as the kinetic energy. The pressure at the inlet of the reaction turbine is much higher than the pressure at the outlet end.

(c) Francis Water Turbine

This is an inward mixed-flow reaction turbine. In this case, water under pressure enters the runner from the guide vanes towards the centre in the radial direction and discharges out of the runner axially. The runner of a Francis turbine consists of a series of curved vanes evenly arranged around the circumference of the runner. The vanes are so shaped that water enters the runner radially at the outer periphery and leaves it axially at the inner periphery. The change in the direction of the flow of water from the radial to the axial direction as it passes through the runner, produces a circumferential force on the runner that makes it rotate and thus contributes to the useful output of the turbine. The water after passing through the runner flows to the tail race through a draft tube. A draft tube is a pipe or passage of gradually increasing cross-sectional area that connects the runner exit to the tailrace. The outline of Francis turbine is shown in Figure 6.9.

Work Done and Efficiency of Francis Turbine

If ρQ kg of water per second strikes the runner, then the work done per second on the runner in case of radial flow can be written as

$$\text{The work done per second } (P_j(t)) = \rho Q(u_{jw}u_t \mp u_{jw1}u_{t1})$$

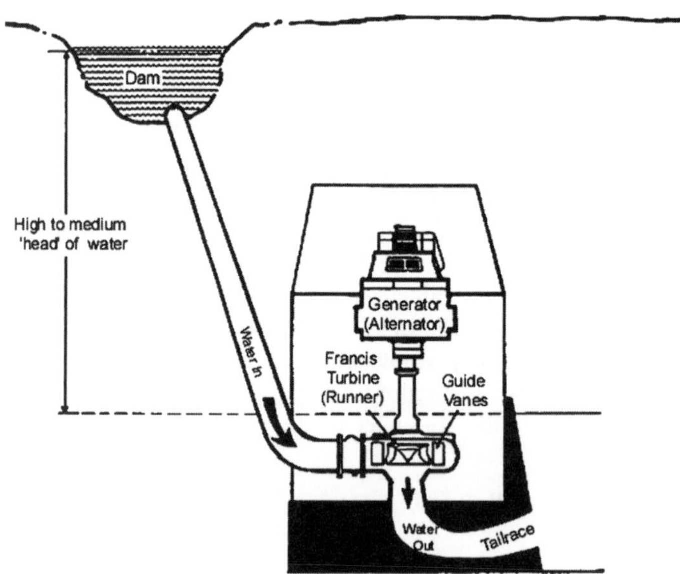

Figure 6.9 A schematic diagram of a Francis water turbine.

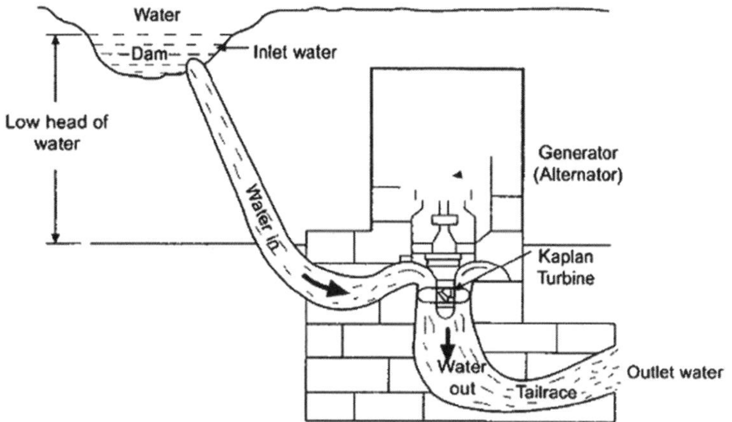

Figure 6.10 A schematic diagram of a Kaplan water turbine.

But, in the case of a Francis turbine, the jet leaves in the same direction as that of the motion of the wheel (in axial direction), then the sign between u_{jw} and u_{jw1} becomes negative. Therefore, the above equation reduces to

$$P_j(t)\,\text{for Francis turbine} = \rho Q(u_{jw}u_t - u_{jw1}u_{t1}) \tag{6.24}$$

(ii) Kaplan Water Turbine

This is an axial flow reaction turbine. It is named after the Austrian engineer, V. Kaplan. It is suitable for relatively low heads and hence requires a large quantity of water to develop a large amount of power, as it is a reaction type of turbine. It operates in an entirely closed conduit from the head race to the tailrace. The outline of a Kaplan turbine is shown in Figure 6.10.

In a Francis turbine, water enters radially, while in a Kaplan turbine, it strikes the blade axially. The number of blades (vanes) in the runner of a Francis turbine is about 16 to 24, but in a Kaplan turbine, the number is only 3 to 6. As a result, the contact surface with the water is reduced causing less frictional resistance between the flowing water and the vane. Also, the blades of the Kaplan runner can be turned about their own axis, so that their angle of inclination may be adjusted while the turbine is in motion. This is done automatically by means of a servomotor operating inside the hollow coupling of the turbine and the generator shaft. Due to adjustment of the runner-blade inclination, high efficiency can be achieved over a wide range of operating conditions for a Kaplan turbine.

The expression for work done per second (power developed) by a Kaplan turbine is the same as that for a Francis turbine.

6.3 WATER-JET VELOCITY AND NOZZLE SIZE IN PELTON WHEEL TURBINE

The pressures of both water in the reservoir and the water jet at the inlet end are atmospheric. By applying Bernoulli's theorem between the reservoir and the jet at the inlet end, one can have the following expression for u_j as,

$$u_j^2 = -2gH_t, \text{without friction in the pipe}$$

and

$$u_j^2 = 2gH_t, \text{with friction in the pipe} \tag{6.25a}$$

where H_t, is the total head of water flow and H_a is the available head at water jet at inlet end.

If there are n_j nozzles having cross-sectional area a_j of each nozzle, then, total water flow through the water jet can be given as

$$Q = n_j a_j u_j$$

The maximum power output of the Pelton wheel turbine is given by

$$P_j = \frac{1}{2}\rho Q u_j^2 \tag{6.25b}$$

If η_{jm} is the actual mechanical efficiency of the Pelton wheel turbine, then the mechanical output from the water turbine is

$$\begin{aligned} P_m = \eta_{jm} P_j = \frac{1}{2}\eta_{jm}\rho Q_j^2 &= \frac{1}{2}\eta_{jm} n_j a_j \rho u_j^3 \\ &= \frac{1}{2}\eta_{jm} n_j a_j \rho (2gH_a)^{3/2} \end{aligned} \tag{6.26}$$

Equation (6.26) shows the relationship between the mechanical power obtained from the water turbine and the available head between the reservoir and the turbine. If a_j is increased, then the size of the turbine also increases, which is not at all desirable. It is usually easier to increase the number of nozzles than to increase the overall size of the turbine.

If the wheel of a Pelton turbine has radius "R" and turns at angular velocity ω, then

Power developed = Force exerted by the fluid on the turbine × tangential velocity of wheel

or,

$$P = F \times u_t = F \times R\omega \tag{6.27}$$

Thus, for a given power output (P), the larger is the size of the wheel (R), the smaller is its angular velocity (ω).

For maximum power output (P) at fixed $u_j, \dfrac{u_t}{u_j} = 0.5$, then

$$u_t = R\omega \quad \text{or} \quad R = \frac{u_t}{\omega} = \frac{0.5u_j}{\omega} = \frac{0.5(2gH_a)^{1/2}}{\omega} \tag{6.28a}$$

The nozzles usually give circular jets of radius r_j then $a_j = \pi r_j^2$ and putting the value of a_j in Eq. (6.26)

$$P_m = \frac{1}{2}\eta_{jm}\,\eta_j \pi r_j^2 \rho (2gH_a)^{3/2} = \sqrt{2}\eta_{jm}\eta_j\,\pi r_j^2 \rho (gH_a)^{3/2} \tag{6.28b}$$

or,

$$r_j^2 = \frac{P_m}{\sqrt{2}\eta_{jm}\eta_j\,\pi r_j^2 \rho (gH_a)^{3/2}} \tag{6.28c}$$

or,

$$r_j = \frac{P_m^{0.5}}{2^{1/4}(\eta_{jm}\eta_j)^{0.5}\pi^{0.5}\rho^{0.5}(gH_a)^{3/4}}$$

or,

$$r_j = 2^{-1/4}(\eta_{jm}\eta_j)^{-0.5}\pi^{-0.5}\frac{P_m^{\ 0.5}}{\rho^{0.5}(gH_a)^{3/4}} = 0.0085(\eta_{jm}\eta_j)^{-0.5}\left(\frac{P_m^{\ 0.5}}{H_a^{\ 3/4}}\right) \tag{6.28d}$$

Now, dividing r_j by R by using Eqs. (6.28a) and (6.28b), one gets

$$\frac{r_j}{R} = 2^{-1/4}\pi^{-0.5}(\eta_{jm}\eta_j)^{-0.5}\frac{P_m^{\ 0.5}}{\rho^{0.5}(gH_a)^{3/4}}\frac{\omega}{(0.5)(2gH_a)^{0.5}}$$

or,

$$\frac{r_j}{R} = \frac{2^{-1/4}\pi^{-0.5}}{2^{1/2}0.5}(\eta_{jm}\eta_j)^{-0.5}\frac{\omega P_m^{\ 0.5}}{\rho^{0.5}(gH_a)^{5/4}} = 0.68(\eta_{jm}\eta_j)^{-0.5}\Psi \tag{6.29}$$

where, Ψ is the nondimensional parameter known as the shape number of the turbine

$$= \frac{\omega P_m^{\ 0.5}}{\rho^{0.5}(gH_a)^{5/4}}$$

Hence,

$$\omega = \Psi\,\rho^{0.5}(gH_a)^{5/4}P_m^{\ -0.5} \tag{6.30}$$

Equation (6.29) relates the shape of a Pelton wheel (measured by the nondimensional parameter r_j/R and n_j) to the Ψ characterising the operating conditions and hence to the efficiency n_{jm} under these condition. For an impulse turbine (Figure 6.7) $r_j/R<1$; the pressure is created at the cup and for a reaction turbine (Figure 6.8) $r_j/R > 1$; the pressure is created around the shaft (axis).

EXAMPLE 6.3

For a Pelton wheel cup, the exit flow makes an angle θ (Figure 6.9) with the incident water jet. u_t, is the tangential velocity of cup measured in the laboratory frame. The energy lost by friction between the water and the cup can be measured by a loss coefficient k such that

$$u_{r1}^{\ 2} = u_{r2}^{\ 2}(1+k)$$

Prove that the power transferred $(P) = \rho Q u_t(u_j - u_t)\left[1 + \frac{\cos\theta}{\sqrt{1+k}}\right]$

Solution

Here,

$$u_{r2} = \frac{u_{r1}}{\sqrt{1+k}} = \frac{(u_j - u_t)}{\sqrt{1+k}} \text{ with } u_{r1} = u_j - u_t$$

From Eq. (6.22),

$$F = \rho Q[u_{r1} - u_{r2}\cos(180^\circ - \theta)]$$

$$= \rho Q[u_{r1} + u_{r2}\cos\theta]$$

Also, power transferred

$$(P) = F \times u_t = \rho Q u_t [u_{r1} + u_{r2} \cos\theta]$$

Substituting the expression for u_{r2} from above one gets

$$P = \rho Q u_t \left[(u_j - u_t) + \frac{(u_j - u_t)}{\sqrt{1+k}} \cos\theta \right]$$

$$= \rho Q u_t (u_j - u_t) \left[1 + \frac{\cos\theta}{\sqrt{1+k}} \right]$$

Hence, the question is proved.

EXAMPLE 6.4a

Determine the dimensions of a single-jet Pelton wheel to develop 200 kW under a head of (i) 90 m and (ii) 10 m. What is the angular speed at which these wheels will perform best?

Assumptions

(a) Water is the working fluid
(b) $\eta_{jm} = 0.85$

Solution

From Eq. (6.29) $\Psi = 0.13$

(i) From Eq. (6.30), the angular speed for best performance is

$$\omega_1 = \Psi \rho^{1/2} (g H_a)^{5/4} P^{-1/2}$$

or,

$$\omega_1 = \frac{0.13(10^3\ \mathrm{kgm^{-3}})^{0.5}[(9.8\ \mathrm{ms^{-2}})(90\ \mathrm{m})]^{5/4}}{(20 \times 10^4\ \mathrm{W})^{1/2}}$$

$$\omega_1 = 44\ \mathrm{rad\ s^{-1}}$$

Again,

$$u_j = (2 g H_a)^{1/2} = (2 \times 9.8 \times 90)^{1/2} = 42\ \mathrm{ms^{-1}}$$

Therefore,

$$R = \frac{1}{2} \frac{u_j}{\omega} = \frac{1}{2} \frac{42}{44} = 0.47\ \mathrm{m}$$

(ii)

$$\omega_2 = \omega_1 \left(\frac{10}{90} \right)^{5/4} = 2.8\ \mathrm{rad\ s^{-1}}$$

$$u_j = 14$$

$$R = 2.5$$

The above problem can be solved as follows:
Substitute the expression of ω from Eq. (6.29) into Eq. (6.28)

$$R = 0.5(2gH_a)^{1/2} \cdot \frac{P_m^{0.5}}{\Psi\rho^{0.5}(gH_a)^{5/4}}$$

$$= 0.5\sqrt{2}(gH_a)^{1/2} \cdot \frac{P_m^{0.5}}{\Psi\rho^{0.5}}(gH_a)^{-5/4}$$

$$= \frac{1}{\sqrt{2}} \cdot \frac{P_m^{0.5}}{\Psi\rho^{0.5}}(gH_a)^{-3/4} = \frac{1}{\sqrt{2}} \cdot \frac{P_m^{0.5}}{\Psi\rho^{0.5}(gH_a)^{3/4}}$$

For

$$P_m = 200 \times 10^3\,\text{W},\ \Psi = 0.13,\ \rho = 10^3\,\frac{\text{kg}}{\text{m}^3} \text{ and } H_a = 90$$

$$R = \frac{1}{\sqrt{2}} \cdot \frac{(200 \times 10^3)^{1/2}}{0.13(10^3)^{1/2}(9.8 \times 90)^{3/4}} = 0.475\,\text{m}$$

Similarly, for

$$H_a = 10\,\text{m},\ \text{R} = 2.5\,\text{m}$$

EXAMPLE 6.4b

Determine the radius of water jet in Example 6.4a by using Eq. (6.28d) for $\eta_{jm} = 0.5$ and 1.0.

Solution

From Eq. (6.28d), one has the expression for radius of water jet as

$$r_j = 0.0085(\eta_{jm} \cdot \eta_j)\left(\frac{P_m^{0.5}}{H_a^{3/4}}\right)$$

Given:

$$\eta_{jm} = 1 \text{ and } 0.5,\quad P_m = 200 \times 10^3\,\text{W}, H_a = 90\,\text{m}$$

So, after substitution of appropriate value, one gets

$$r_j = 0.0085(0.5 \times 1)\frac{(200 \times 10^3)^{0.5}}{90^{3/4}} = 0.065\,\text{m}$$

Similarly, for $\eta_{jm} = 1$, $r_j = 0.13\,\text{m}$

This indicates that the radius of nozzle for a single jet increases with increase of mechanical efficiency.

Now, from Eq. (6.29), the radius of a water turbine is given by

$$\frac{r_j}{R} = 0.68(\eta_{jm}\eta_j)^{-1/2}\Psi$$

Here, substituting $\eta_{jm} = 0.5,\ r_j = 0.065\,\text{m}$ and $\Psi = 0.13$ in the above equation, one gets

$$\frac{0.065}{R} = 0.68(0.5 \times 1)^{-1/2} \times 0.13$$

$$R = 0.5\ \text{m}$$

Similarly, for $\eta_{jm} = 1.0,\ r_j = 0.13\,\text{m},\ \eta_j = 1$ and $\Psi = 0.13$, we have

$$R = 1.47\,\text{m}$$

This shows that the radius of turbine also increases with increase of mechanical efficiency of turbine (η_{jm}) for a given rated power of the turbine.

EXAMPLE 6.5

A Pelton wheel water turbine is to be installed at a site with $H = 20$ m and $Q_{min} = 0.05\,\text{m}^3\,\text{s}^{-1}$

(a) By neglecting friction in the cup, find (i) the water-jet velocity (ii) the maximum power available (iii) the radius of the nozzles for $n_j = 2$.

(b) Assuming further that the wheel has shape number

$$\psi = \frac{\omega P_1{}^{0.5}}{\rho^{0.5}(gH)^{5/4}} = 0.1$$

where P_1 is the power per nozzle, find (iv) the number of cups (v) the diameter of the wheel (vi) angular speed of the wheel in operation.

Solution

(a) (i) From Eq. (6.25a), we have the expression for the water-jet velocity as

$$u_j{}^2 = 2gH_a \text{ or } u_j = \sqrt{2gH_a} = (2 \times 9.8 \times 20)^{1/2} = 19.79\,\text{ms}^{-1}$$

(ii) From Eq. (6.25b), we have the expression for maximum power by

$$P_m = \frac{1}{2}\rho Q u_j{}^2 = \frac{1}{2} \times 10^3 \times 0.05 \times 392 = 9.8\ \text{kW}$$

(iii) From Eq. (6.28c), we have the following expression

$$r_j{}^2 = \frac{P_m}{\eta_{jm}\rho n_j \pi (gH_a)^{3/2}\sqrt{2}}$$

Let $\eta_{jm} = 0.9$, we have

$$r_j^2 = \frac{9800}{0.9 \times 1000 \times 2 \times \pi \times (9.8 \times 20)^{3/2}\sqrt{2}}$$

or,

$$r_j = 0.02\,\text{m}$$

(b) For $a_j = \pi r_j^2 = 1.25 \times 10^{-3}\,\text{m}^2$ and using this value in Eq. (6.26), one gets

$$n_j = \frac{P}{\sqrt{2}\eta_{jm} a_j \rho (gH_a)^{3/2}} = \frac{9800}{\sqrt{2} \times 0.9 \times 1.25 \times 10^{-3} \times 10^3 (9.8 \times 20)^{3/2}}$$

From Eq. (6.28a), we have the expression for the diameter of the wheel as

(iv) Diameter of wheel $R = 0.5(2gH_a)^{0.5}/\omega = \dfrac{0.5(2 \times 9.8 \times 20)^{1/2}}{23.42} = 0.42\,\text{m}$

(v) From Eq. (6.30), we have

$$\omega = \Psi \rho^{1/2} (gH_a)^{5/4} \times P_m^{-1/2} = 0.1 \times 1000^{1/2} \times (9.8 \times 20)^{5/4} \times 9800^{-1/2}$$

$$= 23.42\,\text{rad/s}$$

6.4 HYDROELECTRIC SYSTEM

6.4.1 Principle

The power of water available at height "H" can be obtained from the following equation

$$P_0 = \rho Q g H = \frac{mgH}{t} = \frac{force \times height}{t} \tag{6.31}$$

where P_0 is the power of water per second (or power in Watt), g is the acceleration due to gravity (m/s^2), ρ is the mass density of water (kg/m^3) and Q is the volume of water falling down a slope per second (m^3/s). Therefore, for more power output from the system it is cheaper to increase H.

In a hydropower system, it is thus the head H that is very important. During the conveyance of water in the pipeline, a portion of the head is lost due to friction. There are various types of water heads in hydroelectric systems, which are given as follows.

6.4.2 Water Head

Water head is the difference in the elevation between two levels of water.

(i) **Gross head:** This is the difference in elevation of head race level at the intake end and the tail race level at the discharge side when no water flows.

(ii) **Net or effective head:** The head of water available for doing work on the turbine, *i.e.* the difference between the gross head and all the losses occurring in carrying water from the head-race level to the entrance of the turbine.

(iii) **Total head:** This is the sum total of velocity head, pressure head and potential head.

(iv) **Rated head:** The head at which the turbine produces the rated output at the rated speed.

6.4.3 Water Power

The classification of water power is given below

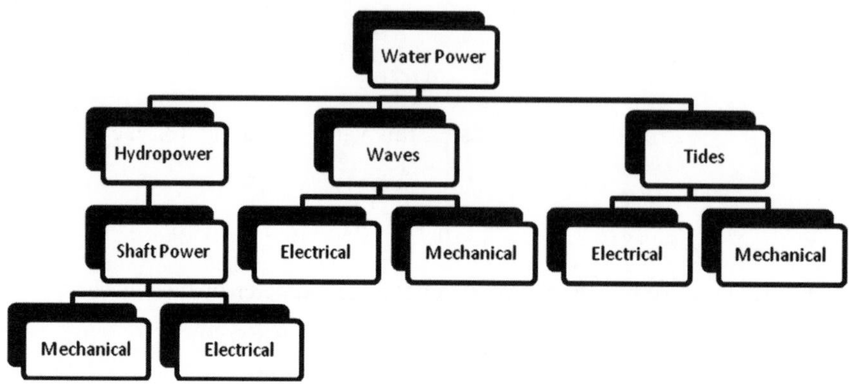

The brief discussion about waves and tidal power will be developed in Chapter 9.

6.4.4 Classification of Hydropower Plant

Hydropower plants are classified on to the following basis.

(a) Based on Head

(i) low head: working under a heads less than 30 m;
(ii) medium head: working under heads between 30–300 m;
(iii) high head: working under heads more than 300 m.

(b) Based on Capacity

(i) micro: capacity up to 100 kW;
(ii) mini: capacity from 101 to 1000 kW;
(iii) small: capacity from 1001 to 6000 kW;
(iv) medium: capacity from 6001 to 10 000 kW;
(v) high: capacity more than 10 000 kW.

(c) On the Basis of Storage Being Provided:

(i) **Run-of-river plant:** This utilises the flow of water without storage being provided. Such plants would be based only on such rivers that have a minimum dry weather condition that makes the development worthwhile. A weir or barrage may be constructed across the river close to the power plant to maintain a given water level. These are generally low-head plants.
(ii) **Reservoir plant:** This utilises the flow from large storage reservoir developed by constructing dams across the rivers. These plants utilise the water for producing electricity according to the requirements throughout the year.
(iii) **Pumped storage plant:** This utilises the peak load of power station. They store water by pumping a portion of the water from the tailrace back to the reservoir during off-peak hours. The pumping is done by the same turbine generator (producing power) which now acts as a pump–motor set (consuming power). It is expensive.

(d) On the Basis of Load Capacity

(i) **Base-load plants:** These plants cater for the base load of the system. Such plants are required to supply a constant power when connected to the grid. Thus, they run without stopping and are often remotely controlled, so fewer staff are required for such plants.
(ii) **Peak-load plants:** These plants supply power during peak loads. Pumped storage plants are peak-load plants. Run-off-river plants with pondage can operate both as peak-load and base-load plants as river flow permits.

6.4.5 Measurement of Water Head

The available vertical fall "H". (a) vertical fall: trigonometric methods are most suitable, (b) sloping site: the head is measured with the help of dumpy levels and a theodolite and (c) For high-head measurement: an altimeter with good accuracy is used.

6.4.6 Measurement of Water Flow Rate (Q)

Flow rate is also an important parameter to measure the shaft power in the water turbine. The flow rate of the water stream (Q) is measured with the following equation.

$$Q = (\text{volume of fluid passing in time } \Delta t)/\Delta t \tag{6.32a}$$

$$= (\text{mean speed of fluid } u)(\text{cross sectional area A}) \tag{6.32b}$$

$$= \int u \cdot \hat{n}\, dA \tag{6.32c}$$

where $\hat{n}$ is the unit vector normal to the elemental area dA.

The following methods to measure flow rate (Q) are used

(i) **Bucket method:** This is shown in Figure 6.11a. It is a simple way of measuring flow in very small streams. The entire flow is diverted into a bucket or barrel and the time for the container to fill is recorded. The flow rate is obtained by simply dividing the volume of the container with the filling time.
(ii) **Floating method:** In this case, the flow rate is measured by multiplying the mean velocity of flow by the cross-sectional area. The flow speed varies from the bottom to the top of the channel. The flow speed is zero on the bottom of the stream because of viscous

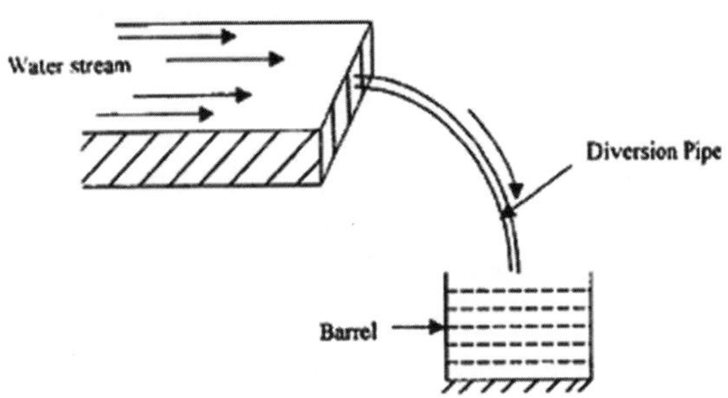

Figure 6.11a Measurement of volume flow rate by bucket method.

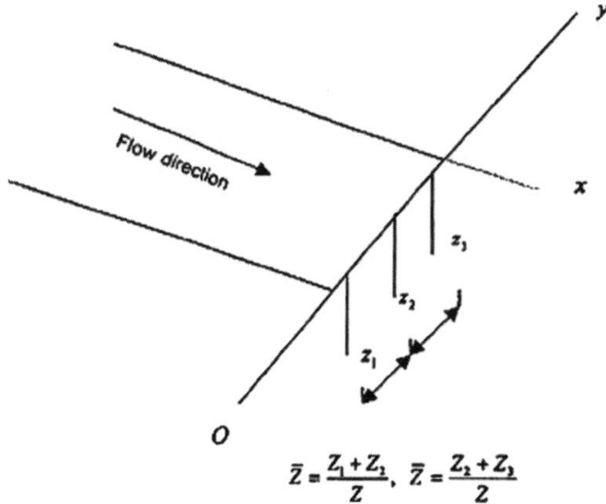

Figure 6.11b Measurement of flow by float method.

friction. Hence, the mean speed is slightly less than the speed u_s on the top surface. But for a rectangular cross section, it is assumed that $\bar{u} = 0.8\,u_s$. Then, u_s can be measured by simply placing a float (piece of wood, leaf, *etc.*) on the surface and measuring the time it takes to move a certain distance along the stream. The cross-sectional area A can be estimated by measuring the depth at several points and corresponding lengths of each section, as shown in Figure 6.11b. The total cross-sectional area is obtained by summing the areas of all sections as follows:

$$A = \frac{1}{2}y_1 z_1 + \frac{1}{2}(y_2 - y_1)(z_1 + z_2) + \frac{1}{2}(y_3 - y_2)(z_2 + z_3) + \frac{1}{2}(y_4 - y_5)z_3$$

Finally, the flow rate is measured by multiplying A by u_s.

(iii) **Current meter:** A current meter is an instrument for measuring the velocity of a flow channel. It consists of a shaft with a revolving element containing cup and a tail on which flat vanes are fixed. The current meter is suspended by means of a cable and is held vertically immersed in the stream of water to the required depth such that the revolving element is facing towards the upstream direction. The revolving element is free to rotate and the speed of rotation is related to the stream velocity. A simple mechanical counter records the number of revolutions of the revolving element of the current meter placed at a desired depth. By averaging the observations taken throughout the cross section, the average speed of the flow is determined. Then by multiplying the average speed by the cross-sectional area the rate of water flow Q is obtained.

(iv) **Use of a weir:** A weir is similar to a small dam constructed across a river. By measuring the height of upstream water surface, the rate of flow can be determined.

Let L be the length of a crest and H be the height of water in a weir over which water flows is known as crest. For computing the discharge of flowing water over the weir, let us consider an elementary horizontal strip of water of thickness dh and length L at the depth h below the water surface. The area of the strip is $(L \times dh)$ and the theoretical velocity of water flowing through the strip will be $\sqrt{2gh}$. Thus, if dQ is the discharge (flow rate) through the strip, then

$$dQ = c_d \times L \times dh \times \sqrt{2gh} \tag{6.33a}$$

where c_d is the coefficient of discharge and g is the acceleration due to gravity. The total discharge Q for the entire weir may be determined by integrating the above expression within limits 0 to H.

So,

$$Q = \int_0^H C_d L \sqrt{2gh}\, dh.$$

Assuming the coefficient of discharge c_d to be constant for the entire weir, one can obtain

$$Q = \frac{2}{3} c_d \sqrt{2g}\, L\, H^{3/2}$$

The contractions and losses of flow through the weir reduce the actual discharge to about 62 per cent of the theoretical value. Substituting the value of g and c_d in the above expression

$$Q = 1.84\, LH^{3/2} \tag{6.33b}$$

The discharge of the flow can be measured by only knowing head H and L from the above equation.

6.5 ESSENTIAL COMPONENTS OF HYDROELECTRIC SYSTEMS

The various components of hydroelectric system include:

Water source: The water available from a catchment's area is stored in a reservoir. It can be utilised to run turbine for generation of electricity. The storage reservoir may be natural or artificial. Natural reservoirs or lakes may be found in high and large mountains. Water is taken off from one end of the reservoir through a tunnel built by cutting the mountain. An artificial reservoir is made by constructing dam across a river at a suitable site for storage of water to create a head.

(a) **Water ways:** In this case, water is carried from the storage reservoir to the power house. It may consist of a tunnel, channel or penstock. A tunnel is a water passage made by cutting the mountain to save distance. Penstocks are the pipes of large diameter usually made of steel, wood or reinforced concrete. This carries water under pressure from the storage reservoir to the turbine.

(b) **Flow control:** The flow-control structures are mainly trash racks, gates, forebay, and surge tank. A trash rack is a grating made of a series of steel bars of rectangular cross section set parallel to each other and placed across the entire intake opening to prevent debris entering into the water passage of the hydropower plant. Gates and valves of different types are provided to control the flow of water inside the pen stock as well as a streamlined transition to ensure smooth flow. A forebay is essentially a storage reservoir provided just in front of the penstock. The purpose of a forebay is to temporarily store water when it is not required by the turbine and to supply the same when required. When the powerhouse is located just at the base of the dam no forebay is required since the reservoir itself serves the same purpose.

(c) **Power house:** This is a building to house the turbines, governors, generators and other accessories for operating the machine.

(d) **Tailrace:** The tailrace is a reservoir to conduct the water discharged from the turbine to a suitable point. It can be safely disposed of or stored to be pumped back into the original reservoir.

(e) **Generation and transmission of electric power:** This consists of electrical generating machines, transformers, switching equipments and transmission lines.
(f) **Drive system:** The drive system transmits power from the turbine shaft to the generator shaft or the shaft powering another device. It has also the function of changing the rotational speed from one shaft to the other. The turbine speed is different from the required speed of the alternator or other device.

The following are the drive systems considered in hydropower plants.

(i) direct drive (uses flexible couplings to join two shafts together directly);
(ii) flat belt and pulleys;
(iii) V-belt and pulleys;
(iv) chain and sprocket;
(v) gearbox.

6.6 HYDROPOWER SYSTEM EFFICIENCY

In hydropower systems, even though the efficiency of each individual step is high, there is still a substantial energy loss in passing from the original power P_0 of the stream to the electrical output P_e, from the generator. Hence, the overall efficiency of the system becomes

$$\frac{P_e}{P_o} = \frac{P_j}{P_o} \times \frac{P_m}{P_j} \times \frac{P_e}{P_m} = \frac{H_a}{H_t} \eta_{jm} \eta_{me}$$

where η_{jm} is mechanical efficiency of the turbine.
η_{me} is the electrical efficiency of the generator
P_m is the mechanical power output from turbine
P_j is the power available in the jet of fluid
P_o is the power of water available in supply source
P_e is the electrical output for the generator.
H_a is the available head at the entrance of the turbine
H_t is the total head of the stream.

For $n_{jm} = \eta_{m\theta} = \frac{H}{H_t} = 0.9$, the overall system efficiency will be obtained as

$$\eta = 0.9 \times 0.9 \times 0.9 = 0.73$$

6.7 ADVANTAGES AND DISADVANTAGES OF HYDROELECTRIC SYSTEMS

These are discussed in brief as follows:

6.7.1 Advantages

(i) Hydroelectric systems use an infinitely available renewable energy source, *i.e.* water
(ii) The system is purely nonpolluting and no heat is released during the process.

(iii) No additional fuel is required and hence running costs are minimised as compared to thermal power plants.
(iv) The efficiency of changing stored water into electricity is normally over 80% compared with 30–40% for either fossil fuels or nuclear plants.
(v) Due to the presence of a storage facility, flood water is retained and is better utilised for agricultural production, river regulation, and wildlife protection.
(vi) Equipments have long life and their malfunctions are rare as no heat is involved in the system.
(vii) The system requires limited maintenance. A few operating personnel are required for its functioning.
(viii) They are constructed on locally available material and labour.
(ix) There is a good opportunity for power generation in remote areas.
(x) Long-lasting and robust technology can last for 50 years or more without major new investment.

6.7.2 Disadvantages

(i) Many people need to be evacuated to make room for the installation of a hydropower plant.
(ii) Landscape clearing from the site of a hydroelectric plant results in soil erosion and landslides.
(iii) Water reservoirs have limited natural flow in summer periods. Stratification of water (warm water at the surface and cooler water at the bottom) results in favorable living conditions for pathogens in the bottom of the reservoir.
(iv) The dissolved oxygen level is reduced causing adversely the health of fish due to increase of temperature of the water in the reservoir.
(v) Silt is naturally transported downstream by the flow of river. It is captured by the reservoir resulting in the decrease in fertility of downstream plains.
(vi) Fish are killed in the turbines, especially young ones swimming downstream.

6.8 HYDRAULIC MACHINES

There are several hydraulic machines.[2] They are employed for either storing the hydraulic energy or then transmitting it when required or doing pumping work. In this section, hydraulic presses and hydraulic ram pumps are described.

(1) Hydraulic Press

In this case, mechanical advantage is obtained by a practical application of Pascal's Law of transmission of fluid pressure. An hydraulic press was first built in 1795 by Joseph Bramah. Since then, hydraulic power has been in use for heavy operations in cranes and lifts.

Principle

Two pistons of different size operate inside two cylinders suitably connected with a pipe so that the pressure in each is the same. A multiplication of force can be achieved by application of fluid

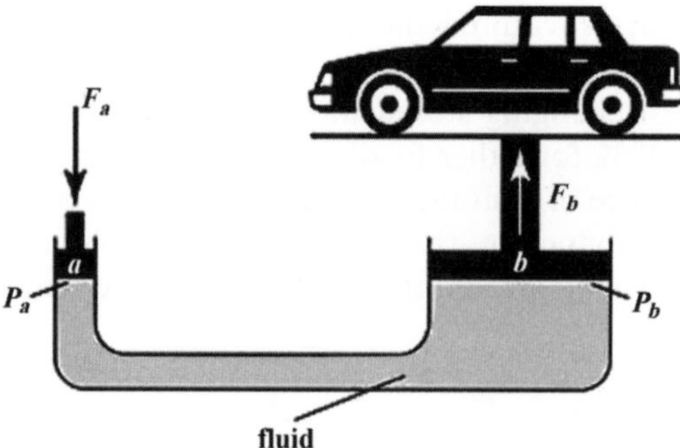

Figure 6.12a Lifting of a large weight by a small force by the application of Pascal's principle.

pressure according to Pascal's principle, which states that pressure is transmitted undiminished in an enclosed static fluid. This allows the lifting of a heavy load with a small force. P is the pressure and A_1 and A_2 are the area of the Pistons as shown in Figure 6.12a. If F_1 is the force applied to the smaller piston, a large force F_2 will be developed in the large piston.

$$P = \frac{F_1}{A_1} = \frac{F_2}{A_2}$$

or,

$$F_2 = \frac{A_2}{A_1} F_1$$

The mechanical advantage of an hydraulic press $= \frac{A_2}{A_1} > 1$.

In an hydraulic press, large force (F_2) is obtained with the application of smaller force (F_1). The principle of an hydraulic press is employed to perform many odd jobs requiring tremendous pressures, like

(a) metal presswork: employed to press sheet metal to any desired shape;
(b) bakelite press: used to prepare moulds and casting of bakelite;
(c) metal-pushing presses;
(d) forging presses, *etc.*

(2) Hydraulic Ram Water Pump

This is a device to raise a portion of a large amount of the water available at some height to a greater height. The momentum of the stream flow is used to pump some of the water to a considerably higher level. This is employed when some natural source of water like a spring or a stream is available at some altitude, especially in hilly regions. The potential energy of the supply water is first converted into kinetic energy and subsequently into potential energy at the higher level. No external power is required to operate the pump. The working of an hydraulic ram pump is based on the principle of a water hammer or inertia pressure developed in the supply pipe. A general layout of a pumping system using a hydraulic ram is shown in Figure 6.12b.

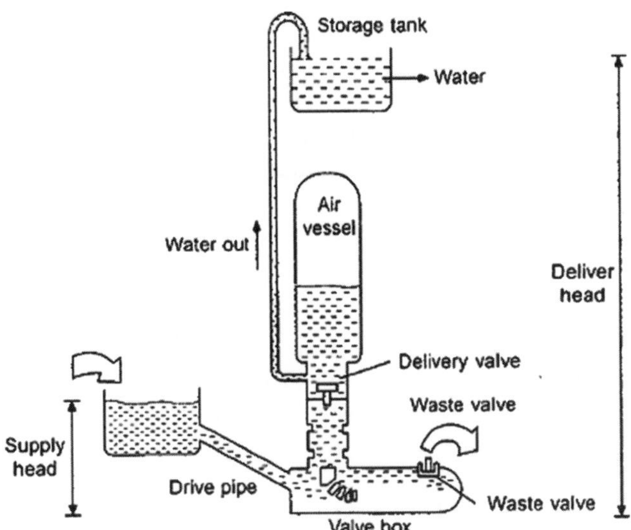

Figure 6.12b A general layout of a hydraulic ram water pump.

The components of the hydraulic ram are (i) supply source (ii) drive pipe (iii) valve box (iv) waste valve (v) delivery valve (vi) air vessel (vii) storage tank. Initially, water flows down the drive pipe into the valve box, the waste valve being open and the delivery valve being closed. Water flows through the waste valve to the waste water channel. As a result, the rate of discharge through the waste valve increases and thereby the flow of water in the drive pipe accelerates. Due to acceleration of a water column in the drive pipe, there occurs an appreciable increase of pressure in the valve box. The pressure (static) in the valve box rapidly increases to such a value that it overcomes the weight of the waste valve and then closes rapidly. The instantaneous closing of the waste valve brings the water in the drive pipe suddenly to rest, causing a further increase of pressure in the valve box due to the development of inertia pressure. Due to the increase of the pressure in valve chamber (closure of waste valve), the delivery valve is forced to open. The water then flows from the supply source through the delivery valve into the air vessel and delivery pipe. Thus, some of the water flowing through the delivery valve is directly supplied to the delivery tank and some of it is stored in the air vessel. The water flowing in the air vessel compresses the air inside it, which pushes some of the water in the delivery pipe even when the delivery valve is closed. Thus, an air vessel of a hydraulic ram assists in providing a continuous delivery of water at a more or less uniform rate. The flow of water in the delivery valve continues until the pressure in the valve chamber is reduced, the delivery valve then closes and the waste valve opens, causing the water again to flow from the supply source to wastewater channel. This constitutes one cycle of operation. The same cycle is then repeated.

Advantages

(a) An hydraulic ram can work continuously for 24 h,. thus providing a regular water supply.
(b) No motive power is required.
(c) Maintenance expenses are low and almost no supervision is required.

EXAMPLE 6.6

The water in a large lake to be used to generate electricity by installing a hydraulic turbine generator at a location where the depth of water is 60 m as shown in the figure. Water is to be supplied at a rate of 4000 kg/s. If the electric power generated is measured as 1800 kW and the generator efficiency is 90 per cent, determine (a) the overall efficiency of the turbine-generator (b) the mechanical efficiency of the turbine (c) the shaft power supplied by the turbine to the generator.

Solution

We can take the bottom of the lake as the reference level for convenience. Also, it is assumed that the mechanical energy of water at the turbine exit is negligible. Hence, in the entrance of the turbine, the mechanical energy of water consists of pressure energy, *i.e.* P/ρ

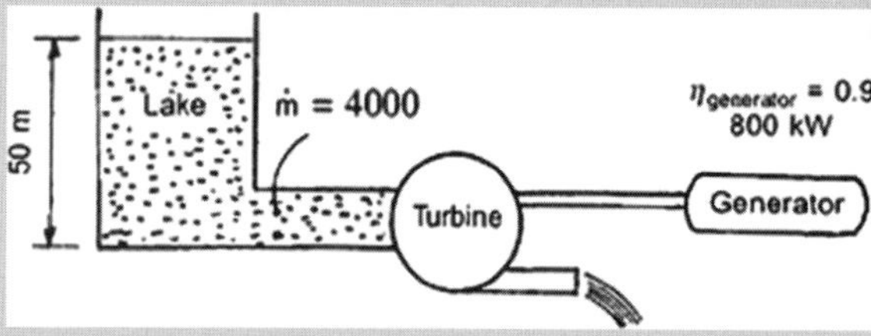

where P is the pressure at the depth of 60 m of the lake and P/ρ is the pressure energy.

$$e_{mech,in} - e_{mech,out} = \frac{P}{\rho} - 0 = \frac{P}{\rho} = gh = (9.81\ \text{m/s}^2)(60\ \text{m})$$

e_{mech} represents mechanical energy = 0.589 kJ/kg

The rate at which mechanical energy is supplied to the turbine by the fluid is

$$\left(0.589\frac{\text{kJ}}{\text{kg}}\right)\left(4000\frac{\text{kg}}{\text{s}}\right) = 2356\,\text{kW}$$

$$\eta_{overall} = \eta_{turbine-generator} = \frac{1800\ \text{kW}}{2356\ \text{kW}} = 0.76$$

(b) Knowing the overall and generator efficiencies, the mechanical efficiency of the turbine is determined from

$$\eta_{turbine-generator} = \eta_{turbine} \cdot \eta_{generator} \Rightarrow \eta_{turbine} = \frac{\eta_{turbine-generator}}{\eta_{generator}}$$

$$= \frac{0.76}{0.9} = 0.84$$

Hence,

$$\eta_{turbine} = 0.84$$

(c) The shaft power output is the product of $\eta_{turbine}$ and the mechanical energy supplied to the turbine

$$\text{Shaft power} = 0.84 \times 2356 = 1979\ \text{kW}$$

From the above calculation, it is clear that lake supplies 2356 kW energy to the water turbine that converts 1979 kW of it to shaft work that drives the generator, which generates 1800 kW of electric power.

EXAMPLE 6.7

In a hydroelectric power plant, 150 m^3/s of water flows from an elevation of 150 m to a turbine, where electric power is generated as shown below. The total head loss in the system from point 1 to point 2 (excluding turbine unit) is determined to be 40 m as shown in the figure. If the overall efficiency of the turbine-generator is 85 per cent, estimate the electrical power output.

Solution

In this case, the water levels at the reservoir and the discharge site remain constant. The mass flow rate of water through the turbine is

$$\dot{m} = \rho Q = \left(1000\frac{\text{kg}}{\text{m}^3}\right)\left(150\frac{\text{m}^3}{\text{s}}\right) = 15 \times 10^4 \frac{\text{kg}}{\text{s}}$$

Let us take point (2) as the reference level for energy calculation. Both points (1) and (2) are open to the atmosphere ($P_1 = P_2 = P_{atm}$). Also, it is assumed that the flow velocity is negligible at both points, *i.e.* $v_1 = v_2 = 0$.

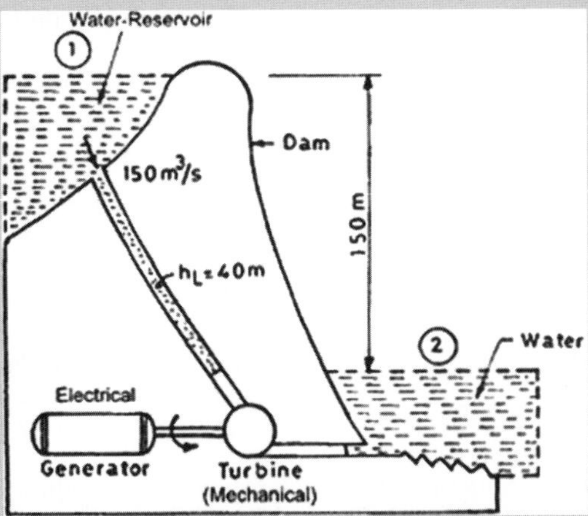

By applying Bernoulli's equation to the system for incompressible flow,

$$h_{turbine} = z_1 - hL = 150 - 40 = 110 \text{ m}$$

Hence, the mechanical energy available to the turbine $\dot{m}gh$

$$= (15 \times 10^4 \text{ kg})(9.81 \text{ m/s}^2)(110 \text{ m}) = 1\,61\,865 \text{ kW}$$

The turbine-generator efficiency is 85 per cent, then the electrical power output will be 0.85 × 1, t r 865 = 137 585 kW = 137 MW. Therefore, the electrical power output in this system is estimated to be 137 MW.

6.9 STATUS OF HYDROPOWER SYSTEMS

In the present scenario, major hydropower installations of larger capacity belong to the category of conventional energy sources. However, a microhydropower system is generally considered in the renewable energy power sector. In India there is about 25% contribution to power required based on hydro. The present status of hydropower system is discussed in the following sections:

6.9.1 Potential of Hydropower in India[1]

India celebrated 100 years of hydropower development in 1997. Around that time, there were about 225 hydroelectric power stations (excluding mini hydro) in the country, with an installed capacity of about 23 000 MW and generating about 80 billion units a year. By the turn of the century, the installed hydropower capacity rose to about 25 000 MW. This was about 25 per cent from all sources. Major hydroelectric projects (above 750 MW): Bhakra, Dehar, Koyna, Nagarjunasagar, Srisailam, Sharavarthy, Kalinadi and Ldukki.

Table 6.1 India's hydroelectric power plants (100 MW and greater).

Power Plant	Owner	Location		Total Capacity (MW)
		River(s)	State	
Koyna l-lV	Maharashtra SEB	Koyna	Maharashtra	1920
Sharavathi	KPCL	Sharavathi	Karnataka	1035
Dehar	BBMB	Beas; Satluj	Rajasthan	990
Kalinadi Nagjhari	KPCL	Kalinadi	Kamataka	840
Nagarjuna Sagar	Andhra Pradesh Power Generation Corp.	Krishna	Andhra Pradesh	810
Idduki	Kerala SEB	Ldduki	Kerala	780
Srisailam Right Bank	Andhra Pradesh Power Generation Corp.	Krishna	Andhra Pradesh	770
Bhakra Right Bank	BBMB	Satluj	Rajasthan	710
Salal	NHPC	Chenab	Jammu & Kashmir	690
Ranjit Sagar	Punjab SEB	Ranjit	Punjab	600
Upper Indravati	OHPC	Lndravati	Orissa	600
Kundah	Tamil Nadu SEB	Kundah	Tamil Nadu	555
Bhakra Left Bank	BBMB	Satluj	Rajasthan	540
Chamera I	NHPC	Ravi	Himachal Pradesh	540
Uri I	NHPC	Jhelum	Jammu & Kashmir	480
Lower Sileru	Andhra Pradesh Power Generation Corp.	Godavari	Andhra Pradesh	460
Srisailam Left Bank	Andhra Pradesh Power Generation Corp.	Krishna	Andhra Pradesh	450
Ranganadi I	NEEPCO	Ranganadi; Dikrong	Arunachal Pradesh	450
Kadampari	Tamil Nadu SEB	Cauvery	Tamil Nadu	400
Koteshwar	THDC	Bhagirathi	Uttar Pradesh	400
Balimela	OHPC	Machkund	Orissa	360
Pong	BBMB	Beas	Himachal Pradesh	360
Upper Kolab	OHPC	Kolab	Orissa	320
Bansagar	Madhya Pradesh SEB	Sone	Madhya Pradesh	315
Hirakud	OHPC	Mahanadi	Orissa	308
Ukai	Gujarat SEB	Tapti	Gujarat	305
Rihand	UPJVNL	Rihand	Uttar Pradesh	300
Sabarigiri	Kerala SEB	Anathodu; Pamba	Kerala	300
Rengali	OHPC	Brahmani	Orissa	250

Table 6.1 (*Continued*)

Power Plant	*Owner*	*Location*		*Total Capacity (MW)*
		River(s)	*State*	
Chibro	UJVNL	Yamuna	Uttaranchal	240
Kadana	Gujarat SEB	Mahi	Gujarat	240
Upper Sileru	Andhra Pradesh Power Generation Corp.	Godavari	Andhra Pradesh	240
Varahi	Karnataka Power Corp. Ltd.	Varahi	Karnataka	230
Mukerian	Punjab SEB	Beas	Punjab	207
Kopili	NEEPCO	Umrong	Assam	200
Mettur Tunnel	Tamil Nadu SEB	Cauvery	Tamil Nadu	200
Ramganga	UJVNL	Ramganga	Uttaranchal	198
Baira Siul	NHPC	Siul	Himachal Pradesh	180
Gerusupa	KPCL	Sharvathi	Karnataka	180
Lower Periyar	Kerala	Periyar	Kerala	180
Rana Pratap Sagar	Rajasthan State Electricity Corp.	Chambal	Rajasthan	172
Pench	Madhya Pradesh SEB	Narmada	Madhya Pradesh	160
Kadra	Karnataka Power Corp. Ltd.	Bethi; Kalinadi	Karnataka	150
Chilla	UJVNL	Ganga	Uttaranchal	144
Mahi Bajaj Sagar	Rajasthan State Electricity Corp.	Mahi	Rajasthan	140
Periyar	Tamil Nadu SEB	Periyar	Tamil Nadu	110
Subernrekha	Bihar SEB	Subernrekha	Jharkhand	130
Upper Sindh	Jammu & Kashmir SEB	Sindh	Jarnmu & Kashmir	127
Kuttiadi	Kerala SEB	Kuttiadi	Kerala	125
Jog	KPCL	Sharvathi	Karnataka	120
Khodri	UJVNL	Tons	Uttaranchal	120
Kodasalli	KPCL	Kalinadi	Karanatka	120
Lower Mettur	Tamil Nadu SEB	Cauvery	Tamil Nadu	120
Sanjay Bhaba	Himachal Pradesh SEB	Bhaba Khad	Himachal Pradesh	120
Tanakpur	NHPC	Sarda	Uttaranchal	120
Gandhi Sagar	Madhya Pradesh SEB	Chambal	Madhya Pradesh	115
Machkund	Andhra Pradesh Power Generation Corp.	Machkund	Andhra Pradesh	115
Umiam	Meghalaya SEB	Umiam	Meghalaya	114
Shanan (Uhl I)	Punjab SEB	Uhl	Punjab	110
Loktak I	NHPC	Leimatak	Manipur	105
Lower Jhelum	Jammu & Kashmir SEB	Jhelum	Jammu & Kashmir	105
Kalinadi Supa	KPCL	Kalinadi	Karnataka	100
Kodayar	Tamil Nadu SEB	Kodayar	Tamil Nadu	100
Total				**21 960**

Sources: Water and Dam Construction, India, 2000.

6.9.2 Global Hydropower Development and Potential

Table 6.2 gives a glimpse of hydropower systems in the world. It is inferred from this table that installed capacity of hydropower systems in Asia is more (225 GW) in comparison with other regions. However, the planned hydro capacity for South America is 35 480 MW hydro-based power plants in most countries is more than 50%, This indicates that India should have more hydro-based power plant in the future.

6.9.3 Comparison with Other Sources

A comparison of hydropower-based power plants with other source-based power plant is given in Table 6.3. The table indicates that natural gas-based power plants have the maximum capacity.

6.9.4 Economics of Hydropower Systems

Table 6.4a gives the cost of a turbine per kW, lifecycle cost per kWh and project cost per kW for small-scale (<10 MW) units. Further, the project cost breakdown is also given along with the construction time in Table 6.4b.

Table 6.2 Overview of world hydropower development and potential.

Africa
>2403 MW of hydro under construction in at least 18 countries
>76 000 GWh/yr produced from hydro plants (20.3 GW)
Feasible potential
Hydro supplies >50 per cent of electricity in 25 countries
Planned hydro capacity: >60 GW

Asia
>84 400 MW of hydro under construction in 27 countries
754 000 GW h/yr produced from hydro plants (225 GW)
6800 TWh/yr of technically feasible potential
Hydro supplied >50 per cent of electricity in nine countries
Planned hydro capacity: >156 GW

Australasia
42 200 GWh/yr produced from hydro plants (13.1 GW)
270 TWh/yr of technical feasible potential
Hydro generated in at least eight countries (contributes >50 per cent in 4)
Planned hydro capacity: about 50 MW

Europe
>221I MW of hydro under construction in 23 countries
567 000 GWh/yr produced from hydro plants (173.2 GW)
1225 TWh/yr of technically feasible potential
Hydro supplies >50 per cent of electricity in seven countries
Planned hydro capacity: >8000 MW

North and Central America
>1236 MW of hydro under construction in five countries
702 500 GWh/yr produced from hydro plants (157.2 GW)
1660 TWh/yr of technical feasible potential
Hydro supplied >50 per cent of electricity in six countries
Planned hydro capacity: >9358 MW

South America
14 792 MW of hydro under construction in seven countries
512 238 GWh/yr produced from hydro plants (108.2 GW)
2665 TWh/yr of technically feasible potential
Hydro supplies >50 per cent of electricity in 10 countries
Planned hydro capacity 35 480 MW
Continental totals are generally based on information for (Hydropower & Dams, 2001) world

Survey: where new figures were not available, the latest available data have been taken.

Table 6.3 Thermal equivalents to hydropower generation.[3]

Global region	*Hydropower electrical generation (billion kWh)*	*Equivalent coal (million tons)*	*Equivalent oil (million barrels)*	*Equivalent natural gas (billion cu. Ft.)*
North America	672	291	1167	7000
Central and South America	521	226	904	5423
Western Europe	520	225	901	5406
Eastern Europe and former USSR	261	113	453	2718
Middle East	16	7	27	164
Africa	63	27	109	656
Far East and Oceana	514	223	890	5342
World total	**2567**	**1112**	**4451**	**26 709**

Assumptions: 1 kWh = 10 400 BYU; 1 ton/coal = 24 000 000 BTU; 1 ft^3/natural gas = 1000 BTU; 1 barrel/oil = 6 000 000 BTU.

Table 6.4a Cost of small-scale hydro power.

S. No.	*Component*	*Cost*
1	Project cost	US \$1000–\$5000/kW
2	Turbine cost	US \$450–\$600/kW
3	Life-cycle cost	US \$0.05–\$0.15/kWh
4	Construction time	2–3 yr
5	Life of the system	20–30 yr

Table 6.4b Cost breakdown of the project.

S. No.	*Component*	*Percentage share (%)*
1	Civil work	21
2	Development	15
3	Infrastructure	13
4	Equipment	45

ADDITIONAL EXERCISES

Exercise 6.1: Determine the condition for maximum power for fluid jet striking a moving vertical plate.

Solution

The power for fluid water jet striking a moving vertical plate is given by:

$$P(u_t) = F \times u_t$$

$$= \frac{\omega_a (u_j - u_t)^2}{g}$$

$$= \rho_a (u_j - u_t)^2 u_t \quad (1)$$

Differentiating Eq. (1) w.r.t to u_t and equating to zero one gets,

$$\frac{d}{du_t}[\rho_a(u_j - u_t)^2 u_t] = 0$$

$$\rho_a(u_j - u_t)(u_j - 3u_t) = 0$$

$$(u_j - u_t)(u_j = 3u_t) = 0$$

$$(u_j - u_t) \neq 0$$

$$(u_j - u_t) \neq 0$$

Hence,

$$u_j = 3u_t$$

$$u_t = \frac{u_j}{3}$$

Exercise 6.2: Prove that the normal thrust for a fluid jet striking a moving horizontal plate is equal to zero.

Solution

The force exerted on the plate in the direction normal to the plate per second is equal to:

$$F = \frac{\omega_a}{g}(u_j - u_t)^2 \sin\theta \tag{1}$$

The normal thrust F can be resolved into two components F_x parallel to the direction of jet and F_y normal to the direction of jet.

$$F_x = \frac{\omega_a}{g}(u_j - u_t)^2 \sin^2\theta \tag{2}$$

and,

$$F_y = \frac{\omega_a}{g}(u_j - u_t)^2 \sin\theta \cos\theta \tag{3}$$

For the normal thrust for a fluid jet striking a moving horizontal plate, put $\theta = 0$ in Eqs. (2) and (3), one gets,

$$F_x = \frac{\omega_a}{g}(u_j - u_t)^2 \sin^2(0) = 0$$

$$F_y = \frac{\omega_a}{g}(u_j - u_t)^2 \sin(0)\cos(0) = 0$$

$$F = \sqrt{F_x{}^2 + F_y{}^2} = 0$$

Exercise 6.3: Repeat Exercise 6.1 for Figure 6.4.

Solution

For Figure 6.4:
i.e. for a water jet striking the moving inclined plate the power delivered by the jet is given by

$$P(u_t) = \rho_a(u_j - u_t)^2 u_t \sin^2\theta$$

After differentiating w.r.t. u_t, one gets:

$$\rho_a(u_j - u_t)(u_j - 3u_t)\sin^2\theta = 0$$

or,

$$(u_j - u_t)(u_j - 3u_t) = 0$$

Since,

$$(u_j - u_t) \neq 0$$

Hence,

$$u_j = 3u_t$$

or,

$$u_t = \frac{u_j}{3}$$

Exercise 6.4: Show that the expression for forces along x- and y-directions for an asymmetrical stationary curved vane reduces to an expression for the force for a symmetrically stationary vane.

Solution

The expression for forces along the x- and y-directions is given by

$$F_x = \rho Q u_j(\cos\theta + \cos\phi)$$

Putting $\theta = 0$; $\phi = \theta$

$$F_x = \rho Q u_j(1 + \cos\theta)$$

The expression for the force exerted by the jet of fluid in the y-direction becomes:

$$F_y = \rho Q u_j(\sin\theta - \sin\phi)$$

$$F_y = \rho Q u_j(-\sin\phi)$$

$$F = \sqrt{F_x{}^2 + F_y{}^2}$$

or,

$$F = \sqrt{(\rho Q u_j)^2(1 + \cos\theta)^2 + (\rho Q u_j)^2(-\sin\theta)^2}$$

or,

$$F = \sqrt{(\rho Q u_j)^2[1 + 2\cos\theta + \cos^2\theta + \sin^2\theta]}$$

or,

$$F = \sqrt{(\rho Q u_j)^2[2 + 2\cos\theta]}$$

or,

$$F = \rho Q u_j \sqrt{[2 + 2\cos\theta]}$$

If $F_y = 0$

$$F = \sqrt{(\rho Q u_j)^2(1 + \cos\theta)^2}$$

or,

$$F = \rho Q u_j(1 + \cos\theta)$$

Exercise 6.5: Prove that the maximum power for a symmetrically moving curved vane (Figure 6.6a) occurs at $u_j = 3u_t$

Solution

The power for a symmetrically moving curved vane is given by:

$$P(u_t) = F \times u_t = \rho_a(u_j - u_t)^2(1 + \cos\theta)u_t$$

Now,

$$\frac{dP_j}{du_t} = \frac{d}{du_t}[\rho_a(u_j - u_t)^2(1 + \cos\theta)u_t] = 0$$

or,

$$\rho a(u_j - u_t)(u_j - 3u_t)(1 + \cos\theta) = 0$$

or,

$$(u_j - u_t)(u_j - 3u_t) = 0$$

Since,

$$(u_j - u_t) \neq 0$$

Hence,

$$u_j = 3u_t$$

Exercise 6.6: What will be the power for fluid jet on a horizontally moving flat plate?

Solution

The power delivered by a flat plate inclined at an angle θ to the fluid jet is given by

$$P(u_t) = \rho_a(u_j - u_t)^2 \sin^2\theta\, u_t$$

For a horizontally moving flat plate, $\theta = 0$

The power developed is given by:

$$P(u_t) = \rho_a (u_j - u_t)^2 \sin^2\theta \, u_t$$

Hence, $P(u_t) = 0$

Exercise 6.7: Determine the condition for the maximum power for a Pelton wheel turbine (Figure 6.9).

Solution

The expression for the power for a Pelton wheel turbine is given by:

$$P = \rho Q (u_j - u_t) u_t (1 + \cos\theta)$$

Now,

$$\frac{d}{du_t} [(\rho Q (u_j - u_t) u_t (1 + \cos\theta)] = 0$$

or,

$$\rho Q (u_j - u_t + u_t(-1))(1 + \cos\theta) = 0$$

or,

$$(u_j - 2u_t) = 0$$

or,

$$u_t = \frac{1}{2} u_j$$

Exercise 6.8: What will be the condition for the maximum power of Exercise 7 for $\theta = 0$?

Solution

$$P = \rho Q (u_j - u_t) u_t (1 + \cos\theta)$$

For, $\theta = 0$

$$P = \rho Q (u_j - u_t) u_t \cdot 2$$

Now,

$$\frac{dP}{du_t} = 2\rho Q[(u_j - u_t) + u_t(-1)]$$

or,

$$2\rho Q (u_j - 2u_t) = 0$$

or,

$$u_t = \frac{u_j}{2}$$

Exercise 6.9: Determine the radius of a single-jet Pelton wheel turbine (dimension) to develop 100 kW power under a head of 10 m, 20 m, 40 m and 60 m, respectively, for $\Psi = 0.1$

Solution

Diameter of wheel

$$R = \frac{1}{2}\frac{(2gH_a)^{1/2}}{\omega}$$

$$D = 2R = \frac{(2gH_a)^{1/2}}{\omega}$$

$$\omega = \Psi\rho^{0.5}(gH_a) \times P_m{}^{-1/2}$$

As $\Psi = 0.1$,

$$\omega_{10} = 0.1 \times 1000^{1/2} \times (9.8 \times 10)^{5/4} \times (100 \times 10^3)^{-1/2} = 1.742\,\text{Rad/s}$$

After substitution of appropriate values, one gets

$$R_{10} = \frac{1}{2} \times \frac{(2 \times 9.8 \times 10)^{1/2}}{1.74} = 4.01\,\text{m}$$

$$R_{20} = \frac{1}{2} \times \frac{(2 \times 9.8 \times 20)^{1/2} \times (10^5)^{1/2}}{0.1 \times (1000)^{1/2} \times (9.8 \times 20)^{5/4}} = 1.35\,\text{m}$$

$$R_{40} = \frac{1}{2} \times \frac{(2 \times 9.8 \times 40)^{1/2} \times (10^5)^{1/2}}{0.1 \times (1000)^{1/2} \times (9.8 \times 40)^{5/4}} = 0.802\,\text{m}$$

$$R_{40} = \frac{1}{2} \times \frac{(2 \times 9.8 \times 40)^{1/2} \times (10^5)^{1/2}}{0.1 \times (1000)^{1/2} \times (9.8 \times 40)^{5/4}} = 0.802\,\text{m}$$

and

$$R_{60} = \frac{1}{2} \times \frac{(2 \times 9.8 \times 60)^{1/2} \times (10^5)^{1/2}}{0.1 \times (1000)^{1/2} \times (9.8 \times 60)^{5/4}} = 0.593\,\text{m}$$

Exercise 6.10: Find the condition for maximum power for Exercise 6 for a given θ.

Solution

Expression for maximum power is given by

$$P(u_t) = \rho_a(u_j - u_t)^2 \sin^2\theta\, u_t$$

Now,

$$\frac{dP(u_t)}{du_t} = \rho_a[2(u_j - u_t)(-1) \cdot u_t + (u_j - u_t)^2]\sin^2\theta = 0$$

or,

$$\rho_a \sin^2\theta[(u_j - u_t)(u_j - u_t - 2u_t)] = 0$$

or,

$$(u_j - 3u_t) = 0$$

or,

$$u_j = 3u_t$$

Exercise 6.11: Prove that for the volume flow rate (use of Weir's method) for $C_d = 1$ (open-flow conditions) is given by

$$Q = \sqrt{\frac{8g}{9}} LH^{3/2}$$

$$Q = \int u \cdot \hat{n}\, dA$$

($\hat{n}$ = unit vector normal to elemental area dA)

Solution

We have,

$$dQ = C_d \times L \times dh \times \sqrt{2gh}$$

Now,

$$Q = \int_0^H C_d L \sqrt{2gh}\, dh$$

$$= \frac{2}{3} C_d \sqrt{2g} LH^{3/2}$$

$$= \frac{2\sqrt{2}}{3} C_d \sqrt{g} LH^{3/2}$$

or,

$$Q = \frac{2\sqrt{2}}{3} \sqrt{g} LH^{3/2} = \sqrt{\frac{8g}{9}}$$

OBJECTIVE QUESTIONS

6.1 Hydropower is stored in
   (a) Water energy in the form of potential energy
   (b) Wind energy in the form of kinetic energy
   (c) Biomass in the form of chemical energy
   (d) All of them

6.2 Hydropower is derived from
   (a) Hydraulic energy
   (b) Wind energy
   (c) Geothermal energy
   (d) All of them

6.3 Hydraulic power consists of
   (a) Turbine and generator
   (b) Water head, turbine and generator

(c) Water head and generator
(d) None of them

6.4 A Pelton wheel turbine works on principle of
(a) Reaction
(b) Impulse
(c) Dynamic force
(d) All of them

6.5 A Pelton wheel turbine gives mechanical power if
(a) Moving water jet and stationary vertical plate
(b) Moving water jet and vertical plate
(c) Stationary water jet and moving vertical plate
(d) None of them

6.6 Maximum mechanical power is derived from a Pelton wheel if
(a) The vertical plate is horizontal ($\theta = 0$)
(b) The vertical plate is inclined ($\theta \neq 0$)
(c) The vertical plate is vertical ($\theta = 90^\circ$)
(d) None of them

6.7 For maximum power from a Pelton wheel turbine
(a) The vertical plate velocity is equal to the water-jet velocity
(b) The vertical plate velocity is less than the water-jet velocity
(c) The vertical plate velocity is greater than the water-jet velocity
(d) The vertical plate velocity is half the water-jet velocity

6.8 The mechanical power from a Pelton wheel turbine for a given capacity depends on
(a) The radius of water jet (r_j)
(b) The length of vertical blade (R)
(c) Velocity of water jet
(d) Velocity of vertical jet

6.9 The maximum mechanical efficiency of vertical plate Pelton wheel turbine is
(a) 100%
(b) 25%
(c) 50%
(d) 75%

6.10 The symmetrically curved vane of a Pelton wheel turbine gives
(a) Less mechanical power in comparison with a vertical plate
(b) Equal mechanical power in comparison with a vertical plate
(c) Higher mechanical power in comparison with a vertical plate
(d) None of them

6.11 The maximum mechanical efficiency of a symmetrically curved vane Pelton wheel turbine is
(a) 100%
(b) 25%
(c) 50%
(d) 75%

6.12 The Pelton wheel turbine gives maximum mechanical power
(a) For friction/rough surface of symmetrically curved vane/vertical plate
(b) For polished surface of symmetrically curved vane/vertical plate
(c) For white surface of symmetrically curved vane/vertical plate
(d) All of them

6.13 The mechanical power of a symmetrically curved vane Pelton wheel turbine is a maximum if
(a) The derivative of the water jet before and after the strike with the vane is 90°

(b) The derivative of the water jet before and after the strike with the vane is 180°
(c) The derivative of the water jet before and after the strike with the vane is parallel
(d) None of them

6.14 The expression for mechanical efficiency of symmetrically curved van Pelton wheel turbine is given by
(a) $\eta_m = (1 + \cos\theta)$
(b) $\eta_m = (1 + \cos\theta)$
(c) $\eta_m = (1 + \cos\theta)^2$
(d) None of them

6.15 For an impulse turbine
(a) $\frac{r_j}{R} = 1$
(b) $\frac{r_j}{R} > 1$
(c) $\frac{r_j}{R} < 1$
(d) $\frac{r_j}{R} \rightarrow \infty$

6.16 For a reaction turbine
(a) $\frac{r_j}{R} = 1$
(b) $\frac{r_j}{R} > 1$
(c) $\frac{r_j}{R} > 1$
(d) $\frac{r_j}{R} \rightarrow \infty$

6.17 The water-jet velocity (u_j) depends on
(a) Water head (H_a)
(b) Jet radius (r_j)
(c) Length of nozzle
(d) All of them

6.18 The volume flow rate (Q) depends on
(a) Jet radius (r_j)
(b) Jet velocity (u_j)
(c) Plate velocity (u_t)
(d) All of them

6.19 The mass flow rate ($\dot{m}$) depends on
(a) Jet radius (r_j)
(b) Jet velocity (u_j)
(c) Plate velocity (u_t)
(d) All of them

6.20 The mechanical power of a Pelton wheel turbine is positive for
(a) $u_t = 0$
(b) $u_t > 0$
(c) $u_t = u_j$
(d) All of them

6.21 The mechanical power of a Pelton wheel turbine is maximum for
(a) Vertical plate vane
(b) Symmetrically curved vane
(c) Unsymmetrically curved vane
(d) None of them

6.22 The volume flow rate of water steam in a dam is
(a) $\propto H$
(b) $\propto H^2$

(c) $\propto H^{3/2}$
(d) None
where H is the height of water surface from crest.

6.23 The available water head (H_a) through a pipe is
(a) Less than the total water head ($H_a < H_t$)
(b) Equal to the total water head ($H_a = H_t$)
(c) More than the total water head ($H_a > H_t$)
(d) None of above

6.24 The available water head for free fall of water is
(a) Less than the total water head ($H_a < H_t$)
(b) Equal to the total water head ($H_a = H_t$)
(c) More than the total water head ($H_a > H_t$)
(d) None of above

6.25 Overall efficiency of hydropower system efficiency (η) depends on
(a) Water head efficiency
(b) Mechanical efficiency (η_{jm})
(c) Electrical efficiency of generator (η_{me})
(d) All of them

6.26 Overall efficiency of hydropower system (η) is
(a) Less than the head efficiency
(b) Less than the mechanical efficiency
(c) Less than the electrical efficiency of a generator
(d) All of them

6.27 A hydropower system is
(a) Renewable
(b) Sustainable
(c) Environmental friendly
(d) All of them

6.28 A hydropower system provides
(a) Clean energy
(b) Sustainable energy
(c) Nonpolluting energy
(d) All of them

6.29 The expression for the maximum mechanical power of a curved-vane Pelton wheel turbine is
(a) $\frac{1}{2}\rho Q u_j^2$
(b) $\frac{1}{2}\rho Q u_j^3$
(c) $\frac{1}{2}\dot{m} u_j^2$
(d) All of them

6.30 The shape number of a Pelton wheel turbine is
(a) $= 1$
(b) > 1
(c) < 1
(d) $= 0$

6.31 For a given shape number of a Pelton wheel turbine, the radius of the turbine (R)
(a) Increases with increase of mechanical efficiency (η_{jm})
(b) Decreases with increase of mechanical efficiency (η_{jm})

(c) Unaffected with increase of mechanical efficiency (η_{jm})
(d) None of the above

6.32 The energy contained in water at height "H" is given by
(a) ρQgH
(b) $\frac{mgH}{2}$
(c) $\dot{m}gH$
(d) All of them

6.33 An hydraulic ram pump is used for
(a) Lifting of water
(b) Storing of water
(c) Irrigation purpose
(d) None of them

6.34 An hydraulic press works on the principle of
(a) Bernoulli's theorem
(b) Pascal's law
(c) Newton's law
(d) All of them

6.35 The diffusion coefficient (Cd) for open-flow conditions is
(a) $C_d = 0$
(b) $C_d > 1$
(c) $C_d < 1$
(d) $C_d = 1$

ANSWERS

6.1 **(a)**; 6.2 **(d)**; 6.3 **(a)**; 6.4 **(c)**; 6.5 **(c)**; 6.6 **(b)**; 6.7 **(a)**; 6.8 **(c)**; 6.9 **(b)**; 6.10 **(d)**; 6.11 **(d)**; 6.12 **(a)**; 6.13 **(d)**; 6.14 **(a)**; 6.15 **(d)**; 6.16 **(a)**; 6.17 **(a)**; 6.18 **(a)**; 6.19 **(a)**; 6.20 **(c)**; 6.21 **(a)**; 6.22 **(d)**; 6.23 **(a)**; 6.24 **(a)**; 6.25 **(a)**; 6.26 **(d)**; 6.27 **(a)**; 6.28 **(a)** & **(c)**; 6.29 **(a)**; 6.30 **(b)**; 6.31 **(a)**; 6.32 **(a)** & **(c)**; 6.33 **(a)** & **(b)**; 6.34 **(d)**; 6.35 **(b)**.

REFERENCES

1. Bartle, A. 2002. *Energy Policy*, **30**, 1231–1239.
2. Lal Jagadish 1985. *Hydraulic Machines*. 6th edn, Metropolitan Book Co. Private Ltd, New Delhi.
3. Frey, G. W. and Linke, D. M. 2002. *Energy Policy*, **30**, 1261–1265.

The Source of Wind Energy is the Day-Night Cycle Due to Temperature Variations. It is Renewable and Provides Clean Mechanical and Electrical Power to Meet the Needs of Human Beings without Disturbing the Ecosystem.

CHAPTER 7

Wind Energy

7.1 INTRODUCTION

In the past, wind energy was used (i) to propel ships (ii) to produce mechanical energy for pulling up underground water from wells and (iii) grinding agriculture products. There is also evidence that suggests that the ancient Egyptians used windmills to pump water for irrigating agricultural lands and to grind grains during 3600 B.C. Wind is simply air in motion that carries kinetic energy with it. The kinetic energy is converted into first mechanical and then electrical energy by generation. The mechanical energy of wind can be used for driving ships, pumping water, grinding grains, *etc.*

The harnessing of electrical power from wind is gaining momentum due to the depletion of fossil fuels and their rising running cost. Moreover, wind energy is considered to be a green/clean power technology. It has minor impacts on the economy and environment. Hence, wind energy, which is a renewable sources of energy, can be harnessed to provide an environmentally friendly and reliable source of energy without producing any air pollutants or greenhouse gases. The kinetic energy of wind is captured by the wind turbine that is mechanically coupled to an electrical generator. The turbine is mounted on a tall pillar to enhance the energy capture. In the 19th century, wind turbines contributed greatly to the economic development of many countries like the Netherlands, Denmark and the USA. The use of wind energy declined very fast everywhere due to the cheap availability and exploitation of coal, oil and gas resources.

The old wind turbines were no longer economically competitive with conventional sources of energy. Therefore, very little research was done to develop new and more efficient wind turbines. Due to the energy crisis during 1973 the development of new and more efficient wind turbines was resumed to generate electricity. As a result, the cost of electricity produced by wind turbines decreased dramatically due to improved technology. Nowadays, the extraction of electrical power with modern turbines from the wind is an established industry. A device for direct mechanical work is often called a windmill or just wind turbines. If electricity is produced, the combination of turbine and generator may be called a wind generator or aerogenerator that is also referred as a **wind energy-conversion system (WECS)**.

Generated electricity from wind has been used in three modes namely (a) small wind electric generators below 4 kW capacity for battery chargers; (b) wind electric generators in the range of 20 to 100 kW in standalone model supplemented by power from diesel generator sets and (c) wind electric generators in the range of 50 to 300 kW capacities have been used in grid-connected wind farms.

Advanced Renewable Energy Sources
G. N. Tiwari and R. K. Mishra

Published by the Royal Society of Chemistry, www.rsc.org

Being a clean and ecofriendly, wind energy has the limitation of its intermittent nature, like solar energy. Wind potential is a localised concept in comparison with solar energy. Power fluctuations due to uncertainty of wind, variations in magnitude and directions of wind velocities, structural instability due to heavy gusts and cyclonic storms are also some of the problems associated with wind energy-conversion systems. It can be harnessed in a clean and inexhaustible manner through the application of technically advanced and efficient systems.

7.2 HISTORICAL DEVELOPMENT

The wind has played a long and important role in the history of human civilisation. Wind power has been harnessed by mankind for thousands of years. Since the earliest recorded history, wind power has been used to move ships, grind grain and pump water. There is evidence that wind energy was used to propel boats along the Nile River as early as 5000 B.C. The first true windmill, a machine with vanes attached to an axis to produce circular motion, may have been built as early as 2000 B.C. in ancient Babylon. By the 10th century A.D., windmills with wind-catching surfaces as long as 16 feet and as high as 30 feet were grinding grain in the area now known as eastern Iran and Afghanistan. The western world discovered the windmill much later. The earliest written references to working wind machines date from the 12th century. These were used for milling grain. It was not until a few hundred years later that windmills were modified to pump water and reclaim much of Holland from the sea. The first horizontal-axis windmill appeared in England around 1150, in France 1180, in Flanders 1190, in Germany 1222 and in Denmark 1259. This fast development was most likely influenced by the Crusaders, taking the knowledge about windmills from Persia to many places in Europe. The people of Holland improved the basic design of the windmill. They gave it propeller-type blades made of fabric sails and invented ways for it to change direction so that it could continually face the wind. Windmills helped Holland become one of the world's most industrialised countries by the 17th century. The first person, who generated in 1891 electricity from wind speed, was the Dane Poul LaCour, who lived in Denmark. He had also received meteorology education and used the wind tunnel for the first time in order to obtain some theoretical formulations. Danish engineers improved the technology during World Wars I and II and used the technology to overcome energy shortages.

In Europe, windmill performance was constantly improved between the 12th and 19th centuries. By 1800, about 20 000 modern European windmills were in operation in France alone. And in the Netherlands, 90% of the power used in industry was based on wind energy. Industrialisation then led to a gradual decline in windmills, but even in 1904 wind energy provided 11% of the Dutch industry energy and Germany had more than 18 000 units installed. American colonists used windmills to grind wheat and corn, pump water, and cut wood. As late as the 1920s, Americans used small windmills to generate electricity in rural areas without electric service. When power lines began to transport electricity to rural areas in the 1930s, local windmills were used less and less, though they can still be seen on some Western ranches. The popularity of windmills in the US reached its peak between 1920 and 1930 with about 600 000 units installed. Various types of American windmills are still used for agricultural purposes all over the world.

In the 1930s and 1940s, hundreds of thousands of electricity producing wind turbines were built in the US. They had two or three thin blades, which rotated at high speeds to drive electrical generators. These wind turbines provided electricity to farms beyond the reach of power lines and were typically used to charge storage batteries, operate radio receivers and power a light bulb or two. By the early 1950s, however, the extension of the central power grid to nearly every American household, *via* the Rural Electrification Administration, eliminated the market for these machines.

The popularity of using the energy in the wind has always fluctuated with the price of fossil fuels. When fuel prices fell after World War II, interest in wind turbines waned. But when the price of oil skyrocketed in the 1970s, so did worldwide interest in wind-turbine generators. The wind-turbine technology R&D that followed the oil embargoes of the 1970s refined old ideas and introduced new ways of converting wind energy into useful power. Many of these approaches have been demonstrated in "wind farms" or wind-power plants – groups of turbines that feed electricity into the utility grid – in the US and Europe. The wind technology has improved step by step since the early 1970s. By the end of the 1990s, wind energy had re-emerged as one of the most important sustainable energy resources.

Today, the lessons learned from more than a decade of operating wind power plants, along with continuing R&D, have made wind-generated electricity very close in cost to the power from conventional utility generation in some locations.

As for the background in Turkey, wind energy has always played an important role in the historical and economical development of Asia Minor and the geographical area covered by the Republic of Turkey today. The earliest documented evidence of this statement goes back to the ancient city of Troia. It is not known when the first windmills were installed in Anatolia. However, they must have been dominant landmarks already in the 14th century. A naval map dated 1389 AC shows windmills as landmarks along with the shallows and sand banks in the Bay of Izmir.[1]

In the 1940s windmills ground corn, pumped water to fields and even powered the first radio sets at the Anatolia countryside. Based on a survey performed by the Turkish Ministry of Agriculture between 1960 and 1961, there were 749 windmills. Of these, 718 were used for water pumping, while 41 were for generating electricity. Two surveys between 1966–1967 and 1978–1979 revealed 309 and 894 units, of which 2 and 23 were electricity-producing turbines with capacities lower than 1 kW, respectively. Since the 1960s, several universities have conducted studies on wind energy. The Turkish Scientific and Technologic Research Institution (TUBITAK) Marmara Research Centre has started with studies towards developing a wind atlas for Turkey since the 1980s. The General Directorate of Electrical Power Resources Survey Administration (EIE) has made some wind measurements. Electricity generation through wind energy for general use was first realised at the Cesme Altinyunus Resort Hotel (The Golden Dolphin Hotel) in Izmir, Turkey in 1986 with a 55-kW nominal wind power capacity. This hotel with 1000 beds consumes about 3 million kWh of electrical energy annually, while the windmill installed produces 130 000 kW h per year approximately. Between 1986 and 1996, there were some attempts to generate electricity from wind, but they were never successful. In 1994, the first build-operate-transfer (BOT) feasibility study for a wind energy project in Turkey was presented to the Ministry of Energy and Natural Resources of Turkey (MENR).[2] Apart from initial high investment costs in harnessing wind energy, lack of adequate knowledge on the wind-speed characteristics in the country is the main reason for the failure to harvest the energy from wind. In terms of generating electricity from wind, the development of wind energy in Turkey started in 1998 when some wind plants were installed at several locations in the country. By January 1998, there were 25 applications for wind energy projects recorded at the MENR. To date, three wind power plants have been installed with a total capacity of about 18.9 MW. Considering the installation of a wind plant with a capacity of 1.2 MW in November 2003, the total installed capacity will reach 20.1 MW. Recently, small wind-turbine systems with capacities ranging from 1.5 to 5 kW have also been installed in some Turkish universities for conducting wind energy investigations as well as for lighting purposes.[3] Wind energy projections made by some institutions and organisations, such as the International Energy Agency, European Commission and BTM Consult ApS, differ from each other, as given in detail elsewhere. These projections are based on the studies performed between 1997 and 2000.[4] According to a forecast of 2003 made by BTM Consult ApS for the period up to 2007, an average growth rate of 11.2% yearly is projected, while for 2003 a growth of 24% over 2001 is expected. Total demand during the 5-year period is estimated to be 51 000 MW. By the end of 2007, 83 000 MW of capacity will be on line, of which 58 600 MW is in Europe. According to long term prediction up to 2012, an annual installation level of 24 000 MW by 2012 is expected. Cumulative installation growth to 177 000 MW is projected to equal a penetration of wind power close to 2% of the world's electricity consumption by 2012.

7.3 BASIC CONCEPTS OF LIFT AND DRAG FORCES

Flowing of fluid air over the solid bodies frequently occurs in practice. It is responsible for numerous physical phenomena namely the lift force development by airplane wings, dust particles in high-wind turbines, *etc.*, and the drag force acting on automobiles, trees, *etc.* The two processes, namely fluid moves over a stationary body and body moves through a quiescent fluid are equivalent to each other. The relative motion between the body and the fluid is important. These motions are referred to as flow over bodies or external flow.

7.3.1 Drag Force (*D*)

When a body is forced to move through a fluid especially through a liquid, it is found that the body meets some resistance. It is well known that it is very difficult to walk in water because of the much greater resistance offered by water to motion as compared to air. It is also felt that the strong push is exerted by the flowing wind on the human body. **The force exerted by a flowing fluid on a body in the flow direction is called drag force**. It can be directly measured by attaching the body subjected to fluid flow to a calibrated spring and measuring the displacement in the flow direction of. Drag is usually an undesirable effect, like friction and is minimised as far as practicable. Reduction of drag is closely associated with the reduction of fuel consumption in automobiles, submarines, aircraft, *etc.* In some cases drag produces a very beneficial effect and is maximised. Friction for example is a life saver in the brakes of automobiles. Likewise, it is the drag that helps people to parachute, pollen to fly to distant locations, *etc.*

7.3.2 Lift Force (*L*)

A stationary fluid exerts only normal forces on the surface of a body immersed in it. A moving fluid, however, also exerts tangential shear forces on the surface of a body. Both of these forces have the components in the direction of the flow. Hence, drag force is due to the combined effects of pressure and wall shear forces in the flow direction. The components of pressure and wall-shear forces in the direction normal to the fluid flow that tend to move the body in that direction are called lift.

In a moving car when a hand is extended out of the window, the hand is moved towards the rear side of the car and due to lift force, it is moved towards the ceiling of the car. Hence, both the lift and drag forces act on a body immersed in a mass of moving fluid, as shown in Figure 7.1.

Let P and τ be pressure (N/m^2) and wall shear stress (viscous force) acting at M, as shown in the Figure 7.2. The θ is an angle of incidence of wind on a body. The net pressure and shear force on a differential area $\mathrm{d}A$ are $P\mathrm{d}A$ and $\tau\mathrm{d}A$, respectively. Let a body held stationary in a stream of air (fluid) moving at a uniform velocity v_0. The pressure force $P\mathrm{d}A$ acts normal to the surface and the shear force $\tau\mathrm{d}A$ acts along the tangential direction of the body. The drag force on the body is therefore the sum of the components of both pressure force ($P\mathrm{d}A$) and shear force ($\tau\mathrm{d}A$) acting over the entire surface of the body in the direction of motion.

The component of the shear force in the direction of flow of fluid is called the friction drag ($\mathrm{d}F_{\mathrm{Df}}$) that may be expressed as

$$\text{Friction drag}\,\mathrm{d}F_{\mathrm{Df}} = \tau\mathrm{d}A\cos\theta$$

Similarly, the components of the pressure forces in the direction of the fluid motion is called the pressure drag $\mathrm{d}F_{\mathrm{DP}}$ and is expressed as

$$\text{Pressure drag}\,\mathrm{d}F_{\mathrm{DP}} = P\mathrm{d}A\sin\theta$$

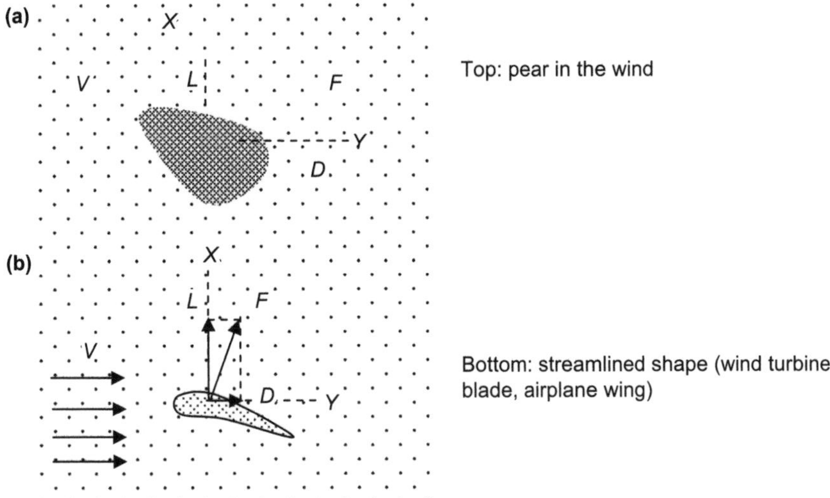

Figure 7.1 View of lift and drag forces in the immersed bodies (V = wind velocity, F = net force exerted by the wind on blade).

The net differential drag force dF_D acting on the body is therefore equal to the sum of the friction drag and the pressure drag. Thus

$$dF_D = dF_{DF} + dF_{DP} = \tau dA \cos\theta + PdA \sin\theta \tag{7.1}$$

The relative magnitude of the total drag depends on the shape and position of the immersed body.

For example:

(i) if a thin flat plate is held immersed in a fluid, parallel to the direction of flow, the pressure drag ($PdA \sin\theta$) is practically equal to zero due to minimum value of dA, hence the differential drag force (dF_D), which is only the friction drag (dF_{DF}) becomes

$$dF_D = \tau dA \cos\theta$$

(ii) if the same plate is held perpendicular to the air flow, the friction drag ($\tau dA \cos\theta$) is equal to zero and the total drag is due to the pressure difference between the upstream and downstream sides of the plate hence the differential drag force (dF_D) becomes

$$dF_D = PdA \sin\theta$$

In between these two extreme cases, there are several shapes of the body for which the contribution of each of the two components to the total drag varies considerably depending on the shape and position of the immersed body and the flow and fluid characteristics.

Similarly, the net lift force (dF_L) on the differential body is given by the sum of the components of shear and the pressure forces acting over the surface of the body (dA) in the direction perpendicular to the direction of the fluid motion. Thus,

$$dF_L = \tau dA \sin\theta - PdA \cos\theta \tag{7.2}$$

The total drag and lift forces acting on the body can be determined by integrating Eq. (7.1) over the entire surface of the body. This is not practical since the detailed distributions of pressure and

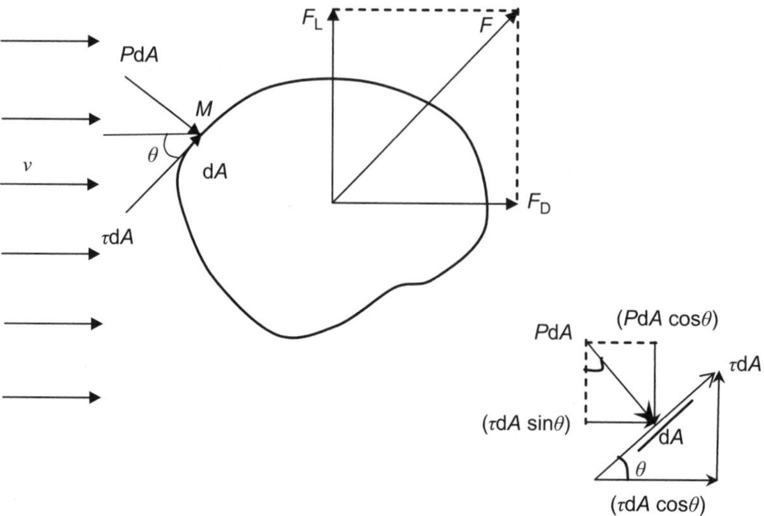

Figure 7.2 View of components of pressure and frictional forces on an element of surface of an immersed body.

shear forces over the entire surface are difficult to obtain by analysis or measurement. In practice, the resultant drag and lift forces acting on the entire body can be measured directly in a wind tunnel.

The airfoil of the wind turbine is shaped specifically to produce the maximum lift force when coming into contact with the flowing air. This is achieved by making the top surface of the airfoil curved and the bottom surface nearly flat. The fluid flowing over the airfoil travels a longer distance to reach the end of the airfoil causing a pressure difference, as mentioned earlier. The pressure difference generates an upward force that tends to lift the airfoil and thereby the rotation of the rotor of the wind turbine. The drag and lift forces depend on the density (ρ) of the fluid, the upstream velocity of the wind (v_0), size, shape and orientation of the body. It is better to work with appropriate dimensionless numbers that represents the drag and lift characteristics of the body. These numbers are the drag coefficients (C_D) and lift coefficients (C_L) and they are defined as

$$\text{Drag coefficient, } C_D = \frac{F_D}{\frac{1}{2}\rho v_0^2 A} \tag{7.3}$$

$$\text{Lift coefficient, } C_L = \frac{F_L}{\frac{1}{2}\rho v_0^2 A} \tag{7.4}$$

where A is the frontal area projected on a plane normal to the direction of the flow of the body. The frontal area of a cylinder of diameter D and length L, for example is $A = LD$.

For an airfoil of width (or span) b and chord length c (the length between the leading edge and trailing edge), the platform area $A = bc$. In most cases the drag and lift coefficients are functions of the shape and roughness of the body. The term $\frac{1}{2}\rho v_0^2$ is the dynamic pressure due to wind.

It is desirable for airfoils to generate the most lift force while producing the least drag. Therefore, a measure of performance of airfoil is the lift to drag ratio, which is equivalent to the ratio of lift to drag coefficients C_L/C_D. This information is provided by either plotting C_L versus C_D for different values of the angles of attack as shown in Figure 7.3. The ratio C_L/C_D increases with the angle of attack, until the airfoil stalls. Similarly, the variations of C_D as well as C_L/C_D with the angle of attack for an airfoil NACA 0012 are shown in Figures 7.4 and 7.5. From these figures it is seen that

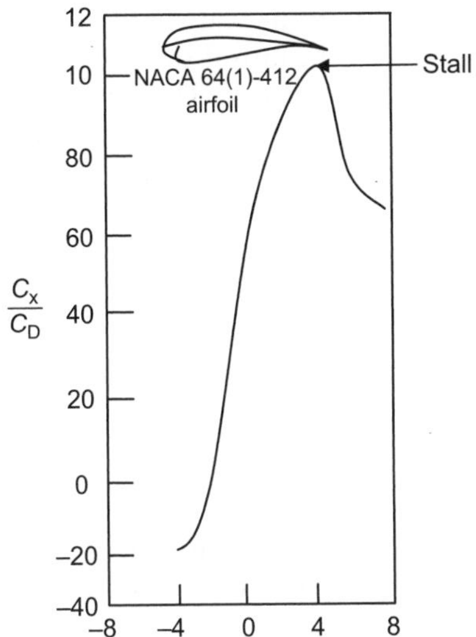

Figure 7.3 Effect of the angle of attack of an airfoil on lift-to-drag ratio.

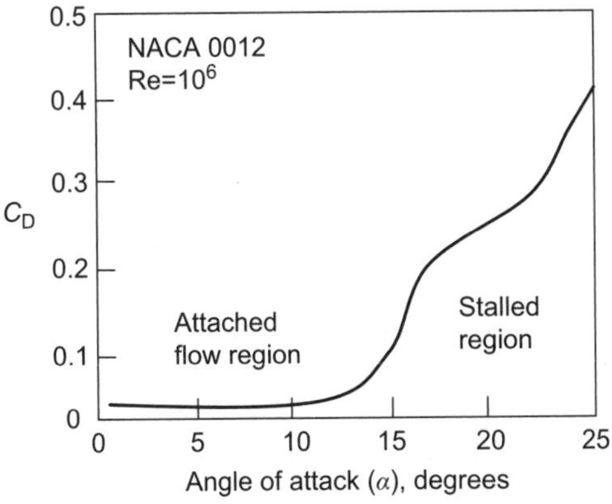

Figure 7.4 Effect of the angle of attack of an airfoil on C_D for the NACA 0012 aerofoil.

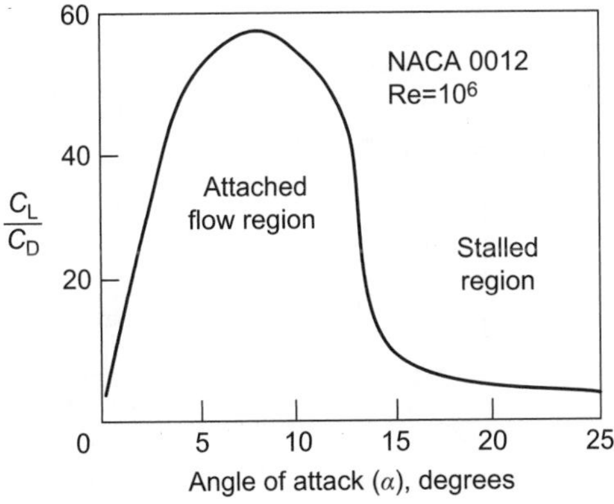

Figure 7.5 Effect of the angle of attack of an airfoil on Lift/Drag ratio for the NACA 0012 aerofoil.

there is no significant variation of C_D with the increase of the angle of attack. C_D increases abruptly when the angle of attack exceeds 13° and the airfoil stalls. The values C_L/C_D from Figure 7.5 are seen to be highest at around a 10° angle of attack. Therefore, the lift/drag ratio has a significant effect upon the efficiency of a wind turbine and it is desirable that a turbine blade must operate at the maximum ratio for better energy extraction from the wind.

7.4 TYPES OF TURBINE

There are two types of wind energy-conversion devices namely, aerodynamic lift and aerodynamic drag, which is also referred to as wind turbines (rotors).

7.4.1 Lift Type Wind Turbine

A high-speed turbine depends on lift forces to move the blades of the wind turbine. The linear speed of the blades is usually several times higher than the wind speed. The torque of lift force is low as compared to the drag type.

7.4.2 Drag Type Wind Turbine

Low-speed turbines are slower than the wind. They are mainly driven by the drag force. The torque at the rotor shaft is relatively high. Wind turbines are classified as (a) horizontal axis and (b) vertical axis wind turbines. In horizontal-axis turbines the axis of rotation is horizontal with respect to the ground. In this case, the rotating shaft is parallel to the ground and the blades are perpendicular to the ground. In vertical-axis turbines, the axis of rotation is vertical with respect to the ground. The configurations of horizontal- and vertical-axis turbines are shown in Figure 7.6. Horizontal-axis or propeller-type turbines are more common and highly development than the vertical-axis designs.

Vertical-axis machines operate independently in the wind direction. Gear box and generating machinery can be directly coupled to the axis at the ground level, as shown in Figure 7.6. Vertical-axis machines have the following disadvantages: they are not self-starting, the speed regulation in high winds is difficult; the torque fluctuates with each revolution as the blades move into and away

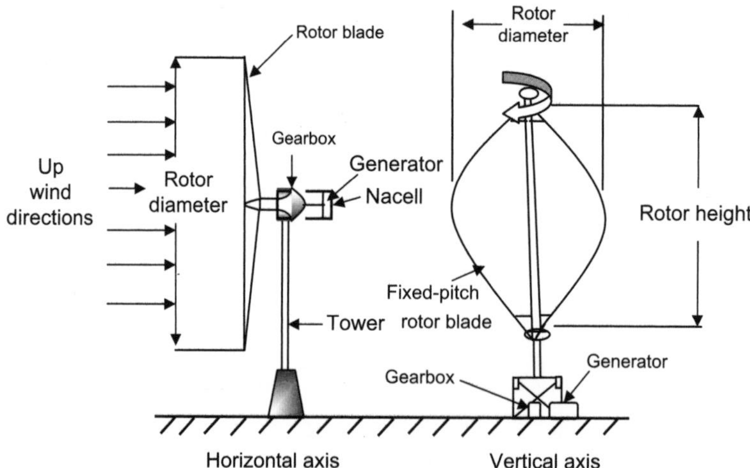

Figure 7.6 View of horizontal-axis and vertical-axis wind turbine.

from the wind. Due to these disadvantages, most working machines are of the horizontal-axis type. The various types of vertical-axis wind turbines are as follows:

(a) **Anemometer:** This rotates by drag force. Due to the small size of the cup, there is a linear relationship between rotational frequency and wind speed ($v = r\omega$).
(b) **Savonius rotor (turbo machine):** There is a complicated motion of wind through and around the two curved sheet airfoils rotates by drag force, has a simple construction and is inexpensive.
(c) **Darrieus rotor:** This rotator consists of two or three curved blades; the driving forces are lifting forces. The maximum torque occurs when a blade is moving across the wind of a speed much high than the wind speed. Initial movement may be initiated with the electrical generator used as a motor.
(d) **Musgrove rotor:** In this rotor, the blades are vertical for normal power generation. This rotor has an advantage of fail-safe shut down in strong winds.
(e) **Evans rotor:** Vertical blades twist about a vertical axis speed for control and a fail-safe shut down.

Rotors can also be classified on the basis of their movement at variable speed or constant speed. For water pumping and small-battery operation, it is desirable to allow the rotor speed to vary. However, for the large scale generation of electricity, it is common to operate wind turbines at constant speed. This allows the use of simple generators whose speed is fixed by the frequency of the electrical network. Variable-speed wind turbines are sometimes used for electricity generation but a power electronic frequency converter is then required to connect the variable frequency output of the wind turbine to the fixed frequency of the electrical system.

7.5 AERODYNAMICS OF WIND TURBINE

In wind turbines, aerodynamics deals with the relative motion between the moving air and the stationary airfoil. The airfoil is the cross section of the blade of the wind turbine. It is the shape designed to create maximum lift force when air flows over it. In the wind turbine, linear kinetic energy associated with the wind is converted into the rotational motion that is required to turn the electrical generator for power generation. This change is accomplished by a rotor that has one, two or three blades or airfoils attached to the hub. The wind flowing over the surfaces of these airfoils generates the forces that cause the motor to run. The basic principle of aerodynamics of a horizontal-axis wind turbine is shown in Figure 7.7a.

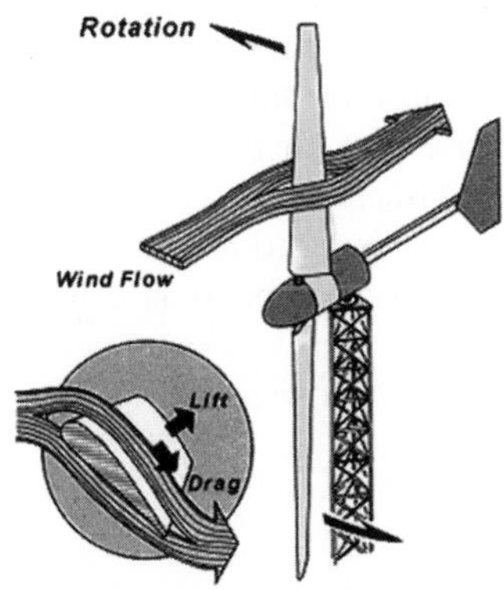

Figure 7.7a View of wind turbine aerodynamic lift.

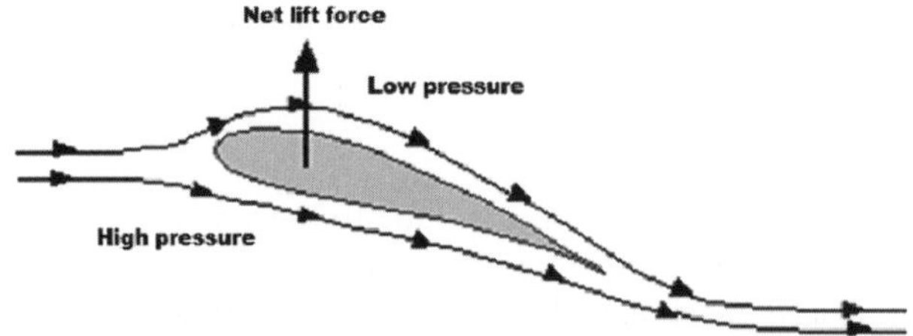

Figure 7.7b View of aerodynamic lift force on an airfoil section of wind turbine.

Wind passes more rapidly over the longer (upper) path of the airfoil in comparison to the shorter (lower) path, as shown in Figure 7.7b. High- and low-pressure regions can be identified by using Bernoulli's equation (Chapter 1). The pressure is low at locations where the flow velocity is high and the pressure is high at locations where flow velocity is low. Therefore, low pressure is created in the upper surface of the airfoil and high pressure in its lower surface. The pressure difference between the top and bottom surfaces of the airfoil results in a force called the aerodynamic lift as air moves from the high-pressure region to the low-pressure region. The upward force due to aerodynamic lift pushes the blades to move up. Since the blades of the wind of the wind turbine are constrained to move up with the hub at its centre, the lift force causes the rotation of the blade about the hub. Air flowing smoothly over an airfoil produces two forces; the force perpendicular to the air flow and drag, which acts in the direction of flow.

In wind turbines, the drag force perpendicular to the lift force also acts on the blade causing the impediment or rotor rotation. The prime objective in wind-turbine design is for the blade (airfoil-shaped) to have a relatively high lift to drag ratio. This ratio can be varied along the length of the blade to optimise the output energy of the turbine at various wind speeds. Hence, in aerodynamic analysis of wind turbines, both lift and drag forces are important for their optimisation in efficient design. In the next section, the basic concept of lift and drag forces are discussed.

7.6 MOMENTUM THEOREM

The simplest model for wind-turbine aerodynamics is the momentum theorem. In this model, the rotor is approximated by an actuator disk. The kinetic energy of the wind is extracted in the wind turbines. After extraction of kinetic energy, the velocity of flowing wind decreases. Further, the affected mass of air due to the rotor disk remains separate from the air that does not pass through the rotor disk. Pressure energy can be extracted in a step-like manner. Hence, the model assumes a sudden drop in pressure at the rotor disk. The variation in velocity is assumed to be smooth from far upstream. The air then proceeds downstream with reduced speed and static pressure. This region of flow is called the **wake**. Eventually, the static pressure in the wake becomes the atmospheric level at far downstream. The rise of static pressure is at the expense of the kinetic energy and thereby wind speed further decreases. Thus, between the far-upstream and far-wake conditions, no change in static pressure exists, but there is a reduction in kinetic energy.

The device that carries out the extraction process is called an actuator disk, as shown in the Figure 7.8.

The assumptions for this model are as follows:

- flow is regarded as an incompressible fluid;
- sudden drop in pressure occurs at the rotor disk;
- the flow velocity varies smoothly and in a continuous manner from far upstream to far downstream;

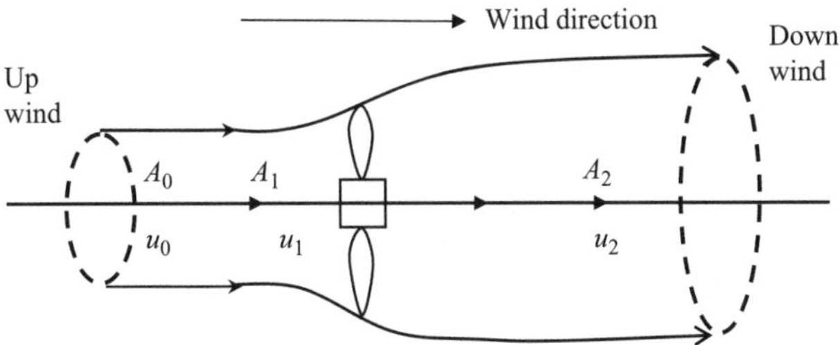

Figure 7.8 Flow diagram of energy extraction actuator disk of wind turbine.

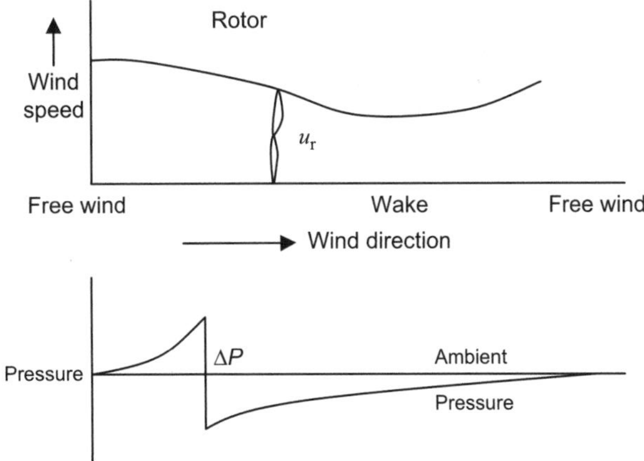

Figure 7.9 Conditions in traversing a wind turbine.

- the rotation of wake after rotor is ignored;
- the flow is steady;
- the mass flow rate is constant at far upstream, at the rotor and at far downstream.

The variations of wind speed as well as pressure in the actuator disk (rotor) are shown in Figure 7.9.

7.6.1 Energy Extraction

The kinetic energy of the air stream passing through the turbine rotor as shown in Figure 7.10 of cross-sectional area A per unit time is expressed as

$$\text{kinetic energy per unit time} = \frac{1}{2}\left(\frac{m}{t}\right)v_0^2 = \frac{1}{2}(\rho A_1 v_0)(v_0^2) = \frac{1}{2}\rho A_1 v_0^3$$

Here, ρ is the air density and v_0 is the free wind speed in the upstream side of the airflow. The above expression also represents the power in the wind at the speed v_0.

Hence, the initial power associated with wind is given by

$$P_0 = \frac{1}{2}\rho A_1 v_0^3 \tag{7.5}$$

For $\rho = 1.2$ kg/m^3 at sea level and
(i) $v_0 = 10$ m/s, $P_0 = 600$ W/m^2 and (ii) $v_0 = 25$ m/s, $P_0 = 10\ 000$ W/m^2

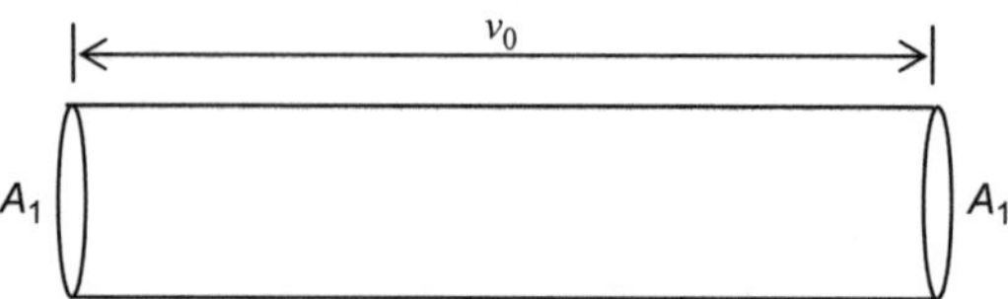

Figure 7.10 Cross-sectional view of the wind-turbine rotor.

In case (i), the input power in 600 W/m^2, which is equivalent to the average daily insolation on horizontal surface, *i.e.* 500 W/m^2. Hence, the optimum wind velocity for power extraction should be 10 m/s.

In Figure 7.8, A_1 is the rotor swept area and A_0 and A_2 are area of upwind and downwind enclose the stream of constant air mass passing through A_1.

The force (F) or thrust on the wind turbine (actuator) is the rate of change of momentum from the air mass flow rate $\dot{m}$, which is given by

$$F = \dot{m}v_0 - \dot{m}v_2 \tag{7.6}$$

The power extracted by the wind turbine is obtained by multiplying force/thrust and v_1 as

$$P_T = Fv_1 = (\dot{m}v_0 - \dot{m}v_2)v_1 \tag{7.7}$$

The rate of loss of kinetic energy of wind flowing across the wind turbine is given by

$$P_w = \frac{1}{2}\dot{m}(v_0^2 - v_2^2) \tag{7.8}$$

By equating Eqs. (7.7) and (7.8), one gets

$$(v_0 - v_2)v_1 = \frac{1}{2}(v_0^2 - v_2^2) = \frac{1}{2}(v_0 - v_2)(v_0 + v_2) \tag{7.9}$$

or,

$$v_1 = \frac{v_0 + v_2}{2} \tag{7.10}$$

The mass of air flowing through the disk per unit time is given by

$$\dot{m} = \rho A_1 v_1 \tag{7.11}$$

Substituting the expression for $\dot{m}$ from Eq. (7.11) in Eq. (7.12) one gets,

$$P_T = \rho A_1 v_1^2 (v_0 - v_2) \tag{7.12}$$

Now, substituting the expression for v_2 from Eq. (7.10) in Eq. (7.12), one gets

$$P_T = \rho A_1 v_1^2 \{v_0 - (2v_1 - v_0)\} = 2\rho A_1 v_1^2 (v_0 - v_1) \tag{7.13}$$

If the interference factor or perturbation factor (a) is defined as the fractional decrease of wind speed at the wind turbine as,

$$a = \left(\frac{v_0 - v_1}{v_0}\right) \Rightarrow v_1 = (1-a)v_0 \tag{7.14a}$$

Sometimes $b = \frac{v_2}{v_0}$ is also referred to as the interference factor (7.14b)

Now, using Eq. (7.10) in Eq. (7.13),

$$a = \left(\frac{v_0 - v_2}{2v_0}\right) \Rightarrow v_2 = (1-2a)v_0 \tag{7.15a}$$

From Eqs. (7.14b) and (7.15a) one can get

$$b = 2a \tag{7.15b}$$

Now, using Eq. (7.14) in Eq. (7.13)

$$P_T = 2\rho A_1(1-a^2)v_0^2[v_0 - (1-a)v_0]$$

or,

$$P_T = \frac{1}{2}\rho A_1 v_0^3\left[4a(1-a)^2\right] \tag{7.16}$$

or,

$$P_T = C_P P_0 \tag{7.17}$$

where,

$$C_P = 4a(1-a)^2 \tag{7.18a}$$

P_0 is the power of free wind and C_P, the fraction of power extracted, also called the power coefficient.

In order to maximise the value of C_P, differentiate Eq. (7.18a) with respect to "a", we have

$$\frac{dC_p}{da} = 1 + 3a^2 - 4a = 0$$

Thus, $a = 1$ or $1/3$, $a \neq 1$ as $C_p = 0$ when $a = 1$ and hence $a = 1/3$

Putting the value of $a = 1/3$ in Eq. (7.18a) one can get

$$C_{p\max} = \frac{16}{27} = 0.59$$

With the help of Eq. (7.14b), an expression for C_p can also be written as

$$C_P = \frac{1}{2}(1-b)(1+b)^2 \tag{7.18b}$$

EXAMPLE 7.1

By using Eq. (7.18b), find the condition for maximum "C_p".

Solution: From Eq. (7.18b), one has

$$C_P = \frac{1}{2}(1-b)(1+b)^2$$

For the maximum value of "C_p", differentiate the above equation with respect to "b" and equate it to zero, one gets

$$\frac{dC_p}{dt} = \frac{1}{2}\left[(1-b)\times 2(1+b) + (1+b)^2 \times (-1)\right]$$
$$= \frac{1}{2}(1+b)(1-3b) = 0 \Rightarrow b = -1 \text{ or } \frac{1}{3}$$

Since v_2 is not negative, condition $b=1/3$ is valid.

For $b=1/3, C_p = \frac{1}{2}\times\frac{2}{3}\times\frac{4}{3}\times\frac{4}{3} = \frac{16}{27} = 0.59$

EXAMPLE 7.2

Plot the curve between "C_p" and "b" (Example 7.1) and compare it with Figure 7.15.

Solution: The variation of "C_p" and "b" can be obtained by using Eq. (7.18b). The variation of "C_p" and "b" is shown below:

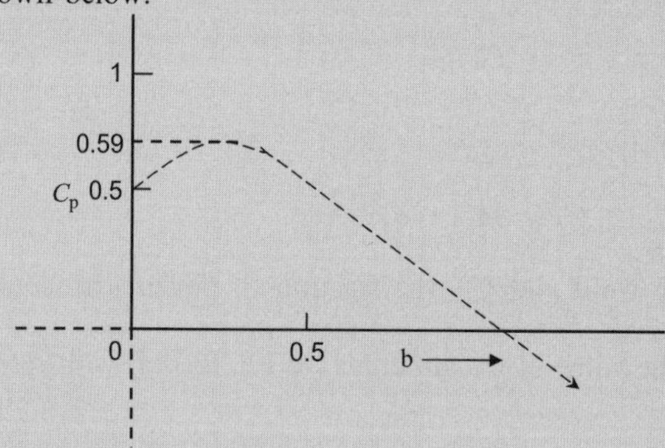

The above figure shows that

(i) The C_p is maximum at $b=1/3$ and the same at $a=1/3$.
(ii) The C_p is zero at b = 1 or $v_2 = v_0$ as expected.
(iii) The turbine loses power for $b>1$.

7.6.2 The Betz Limit

The Betz limit is the maximum achievable value of the power coefficient. This limit is applied to all types of wind turbine set in an extended fluid stream. This is also applied to power extraction from tidal and river sources. With conventional hydropower system, water reaches the turbine from a

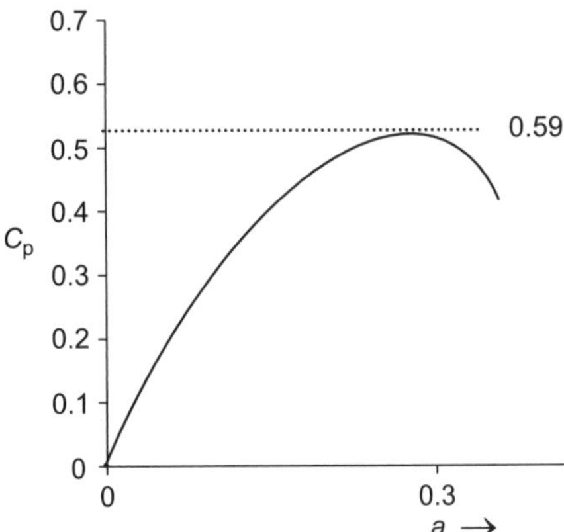

Figure 7.11 Effect of interference factor 'a' on power coefficient (C_p).

pipe and therefore this limit is inapplicable for a wind turbine which is not in extended flow condition. The variation of "C_p" with "a" has been shown in Figure 7.11.

Modern designs of wind turbines for electricity generation operate at C_p values of about 0.4. The major losses in efficiency for a real wind turbine arise from the viscous drag on the blades. While deriving C_p, the power coefficient, the extraction of power has to be taken from the mass of air in the stream to be through the activator disk of area A_1.

The power extracted per unit area of cross section equal to A_0 in the upstream side is greater than the power extracted per unit area of A_1. This is due to the fact that $A_0 < A_1$ and $v_0 > v_1$.

For maximum power extraction, we have $a = 1/3$

$$a = \frac{v_0 - v_1}{v_0} = \frac{1}{3} \Rightarrow 2v_0 = 3v_1 \Rightarrow v_0 = \frac{3}{2}v_1 \Rightarrow \frac{v_0}{v_1} = \frac{3}{2} \Rightarrow v_0 = \frac{3}{2}v_1$$

The mass flow rate is constant throughout, *i.e.*

$$\rho A_0 v_0 = \rho A_1 v_1 = \rho A_2 v_2 = \dot{m} \Rightarrow \frac{v_0}{v_1} = \frac{A_1}{A_0}$$

After substituting the value of $\frac{v_0}{v_1}$ in the above equation we get

$$\frac{A_1}{A_0} = \frac{v_0}{v_1} = \frac{3}{2} \Rightarrow A_1 = \frac{3}{2}A_0$$

For $C_p = \frac{16}{27}$ the maximum mechanical power output $= \frac{16}{27}\left(\frac{1}{2}\rho A_1 v_0^3\right)$ (Eq. (7.17))

The power of the wind at the upstream side is $\left(\frac{1}{2}\rho A_1 v_0^3\right)$

Thus, an overall mechanical efficiency of wind turbine is given by

$$\eta_m = \frac{\textit{output maximum mechanical power}}{\textit{input power}} = \frac{\left(\frac{16}{27}\right)\left(\frac{1}{2}\rho A_1 v_0^3\right)}{\left(\frac{1}{2}\rho A_0 v_0^3\right)} = \frac{16}{27}\left(\frac{A_1}{A_0}\right)$$

or,

$$\eta_m = \left(\frac{16}{27}\right)\left(\frac{3}{2}\right) = \frac{8}{9} = 0.89 \tag{7.19}$$

Therefore, the maximum mechanical power extraction per unit area of A_0 is 8/9 of the power in the wind.

7.6.3 Thrust on Turbine

Bernoulli's equation can be used for evaluating the thrust on a wind turbine (actuator disk) in a stream line flow, as shown in Figures 7.12a and b, respectively.

Referring to Figure 7.12a, Bernoulli's equation for no power extraction can be written for the above system per unit mass as follows:

$$\frac{P_0}{\rho_0} + gz_1 + \frac{v_0^2}{2} = \frac{P_2}{\rho_0} + gz_2 + \frac{v_2^2}{2} \tag{7.20}$$

Since the change in z and ρ with respect to height and temperature are negligible as compared to other terms, the static pressure difference across a wind turbine is given by

$$\Delta P = P_2 - P_0 = \frac{(v_0^2 - v_2^2)\rho}{2} \tag{7.21}$$

The term $v^2\rho/2$ is the **dynamic pressure**. The maximum value of static pressure difference occurs as v_2 tends to zero.

Therefore,

$$\Delta P_{\max} = \frac{\rho v_0^2}{2} \tag{7.22}$$

The maximum thrust on the turbine can be obtained by multiplying the above equation area of wind turbine (A_1)

$$F_{\max} = \rho A_1 v_0^2/2 \tag{7.23}$$

On a horizontal-axis machine, the thrust is centred on the turbine axis and is called the axial thrust F_{A}.

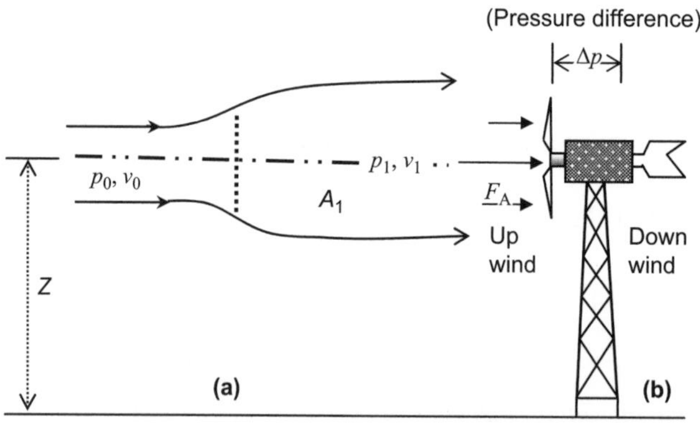

Figure 7.12 Force on turbine (a) air flow speed v, pressure p, height z, (b) axial force F_A, and pressure.

The thrust is also equal to the rate of loss of momentum of the air stream across the wind turbine and hence,

$$F_A = \dot{m}\,(v_0 - v_2)$$

Using Eqs. (7.11), (7.14) and (7.15), the above equation reduces to

$$F_A = (\rho A_1 v_1)\,(2v_0 a) = 2\rho A_1 v_0 (1 - a) a v_0$$

$$F_A = \frac{\rho A_1 v_0^2}{2} 4a(1 - a) \tag{7.24}$$

or,

$$F_A = C_F \frac{\rho A_1 v_0^2}{2} \tag{7.25}$$

The term $\frac{\rho A_1 v_0^2}{2}$ is the maximum force given by the actuator disk model (Eq. (7.23)) for wind hitting a solid disk. C_F is the axial force coefficient, *i.e.* the fraction of this force experienced by the actual turbine

$$C_F = 4a(1 - a) \tag{7.26}$$

The maximum value of C_F would be 1 when $a = 1/2$, equivalent to $v_2 = 0$ from Eq. (7.19).

EXAMPLE 7.3

Determine the condition for maximum axial force coefficient (C_F) by using "b" as an interference factor.

Solution: From Eq. (7.26), we have an expression for an axial force coefficient as

$$C_F = 4a(1 - a)$$

Substituting $b = (1 - a)/2$ in above equation one gets

$$C_F = 4 \times \frac{1 - b}{2}\left[1 - \frac{1 - b}{2}\right] = (1 - b)\,(1 + b)$$

Differentiating "C_F" with respect to "b" and equating it to zero, one gets

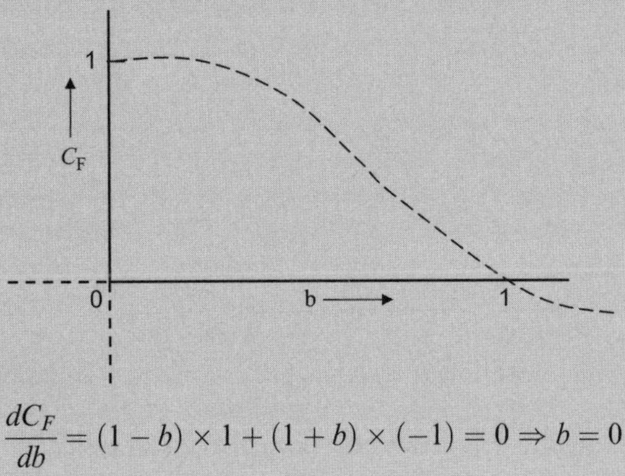

$$\frac{dC_F}{db} = (1 - b) \times 1 + (1 + b) \times (-1) = 0 \Rightarrow b = 0$$

EXAMPLE 7.4

Determine the condition for maximum "C_F" with respect to "a".
Solution: From Eq. (7.26) $C_F = 4a(1 - a)$
Differentiating "C_F" with respect to "a" and equating it to zero, one gets

$$\frac{dC_F}{da} = 4(1 - 2a) = 0 \Rightarrow a = \frac{1}{2}$$

Substituting the value of $a = 1/2$ in the expression of C_F, $C_F = 1$

EXAMPLE 7.5

Calculate C_F value of C_p at $a =$ 1/2 and value of C_F at $a = 1/3$.
Solution: From Eq. (7.18a) $C_P = 4a(1 - a)^2$
Hence,

$$C_p \,\text{at}\, a = \frac{1}{2} = 4 \times \frac{1}{2} \times \left(1 - \frac{1}{2}\right)^2 = 4 \times \frac{1}{2} \times \frac{1}{4} = \frac{1}{2}$$

From Eq. (7.26) $C_F = 4a(1 - a)$
Hence,

$$C_F \,\text{at}\, a = \frac{1}{3} = 4 \times \frac{1}{3} \times \left(1 - \frac{1}{3}\right) = 4 \times \frac{1}{3} \times \frac{2}{3} = \frac{8}{9}$$

EXAMPLE 7.6

Derive an expression for the power coefficient for two activator disks (WECS) connected in series.
Solution: The two activator disks (WECS) connected in series are shown in the figure below:

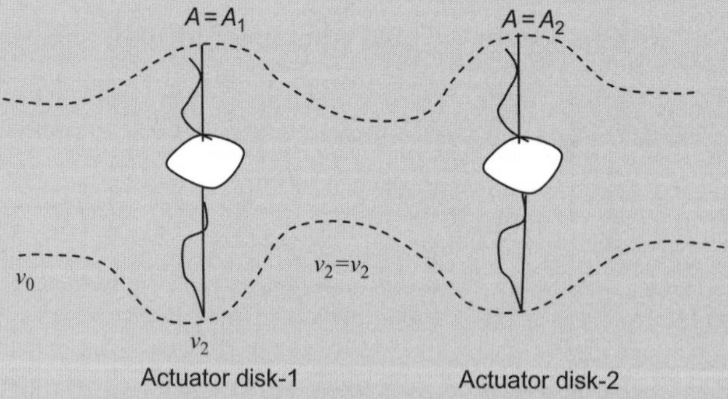

Let us consider the prime notation for activator disk-1 and double prime notation for activator disk-2.

The overall power extraction from two activator disks connected in series as shown in the figure above, can be written as

$$P = P' + P'' = \frac{1}{2}\rho A v_0^3 C'_P + \frac{1}{2}\rho A {v'_2}^3 C''_P$$
$$= \frac{1}{2}\rho A v_0^3 \left[C'_P + \left(\frac{v'_2}{v_0}\right)^3 .C''_P \right] = \frac{1}{2}\rho A v_0^3 .C_{Peff}$$

where,

$$C_{Peff} = C'_P + \left(\frac{v'_2}{v_0}\right)^3 .C''_P \text{ with } C'_P = 4a'(1-a')^2$$

From Eqs. (7.14b) and (7.15b), $\frac{v'_2}{v_0} = b' = 1 - 2a'$, so we have

$$C_{Peff} = C'_P + (1-2a')^3 .C''_P$$

For
$$C''_P = \frac{16}{27},\ C_{Peff} = 4a'(1-a')^2 + (1-2a')^3 .\frac{16}{27}$$

For the maximum value of C_{Peff}, differentiating C_{Peff} with respect to a' and equating to zero.

$$\frac{dC_{Peff}}{da'} = 0 \Rightarrow a' = 0.2$$

For $a' = 0.2$, $C_{Peff} = 0.64$

Here, we see that there is a small increase (0.04) in the value of C_{Peff} and hence it is not economical to have two disks connected in series.

EXAMPLE 7.7

Plot the curve between an axial force coefficient (C_F) and "a". Compare the results obtained in Example 7.3.

Solution: From Eq. (7.26) an expression for C_F can be written as

$$C_F = 4a(1-a)$$

The variation of C_F with "a" is shown below:

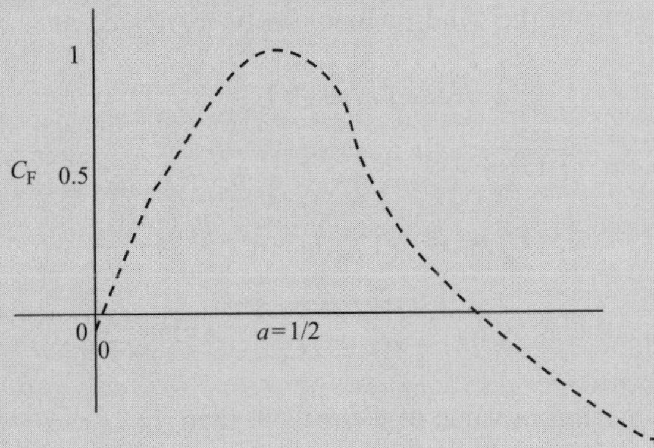

7.6.4 Torque on Turbine

The maximum torque (Γ) on a wind turbine rotor occurs when the maximum force is applied at the blade tip farthest from the axis. For a parallel turbine of radius R, the maximum torque (Γ_{max}) is given by

$$\Gamma_{max} = F_{max} R$$

From Eq. (7.23), one has

$$F_{max} = \rho A_1 \frac{v_0^2}{2}$$

After substituting the expression of F_{max} from the above equation in expression for $\Gamma_{max,}$ we have,

$$\Gamma_{max} = \rho A_1 v_0^2 R/2 \tag{7.27}$$

For a working wind machine producing a shaft torque Γ proportional to Γ_{max}

$$\Gamma \propto C_\Gamma \Gamma_{max}$$

or,

$$\Gamma = C_\Gamma \Gamma_{max} \tag{7.28}$$

where C_Γ is the torque coefficient.

The tip-speed ratio (λ) is defined as the ratio of the outer blade tip speed (v_t) to the unperturbed (free) up wind speed v_0, *i.e.*

$$\lambda = \frac{v_t}{v_0} = \frac{R\omega}{v_0} \Rightarrow R = \frac{\lambda v_0}{\omega} \tag{7.29}$$

where R and ω are the outer blade radius and the rotation frequency of the rotor.

Substituting the expression of R from Eq. (7.29) into Eq. (7.27), one gets

$$\Gamma_{max} = \rho A_1 v_0^2 \left(\frac{v_0 \lambda}{2\omega}\right) = P_0 \frac{\lambda}{\omega} \tag{7.30}$$

where $P_0 = \frac{1}{2}\rho A_1 v_0^3$ is the maximum power from the wind turbine.

Further, the shaft power from the wind turbine can be expressed as

$$P_T = \Gamma\omega = C_\Gamma \Gamma_{max}\omega \tag{7.31}$$

Since, $P_T = C_P P_0$ from Eq. (7.17)

Then,

$$C_P P_0 = C_\Gamma \Gamma_{max}\omega = C_\Gamma \frac{P_0 \lambda}{\omega}\omega$$

or,

$$C_P = \lambda C_\Gamma \tag{7.32}$$

For the Betz limit, the maximum value of C_P is 0.59, then

$$(C_\Gamma)_{max} = \frac{0.59}{\lambda} \tag{7.33}$$

7.7 WIND ENERGY SOURCES

Before the analysis of the possible contribution of wind energy to the energy supply, the characteristics of wind energy sources need to be investigated.

7.7.1 The Origin of Wind

Wind is produced by the uneven heating of the Earth's surface by energy from the Sun. Since the Earth's surface is made of very different types of land and water, it absorbs the Sun's radiant energy at different rates. Much of this energy is converted into heat as it is absorbed by land areas, bodies of water, and the air over these formations. On a global scale, the nonuniform thermal effects combine with the dynamic effects from the Earth's rotation to produce prevailing wind patterns. There are also minor changes in the flow of the air as a result of the differential heating of sea and land. The nature of terrain ranging from mountain and valleys to more local obstacles such as buildings and trees also has an important effect on the origin of wind. Generally, during the day time the air above the land mass tends to heat up more rapidly than the air above water. In coastal regions this manifests itself in a strong onshore wind. In the night time the process is reversed because of the air cools down more rapidly over the land and the breeze therefore blows offshore, as shown in Figures 7.13a and b. A similar process occurs in mountains and valleys, creating local wind. The speed of wind is affected by the surface over which it blows. Rough surfaces such as areas with trees and buildings produce more friction and turbulence than smooth surface such as lakes or open cropland. The greater friction means the wind speed near the ground is reduced, as shown in Figure 7.14.

7.7.2 The Characteristics of the Wind

The power in the wind is proportional to the cube of the wind speed, as given by Eq. (7.5). It is well known that the highest wind velocities are generally found on hill tops, exposed coasts and out at sea. Various parameters need to be known regarding wind, including the mean wind speed directional data, variations about the mean in the short-term (gusts), daily, seasonal and annual variations as well as variations with height. These parameters are highly site specific and can only be determined with sufficient accuracy by measurements at a particular site over a sufficiently long period. They are used to assess the performance and economics of a wind energy-conversion system. General meteorological statistics may overestimate wind speed at a specific site. Therefore, not only the mean wind speed published by the meteorological organisations but also the wind-frequency

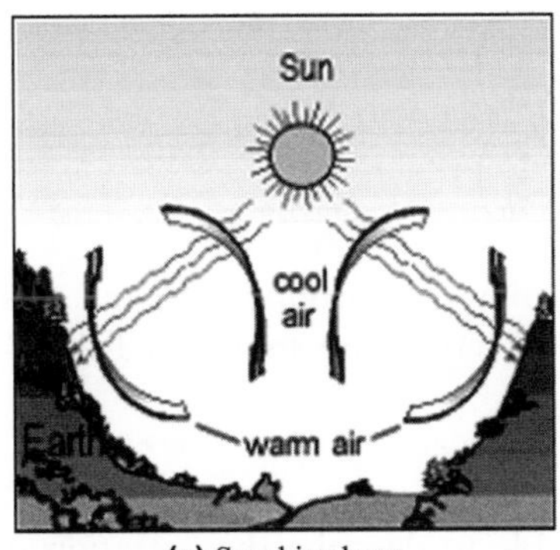

(a) Sunshine hour

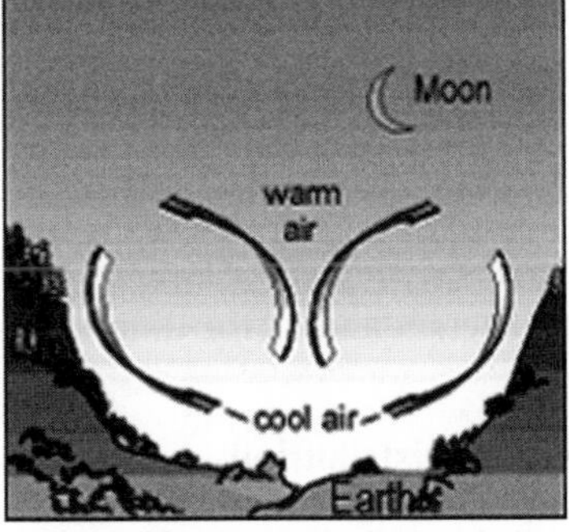

(b) Off-sunshine hour

Figure 7.13 Generation of wind sources (a) during day and (b) night time.

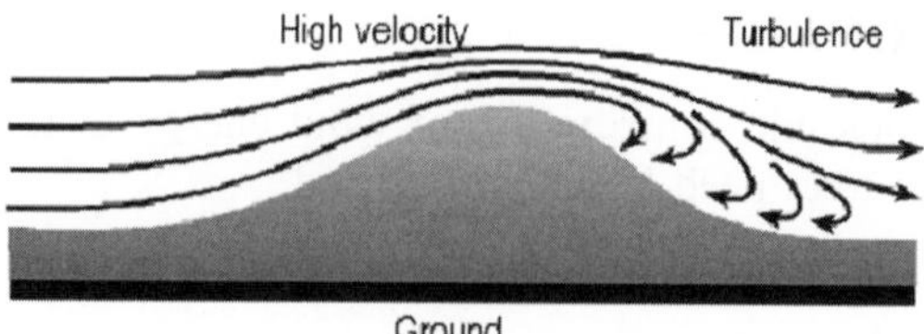

Figure 7.14 Wind sources due to the different ground surface conditions.

distribution, described by a Weibull distribution have to be taken into account in order to calculate the amount of electricity that can be produced by wind turbines in a certain region. A standard meteorological measurement of wind speed measures the speed of wind at a height of 10 m. But the height of the hub in a wind turbine is generally kept at more than 10 m. In that condition, the variations in speed of the wind with height are to be incorporated for predicting the energy available in the wind.

7.7.3 Vertical Wind Speed Gradient

The wind speed varies with the height above the ground. This is called the **wind shear**. The wind speed at the surface due to the friction between the air and the surface of the ground is zero. The wind speed increases most rapidly near the ground with height, increasing less rapidly with greater height. The change in wind speed becomes nearly zero at a height of about 1 km above the ground. The vertical variation of the wind speed and the wind-speed profile can be expressed by many functions. The two more common functions that have been developed to describe the change in mean wind speed with height are based on experiments. This is explained below:

Power exponent function: The power exponent function is given by

$$\mathrm{V}(z) = \mathrm{V}_r\left(\frac{z}{z_r}\right)^{\alpha} \tag{7.34}$$

where z is the height above the ground level, V_r is the wind speed at the reference height z_r above the ground level. A typical value of α is 0.1.

Although the power exponent law is a convenient approximation as given in Eq. (7.34), has no theoretical basis. Instead, aerodynamic theory suggests that when the atmosphere is thermally natural, which is the case on a cloudy day for the strong wind, air flow within the boundary layer varies logarithmically with height, so that the logarithmic function can be expressed as[5]

$$\frac{\mathrm{V}(z)}{\mathrm{V}_r} = \frac{\ln\left(\frac{z}{z_0}\right)}{\ln\left(\frac{z_r}{z_0}\right)} \tag{7.35}$$

where V_r is the wind speed at the reference height z_r above the ground level and z_0 is the roughness length. The parameters α and z_0 for different types of terrain are shown in Table 7.1.

7.7.4 Wind Speed Distribution

For the wind industry it is very necessary to be able to describe the variation of wind speeds. Turbine designers need the information to optimise the design of their wind turbines in order to minimise the generator cost. Turbine investors need the information to estimate their income from

Table 7.1 Wind-speed parameters for calculating a vertical profile.

Types of terrain	*Roughness class*	*Roughness length z_0 (m)*	*Exponent (α)*
Water areas	0	0.001	0.01
Open country, few surface features	1	0.12	0.12
Farmland with building and hedges	2	0.05	0.16
Farmland with many trees, forests, villages	3	0.3	0.28

electricity generation. Wind-speed variations and their distributions at any given site can be described in terms of a Weibull distribution.

The probability that the wind will blow at some wind speed including zero must be 100%, since the area under the curve is always 1. The Weibull function can be expressed as

$$P(\mathrm{V}) = \frac{k}{\mathrm{V}}\left(\frac{\mathrm{V}}{C}\right)^{k-1} \exp\left\{-\left(\frac{\mathrm{V}}{C}\right)^{k}\right\} \tag{7.36}$$

where $P(V)$ is the frequency of the wind at wind speed V, C is the scale parameter or characteristic wind speed and k is the shape parameter.

The Weibull distribution also expresses the proportion of time for which the wind speed exceeds the value V and is expressed as

$$P(\mathrm{V}) = \exp\left\{-\left(\frac{\mathrm{V}}{C}\right)^{k}\right\} \tag{7.37}$$

For a typical value of $k = 1$, the distribution is called a **cumulative Rayleigh distribution**.

The probability of wind distribution between V_1 and V_2 can be given as

$$P(\mathrm{V}_1 < \mathrm{V} < \mathrm{V}_2) = \exp\left\{-\left(\frac{\mathrm{V}_1}{C}\right)^{k}\right\} - \exp\left\{-\left(\frac{\mathrm{V}_2}{C}\right)^{k}\right\} \tag{7.38}$$

The plot of ln V versus ln $\{-(\ln V)\}$ gives the parameters C and k for the Weibull frequency distribution. The graph gives a straight line in which the slope is equal to k and parameter C is equal to exp (ln V) or V, where ln $\{-(\ln V)\}$ is zero.

7.8 COMPONENT OF WIND ENERGY-CONVERSION SYSTEM

The main components of the turbine are:

- nacelle;
- rotor that is the assembly of blades, hub and shaft;
- transmission system that includes a gearbox and a breaking mechanism;
- electric generator;
- yaw and control system;
- tower to support the rotor system.

Nacelle: This includes the gearbox, low- and high-speed shafts, generator controller and brake. It is placed on the top of the tower and is connected to the rotor.

Rotor: Most of the horizontal-axis wind turbines use two or three blades in an upwind design. Blades are manufactured form fibreglass-reinforced polyester (FRD), wood laminates, steel or aluminium. A FRD blade is comparatively lighter and exerts less stress on bearing and rotor hubs. Hence, it is used by most of the wind-turbine manufacturers. Other manufacturers use steel blades because of the ease of fabrication, greater strength and lower cost. Sometimes, wood laminates blades are also used due to their excellent fatigue resistance properties. Vertical-axis wind-turbine manufacturers often use extruded aluminium blades.

Rotor power control: While designing a wind turbine one of the important issue is to limit the power output at high wind speed. There are two options for constant speed machines (i) stall regulation and (ii) pitch control.

(i) For **stall-regulated wind turbines**, the pitch angle distribution along the blades is constant for all wind speeds. The angle of attack of the airfoil over the blades increases until flow separation (stall) occurs at high wind speeds. This results in a loss of lift but drag forces rise. The effect of this process can be influenced by appropriate choice of blade profile, the thickness and chord distribution and the blade twist. The great advantage of stall regulation is its simplicity and relatively low cost.
(ii) For **pitch-regulated wind turbines**, the blades can be rotated about their radial axis during the operation as the wind speed changes. It is therefore possible to have an optimum pitch angle at all wind speeds and a relatively low cut-in wind velocity. The pitch angle changes in order to decrease the angle of attack at high wind speeds. This ensures that the power output from the rotor is limited to the rated power of the generator. Pitch regulation is more expansive and it requires a relatively complicated control system. It is more efficient than stall regulation.

Transmission system: The mechanical power generated by the wind turbine (rotor blades) is transmitted to the electric generator by a transmission system located in the nacelle. The transmission system contains a gearbox, a clutch and a braking system to stop the rotor in an emergency. The purpose of the gearbox is to increase the speed of the rotor typically from 20 to 50 revolutions per minutes (rpm) or from 1000 to 1500 rpm which is required for driving most types of electric generators. There are two main types of gearbox (i) a planetary shaft and (ii) a parallel shaft. The transmission system must be designed for high dynamic torque loads due to the fluctuating power output from the rotor. Some designers have made an attempt to control the dynamic loads by adding (a) mechanical compliances and (b) damping into the drive train. This is particularly important for very large wind turbines.

Electric generator: There are two main options for the generator used in constant-speed wind turbines namely asynchronous (induction) or synchronous generators. Most of the grid-connected wind turbines installed so far use induction generators. These turbines have to be connected to the electricity grid before they can generate electricity. The generator is sometimes used as a motor to run the wind turbine up to synchronous speed, a feature that is utilised by stall-regulated wind turbines. The major disadvantage of the induction generators is that they draw reactive power from the grid system. Synchronous generators do not require reactive power so they are favoured by utilities. Wind turbines driving electrical generators operate at either variable or constant speed. In variable-speed operations, rotor speed varies with wind speed. In constant speed machines, rotor speed remains constant despite changes in the wind speed.

Yaw control: This is used to continuously orient the rotor in the direction of the wind. The horizontal-axis wind turbine has a yaw control system that turns the nacelle according to the actual wind direction, using a rotary actuator attached to the gear ring at the top of the wind tower. The wind direction must be perpendicular to the swept rotor area during normal operation of the wind turbine. A slow closed-loop control system is used to control the yaw drives. A wind vane mounted on the top of the nacelle, senses the relative wind direction and the wind-turbine controller then operates the yaw drives.

Wind tower: The most common types of wind tower are the lattice or tubular types constructed from steel or concrete. The towers are designed to withstand wind loads and gravity loads. The wind tower has to be mounted to a strong foundation in the ground. It is designed so that either its resonant frequencies do not coincide with induced frequencies from the rotor or they can be damped out.

7.9 WIND TURBINE GENERATOR CLASSIFICATION

Table 7.2 gives the classification of the wind-turbine electricity system. The classification depends on the relative size of the aerogenerator capacity (P_T) and other electricity generators capacity (P_G) connected in parallel with it.

Class A: For $P_T \geq 5P_G$, an aerogenerator represents a single autonomous standalone machine without any form of grid linking.

The control of output is very important for efficient cost effective systems. One choice is to have little control so that the output is of variable voltage for use as heat or rectified power, as shown in Figure 7.15a. Such a type of power supply is very useful. The relatively small amount of power that usually has to be controlled at say 240 V/50 Hz or 110 V/60 Hz can be obtained from batteries by inverters. However, it is preferred to have electricity at constant frequency. Two extreme options are:

(a) **Mechanical control of the turbine blades:** With change of wind speed, the pitch of the blades is adjusted to control the frequency of wind-turbine rotation. The power in the wind is wasted. The control method can be expensive and unreliable. This control system is shown in Figure 7.15b.

(b) **Mechanical control of the turbine blades:** With change of wind speed, the pitch of the blades is adjusted to control the frequency of wind-turbine rotation. The power in the wind is wasted. The control method can be expensive and unreliable. This control system is shown in Figure 7.15b.

Class B: $P_T \sim P_G$. This type of aerogenerator classification is common for remote areas and small grid systems. It is considered that "other generators" of capacity P_G are powered by a diesel engine. The basic aim of the wind turbine is to be a diesel saver. The diesel generator supplies power in windless periods. In this case, there are two extreme modes of operation.

(a) **Single-mode electricity supply distribution:** This is usually a three-phase supply that takes single phase to domestic dwellings with a single set of distribution cables, the system operates in a single mode at fixed voltage. A 24-h maintained supply without load-management control depends usually on diesel generation due to nonavailability of wind. The diesel is either left on continuously or switched off when excessive wind blows. Generally, a

Table 7.2 Classification of wind-turbine electricity generator system.

A	*B*	*C*
$P_T \gg P_G$ autonomous (a) blade pitch (b) load matching	$P_T \sim P_G$ wind/diesel (a) wind or diesel (b) wind and diesel together	$P_T \ll P_G$ grid stand (a) direct induction generator (b) to DC then AC

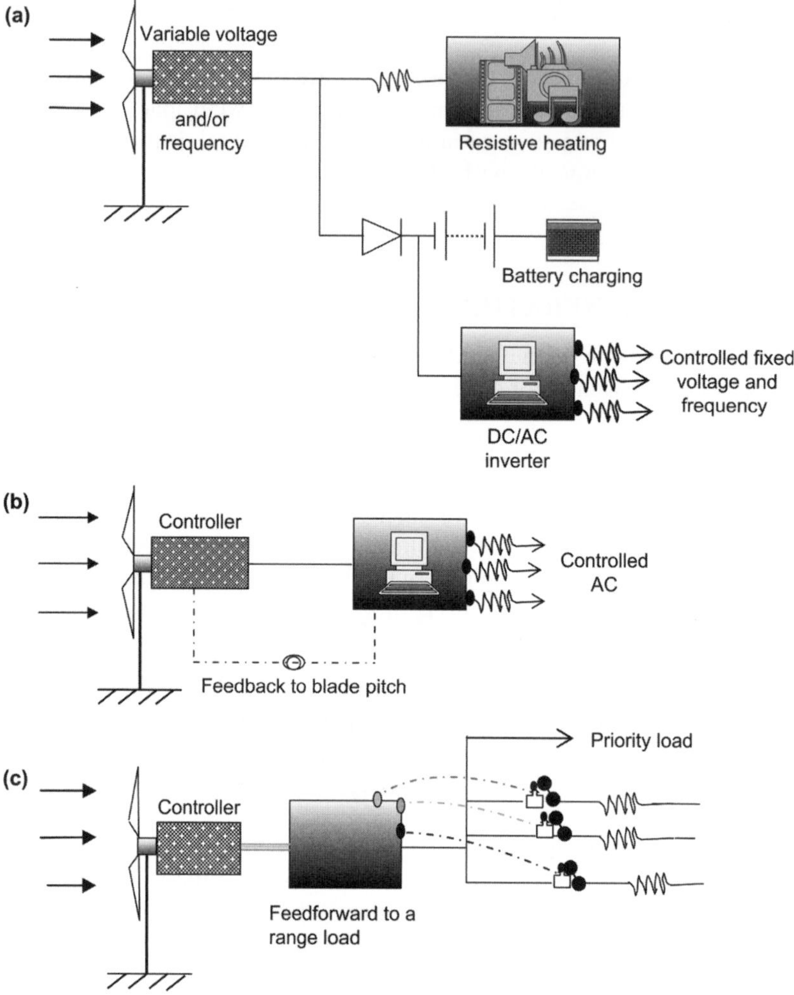

Figure 7.15 Classification of electricity supply options with the aerogenerator.

large amount (over 70%) of the wind generated power is to be dumped into an outside resistor banks owing to the mismatch of supply and demand in windy conditions. This is shown in Figure 7.16a.

(b) **Multiple-mode distribution:** In this case, every effort is made to use all the wind-generated power. Economic electricity for many applications in windy conditions is offered. The economic service loads are automatically switched off to decrease the demand due to wind drop. Only the loads on the expensive supply are enabled for supply by the diesel generator without availability of wind. The economic advantage of multiple mode operation is used at all times. This is shown in Figure 7.16b.

Class C: $5P_T \leq P_G$ (grid linked). This is common arrangement for (i) large capacity ($\sim$3 MW), (ii) medium capacity ($\sim$250 kW) and (iii) small capacity ($\sim$50 kW) machines. The owner of the wind machine uses the wind power directly and sells any excess electric power to the grid. Electricity is purchased from the grid at periods of low or no wind. This is depicted in Figure 7.17.

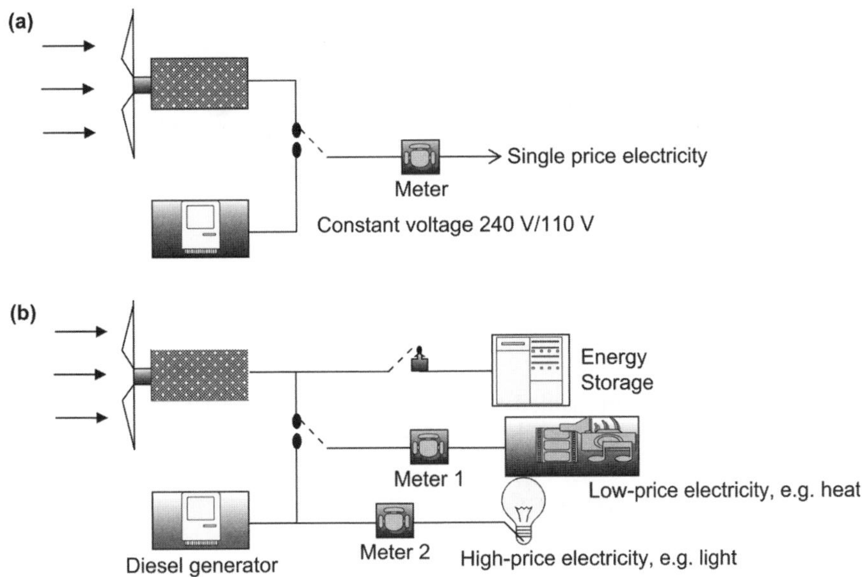

Figure 7.16 Wind/diesel supply modes (a) single-mode supply (b) multiple-mode supply.

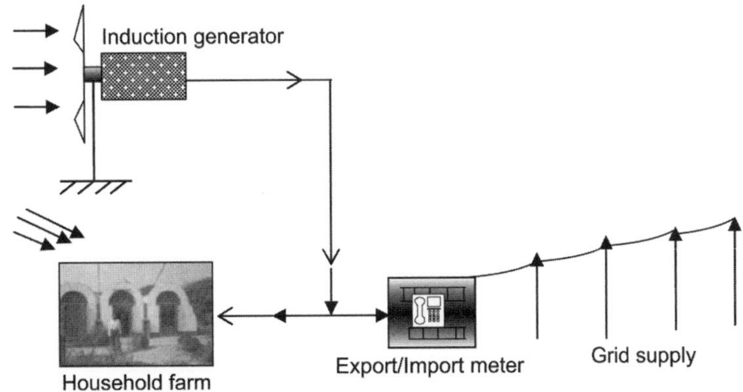

Figure 7.17 Grid-linked aerogenerator slaved in a large system.

7.10 APPLICATION OF WIND TURBINES

The followings are the main application of wind turbines

(a) Drag Machine

This consists of a device with wind-driven surfaces or flaps moving parallel to the undisturbed wind having wind speed v_0 as shown in Figure 7.18.

The pressure difference (ΔP) across a stationary flap held perpendicular to the wind velocity is given by Eq. (7.19). The maximum driving drag force for a loop can be given as

$$F_{max} = \rho A(v_0 - v^2)/2 \tag{7.39}$$

where A and v are the cross-sectional area and moving speed of the flap. Here, v_0 and v are in the same direction.

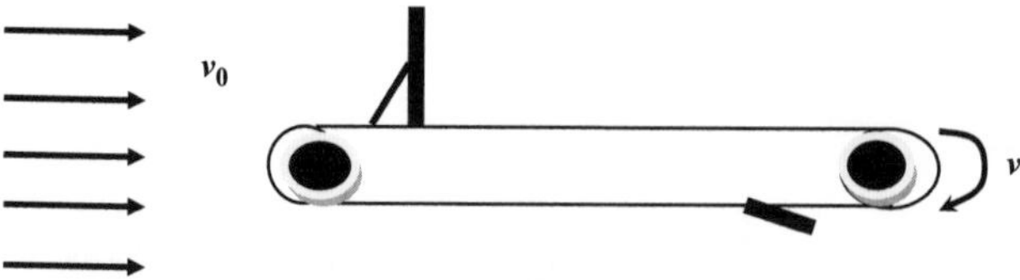

Figure 7.18 View of drag machine with hinged flaps on rotating belt.

The drag force can also be written as,

$$F_D = C_D \rho A(v_0 - v^2)/2 \tag{7.40}$$

where C_D is a dimensional quantity called the drag coefficient.

The power transmitted to the flap can be given as

$$P_D = F_D.v = C_D \rho A(v_0 - v^2)v/2 \tag{7.41}$$

For maximum transmitted power, differentiating Eq. (7.41) w.r.t. v and making it equal to zero one can get,

$$\frac{dP_D}{dv} = 0 \Rightarrow v = \frac{v_0}{3}$$

After substituting the value of v form above equation in Eq. (7.41) and using Eq. (7.17), one gets

$$P_{D,max} = \frac{4}{27} C_D \frac{\rho A v_0^2}{2} = C_P \frac{\rho A v_0^2}{2}$$

so,

$$C_{p,max} = \left(\frac{4}{27}\right) C_D$$

$C_D \rightarrow$ 0 for a pointed object

$C_D \rightarrow$ 1.5 (maximum) foe a concave shape used in standard anemometers.

Thus, the maximum power coefficient for a drag machine can be given as

$$C_{p,max} = \left(\frac{4}{27}\right)(1.5) = \frac{6}{27} = 22\%$$

The maximum value of C_p for a wind turbine in an ideal case is 59%, as per the Betz limit. Hence, $C_{p,max}$ for a drag machine is less than $C_{p,max}$ for a lift forced wind turbine.

$C_{p,max}$ for a drag machine can be increased by incorporating more flaps or by arranging concentrated air flows.

The ratio of $C_{p,max}$ between the drag machine and the lift forced machine is

$$\frac{Drag\,machine}{Lift\,turbine} = \frac{22}{59} = 37\%$$

Therefore, drag-only devices have power extraction coefficients of only about 37% that of lift forced turbines for the same cross-sectional area.

(b) Wind Pump

Water pumping is the one of the main applications of wind energy. The most widely used type of wind pump is constructed of a steel, multibladed, high solidity, fan-like rotor that drives a reciprocating pump linkage usually *via* a reduction gearing that is directly connected with a piston pump located in a borehole directly under it.[6]

The design of wind turbines for water pumping is relatively simple compared to electricity-generating turbines. The mechanical power at the rotor shaft is used directly to drive a pumping device. Turbines with high starting torque are suitable for pumping and this requires high solidity rotors operating at a low tip ratio of 2 or less.

The most prevalent wind turbines for water pumping are of the horizontal-axis type and typically have rotors blades usually made from curved sheet metal and need not be of a complex aerofoil section.

The important components of a wind pump are as follows:

(i) Rotor: This can vary widely in both size and design. Diameters range from less than 2 m to 7 m. The number of blades can vary from 6 to 24. A rotor with more blades runs slower but is able to pump with more force.
(ii) Tail: This keeps the rotor pointing into the wind like a weather vane. The whole top assembly pivots on the top of the tower. It allows the rotor to face in any direction.
(iii) Transmission system: This turns the rotation of the rotor into reciprocating motion (up and down) in the pump rod. Normal types use a gearbox or are direct drive. In direct drive, the pump rod moves up and down once for each turn of the rotor.
(iv) Pump rod: This transmits the motion from the transmission at the top of the tower to the pump at the bottom of the well. The motion of the pump rod is reciprocating (up and down) and the distance it travels (called the stroke) is typically of about 30 cm, depending on the pump. It is usually made of steel.
(v) Pump: This is normally submerged below the water level on the downward stroke, the cylinder is filled with water and on the upward stroke, the water is lifted by the piston up the drop pipe. The pump hangs on the drop pipe.
(vi) Drop pipe: This is the pipe through which the water is pumped and also encloses the pump rod.
(vii) Well: The source of water pumped by the wind pump.
(viii) This is normally of galvanised steel with three or four legs. Its height varies from 5 m to 20 m. The bases of the legs are fixed, often bolted, to a concrete foundation.

The size of a pump driven by the wind turbine is a function of the pump head, the required water flow rate and mean wind speed. The three main parameters that are needed are the total pumped head H (m), the pumped volume flow rate Q (m^3/s) and the expected mean wind speed v_0 (m/s). The actual power delivered by the rotor must be equal to the required hydraulic power.

$$C_p\left(\frac{1}{2}\rho A v_0^3\right) = \rho_w g H Q \tag{7.43}$$

where C_p is the power coefficient of wind turbine, ρ is the density of air, A is the swept area and v_0 is the mean velocity of air, $\rho_w = 1000$ kg/m^2 (the density of water) and $g = 9.81$ m/s^2 is the acceleration due to gravity.

Rearranging Eq. (7.43),

$$A = \frac{1000 \times 9.81 HQ}{0.06 c_p v_0^3} \tag{7.44}$$

The rotor diameter can be given as

$$D = \sqrt{\frac{4A}{\pi}}$$

7.11 ADVANTAGES/DISADVANTAGES OF WIND ENERGY SYSTEMS

The following are the advantages and disadvantages of wind turbine systems:

Advantages

(i) Wind is available free of cost.
(ii) Wind is helpful in supplying electric power to remote areas.
(iii) Wind power generation is cost effective and reliable.
(iv) Pollution free.
(v) Economical.
(vi) Reliable.

Disadvantages

(i) Wind energy has low energy density but is favorable in many geographical locations from cities and forests.
(ii) Requires energy storage batteries that indirectly and substantially contribute to environmental pollution.
(iii) Turbine rotors are not very efficient as they extract only 10 to 40% of the available wind energy.
(iv) Wind energy is capital intensive.

ADDITIONAL EXERCISES

Exercise 7.1 By using the Betz limit, plot the curve between the torque coefficient (C_Γ) and the tip speed ratio (λ)

Solution:
We have, for the Betz limit, the maximum value of C_Γ is 0.59, then

$$(C_\Gamma)_{\max} = \frac{0.59}{\lambda}$$

$$\Rightarrow \lambda = 5: \; C_{\Gamma,max} = \frac{0.59}{\lambda} = \frac{0.59}{5} = 0.118$$

$$\Rightarrow \lambda = 10: \; C_{\Gamma,max} = \frac{0.59}{\lambda} = \frac{0.59}{10} = 0.059$$

$$\Rightarrow \lambda = 15: \; C_{\Gamma,max} = \frac{0.59}{\lambda} = \frac{0.59}{15} = 0.039$$

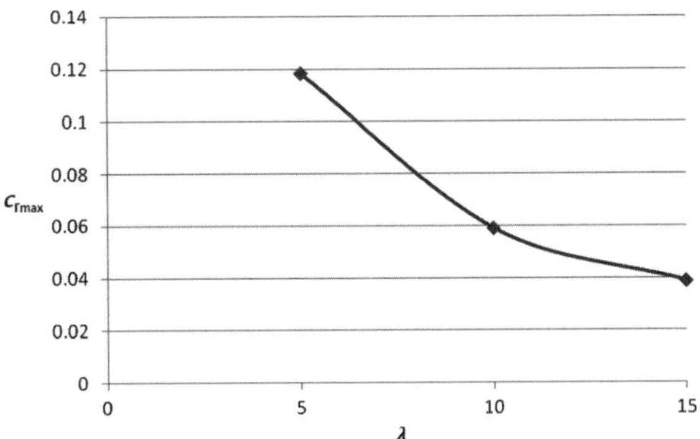

Statement: The value of C_Γ varies linearly with λ, if λ increases, C_Γ decreases, means that if wind turbine blades move faster, the torque coefficient will decrease. For modern high-speed turbines designed for electricity as low torque.

Exercise 7.2 Plot the curve between "C_p" and "C_F" w.r.t "a".
Solution:
The power coefficient C_p can be given in terms of "a" as

$$C_p = 4a(1 - a)^2$$

For C_{pmax}, differentiate the above expression w.r.t "a" and equate it to 0, one gets

$$\frac{dC_p}{da} = \frac{d}{da}[4a(1 - a)^2]$$

$$\frac{d}{da}[4a(a^2 - 2a + 1)] = 0$$

$$\frac{d}{da}[4a + 4a^3 - 8a^2] = 0$$

$$4 + 12a^2 - 16a = 0$$

$$1 + 3a^2 - 4a = 0$$

$$a = 1, \frac{1}{3}$$

$$\text{At } a = 1,\ C_p = 0$$

$$\text{At } a = \frac{1}{3},\ C_p = \frac{4}{3}\left(1 - \frac{1}{3}\right)^2 = \frac{16}{27} = 0.592$$

$$C_{p,\max} = 0.592 \equiv \mathit{Betz\ Criteria}$$

Now, the curve between "C_p" and "C_F" w.r.t. "a" can be plotted as

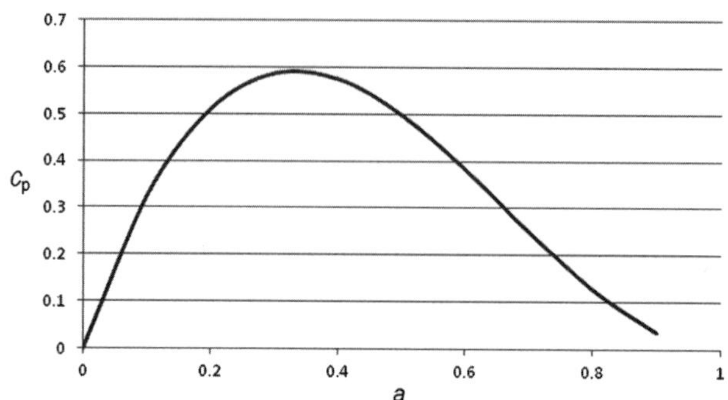

Statement: No wind turbine can generate power more than 59.2%, the value of a lies between 1/3 to 1. If we increase the value of a to more than 1/3, the C_p will decrease and at $a = 1$, it is 0.

Exercise 7.3 Plot the problem in Exercise 2 w.r.t. "b".

Solution:

We have an expression of "C_p" in terms of 'b' can be given as

$$C_p = 4a\,(1 - a)^2 \tag{i}$$

a = interference factor

$$a = \frac{(u_0 - u_1)}{u_0} \text{ and } u_1 = \frac{(u_0 + u_2)}{2} \tag{ii}$$

$$\Rightarrow u_1 = (1 - a)\,u_0 \tag{iii}$$

$b = u_2/u_0$ is also referred to as an interference factor

$$\Rightarrow u_1 = (1 - a)\,u_0$$

$$\Rightarrow \frac{(u_0 + u_2)}{2} = (1 - a)\,u_0$$

$$\Rightarrow (1 - a) = \frac{u_0 + u_2}{2u_0}$$

$$\Rightarrow a = 1 - \frac{(u_0 + u_2)}{2u_0} = \frac{(u_0 - u_2)}{2u_0} = \frac{1}{2}\left(1 - \frac{u_2}{u_0}\right)$$

$$a = \frac{1}{2}(1 - b)$$

$$C_p = 4a(1-a)^2$$

$$C_p = \frac{4}{2}(1-b)\left[1 - \frac{1}{2}(1-b)\right]^2$$

$$C_p = 2(1-b)\left[\frac{1+b}{2}\right]^2$$

$$C_p = 2(1-b)\left[\frac{1+b}{2}\right]^2$$

$$C_p = \frac{1}{2}(1-b)(1+b)^2 \tag{iv}$$

For C_{pmax}, differentiate Eq. (iv) w.r.t. b and equate to 0.

$$\frac{dC_p}{db} = \frac{d}{db}\left[\frac{1}{2}(1-b)(1+b)^2\right]$$

$$\frac{d}{db}\left[\frac{1}{2}(1-b)(1+b)^2\right] = 0$$

$$\frac{d}{db}\left[\frac{1}{2}(1-b)(b^2+2b+1)\right] = 0$$

$$\frac{d}{db}\left[\frac{1}{2}(1+b^2+2b-b-b^3-2b^2)\right] = 0$$

$$\frac{d}{db}\left[\frac{1}{2}(1+b-b^2-b^3)\right] = 0$$

$$\Rightarrow 0 + 1 - 3b^2 - 2b = 0$$

$$\Rightarrow 3b^2 + 2b - 1 = 0$$

$$b = -1, \frac{1}{3}$$

$b = -1$ is not possible, as u_2 is not negative so it is invalid.

For $b = \frac{1}{3}$, $C_p = \frac{1}{2}\left(1 - \frac{1}{3}\right)\left(1 + \frac{1}{3}\right)^2 = 0.592$

$$C_{p,max} = 0.592$$

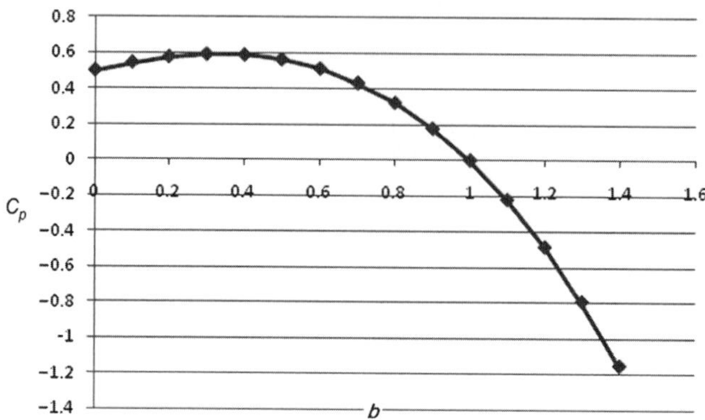

Statements:

1. C_p is maximum at $b = 1/3$ or 0.33, same at $a = 1/3$.
2. C_p is zero at $b = 1$,

At $b > 1$, C_p is downfall (negative) means that turbine losses are present.

OBJECTIVE QUESTIONS

7.1 The drag force (C_D) in an immersed body in a fluid acts along the
 (a) Horizontal direction
 (b) Vertical direction
 (c) Inclined surface at 45°
 (d) None

7.2 The lift force (C_L) in an immersed body in a fluid acts along the
 (a) Horizontal direction
 (b) Vertical direction
 (c) Inclined surface at 45°
 (d) None

7.3 Pressure is defined as
 (a) Force per unit area
 (b) Work per unit area
 (c) Velocity
 (d) None

7.4 The shear force (tangential force) and pressure has the units of
 (a) N/m^2
 (b) W/m^2
 (c) N/m
 (d) N/m^3

7.5 The value of drag coefficient (C_D) is always
(a) <1
(b) >1
(c) $=1$
(d) $=0$
7.6 The ratio C_L/C_D has values of
(a) + ve to −ve
(b) +ve only
(c) −ve only
(d) zero only
7.7 The lift coefficient (C_L) is always
(a) Higher than the drag coefficient (C_D)
(b) Higher than the drag coefficient (C_D)
(c) Equal to the drag coefficient (C_D)
(d) None of the above
7.8 The horizontal-axis turbine works due to
(a) Lift force
(b) Drag force
(c) Both
(d) All
7.9 The vertical-axis turbine works due to
(a) Lift force
(b) Drag force
(c) Both
(d) All
7.10 The angle of attack will be
(a) Only positive
(b) Only negative
(c) Positive and negative
(d) None
7.11 The value of the lift coefficient is always
(a) $<C_D$
(b) $>C_L$
(c) $C_D=C_L$
(d) none
7.12 The numerical value of dynamic pressure ($\frac{1}{2}\rho v^2$) is
(a) $>C_D$
(b) $>C_L$
(c) $>C_D$ and C_L
(d) all true
7.13 The power associated with wind is
(a) $\propto v^2$
(b) $\propto v^3$
(c) $\propto v$
(d) none
7.14 The optimum wind speed for $P_0=600$ W/m^2 is
(a) 5 m/s
(b) 15 m/s
(c) 10 m/s
(d) None

7.15 The wind turbine gives
- (a) Mechanical output
- (b) Electrical output
- (c) Chemical output
- (d) None

7.16 The wind turbine gives maximum mechanical power if
- (a) Wind is streamline
- (b) Wind is turbulent
- (c) Constant wind speed
- (d) None

7.17 The power coefficient of horizontal axis wind turbine (C_p) is given by
- (a) $4a(1-a)$
- (b) $a(1-a)$
- (c) $4a(a-a)^2$
- (d) None

7.18 The maximum value of C_p for a horizontal-axis wind turbine is
- (a) 0.49
- (b) 0.59
- (c) 0.39
- (d) 0.29

7.19 The interface factor for a horizontal-axis wind turbine is given by
- (a) $a = \frac{v_0 - v_2}{2v_0}$
- (b) $a = \frac{v_2}{v_0}$
- (c) $a = \frac{v_2}{v_0}$
- (d) None

7.20 The value of C_p is maximum at
- (a) $a = 1/2$
- (b) $a = 3/4$
- (c) $a = 1/3$
- (d) None

7.21 The value of C_p is maximum at
- (a) $a = 1/3$
- (b) $b = 1/3$
- (c) $a = b = 1/3$
- (d) All of them

7.22 The power extracted by a horizontal-axis wind turbine depends on
- (a) The rate of change of momentum
- (b) The rate of kinetic energy of wind flow across the turbine
- (c) The wind turbine speed
- (d) All of them

7.23 The swept area (A_1) is related to unwind cross-sectional area (A_0) by
- (a) 3/2
- (b) 2/3
- (c) 1
- (d) None

7.24 The maximum power extraction per unit area of A_0 is
- (a) 9/8
- (b) 7/8
- (c) 8/9
- (d) 1

7.25 The wind turbine of a horizontal-axis wind turbine loses power for
   (a) $b>1$
   (b) $b<1$
   (c) $b=0$
   (d) $b=1$
7.26 The expression for the axial force coefficient of a horizontal-axis wind turbine (C_F) is given by
   (a) $4a\ (1-a)2$
   (b) $4a\ (1-a)$
   (c) $4a\ (1-a)3$
   (d) $4a$
7.27 The value of the maximum axial force coefficient of a horizontal-axis wind turbine (C_F) is
   (a) 1/3
   (b) 2/3
   (c) 1/2
   (d) 1
7.28 The maximum value of solidity of wind turbine is
   (a) 10
   (b) 1
   (c) 0
   (d) None
7.29 The maximum axial force coefficient occurs at
   (a) $a=1/2$
   (b) $b=0$
   (c) $a=1/3$
   (d) $b=1$
7.30 The origin of wind energy is
   (a) Earth
   (b) Ocean
   (c) Sun
   (d) None of them
7.31 The percentage of solar radiation converted into wind energy is
   (a) 0.50%
   (b) 0.25%
   (c) 1%
   (d) None
7.32 In which country was the first wind turbine connected to a grid?
   (a) USA
   (b) Denmark
   (c) India
   (d) Germany
7.33 Which of the following sources emits fewer pollutants?
   (a) Coal
   (b) Petroleum
   (c) Charcoal
   (d) Wind
7.34 The maximum energy conversion efficiency of a wind turbine for a given swept area is
   (a) 25.1%
   (b) 50.4%
   (c) 59.3%
   (d) 99.3%

7.35 If the velocity of was wind is doubled, then the power output will increase by
- (a) 10 times
- (b) 8 times
- (c) 2 times
- (d) 6 times

7.36 Windmills work on the principle of
- (a) Rotation
- (b) Momentum
- (c) Gravitation
- (d) Collision

7.37 Which of the following forces act on the blade of the wind-turbine rotors?
- (a) Lift force
- (b) Drag force
- (c) Both (a) & (b)
- (d) None of them

7.38 During the day the surface wind flows
- (a) From sea to land
- (b) From land to sea
- (c) On the surface of the sea
- (d) On the surface of the land

7.39 During the night, the direction of was wind reverses from was land surface to was sea surface because the
- (a) Land surface cools faster than water
- (b) Water surface cools faster than land
- (c) Water surface remains hot
- (d) None of the above

7.40 Wind-turbine conversion devices based on drag force
- (a) Move faster than was wind
- (b) Move slower than was wind
- (c) Move with equal velocity as was wind
- (d) Do not depend on the velocity of the wind

ANSWERS

7.1 **(a)**; 7.2 **(b)**; 7.3 **(a)**; 7.4 **(a)**; 7.5 **(a)**; 7.6 **(a)**; 7.7 **(a)**; 7.8 **(a)**; 7.9 **(b)**; 7.10 **(c)**; 7.11 **(b)**; 7.12 **(d)**; 7.13 **(b)**; 7.14 **(c)**; 7.15 **(a)**; 7.16 **(a)**; 7.17 **(c)**; 7.18 **(b)**; 7.19 **(a)**; 7.20 **(c)**; 7.21 **(d)**; 7.22 **(a)**; 7.23 **(c)**; 7.24 **(a)**; 7.25 **(b)**; 7.26 **(d)**; 7.27 **(b)**; 7.28 **(a) & (b)**; 7.29 **(a)**; 7.30 **(b)**; 7.31 **(a)**; 7.32 **(d)**; 7.33 **(d)**; 7.34 **(c)**; 7.35 **(b)**; 7.36 **(b)**; 7.37 **(c)**; 7.38 **(a)**; 7.39 **(a)**; 7.40 **(b)**

REFERENCES

1. Gipe, P. 1995. *Wind Energy Comes of Age*. John Wiley and Sons, New York.
2. Putnam, G. C. 1948. *Power from the Wind.* Van Nostrand Rheinhold, New York, USA.
3. CEU, 1997. Energy for the future, renewable sources of energy – White Paper for a community strategy and action plan. COM (97) 559 final.
4. Zervos, A. 2000. European targets, time to be more ambitious? Wind directions. European Wind Energy Association. 18–19. (www.ewea.org).
5. Lumley, J. L. and Panofsky, H. A. 1964. *The Structure of Atmospheric Turbulence*, Interscience, London.
6. Vendot, L. 1957. *Water Pumping by Windmills*, La Huille Blanche No.4, Grenoble.

The Geothermal Thermal Energy, Stored in the Interior of the Earth, can be Harnessed Either in the Form of Thermal or Electrical Energy to Achieve the Energy Security of Both Developing and Underdeveloped Countries.

CHAPTER 8

Geothermal Energy

8.1 INTRODUCTION

Geothermal energy is the thermal energy contained in the Earth's interior. The literal meaning of the word "geothermal" is from geo (Earth) and thermal (heat energy). The origin of this thermal energy (heat) is linked with the internal structure of our planet. Volcanoes, geysers, hot springs and boiling mud pots are the visible evidence of heat available within and beneath the Earth's crust. The amount of thermal energy within the Earth is very large and in practically inexhaustible quantities in the Earth's crust. Useful geothermal energy is limited to certain sites. It is unevenly distributed, seldom concentrated and often at depths too great to be exploited industrially. The evolution of life began on Earth billions of years ago. The Earth was a ball of fire. Gradually, its outer surface cooled to allow life to begin. The heat is transferred from the depths to subsurface regions first by conduction and then by convection and radiation.

Geothermal energy is not a new energy source. People have been using hot springs for bathing and washing clothes in many parts of the world since the dawn of civilisation. Geothermal energy was first harnessed on a large scale for space heating, industry and electricity generation in the 20th century. Electric power was generated with geothermal steam at Larderello, Tuscany, Italy in 1904 by Prince Piero Ginori Conti. Commercial production of electricity started in Larderello in 1913. Geothermal energy has been produced commercially for about 90 years. Geothermal resources had been identified in over 80 countries in the year 2000. The worldwide use of geothermal energy amounted to about 49 TWh/annum of electricity in 2002 and 53 THw/annum for direct use ($1\ T = 10^{12}$).[1,2]

8.2 STRUCTURE OF THE EARTH

The internal structure of the Earth consists of crust, mantle and core has been shown in Figure 8.1.

8.2.1 Crust

The Earth's crust is similar to the skin of an apple. The thickness of the crust is 7 km on average under the ocean basins, and 20–65 km under the continents. This is insignificant compared to the rest of the Earth having an average radius of 6370 km. Wells give us direct access to heat in up to 10 km of the crust. It is assumed that the two types of crust (namely, continental and ocean) are made

Advanced Renewable Energy Sources
G. N. Tiwari and R. K. Mishra

Published by the Royal Society of Chemistry, www.rsc.org

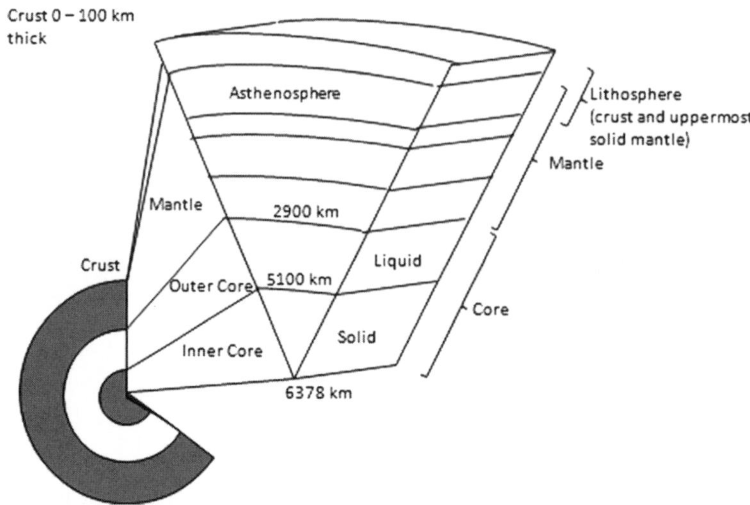

Figure 8.1 Internal structure of the earth.

up of different kinds of rock. The oceanic crust is made of basalt and the continental crust is referred to as being largely granite. The Earth's crust is thinner beneath the oceans than beneath the continents. Seismic waves travel faster in oceanic crust than in continental crust. Seismic waves are introduced into the Earth by detonating an explosive charge in a shallow bore hole. Returns of the seismic waves are measured at the surface. Seismic waves also originate naturally from earthquakes as well as microearthquakes. These waves can also be detected at the surface. Interpretation of the seismic information can also provide data on the location of active faults. It can channel hot fluids towards the surface. The temperature in the crust increases proportionally with depth at a rate of about 30 °C/km. The temperature at the base of the crust (the top of the mantle) levels off at a value of about 1000 °C. However, it is not possible to extract the thermal energy of the Earth's crust at such a higher depth for effective utilisation.

8.2.2 Mantle

The Earth's mantle lies closer to the Earth's surface beneath the ocean at a depth of 7 km. It is 20–65 km beneath the continents. The composition of the mantle consists of ultrabasic rock that is very rich in iron and magnesium and peridotite, which is a heavy igneous rock, made up chiefly of ferromagnesian minerals.

The Earth's crust and uppermost mantle together form the lithosphere. The outer shell of the Earth is relatively rigid and brittle, as shown in Figure 8.1. The lithosphere is split into a number of large blocks at the continental scale. These are called lithospheric plates in plate-tectonic theory. Its lower boundary inside the mantle is marked by a low-velocity zone. Seismic waves slow down in this zone. The zone extending up to a depth of perhaps 200 km from the surface of Earth is known as the asthenosphere. In the asthenosphere rocks may be closer to their melting point than rocks above or below this zone. Mantle rocks in the asthenosphere are weaker than in the overlying lithosphere. This causes the asthenosphere to deform easily by plastic flow. Convection takes place within the asthenosphere as well as within the lower mantle.

The lithosphere seems to be in continual movement/ floating condition on the mantle. The plates of brittle lithosphere probably move easily over the asthenosphere that acts as a lubricating layer below.

8.2.3 Core

The Earth's core extends from 2900 to 6370 km at the Earth's centre. Its thickness/radius is 3470 km. The temperature in the core and pressure at the Earth's centre are around 4000 °C and 360 000 MPa, respectively. The core consists of an inner and an outer core. It has the greatest average density of 10 g/cm^3. The most likely constituent of the core is an iron-nickel alloy with dense and the presence of some lighter material. Silicon has been proposed as an alloying element in the core. Sulfur is another light element present in the core. The present structure of the core is well established from seismological evidence. The outer part is molten. It does not transmit shear waves. The inner part of the core shows higher velocities leading to the suggestion of the inner core being solid.

The earliest record of space heating dates back to 1300 A.D. in Iceland. The earliest residential heating was in Chaude Aigues (France) in the 14th century in the world by geothermal water. The first mechanical conversion was in 1897 by the steam of the geothermal field at Larderello, Italy to heat a boiler for producing steam. This drove a small steam engine. The first attempt to produce electricity also took place at Larderello in 1904 with an electric generator. It powered four light bulbs. This was followed in 1912 by a condensing turbine. By the year 1914, 8.5 MW of electricity was being produced. 127 MW of electricity was produced by 1944 in the same geothermal power plant. The plant was destroyed towards the end of World War II. It was further rebuilt and expanded and the power production reached 360 MW in 1981.

Geothermal water was first used in greenhouse heating in Iceland in the 1920s. Now hundreds of hectares of greenhouses are operating throughout the world. Similarly, the first municipal district heating system was set up in Reykjavik, Iceland in 1930 using geothermal water. At present, 90% of the total population of Iceland lives in houses heated by geothermal energy. Large-scale district heating systems using geothermal water have been built in many countries (France. Russia, Georgia, China, Italy, Turkey and the USA). Geothermal heat has also been used on a large scale in animal husbandry, fish farming, crop drying and soil heating. Air conditioning using geothermal steam was first developed in a hotel in Rotorua, New Zealand in the late 1960s.

Commercial generation of electricity from geothermal steam began in Larderello, Tuscany, Italy in 1914. Other countries have followed the Italian example. Electricity is generated from geothermal energy in 21 countries all over the world. The world-wide geothermal installed electrical capacity is shown in Figure 8.2. Geothermal utilisation is divided into two categories, *i.e.* electric energy production and direct uses. In the United States, geothermal heat is used in some of their houses and commercial buildings. These are in California, Colorado, Boise, Idaho Oregon and Klamath Falls. Some of these were constructed as early as 1890.

8.3 THE PLATE-TECTONIC THEORY

The plate tectonic theory is currently accepted by most geologists. It is a unifying theory.[3] According to this theory, the rigid outer shell of the Earth or lithosphere is divided into separate blocks or plates. They are known as lithospheric plates. The location of the plate boundaries also corresponds to regions of active volcanoes. These plates move across the Earth's surface, at a speed of a few centimeters per year. The plates comprise both continents and sea floors. The plate tectonics concept means that the continents and the sea floor are moving and sliding on top of the underlying plastic asthenosphere. These plates either pull away from each other, slide past each other or move towards each other.

The boundaries between plates are of three types:

(i) **Diverging plate boundaries** (or spreading centres or ocean ridge). In this case two plates are moving apart. This permits the upwelling of magma from the asthenosphere to form new

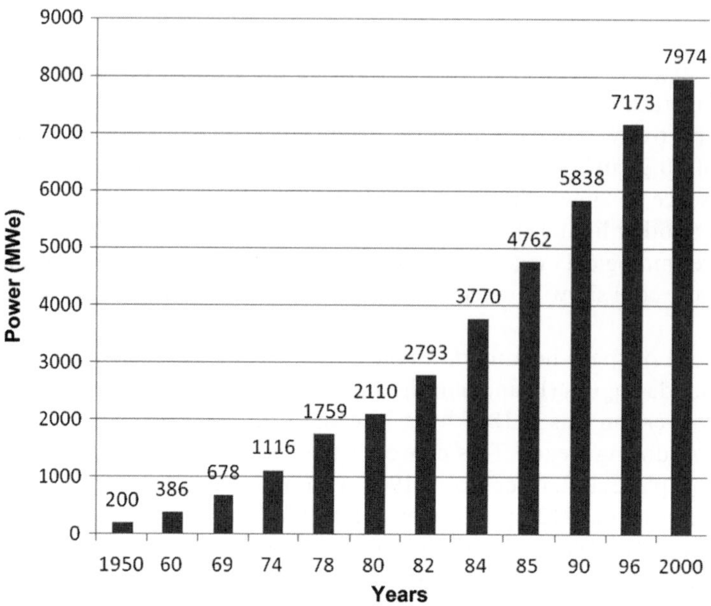

Figure 8.2 World wide electrical geothermal installed capacities to 2000.

lithosphere. Most spreading centres coincide with the crest of submarine mountain ranges, called midoceanic ridges. They rarely rise above sea level, *i.e.* Iceland, the Azores islands and the Afar depression in Ethiopia. There are also spreading ridges on the continents namely (i) the East Pacific rise, with the geothermal fields of Cerro Prieto (Mexico) and (ii) the Imperial Valley (USA) and the East African Rift with the fields of Langano (Ethiopia) and Olkaria (Kenya). The elongated elevated zones created in the centre of the sea-flow spreading are called midridges. They occupy central positions in the Atlantic and Indian Oceans.

(ii) **Converging plate boundaries:** These correspond to oceanic trenches. In this case, two plates converge and collide so that one plate slips and sinks below the other plate. This is eventually reabsorbed into the mantle and destroyed, *e.g.* the Nazca plate in the eastern Pacific Ocean. Convergence occurs at the boundary of oceanic crust and continental crust. In this case, two plates move towards each other. The less dense, more buoyant continental plate will override the denser plate. If the oceanic plate sinks it is known as a subduction zone. If an oceanic plate descends beneath the mantle it is an overriding plate. The entire oceanic plate becomes hotter as it descends deeper into the Earth's interior and it melts down. Melting of down thrust crust produces pods of magma. These rise into the upper plate and act as heat sources for geothermal reservoirs.

The collisions of converging plates occur in three ways, as shown in Figure 8.3.

(a) **Between oceanic lithospheric plates:** In this case two oceanic plates collide, one of them sinks and goes back to the mantle (Figure 8.3a).

(b) **One oceanic and other continental plate:** In this case the oceanic plate is consumed in the subduction zone (Figure 8.3b).

(c) **Between two continental lithospheric plates:** The continental lithosphere is not easily forced down and consumed in the subduction zone, as it is less dense. Therefore, when two continental plates collide, neither of them is consumed and buckling and crumpling take place resulting in great mountain ranges, such as the Himalayas (Figure 8.3c).

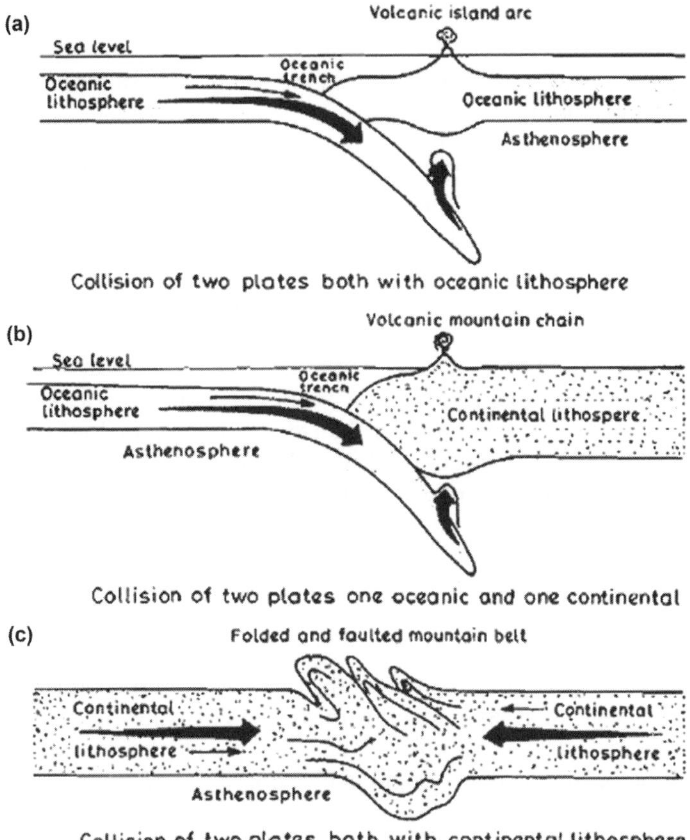

Figure 8.3 Converging plates and the resultant physiographic features.

Oceanic trenches are distinct topographic features associated with the subduction zones of the oceanic lithosphere. The oceanic plate bends down over a distance of the order of 100 km as it enters the subduction zone. The contact between the overlying and the sinking plate causes the trench at the bend.

(iii) **Conservative plate boundaries:** In this case, two plates slide past each other, so that no lithosphere is either created or destroyed. The direction of relative motion of the two plates is parallel to the fault conservative plate margins. This occurs within both the oceanic and continental lithosphere. The most common conservative plate margins are oceanic transform faults (*i.e.* the San Andreas fault in California).

The different types of plate boundaries were originally distinguished on the basis of their seismicity. Earthquakes commonly occur at the plate boundary. There is a close correspondence between plate boundaries and earthquake belts. Geothermal resources are associated with tectonic activity. It allows ground water to come in contact with deep subsurface heat sources.

8.4 THE EARTH'S THERMAL ENERGY

The thermal energy of the Earth is released into space from the interior. It is measured in milliwatt per square meter. It varies from place to place on the surface. It has varied with time at any particular place. The Earth's heat flow originates from the original heat. It is the heat generated during the Earth's formation by the decay of long-lived radioactive isotopes. All radioactive isotopes generate heat as they decay. Only isotopes that are relatively abundant and have half-lives comparable to the age of the Earth (4.5 billion years) are significant heat producers throughout geological time and remain so at present. Four long-lived radioactive isotopes ^{40}K, ^{232}Th, ^{235}U and ^{238}U are important heat producers.

The average heat flow from the continental crust (granite) is **57 mW/m^2**. Through the oceanic crust (basalt) it is **99 mW/m^2**. **The Earth's average heat flow is 82 mW/m^2** and the total global output is over 4×10^{13} W.[4] It is four times more than the present world energy consumption (10^{13}) W.[5] Continental heat flow is derived from radiogenic decay within the upper crust. In the oceanic crust, the concentration of radioactive isotopes is so low that radiogenic heating is negligible. The heat flow largely derives from heat flowing from the mantle below the lithosphere.

The heat flow in the continental crust at the surface is highest in areas experienced magmatic activity more recently than 65 million years (77 mW/m^2). Heat flow decreases to a constant value of about 46 mW/m^2 in crust older than 800 million years. In the young oceanic crust ($<$65 million year), heat flow is higher and variable (70–170 mW/m^2). It has lower and more constant heat flow (about 50 mW/m^2). The heat flow decreases with the age of the oceanic crust.[4]

8.4.1 Heat Transfer

The conductive heat flow inside the Earth is the product of the geothermal gradient and the thermal conductivity of rocks. The geothermal gradient is measured in shallow holes. The conductivity of rocks is measured in the laboratory on samples taken from that part of the well. Two forms of heat transfer occur within the Earth; conduction and convection.

Conduction involves the transfer of random kinetic energy between molecules. Moving molecules strike neighbouring molecules, to vibrate faster and thus transfer heat energy, Conduction is the primary heat transfer mode in solids. Metals are very good conductors of heat. Rocks are relatively poor conductors.

Convection is the common heat transfer process in liquids or gases. It consists of the movement of hot fluid (that is, a liquid or a gas) from one place to another. Due to the motion of material convection is a more efficient process of heat transfer than conduction.

8.4.2 Geothermal Gradient and Thermal Conductivity

The average temperature gradient near the surface is about **30 °C/km** within a few km. Values as low as about 10 °C are found in ancient continental crust and very high values ($>$ 100 °C/km) are found in the areas of active volcanism. After determining the temperature gradient it can be used to determine the rate at which the heat is moving upwards to the upper part of the Earth.

As the heat generally moves upwards through solid impermeable rock, the principal mechanism of heat transfer must be conduction. The amount of heat flowing by conduction through a unit area of 1 m^2 of solid rock in a given time (the rate of heat flow) is proportional to the geothermal gradient. The constant of proportionality is known as the thermal conductivity of rocks. The thermal conductivity is defined as the amount of heat conducted per unit area and per unit time for the temperature gradient of 1 °C/m perpendicular to that area. Its unit is W/m K. The temperature gradient is measured in wells with electrical (platinum-resistance) thermometers. The thermal conductivity of rock is best measured in the laboratory. If the gradient is expressed in °C/km and conductivity in W/m °C, then heat flow will be in mW/m^2 (milliwatt per square meter).

8.4.3 Renewability

Geothermal energy has its origin in the molten core of the Earth having temperatures of about 4000 °C. For this reason, some scientists refer to geothermal energy as a form of "fossil nuclear energy". The interior of the Earth is thought to consist of a central molten core surrounded by a region of semifluid material called the mantle. There are regions in which hot molten rock of the mantle, known as magma, is pushed up through faults and cracks to near the surface. This creates "hot spots" within 2 to 3 km of the surface. The evidence of such activity in volcanic eruptions, geysers and bubbling mud holes are generally observed. Geysers (hot water and steam) are periodically ejected at the surface in a volcanic area. This is caused by heating of ground water by subsurface magma. Similarly, bubbling mud holes are the hot spring for produce boiling mud. The zone of geothermal sites corresponds to the regions of earthquake and volcanic activity. These regions are at the junctions of tectonic plates (Section 8.3). At these plate boundaries, heat travels most rapidly from the interior *via* subsurface magma to surface volcanoes. Most of the world's geothermal sites today are located near the edges of the Pacific plate, the so-called "ring of fire".

Geothermal energy is generally referred to as a renewable energy resource. The average heat flow from the centre of the Earth to surface is **0.04 to 0.06 W/m^2**. This is small compared with the rate of extraction required for economic operation. Geothermal resources are renewable only if the heat extraction rate is less than the reservoir replenishment rate. Exploitation through wells and downhole pumps in the case of nonelectrical uses, lead to the extraction of very large quantities of fluid. As a result, the geothermal resource in the place is reduced.

In electrical uses, steam condenses into hot water. It is often rich in salts. This polluting hot water waste is disposed of. If disposed hot water is reinjected into the reservoir, then extracted fluid from the geothermal field is replenished. Moreover, the Earth's interior heat is so vast that it will take an infinite time for its exhaustion. The reinjection process heat may compensate for the part of fluid extracted by production. It will to a certain limit prolong the commercial lifetime of the field. Geothermal energy has practically no intermittency. It has the highest energy density. It is therefore to some extent a renewable energy source.

8.5 GEOTHERMAL RESOURCES

Geothermal resources are the thermal energy that could reasonably be extracted at a cost competitive with other forms of energy at some specified future time. This definition was given by Muffler and Cataldi.[7] According to them, the "accessible" geothermal sources is all of the thermal energy stored between the Earth's surface and a specified depth in the crust, beneath a specified area and measured from a local mean annual temperature. The accessible resource includes the useful accessible resource (= Resource); that part of the accessible resource that could be extracted economically and legally at some specified time in the future (less than a hundred years). The identified economic resource (= Reserve) is the part of the accessible resource of a given area that can be extracted legally at a cost competitive with other commercial energy sources and is known and characterised by drilling or by geochemical, geophysical and geological evidence.

The most common criterion for classifying geothermal sources is, however, that based on the enthalpy of the geothermal fluids that act as the carrier transporting heat from the deep hot rocks to the surface. Enthalpy, which can by and large be considered proportional to temperature, is used to express the heat (thermal energy) content of the fluid and gives a rough idea of their value. The sources are divided into low-, medium- and high-enthalpy geothermal sources according to Table 8.1.

Geothermal resources are generally confined to areas of the Earth's crust where heat-flow higher than in the surrounding areas heats the water contained in permeable rocks (reservoirs) at depth. The resources with the highest energy potential are mainly concentrated on the boundaries between

Table 8.1 Classification of geothermal resources (°C).

	(a)[7]	(b)[8]	(c)[9]	(d)[6]
Low-Enthalpy Resources	<90	<125	<100	
Intermediate-Enthalpy Resources	90–150	125–225	100–200	–
High-Enthalpy Resources	>150	>225	>200	>150

tectonic plates where visible geothermal activity frequently exists. By geothermal activity, we mean hot springs, fumaroles, steam vents and geysers. Active volcanoes are also a kind of geothermal activity, on a particularly and more spectacular large scale.

Geothermal activity in an area is certainly the first significant indication that subsurface rocks in the area are warmer than the normal value. The local heat source could be a magma body at 600-1000 °C, intruded within a few kilometers of the surface. However, geothermal fields can also form in regions unaffected by recent shallow magmatic intrusions. The anomalous higher heat flow may be due to particular tectonic situations, for example, due to thinning of the continental crust, which implies the upwelling of the crust/mantle boundary and consequently higher temperatures at shallower depths.

However, we need more than a thermal anomaly to have a productive geothermal resource. We also need a reservoir, which is a sufficiently large body of permeable rocks at a depth accessible by drilling. This body of rock must contain large amounts of fluids, water or steam that carry the heat to the surface. The reservoir is bounded by cooler rocks hydraulically connected to the hot reservoir by fractures and fissures, which provide channels for rainwater to penetrate underground. These cooler rocks occur at the surface where they represent the so-called recharge areas of the geothermal reservoir. Thermal water or steam is in fact mainly rainwater that infiltrates into the recharge areas at the surface and proceeds to depth, increasing in temperature while penetrating the hot rocks of the reservoir (Figure 8.1).

Water moves inside the reservoir by convection, due to density variations caused by temperature, transferring heat from the lowest parts of the reservoir to its upper parts. The result of the convection process is that the temperature in the upper parts of the reservoir is not much lower than that of its deeper parts, so that the lowest values of the geothermal gradient are actually found inside the reservoir. Heat is transferred by conduction from the magma body towards the permeable reservoir rocks, the reservoir, filled with fluids. Hot fluids often escape from the reservoir and reach the surface, producing the visible geothermal activity as described earlier.

However, it can be seen that the geothermal resource is of two kinds, that originating from the magma itself, called the magmatic resource, and that from ground water heated by the magma is called the meteoritic resource. The latter is the largest geothermal resource. Not all geothermal sources produce hot fluids. Some receive no ground water at all and contain only hot rocks. Geothermal sources are therefore of four basic types: **(i) hydrothermal (ii) geopressured (iii) hot dry rocks (iv) magma**. The energy potentially recoverable from the last three reservoirs (*i.e.* geopressured, hot dry rocks and magma) is immense, but the technology to exploit it is still being developed. Of the geothermal resource types, hydrothermal energy is the most advanced and cost competitive and the only one being presently used commercially. Geopressured, hot dry rock and magma resource systems are still in experimental stages. They may be exploited industrially in future after more technological development.

8.5.1 Hydrothermal Systems

The heat source, the reservoir, the recharge area and the connecting paths through which cool superficial water penetrates the reservoir and in most cases, escapes back to the surface, compose

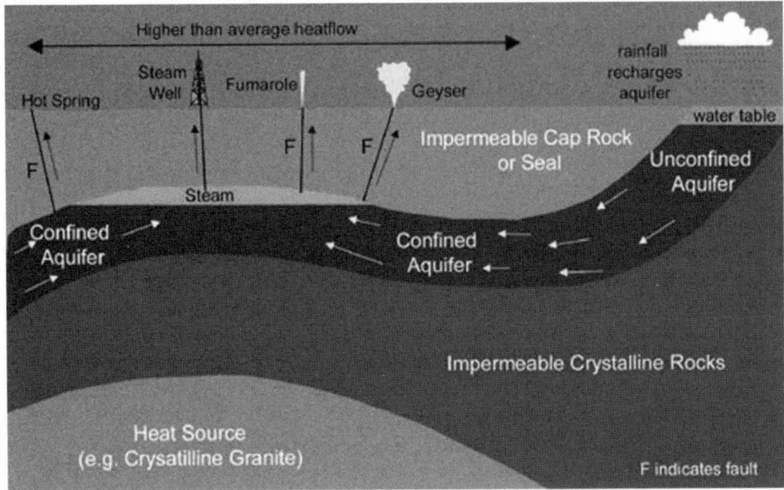

Figure 8.4 Cross-sectional view of the essential characteristics of a hydrothermal geothermal site.

the hydrothermal system. (Figure 8.4). The heat output of hydrothermal systems varies with time. They generally occupy zones of structural weakness, where repeated magmatism is to be expected. The reservoir is the most important part of the system, from the point of view of energy utilisation and in fact we define the reservoir "as the hot part of the geothermal system that can be exploited either by extracting the fluid contained (water, steam or various gases) or using anyhow its heat".[10]

The existence of a hydrothermal system will not necessarily ensure production at industrial levels. Only a part of its rocks may be permeable, constituting a fluid reservoir, so that the system will be able to produce industrially from that part only. That part is called a geothermal field and the geographic name of the locality is usually given to the name of the field (for example, The Geysers geothermal field in California, Tiwi field in the Philippines, Wairakei field in New Zealand and Larderello field in Italy).

Hydrothermal systems (or geothermal reservoirs or fields) are traditionally classified as:

- water dominated; or
- vapour dominated,

The latter have a higher energy content per unit fluid mass. Water-dominated fields are further divided into hot-water fields, producing hot water and fields producing mixtures of water and steam, called wet-steam fields.

A. Water-Dominated Fields

(i) Hot-Water Fields

They are capable of producing hot water at the surface at temperatures up to 100 °C. They are the geothermal fields with the lowest temperature and the reservoir contains water in liquid phase. The thermal aquifers are overlain by confining layers that keep the hot water under pressure. Temperatures in the reservoir remain below the boiling point of water at any pressure because the heat source is not large enough. Surface temperature is not higher than the boiling temperature of water at atmospheric pressure.

These fields may also occur in areas with normal heat flow. On the surface, there are often thermal springs whose temperatures are, in some cases, near the boiling point of water. A hot water field is of economic interest if the reservoir is found at a depth of less than 2 km, if the salt content of water is lower than 60 g/kg and if the wells have high flow rates (above 150 t/h). The best known examples of exploited hot-water fields are those of the Pannonian basin (Hungary), the Paris basin (France), the Aquitanian basin (France), many Russian fields, the Po river valley (Italy), Klamath Falls (Oregon, USA) and Tianjin (China).

(ii) Wet-Steam Fields

These contain pressurised water at temperatures exceeding 100 °C and small quantities of steam in the shallower as well as lower temperature parts of the reservoir. The dominant phase in the reservoir is the liquid one and it is this phase that controls the pressure inside the reservoir. Steam is not uniformly present, occurring in the form of bubbles surrounded by liquid water.

An impermeable cap rock generally exists to prevent the fluid from escaping to the surface, thus keeping it under pressure. This is common but not absolutely necessary. In fact, at any depth below the water table, water bears its own hydrostatic pressure. When the fluid is brought to the surface and its pressure decreases, a fraction of fluid is flashed into steam, while the greater part remains as boiling water. Once a well penetrates a reservoir of this type, the pressurised water rises into the well because the pressure is lower there. The consequence of the pressure drop is the vaporisation of part of the water, with the result that the well eventually produces hot water and steam, with water as the predominant phase. The water–steam ratio varies from field to field and even from one well to the next within the same field. As in many cases only steam is used to produce electrical energy, liquid water must be removed at the surface in special separators.

The surface manifestations of these fields include boiling springs and geysers. The water produced often contains large quantities of chemicals (from 1 to over 100 g/kg of fluid, in some fields up to 350 g/kg). These chemicals may cause severe scaling problems to pipelines and plants. They are mainly chlorides, bicarbonates, sulfates, borates, fluorides and sili*ca*. More than 90% of the hydrothermal reservoirs exploited on an industrial scale are of the wet-steam type. Electricity is their optimal utilisation. One important economic aspect of wet-steam fields is the large quantity of water extracted with the steam (for example 6600 t/h at Cerro Prieto, Mexico). Owing to its generally high chemical content, this water has to be disposed of through reinjection wells drilled at the margins of the reservoir.

Examples of wet-steam fields producing electricity are: Cerro Prieto, Los Azufres and Los Humeros (Mexico), Momotombo (Nicaragua), Ahuachapan-Chipilapa (El Salvador), Ohaaki and Kawerau (New Zealand), Puna (Hawaii), Dieng and Salak (Indonesia), Palinpinon and Bac Man (Philippines), Fang (Thailand) Azores (Portugal), Latera (Italy), Milos (Greece), Cove Fort (Utah). Salton Sea, Coso and Casa Diablo (California) Miravalles (Costa Rica), Zunil (Guatemala), Soda Lake, Steamboat and Brady Hot Springs (Nevada), *etc.*

B. Vapour-Dominated Fields

Vapour-dominated reservoirs (fields) produce dry saturated or slightly superheated steam at pressures above atmospheric. They are geologically similar to wet-steam fields but the heat transfer from depth is certainly much higher. Research suggests that their permeability is lower than in wet-steam fields and the presence of the cap rock is of fundamental importance in this case. Water and steam coexist, but steam is the continuous predominant phase, regulating the pressure in the reservoir and the pressure is practically constant throughout the reservoir. These fields are also called dry or superheated fields. Produced steam is in fact generally superheated with small quantities of other gases, mainly CO_2 and H_2S.

The mechanism governing production in these fields is believed to be the following. When a well penetrates the reservoir and production begins, a depressurised zone forms at the bottom of the well. This pressure drop produces boiling and vaporisation of the liquid water in the surrounding rock mass. A dry area, *i.e.* without liquid water is formed near the bottom of the well and steam flows through this zone. Steam crossing the dry area starts to expand and cool, but the addition of heat from the very hot surrounding rocks keeps steam temperature above the vaporisation value for the pressure existing at that point. As a result, the well produces superheated steam with a degree of superheating that may reach 100 °C, for example with well-head pressures of 5 to 10 bar and a steam outlet temperature of more than 200 °C. Surface geothermal activity associated with vapour-dominated fields, whether dry or superheated is similar to the activity present in wet-steam fields. About half of the geothermal electric energy generated in the world comes from six vapour-dominated fields: Larderello (Italy), Mt. Amiata (Italy), The Geysers (California), Matsukawa (Japan), Kamojang and Daraj at (Indonesia).

8.5.2 Geopressured Reservoirs

Geopressured reservoirs were developed over millions of years by sand deposits that were carried down by rivers and entrapped brine and dissolved natural gas at depths of 4 m to 6 m. Through years of successive faulting, these reservoirs are now overlaid by sedimentary rock formations and have been heated by the Earth's normal heat flow. Reservoirs have temperatures of about 160 °C with an extremely high pressure of nearly 1000 bar. When tapped by wells, they can release three forms of energy: pressure (convertible into mechanical work), heat and natural gas.

Geopressured resources have been investigated extensively in offshore wells in Texas and Louisiana in the US Gulf Coast area (deepest well 6567 m) and pilot projects were operated there for some years to produce geopressured fluid and to extract its heat and methane gas content. Also, the Pannonian Basin in Hungary is at present under evaluation. Electrical energy-conversion experiments have been started in the US, but research has still to confirm the economic feasibility and long-term use of this resource.

8.5.3 Hot Dry Rock Systems

A hot dry rock system is a heated geological formation formed in the same way as hydrothermal resources but containing no water, as the aquifers or fractures required to conduct water to the surface are not present. Hot dry rock (HDR) reservoirs are, instead man-made reservoirs in rocks that are artificially fractured and thus any convenient volume of hot dry rock in the Earth's crust, at accessible depth, can become an artificial reservoir (Figure 8.5).

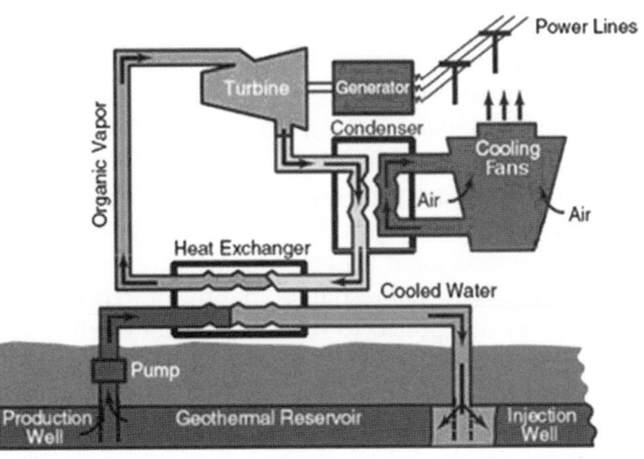

Figure 8.5 View of vapor–turbine cycle.

A pair of wells is drilled into the rock, terminating a hundred meters apart. Water is circulated down the injection well and through the HDR reservoir, which acts as a heat exchanger. The fluid then returns to the surface through the production well and thus transfers the heat to the surface as steam or hot water. The steam is ultimately used to generate electricity. Artificial reservoirs can be made by hydraulically fracturing these rocks and then circulating water through cracks. hot dry rock (HDR) systems are much more common than hydrothermal reservoirs and more accessible, so their potential is quite high.

Experts agree that the following key parameters, representing the lower end of the range are required for a commercially viable HDR reservoir: Production flow rate 50–75 kg/s; effective heat transfer area >2 million square meters; rock volume accessed > 200 million cubic meter; flow losses (% of injection now) < 10%.

A pioneer HDR project at Los Alamos, New Mexico, USA has now reached the threshold of economic viability at a cost of US$ 175 million (1993). Since then, field experiments of various magnitudes have been undertaken in the United Kingdom, France, Germany and Japan and more recently in Sweden and Australia. These experiments have been concerned with the validation of various concepts of HDR exploration.

In order to utilise the HDR system, methods must be found for breaking up impermeable rocks at depth, introducing cold water and recovering the resulting hot water (or steam) for use at the surface. The known temperatures of HDR vary between 150 to 290 °C. This energy, called petrothermal energy, represents by for the largest resource of geothermal energy of any type, as it accounts for a large percentage of the geothermal resource. Much of the HDR occurs at moderate depths, but it is largely impermeable, as stated above, in order to extract thermal energy out of it water will have to be pumped into it and back out to the surface. It is necessary for the heat-transport mechanism to render the impermeable rock into a permeable structure with a large heat-transfer surface. A large surface is particularly necessary because of the low thermal conductivity of the rock. The rock is made permeable by fracturing it. Fracturing methods that have been considered involving drilling wells into the rock and then fracturing by high-pressure water or nuclear explosives. Efforts in this direction are in progress.

At present, the breakthrough and advance of HDR technology in the energy industry is being slowed down by the low price of fossil fuel. But an HDR system is an environmentally benign technology.

8.5.4 Magma Energy

Magma, the naturally occurring molten rock material, is a hot viscous liquid that retains fluidity till solidification. It may contain gases and particles of solid materials such as crystals or fragment of solid rocks. However, the mobility of magma is not much affected until the content of solid material is too large. Typically, magma crystallises to form igneous rocks at temperatures depending upon the composition and pressure, from 700–1200 °C. At its sites of generation, magma is lighter than the surrounding material and consequently it rises as long as the density difference between magma and surrounding cooler rocks persists. Eventually, magma either solidifies or forms reservoirs at some depth from the Earth's surface or it erupts. Magma chambers represent a huge potential energy source, the largest of all geothermal resources, but they rarely occur near the surface of the Earth. Extracting magma energy is expected to be most difficult of all resource types.

The goal of the US Magma Energy Extraction Program was to determine the engineering feasibility of locating, accessing and utilising magma as a viable energy resource. Research is also at present being carried out in Japan. For successful energy extraction, engineering materials need to be selected and tested for compatibility with the magmatic environment. High-temperature drilling and completion technology require development for entry into magma technology.

8.6 DRY ROCK AND HOT AQUIFER ANALYSIS

8.6.1 Analysis of Dry Rock Aquifer

Let us consider a large mass of dry uniform material having density ρ_r specific heat capacity C_r and cross-sectional area A. It extends from near the Earth's surface to deep inside the Earth. There will be a linear increase of temperature with depth.

If x increases downward from the surface at $x=0$ (Figure 8.6), then

$$T = T_0 + \frac{dT}{dx}x = T_0 + Gx \tag{8.1}$$

where, $G = \dfrac{dT}{dx}$ is the temperature gradient, *i.e.* temperature rise per unit depth.

Let the minimum useful temperature be T_1 at depth x_1, so one can have

$$T = T_0 + Gx_1$$

or,

$$x_1 = \frac{(T_1 - T_0)}{G} \tag{8.2}$$

The useful heat content of the rock, δE for temperature $T > T_1$, in an element of thickness δx at depth x is

$$\delta E = (\rho_r A\delta x)C_r(T - T_1) = (\rho_r A\delta x)C_r G(x - x_1) \tag{8.3}$$

The total useful heat content of the rock between x_1 to x_2 becomes, after integration of the above equation between the limit of x_1 and x_2,

$$E_0 = \int_{x=x_1}^{x_2} \rho_r A C_r G(x - x_1)dx$$

$$= \rho_r A C_r G\left[\frac{x^2}{2} - x x_1\right]_{x_1}^{x_2}$$

$$= \rho_r A C_r G\frac{(x_2 - x_1)^2}{2} \tag{8.4}$$

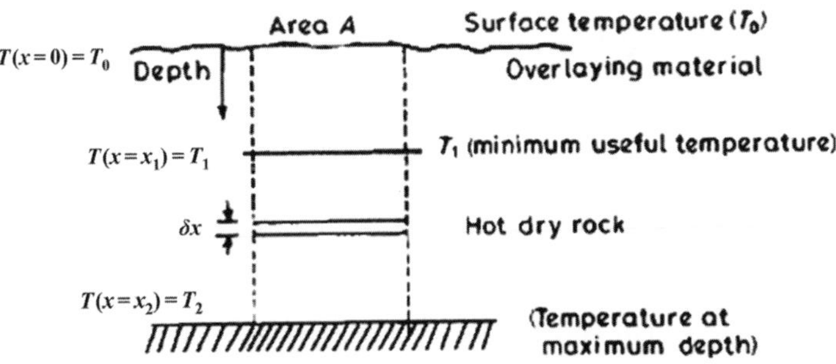

Figure 8.6 Cross-sectional view of hot dry rock system (Density (ρ), specific heat capacity (c), temperature gradient ($dT/dx = G$)).

If $C_r = \rho_r AG(Z - Z_1)$ is the thermal capacity of the rock between x_1 and x_2, the above equation reduces to

$$E_0 = \frac{\rho_r AC_r G(z - z_1)^2}{2} \tag{8.5}$$

Further, let the average available temperature difference that is greater than T_1 be θ,

$$\theta = \frac{(T_2 - T_1)}{2} = \frac{G(x_2 - x_1)}{2}$$

Then, Eq. (8.5) becomes

$$E_0 = C_r \theta \tag{8.6}$$

If heat is extracted from the rock uniformly at volume flow rate V proportion to the temperature greater than T_1 the water will be heated through a temperature difference of $d\theta$ in dt time in the near-perfect heat exchange process, then

$$\dot{V}\rho_w C_w \theta = -\frac{c_r d\theta}{dt} \tag{8.7}$$

where, ρ_w and C_w are the density and specific heat capacity of water.

Equation (8.7) can be rewritten as

$$\frac{d\theta}{\theta} = -\frac{\dot{V}\rho_w C_w}{C_r} dt = -\frac{dt}{\tau}$$

After integration of above equation one gets.

$$\theta = \theta_0 e^{-t/\tau} \tag{8.8}$$

where τ is the time constant given by

$$\tau = \frac{c_r}{\dot{V}\rho_w c_w} = \frac{\rho_r AG(z - z_1)}{\dot{V}\rho_w c_w} \tag{8.9}$$

The useful heat content is $C_r\theta$, and from Eq. (8.6) we have

$$E = C_r \theta_0 e^{-t/\tau} \tag{8.10}$$

Differentiating w.r.t. "t", one gets

$$\frac{dE}{dt} = -\frac{E_0}{\tau} e^{-t/\tau} \tag{8.11}$$

8.6.2 Analysis of Hot Aquifers

In the case of a hot aquifer, the heat resource lies within a layer of water deep beneath the ground surface as shown in Figure 8.7. The thickness of the aquifer (h) is much less than the depth x_2 below the ground level. The water is all at temperature T_2. The fraction of the aquifer containing water has porosity p'. With the remaining space, rock has the density of ρ_r, The minimum useful temperature is T_1 (Section 8.6.1).

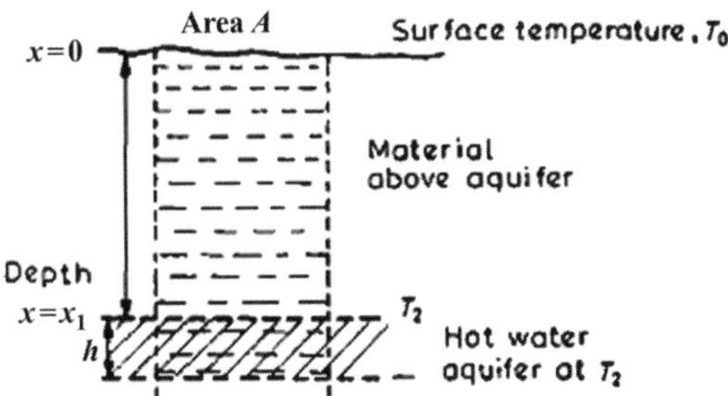

Figure 8.7 Cross-sectional view of hot aquifer system.

If $\frac{dT}{dx}$ is the temperature gradient, then the temperature at $x = x_2$ is given by

$$T_2 = T_0 + \frac{dT}{dx}x = T_0 + Gx \tag{8.12}$$

Following the procedure of Section 8.6.1 for hot dry rock, from Eq. (8.4), we too have

$$\begin{aligned} E_0 &= \frac{\rho_r A C_r G(x_2 - x_1)^2}{2} \\ &= \frac{\rho_r A C_r (x - x_1) G(x_2 - x_1)}{2} \\ &= \rho_r A C_r (x_2 - x_1)\theta \end{aligned}$$

The above equation can be rewritten as,

$$\frac{E_0}{A} = \rho_r A C_r (x_2 - x_1)\theta \tag{8.13}$$

But, in the case of an aquifer (water + rock), $\rho_r C_r$ becomes

$$\rho_r C_r = p' \rho_r C_r + (1 - p')\rho_r C_r \tag{8.14}$$

where p' is the porosity of rock. Equation (8.13), reduces to

$$\frac{E_0}{A} = [p' \rho_r C_r + (1 - p')\rho_r C_r](Z_2 - Z_1)\theta$$

or,

$$\frac{E_0}{A} = C_a \theta \tag{8.15}$$

where,

$$C_a = [p' \rho_r C_r + (1 - p')\rho_r C_r]h \text{ and } Z_2 - Z_1 = h$$

As carried out in Section 8.6.1, one can calculate the removal of heat by a water volume flow rate $\dot{V}$ at θ above T_1 as

$$\dot{V} \rho_w C_w \theta = -C_a \frac{d\theta}{dt} \tag{8.16}$$

The above equation can be solved for θ and the equation for E can be written as

$$E = E_0 e^{-t/\tau_a} \tag{8.17}$$

Further, after differentiating, one gets

$$\frac{dE}{dt} = -\left(\frac{E_0}{\tau_a}\right) e^{-t/\tau_a} \tag{8.18}$$

where

$$\tau_a = \frac{C_a}{\dot{V}\rho_w C_w} = \frac{[p'\rho_r C_r + (1 - p')\rho_r C_r]h}{\dot{V}\rho_w C_w}$$

is a constant in the case of a hot aquifer.

EXAMPLE 8.1

The geysers geothermal site covers an area of 70 km^2. The thickness of the subsurface zone is 2.0 km. In this zone, the temperature is 240 °C and the volumetric specific heat is 2.51 J/cm^3 °C.

(a) Evaluate the heat energy content in joules (at temperatures above $T_0 = 15\,°C$).
(b) Determine the number of years to provide power for a 2000-MWe plant if 1.9% of the thermal energy can be converted to electricity.

Solution

(a) The volume of the zone is the area times the thickness and is given by

$$V = 70\ \text{km}^2 \times 2\ \text{km} = 140\ \text{km}^3$$

The heat content "Q" is given by

$$Q = V(\text{volume}) \times (\text{volumetric specific heat}) \times \Delta T(\text{Temperature difference})$$

Here, $V = 140\,\text{km}^3 \times \dfrac{10^{15}\text{cm}^3}{\text{km}^3} = 1.4 \times 10^{17}\text{cm}^3$, Volumetric specific heat $= 2.51$ J/cm^3 ·°C and $\Delta T = 240\,°\text{C} - 15\,°\text{C} = 225\,°\text{C}$

After substituting the approximate value, one gets

$$Q = 1.4 \times 1017 \times 2.51 \times 225 = 7.9 \times 10^{19}\text{J}$$

(b) For each year of operation, the electrical energy produced is given by

$$E = P \times t = 2000 \times 1 = 2000\,\text{MW} \cdot \text{yr}$$

For an overall efficiency of power plant $= 1.9\%$, the amount of heat energy extracted per year is given by

$$\begin{aligned}\text{Extracted heat energy} &= \frac{2000\ \text{MW} \cdot \text{yr}}{0.019} = 1.053 \times 10^{11}\text{W} \cdot \text{yr} \\ &= 1.053 \times 10^{11}\text{W} \cdot \text{yr} \times (3.15 \times 10^7\text{sec/yr}) \times \frac{1\ \text{J}}{\text{W} \cdot \text{sec}} \\ &= 3.32 \times 10^{18}\ \text{J each year}\end{aligned}$$

In order to obtain the number of years, a 2000-MWe plant can be operated before it is exhausted, divide the available energy by the amount consumed per year, one gets

$$\text{number of years} = \frac{\text{available energy from part (a)}}{\text{amount consumed per year}} = \frac{7.91 \times 10^{19}\ \text{J}}{3.32 \times 10^{18}\ \text{J/yr}} = 23.8 \approx 24\,\text{years}$$

EXAMPLE 8.2

Evaluate the useful heat content per km^2 of dry rock granite to a depth of 7 km. Consider the geothermal temperature gradient $dT = 40\,°C\ km^{-1}$, the minimum useful temperature as 140 K above the surface temperature (T_0) and $\rho_r = 2700\ kg/m^3$.

Solution

(a) At 7 km, the temperature $T_2 = 40 \times 7 = 280$ k above T_0. The minimum useful temperature (T_1) of 140 K above T_0 occurs at 3.5 km. From Eqs. (8.4) and (8.5), one has

$$\frac{E_0}{A} = \frac{\rho_r C_r (Z - Z_1)(T_2 - T_1)}{2}$$

$$= (2.7 \times 10^3\ kgm^{-3})(0.82 \times 10^3 J\,kg^{-1}\,K^{-1})(3.5\,km)(140\,K)$$

$$= 5.42 \times 10^{17} J\,km^{-2}$$

(b) Substituting in Eq. (8.11), one gets

$$\tau = \frac{\rho_r A C_r (Z - Z_1)}{\dot{V} \rho_w C_w}$$

For, density of water $\rho_w = 1000\,kg/m^2$ and specific heat of water $C_w = 4.2\,kJ/kg\,°C$, then one have,

$$\tau = \frac{(2.7 \times 10^3 kg\,m^{-3})(0.82 \times 10^3 J\,kg^{-1}\,K^{-1})(1\,km^2)(3.5\,km)}{1\,m^3\,s^{-1} \times 1 \times 10^3\,kg\,m^{-3} \times 4.2 \times 10^3 J\,kg^{-1}\,K^{-1}}$$

$$= 1.84 \times 10^6\ s = 58\,year$$

(c) From Eq. (8.18), we have

$$\frac{dE}{dt} = -\left(\frac{E_0}{\tau}\right) e^{-\frac{t}{\tau}}$$

At $t = 0$, we have the above equation as

$$\left(\frac{dE}{dt}\right)_{t=0} = \frac{5.42 \times 10^{17} J\,km^{-2}}{1.84 \times 10^9 s} = 294\,MW\,km^{-2}$$

At $t = 10$ years, we have,

$$\left(\frac{dE}{dt}\right)_{t=10\,year} = 294\,e^{\left(-\frac{10}{58}\right)} = 247\,MW\,km^{-2}$$

8.7 EXPLORATION OF GEOTHERMAL ENERGY

The present technology and economic factor restrict extraction of geothermal energy to the upper few kilometers of the Earth's crust, *i.e.* less than 5 km depth.

A strategy for geothermal energy exploitation is as follows: (i) identification of the geothermal region, (ii) use various exploration techniques to locate the most interesting geothermal areas, (iii) identify suitable targets for fluid production, (iv) estimate temperature, reservoir volume and permeability at depth, (v) determine whether wells will produce steam or just hot water, (vi) estimate the chemical composition of the fluid to be produced.

A number of available exploration techniques are as follows:

(a) Inventory and Survey of Surface Manifestations

The knowledge of hot springs, steam vents, fumaroles *etc.* and their physical and chemical is essential and has an important role. This information is extremely useful for subsequent planning of exploration. The surface survey is conducted in two consecutive phases, namely (i) the collation, processing and standardisation of published and recorded data relative to chemistry, temperature, flow rates, *etc.* and (ii) the collection of new data, water samples, gas samples, temperature measurements, *etc.*

(b) Geological and Hydrogeological Surveys

The geological and hydro surveys are not limited to studies of ground waters. These also include geological surveys to provide information on the stratigraphic and structural framework of the area. Geothermal reservoirs are often associated with volcanic regions. Therefore, volcanology also offers many examples of how geological field data give evidence of the location, nature and size of a geothermal resource.

There are some mathematical models that are of great help in hydrogeological surveys. Fluid inclusions may give information on the temperature of deposition of inclusions. It therefore determines the temperature and salinity of geothermal fluids. Fluid inclusions are defects in crystals. They are formed during or after deposition. All crystals have inclusions. Some inclusions are solids, others are empty and a few contain fluids. Inclusions need to be multiphase, namely, liquid and vapour to be most useful.

(c) Geochemical Surveys

Geochemical exploration can start simultaneously with geologic and hydrogeological examination.

This has proved that springs and other geothermal manifestations are available for fluid sampling.

Geochemical studies involve three main steps, namely (i) sample collection (ii) chemical analysis and (iii) data interpretation. The types of samples collected are water samples from hot springs, steam samples from fumaroles, gas samples from hot pools. Geothermometers enable estimation of the temperature of deep reservoirs and calculation of the ratios of certain chemical elements.

The content of tritium and ^{14}C radioisotopes permit evaluation of the age of the geothermal fluids. Geochemical surveys with the use of tracers also offer information (i) on the direction of movement of subsurface groundwater (ii) reinjected fluids, (iii) the type of corrosion and (iv) scaling problems. Hydrogen and oxygen isotopes can be used to identify the recharge surface areas of reservoir.

(d) Geothermometers

Geothermometers are used to estimate the temperature of deep reservoirs to calculate the ratios of certain chemical elements (*i.e.* Na, K, Mg, Ca, *etc.*). They also make some adjustment for the degree of mixing of hot geothermal reservoir water with cooler groundwater in the shallow part of the hydrothermal system.

They are therefore valuable tools in the evaluation of new fields. They are also useful to monitor the hydrology of systems on production. For certain parameters or ratios of parameters, the relationship between temperature and chemical composition is stable and predictable. These parameters (ratios of parameters) are known as geothermometers. The temperature indicated by the geothermometers is not necessarily the maximum temperature of the water.

The three principal indicators of deep-reservoir chemistry are silica, magnesium and sodium/potassium ratios. Silica concentrations are more reliable for hot springs of high discharge in comparison with low discharge. Magnesium is of limited value as a temperature indicator.The majority of gas thermometers require that the gas/steam, and for a hot-water reservoir, the steam/water, ratios are known.

(e) Geophysical Surveys

Classical geophysical techniques (seismic, gravity and magnetic surveys) are defined as indirect methods. These methods are not directly associated with the properties of the hot fluids. They yield information about the attitude and nature of the host rocks. However, other geophysical methods may directly reveal variations in the physical properties of the rocks. This is caused by the presence of hot and saline fluids. These include electrical resistivity, electromagnetic and thermal measurement methods.

(i) **Seismic Surveys**
Elastic waves are transmitted through rocks. Their velocities can be used to determine the structure and properties of rock bodies. Seismic waves are introduced into the Earth by detonating an explosive charge in a shallow borehole. It can also be done by using a large mass striking the surface heavily. Returns of seismic waves are measured at the surface; Seismic waves also originate naturally from earthquakes and microearthquakes. These waves can also be detected at the surface. Interpretation of the seismic information can provide data on the location of active faults that can channel hot fluids towards the surface.

(ii) **Gravity Surveys**
Changes in the density of subsurface rocks cause the variations in the Earth's gravity field. Gravity surveys are rather simple and inexpensive. Gravity anomalies alone are not necessarily indicative of a geothermal region. But they do give valuable information on the type of rocks at depth and their distribution and geometric characteristics.

(iii) **Magnetic Surveys**
The Earth has a primary magnetic field. It induces a magnetic response in certain minerals at and near the Earth's surface. By detecting spatial changes of the magnetic field, the variations in distribution of magnetic minerals may be deduced. This is related to geologic structure. However, each magnetic mineral has a Curie temperature. The mineral loses its magnetic properties above the Curie temperature. For iron, the Curie temperature is 760 °C. Aeromagnetic surveys are much more commonly used in geothermal exploration. The basic principle is to detect zones that are magnetically featureless. This is due to destruction of magnetite in near-surface rocks by hydrothermal alteration.

(iv) **Electrical-Resistivity Surveys**
Most electrical methods are based on measurement of the electrical resistivity of the subsurface. Resistivity is often largely affected by electrical conduction within water occupying the pore spaces in the rock. Consequently, resistivity varies considerably with porosity. Temperature and salinity of interstitial fluids tend to be higher in geothermal reservoirs. The resistivity of geothermal reservoirs is generally relatively low. These techniques are based on injection of current into the ground. As a consequence, measurement of voltage differences is carried out at the ground surface. One of the major drawbacks with electrical methods is the shallow depth of penetration.

(v) **Electromagnetic Surveys**

Electromagnetic methods are a tool to determine the electrical-resistivity distribution in the Earth by means of surface measurements of transient electric and magnetic fields. These fields can be naturally or artificially generated. These methods are more suitable to measure the low resistivities of geothermal reservoirs than electrical-resistivity methods. Furthermore, the surface resistivity is sometimes so high that it prevents current from entering the ground. The electromagnetic methods help to eliminate the screening effect of very resistive surface rocks with a much deeper penetration. Currents of varying frequency (generally from a few to several tens of thousands of Hz) are transmitted into the ground. This is either *via* the electrodes as in the electrical methods or by induction loops. Mobile stations measure the electrical and magnetic fields created by this transmission at several points. Comparison between these fields enables the resistivities of the underlying formations.

Magnetotelluric soundings use the natural oscillations of the Earth's electromagnetic field to determine the resistivity structure of the subsurface. The electromagnetic waves are assumed to be planar. These are vertically incident on the surface of the Earth where they are detected. The lower the frequency, the deeper is the penetration. The longer frequencies take to collect a signal with a satisfactory signal to noise ratio. The depth of penetration of magnetotelluric soundings (MT) surveys is also much greater than direct current (DC) resistivity measurements. It is often possible to achieve a penetration as great as 3–5 km with a reasonable degree of precision. To a reasonable degree of precision, this can be compared to a maximum depth penetration from most (DC) methods of less than 2 km. MT does not require a current source.

(vi) **Thermal-Measurement Surveys**

The traditional geophysical methods had been developed for the oil industry. These are used side-by-side with more specific techniques.

Geothermal prospecting provides (i) information on the thermal conditions of the subsurface, (ii) the distribution of the Earth's heat flow and (iii) the location and intensity of thermal anomalies. Geothermal prospects allow us:

(i) to verify the existence of high-temperature fluids;
(ii) to do the more precise site deep drilling;
(iii) to draw the boundaries of geothermal fields;
(iv) to acquire data for evaluation of the geothermal potential.

Heat-flow measurements are made by drilling small-diameter (10 cm) holes at shallow wells (generally < 300 m). The number of drills depends on local conditions and on the results one wants to achieve. Heat flow is measured every 10–25 km^2. The depth of the wells must be sufficient to avoid the effects of propagation of the annual surface thermal variations. These are negligible beyond 20 m. The thermal disturbance is caused by the circulation of shallow ground water.

The geothermal gradient is obtained from temperatures measured with electric thermometers at various depths along a well. Temperature logging is quick, and relatively inexpensive. The thermal conductivity of the rocks is usually determined by laboratory measurements on core samples. The product of the gradient and conductivity gives the heat flow per unit area.Sometimes, the temperature gradient alone is sufficient to give the required information. However, this is possible, only if the survey is carried out in areas that are lithologically homogeneous at depth for constant thermal conductivity.

(f) Exploratory Wells

The final stage of an exploration survey is well drilling. Usually, the final diameters of these wells are on the order of 8 in (20 cm) or less. This allows the insertion of special logging tools to measure various parameters from the surface to the total depth. Sometimes, it is to carry out fluid production tests. A pump may be lowered into a shallow hot water well at hundreds of metres depth. Compressed air may be injected in deeper hot-water wells.

Most geothermal reservoirs are made up of fluid-filled fractures. It is essential that an exploratory well intersects as many fractures as possible. In some cases, it may be necessary to redrill the well at an angle in order to intersect the natural fracture pattern. Since natural fractures are related to the tectonic activity the siting of exploratory wells is greatly dependent on our geological interpretation at the local structural conditions.

8.8 WELL DRILLING AND FLUID EXTRACTION

8.8.1 High-Temperature Wells (>150 °C)

Drilling for geothermal fluids is similar to rotary drilling for oil and gas. However, geothermal drilling is generally more difficult than in oil and gas operations. This is due to the nature of the rock at higher temperatures and corrosive nature of the fluids. The rock is usually harder, metamorphic or igneous. High temperatures associated with geothermal wells affect the circulation system and the cementing procedures as well as the design of the drill string and casing. Safety devices called blow-out preventers are used to prevent blow-outs of high-temperature wells during drilling.

Mud is generally used as the drilling fluid. The use of air instead of mud makes drilling much faster and cheaper and has been adopted frequently in recent years. One major obstacle in air drilling is its unsuitability in formations carrying excessive water or in formations that tend to collapse.

Directional drilling techniques are used in the surface areas directly above the area that is not the drilling target. The angle is established and maintained by utilising a downhole turbine drill. Directionally drilled wells cost about 25% more than vertically drilled wells due to slower penetration rates.

The hottest geothermal of 500 °C well drilled is located in the Kakkonda field, Japan. at the depth of 3279 m in quaternary granite.[11]

8.8.2 Low-Temperature Wells (<150 °C)

Drilling and casing of a low-temperature well <150 °C is probably the most expensive activity in geothermal projects. This is for nonelectrical uses. Present drilling technology is expensive. The costs of all wells increase exponentially with depth. The cost of drilling is uneconomic for heat production. The cost of drilling includes the costs for land acquisition and geological surveys.

In this case, the technology is similar to that used for groundwater wells.

8.8.3 Extraction of Fluids

The flow rate and the performance of a geothermal reservoir depend on many parameters, namely the volume and type of the fluid, the rate of recharge, the permeability of the rocks, the design of the drilled well and piping and the type of completion equipment utilised.

Depending on the characteristics of the particular reservoir, the fluid may exist at the surface as a liquid (a vapour or a mixture of the two). It may also include various dissolved gases and solid material. Thus, the type of equipment will depend on the enthalpy pressure characteristics of the fluid and its salt content. For low-temperature fluids, it is in nonelectrical applications. We will generally be looking for fluid in the liquid phase.

8.9 UTILISATION

Electricity generation is the most important for high-temperature geothermal resources (> 150 °C). The medium- to low-temperature resources (< 150 °C) are suited to many other types of application including domestic. Therefore, utilisation of geothermal resources is classified as (i) direct and (ii) electrical utilisation. The modified Lindal diagram shows the possible uses of geothermal resources at different temperature ranges (Figure 8.8).[2]

8.9.1 Direct Utilisation

Direct utilisation of geothermal energy refers to the immediate use of the heat energy. Direct heat use is one of the oldest and most versatile. This is the most common forms of utilisation of geothermal energy. The primary form of direct use is swimming pool, bathing and therapeutic use (balneology), and space heating and cooling. It also includes district heating, agriculture (mainly greenhouse heating and some animal husbandry), aquaculture (mainly fish pond and raceway heating), industrial processes and heat pumps (for both heating and cooling).

Direct use of geothermal energy has been on a small scale. Recent developments involve large-scale projects. This includes district heating (Iceland and France), greenhouse complexes (Hungary and Russia) or major industrial use in New Zealand and the US. Heat exchangers are also becoming more applicable to geothermal projects. This allows use of lower temperature water and highly saline fluids.

Many groups like Romans, Chinese, Japanese, Turks and Central Europeans have been using geothermal water for bathing purpose for centuries. The geothermal water in the northwest of Beijing, China has been used for medical purposes for over 500 years. The water of about 50 °C can be used to treat high blood pressure, rheumatism, and skin disease, diseases for the nervous system and generally for recuperation after surgery.

Space conditioning includes both heating and cooling. Buildings heated from individual wells are popular in Klamath Falls, Oregon and Nevada in New Zealand.

District heating originates from a central location. This supplies hot water or steam through a network of pipes to individual dwellings or blocks of buildings. Geothermal district heating systems are in operation in at least 12 countries, including Iceland, France, Poland, Hungary, Turkey, Japan and the US. The Reykjavik, Iceland district heating system is probably the most famous. This system supplies heat for a population of around 145 000 people. In France, production wells in sedimentary basins provide direct heat to more than 500 000 people from 40 projects. These wells provide from 40 to 100 l/s of 60 to 100 °C water from depths of 800 to 1200 m.

Agriculture and aquaculture are particularly attractive due to heating at the lower end of the temperature range. A number of agribusiness applications can be considered for greenhouse heating, aquaculture, animal husbandry, soil warming and irrigation, mushroom culture and biogas generation.

Livestock-raising facilities can encourage the growth of domestic animals by a controlled heating and cooling environment.

Table 8.2 gives the various uses of geothermal energy in terms of capacity and energy utilisation for comparison in the years 1995 and 2000.[2]

Direct applications of geothermal energy can use both high- and low-temperature geothermal resources. It is therefore much more widespread. Direct application is, however, more site specific

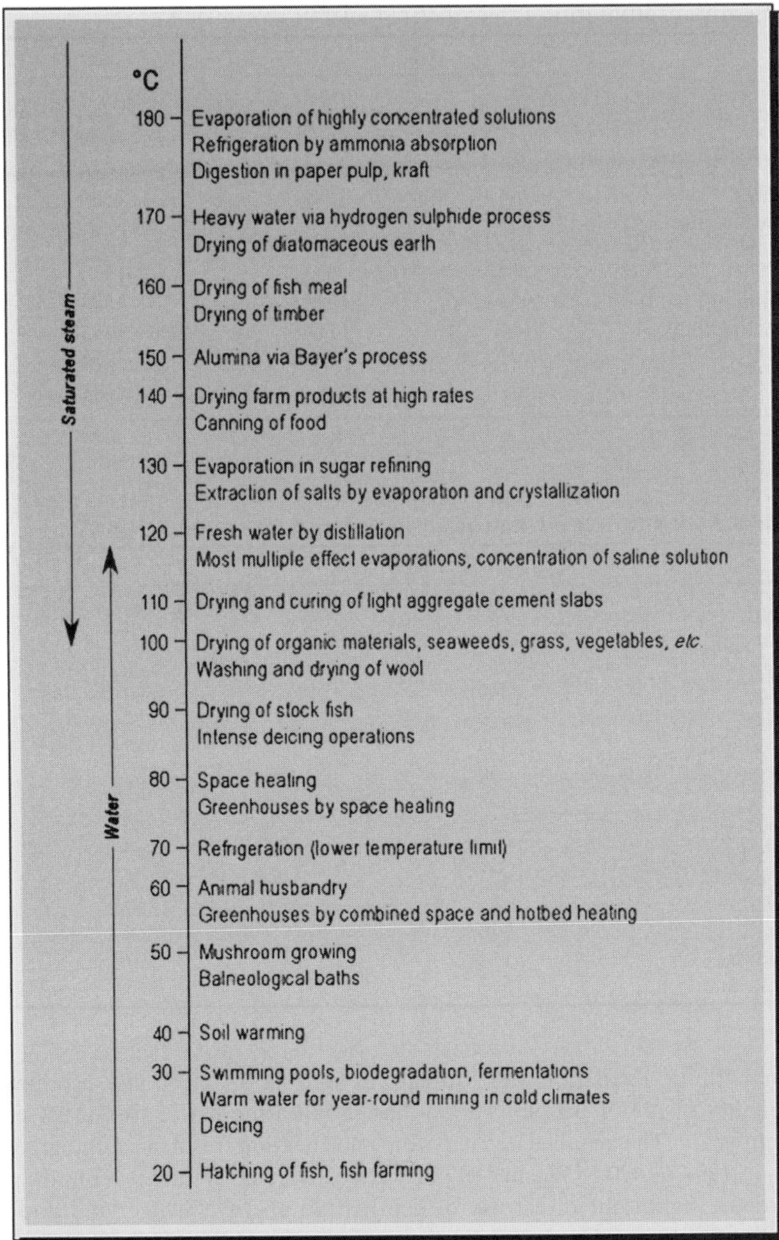

Figure 8.8 The Lindal diagram on typical fluid temperatures for direct applications of geothermal resources.

for the market. Steam and hot water is rarely transported long distances from the geothermal site. The longest geothermal hot water pipeline in the world is 63 km, in Iceland. The production cost/kWh for direct utilisation is highly variable. Table 8.3 shows the installed capacity and produced energy.[12]

China has geothermal water in almost every province. The direct utilisation is expanding at a rate of about 10% per year. It is used mainly for space heating (replacing coal), bathing, and fish farming sectors. Japan is also blessed with every extensive geothermal resource. They are mainly

Table 8.2 Categories of utilisation of geothermal energy worldwide.

Category	Capacity (MW_{th}) 2000	Capacity (MW_{th}) 1995	Utilisation (TJ/yr) 2000	Utilisation (TJ/yr) 1995
Geothermal heat pumps	5275	1854	23 275	14 617
Space heating	3263	2579	42 926	38 230
Greenhouse heating	1246	1085	17 864	15 742
Aquaculture pond heating	605	1097	11 733	13 493
Agricultural drying	74	67	1038	1124
Industrial uses	411	544	10 220	10 120
Bathing and swimming	3957	1085	79 546	15 142
Cooling and snow melting	114	115	1063	1124
Others	137	238	3034	2249
Total	15 145	8664	190 699	112441

Table 8.3 World's top countries using geothermal in direct uses.

	Installed (MW_{th})	Production (GWh/a)
China	2282	10 531
Japan	1167	7482
USA	3766	5640
Iceland	1469	5603
Turkey	820	4377
New Zealand	308	1967
Georgia	250	1752
Russia	308	1707
France	326	1360
Sweden	377	1147
Hungary	473	1135
Mexico	164	1089
Italy	326	1048
Romania	152	797
Switzerland	547	663

(80%) using for bathing, recreation and tourism, as well as for electricity production. This has improved the quality of life of people. Turkey has shot up the list of top direct use geothermal countries in recent years. The installed capacity for space heating (residences and greenhouses) grew from 160 MW_{th} in 1994 to 490 MW_{th} in 1999 in Turkey. Mexico is the first country extending into the tropics to report significant direct use of geothermal energy. Switzerland and Sweden have recently joined the top league through extensive use of ground-source heat pumps. The country with the most extensive use of geothermal energy is Iceland. It obtains 50% of its total primary energy use from geothermal energy, the remainder coming from hydropower (18%). oil (30%) and coal (2%).

8.9.2 Electricity Generation

Geothermal power plants take advantage of a natural, clean energy source from the Earth's interior. The geothermal resources produce steam or hot water. The energy in the form of steam or hot water is tapped by drilling wells into the reservoirs. The piping of the steam or hot water is carried out to power plants that convert the heat into electricity through steam turbines.

Higher-temperature hot water (175 °C or more) can be flashed to steam to drive a turbine in a flash plant. Lower-temperature hot water (less than 175 °C) requires a binary plant. Thus, the hot water is used to boil a working fluid (an organic fluid with low boiling point). The working fluid vapour rotates a turbine and is condensed for reuse. The rotating turbine spins a generator to produce electricity. The geothermal fluids are drawn from the Earth and returned to the Earth and hence there is no environmental effect.

Geothermal power plants are classified as:

(A) Dry-Steam Power Plants

Conventional steam-cycle plants are used to produce electricity from vapour-dominated reservoirs. This is shown in Figure 8.9. Steam is extracted from the reservoir, cleaned to remove entrained solids and is fed directly to a steam turbine. After the turbine, it is passed through a condenser and is converted to water. This improves the efficiency of the turbine. The environmental problems associated with the direct release of steam into the atmosphere can be avoided in case of non-condensing systems. The waste water is then reinjected into the field.

This type of power plant is well developed and it is also commercially available in the 20–120 MWe capacity range. A new trend of installing modular standard generating units of 20 MWe has been adopted in Italy. A dry-steam generation operation is at the Geysers in Northern California. The efficiency and economics of dry-steam plants are affected by the presence of noncondensable gases (carbon dioxide and hydrogen sulfide). The removal of these gases is essential to increase the efficiency of the plants. Vapour-dominated systems are less common in the world. Steam from these fields has the highest enthalpy (energy content), close to 670 kcal/kg. These systems are only available in Indonesia, Italy, Japan and the USA.

(B) Flash-Steam Power Plants

Flash-steam power plants are a common type of geothermal power plant available today. They use the energy from liquid-dominated reservoirs. The temperature is above 175 °C to flash a large proportion of the liquid to steam. Single-flash systems evaporate hot geothermal fluids into steam and direct it through a turbine. In dual-flash systems, steam is flashed from the remaining hot fluid of the first stage. It is separated and fed into a dual-inlet turbine. In both cases the condensate can be used for cooling and the brine is reinjected into the reservoir. Commercially available turbogenerator units are in the range of 10–55 MW_e.

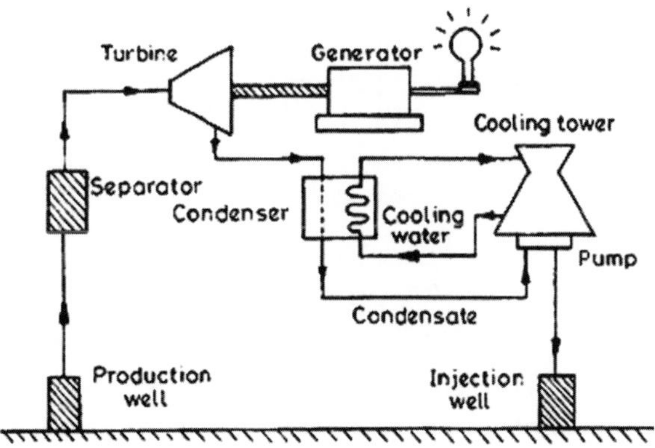

Figure 8.9 View of dry steam system.

(C) Binary-Cycle Power Plants

If the geothermal well produces hot water above 85 °C electricity can still be generated by means of binary cycle plants. These plants operate with low boiling-point working fluid (Freon, isobutane, ammonia, *etc.*) in a thermodynamic cycle. This is known as the organic Rankine cycle. The working fluid is vaporised by the geothermal heat in the vaporiser or heat exchanger. The vapour expands as it passes through the organic vapour turbine, and is coupled to the generator. The exhaust vapour is subsequently condensed in a water-cooled condenser or is air cooled. It is further recycled to the vaporiser by the motive fluid cycle as shown in Figure 8.10. The efficiency of these cycles is between 2.8 and 5.5%. The typical unit size is 1 to 3 MW_e. However, the binary power-plant technology is a cost effective and reliable way to convert large amounts of low temperature geothermal sources into electricity. It is now well known that large low-temperature reservoirs exist at accessible depths almost everywhere in the world.

(D) Geothermal–Fossil Hybrid Power Plants

The concept of hybrid geothermal–fossil fuel systems utilises the relatively low-temperature heat.

There are two arrangements of hybrid plants: (i) geothermal–preheat and (ii) fossil–superheat. Geothermal–preheat arrangement is suitable for low-temperature liquid-dominated systems. The fossil–superheat system is suitable for vapour-dominated and high-temperature liquid-dominated systems.

(i) **Geothermal–Preheat Hybrid Systems:** In this case, the feed water of a conventional fossil-fueled steam plant is heated by low-temperature geothermal energy. Geothermal heat replaces the feed-water heaters, depending upon its temperature. Geothermal heat heats the feed water throughout the low-temperature end prior to an open-type deaerating heater. This is followed by a boiler feed pump and three closed-type feedwater heaters.

(ii) **Fossil–Superheat Hybrid Systems:** The working principle of the fossil–superheat hybrid cycle is shown in Figure 8.11. In this case the vapour-dominated steam (the vapour) obtained from a flash separator in a high-temperature liquid-dominated system is superin a fossil-fired superheater. The system consists of a double-flash geothermal steam system. Steam produced at 4 in the first-stage the flash separator is preheated from 4 to 5 in a regenerator by exhaust steam from the high-pressure turbine at 7. It is then superheated by a fossil fuel-fired superheater to 6, and expands in the high-pressure turbine to 7 at a pressure near that of the second-stage steam separator. It then enters the regenerator, leaves it at 8,

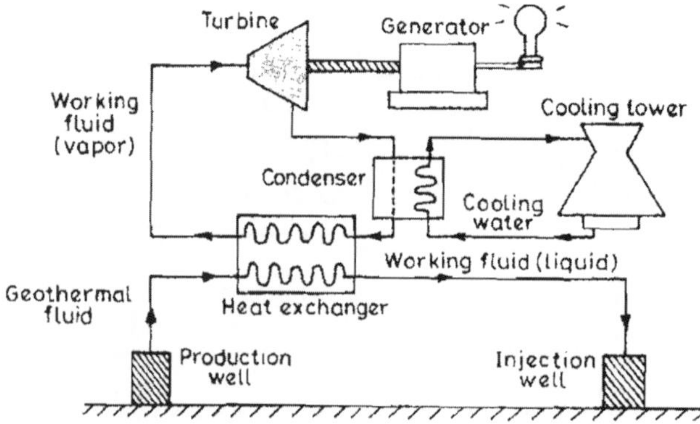

Figure 8.10 View of binary system.

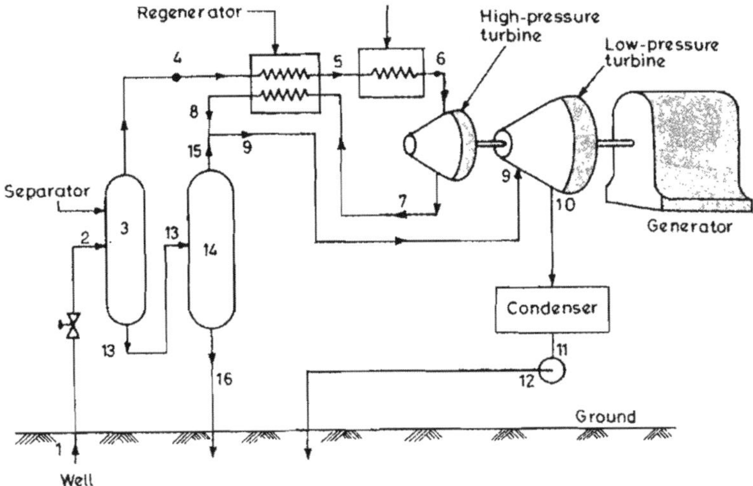

Figure 8.11 Schematic view of a fossil-superheat hybrid system with two-stage flash separator, regenerative and fossil-fired superheater.

where it mixes with the lower-pressure steam produced in the second-stage flash separator at 15, and produces steam at 9, which expands in the lower-pressure turbine to 10. The condensate at 11 is pumped. It is reinjected into the ground at 12. The spent brine from the second-stage evaporator is also reinjected into the ground at 16.

The geothermal electrical capacity installed in the year 2000 was 7974 MWe (Table 8.4) with the generation of 49.3 billion kWh.[1] A further 282 MWe have now been installed.[13] The total electricity produced worldwide from all sources in the year 1998 was 14 411 billion kWh (15 342 billion kWh in 2000).

The efficiency of the generation of electricity from geothermal steam ranges from 10 to 17%. This is three times lower than the efficiency of nuclear or fossil-fuelled plants. Geothermal plants have the lowest efficiency values due to the low temperature of the steam, which is generally below 200 °C. Geothermal power plants require from 6 kg/kWh (if dry steam is available) to 400 kg/kWh of fluid (if hot water is used in binary cycle plants).

8.10 GEOTHERMAL HEAT PUMP

8.10.1 Basic Concept of Heat Pump

As we know, heat flows in the direction of decreasing temperature from high temperature to low temperature. This heat-transfer process occurs in nature without any devices. The reverse process, however, cannot occur by itself. The device that transfers heat from a low temperature to a high temperature is the heat pump. The transfer of heat from a low temperature to a high temperature requires special devices called refrigerators. Refrigerators and heat pumps operate on the same cycle but differ in their objectives. The objective of a refrigerator is to maintain the refrigerated space at a low temperature by removing heat from it. Discharging this heat to a high-temperature is a necessary part of the operation, not the purpose. The objective of a heat pump, however, is to maintain a heated space at a high temperature. This is accomplished by absorbing heat from a low-temperature source, such as well water or cold outside air in winter and supplying this heat to the high temperature such as a house. An ordinary refrigerator that is placed in the window of a house

Table 8.4 Installed geothermal generating capacities in the world in the year 2000.[1]

Country	Installed (MWe)	Generated (GWh)	Percentage of national capacity	Percentage of national energy
Australia	017	0.9	n/a	n/a
China	29.17	100	n/a	n/a
Costa Rica	1425	592	7.77	10.21
El Salvador	161	800	15.39	20
Ethiopia	852	3005	1.93	1.85
France	4.2	246	n/a	2
Guatemala	33.4	2159	368	3.69
Iceland	170	1138	13.04	14.73
Indonesia	589.5	4575	3.04	5.12
Italy	785	4403	1.03	1.68
Japan	546.9	3532	0.23	0.36
Kenya	45	366.47	529	8.41
Mexico	755	5681	2.11	3.16
New Zealand	437	2268	5.11	6.08
Nicaragua	70	583	16.99	17.22
Philippines	1909	9181	n/a	21.52
Portugal	16	94	0.21	n/a
Russia	23	85	0.01	0.01
Thailand	0.3	1.8	n/a	n/a
Turkey	20.4	119.73	n/a	n/a
USA	2228	15 470	0.25	0.4
Totals	797 406	49 261.45		

with its door open to the cold outside air in winter will function as a heat pump, since it will try to cool the outside by absorbing heat from it and rejecting this heat into the house through the coils behind it. The working principle of a heat pump can be very well understood by knowing the fundamentals of an air conditioner.

8.10.2 Air Conditioner

An air conditioner is basically a refrigerator. In this case, the refrigerated space is a room of a building instead of the food compartment of a refrigerator. A window air-conditioning unit cools a room by absorbing heat from the room air and discharging it to the outside. The same air-conditioning unit can be used as a heat pump in winter if it is installed in reverse. In this mode, the unit will pick up heat from the cold outside air and deliver it to the room. An air-conditioning system is equipped with proper controls. A reversing valve operates as an air conditioner in summer and as a heat pump in winter.

The air conditioner is a cyclic device as shown in Figure 8.12a. It is like a heat engine. The most frequently used cycle is the vapour-compression cycle. It involves four main components: a compressor, a condenser, an expansion valve and an evaporator.

The refrigerant enters the compressor as a vapour. It is compressed to the condenser pressure (high pressure). It leaves the compressor at a relatively high temperature and cools down. It condenses as it flows through the coils of the condenser by rejecting heat to the surrounding medium. It then enters a capillary tube where its pressure and temperature drop drastically due to the throttling effect. The low-temperature refrigerant then enters the evaporator where it evaporates by absorbing heat from the refrigerated space. The cycle is completed as the refrigerant leaves the evaporator and re-enters the compressor. A fan blows air both across the condenser as well as evaporator for heat exchange to occur by convection.

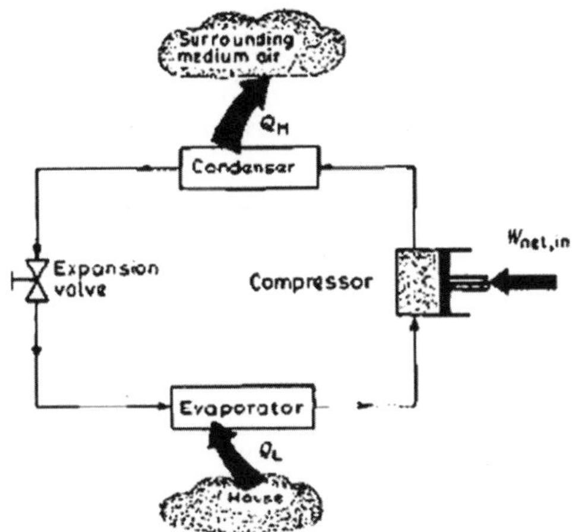

Figure 8.12a Basic Components of a refrigeration system for typical operating system.

8.10.3 Heating and Cooling Mode in Heat Pump

The most common energy source for heat pumps is atmospheric air. Although water and soil are also used. The major problem with air-source systems is frosting. This occurs in humid climates when the temperature falls below 2 to 5 °C. The frost accumulation on the evaporator coils is highly undesirable. It seriously disrupts convective heat transfer. The coils can be defrosted, however, by reversing the heat pump cycle (running it as an air conditioner). This results in a reduction in the efficiency of the system. Water-source systems usually use well water from depths of up to 80 m in the temperature range of 15 to 25 °C. They do not have a frosting problem. They typically have higher COPs (coefficient of performance). This requires ready access to a large body of water, such as underground water. Ground source systems are also involved since they require long tubing placed deep in the ground where soil temperature is relatively constant.

Both the capacity and the efficiency of a heat pump decrease significantly at low temperature. Therefore, most air-source heat pumps require a supplementary heating system. This can be electric-resistance heaters or an oil or gas furnace. Supplementary heating may not be required for water-source or ground-source systems due to constant water and soil temperatures. The heat-pump system must be large enough to meet the maximum heating load.

Heat pumps and air conditioners have the same mechanical components. Therefore, it is not economical to have two separate systems to meet the heating and cooling requirements of a building. One system can be used as a heat pump in winter for heating and an air conditioner in summer for cooling. In heating mode, the compressed refrigerant is directed to the indoor coil first (condenser). This makes releases of heat energy to the inside of the house. The heated air is ducted by the fan to the room. The outdoor coil is used to collect the heat energy from outside. It becomes an evaporator. The reverse is the processes in cooling mode in which the condenser of the heat pump (located indoor) functions as the evaporator in summer. The difference in the two modes is the working of the reversing value.

EXAMPLE 8.3

A heat pump is used to meet the heating requirements of a house to maintain a temperature of 20 °C. If the outside air temperature drops to −1 °C, the house is estimated to lose heat at a rate of 40 000 kJ/h. If the heat pump has a COP (coefficient of performance) of 2.5, evaluate (a) the power consumed by the heat pump (b) the rate of heat is absorbed from the cold outdoor air.

Solution

Under steady operating conditions:
(a) The power consumed by this heat pump is determined from the definition of the coefficient of performance.

$$COP = \frac{Desired\ ouput}{Required\ output} = \frac{Q_H}{W_{net,in}}$$

where Q_H is the magnitude of the heat rejected to the warm environment at temperature T_H and $w_{net,in}$ is the net work input to the refrigerator.

Now,

$$\dot{w}_{net,in} = \frac{\dot{Q}_H}{COP_{Hp}} = \frac{40\,000\,\text{kJ/hour}}{2.5} = 16\,000\frac{\text{kJ}}{\text{hr}}(4.45\,\text{kW})$$

(b) The house is losing heat at a rate of 40 000 kJ/h. If the house is to be maintained at a constant temperature of 20 °C, the heat pump must deliver heat to the house at the same rate. Then, the rate of heat transfer from the outdoor is given by,

$$\dot{Q}_L = \dot{Q}_H - \dot{w}_{net,in} = (40\,000 - 16\,000) = 24\,000\,\text{kJ/hour}$$

where $\dot{Q}_L$ is the rate of heat removed from outside air at temperature T_L.
Therefore, 24 000 of the 40 000 kJ/h heat delivered to the house is actually extracted from the cold outside air. Hence, we are paying only for the 16 000 kJ/h of energy. This is supplied as electrical work to the heat pump. If we were to use an electric-resistance heater instead, we would have to supply the entire 40 000 kJ/h to the resistance heater as electric energy. This would mean a heating bill 2.5 times higher. This explains the popularity of heat pumps as heating systems.

EXAMPLE 8.4

Calculate the ideal coefficient of performance (COP) for an air-to-air heat pump used to maintain the temperature of a house at 20 °C for an outdoor temperature if –1 °C.

Solution

Given,

$$T_H = 20\,°\text{C} = 293\,\text{K}$$

and,

$$T_H = -1\,°\text{C} = 272\ \text{K}$$

Now, the COP is given by,

$$COP = \frac{T_H}{T_H - T_L} = \frac{293}{292 - 272} = 13.9$$

Thus, for every watt of power used to drive this ideal heat pump, 13.9 W is delivered to the hot reservoir (the interior of the house) and 12.9 W is extracted from the cold reservoir (the outside air). In practice, the COP for such a situation would be much less favorable. It would probably have a value of about 2–6.

8.10.4 Heat Pumps with Geothermal Resources

Ground-coupled heat pumps combine a heat pump with a ground heat exchanger. They are also known as geothermal heat pumps (GHP). They use Earth as a heat source in heating mode. A fluid (usually water or a water–antifreeze mixture) transfers the heat from the Earth to the evaporator of the heat pump. They use the Earth as a heat sink in cooling mode. Heat pumps utilising Earth energy are known as geothermal heat pumps (GHP). They transfer heat from the warmer Earth to the building in the winter through underground pipes. They take the heat from the building in the summer and discharge it to the cooler ground. Geothermal heat pumps (GHP) do not create heat. They transfer heat from one area to another.

The components of GHP are as follows:

- geothermal wells;
- circulating pumps;
- heat exchanger;
- disposal area for open loop system;
- compressor.

In heating mode, the heat from the geothermal resources is absorbed by the refrigerant circulating downwards. The heated refrigerant then moves upwards. A heat exchanger is used to transfer the heat from the refrigerant to water. It can then be used directly as hot water circulating through radiators for space heating. Such installations are called water-to-water systems. If heat is delivered to interior spaces through a system of air ducts, then it is referred to as a water-to-air system. In cooling mode, the processes in GHP are reversed. The refrigerant absorbs the heat from the interior spaces and transfers it to the ground. After identification of geothermal resources, a well is drilled either as a vertical borehole or a horizontal trench. The boreholes are spaced around 45 m apart. Trenches are spaced around 1.5 m apart. High-density plastic (polyethylene) pipes are placed vertically in the boreholes or horizontally in the trench and a refrigerant (or water in the case of direct use and water-to-water systems) is pumped through them. In general, a GHP works much like the refrigerator in the room of a building.

8.10.5 Typical GHP Loop Configurations

There are four basic types of ground-loop configurations namely (i) horizontal, (ii) vertical and (iii) pond/lake. These are closed-loop configurations. The fourth type of loop configuration is the open-loop type. These approaches can be used for residential and commercial building applications. The length of the loop depends upon the type of loop configuration, a home's heating and air conditioning load, soil conditions, local climate and landscaping.

(i) **Horizontal Ground Closed Loops:** This configuration is usually the most cost effective. Trenches are easy to dig. Plastic pipes are buried in trenches dug about 6 ft deep with a backhole, as shown in Figure 8.12b. As many as six pipes can be buried in one trench. The pipes may also be curled into a slinky shape in order to fit more of them into shorter trenches. Horizontal ground loops are easiest and economical to install during construction of houses. These arrangements can be made below a house to save extra land for the purpose.

(ii) **Vertical Ground Closed Loops:** For a vertical loop configuration, holes (approximately four inches in diameter) are drilled about 20 feet apart and 100 to 400 feet deep. Each hole contains a single loop of pipe with a U-bend at the bottom. After the pipe is inserted, the hole is backfilled. Each vertical pipe is then connected to a horizontal pipe, which is also

Figure 8.12b View of ground source heat pump horizontal trench.

concealed underground and then connected to the heat pump in the building. The vertical loop configuration is used in the case of insufficient space around the building.

(iii) **Surface Water Pond Closed Loop:** This type of configuration is the most economical and suitable for sites having an adequate water body. It is coiled in a slinky shape to fit more of it into a given amount of space. Pond loops are used in a closed system to have no adverse impacts on the aquatic system.

(iv) **Open-Loop Systems:** Open-loop systems are the simplest to install. These are used in areas where local codes permit. In this case, ground water from the aquifer is piped directly to the building. It transfers its heat to a heat pump. After it leaves the building, the water is pumped back into the same aquifer *via* a second well. This is called a discharge well, located at a suitable distance from the first. The use of an open-loop system can lead to a need for frequent cleaning of the heat exchanger through addition of chemical inhibitors to prevent fouling of loops by organic matter.

The benefits of using ground heat pipes (GHP) are as follows

- reduces energy use by 40–60% in winter and 30–50% in summer;
- operates for 20–40 years with minimum maintenance;
- cost effective;
- does not damage surface vegetation;
- no emission of greenhouse gases (GHS);
- no dependence on fossil fuels;
- provides good indoor-air quality.

8.10.6 Status of GHP

Geothermal heat pumps are gaining increasing momentum mostly in the United States and Europe. The installed capacity is 5275 MW_{th}. The annual energy use is 23 275 TJ/year in 26 countries.[2] The actual number of installed units is around 500 000. The total number of 12-kW units installed is slightly over 570 00. The 12-kW equivalents units are used in homes of the United States and western European countries. The size of individual units, however, ranges from 5.5 kW (Poland and Sweden) for residential use to 150 kW (Germany and the United States) for commercial and institutional installations. In the United States, most units are sized for the peak cooling load. These are oversized for heating with a capacity factor of 0.11. In Europe, units are sized for the heating load. These are designed to provide base load with peaking by fossil fuel. These units may operate from 2000 to 6000 full-load hours per year with a capacity factor of 0.23–0.68. Thus, the geothermal component is 67% of the energy output.

8.11 GEOTHERMAL FIELDS IN INDIA

India has the potential of geothermal fields with more than 340 hot water springs. Many of these have temperatures near the boiling point at their sources. Higher geothermal fields of 140–200 °C are also found.

Important geothermal provinces and places for hot springs occurrence are given in Table 8.5.

(A) Geothermal Energy Sources in the Himalayan Region

A large number of thermal springs occurs in the Himalayan region with temperatures close to 100 °C. Puga and Chumathang of Jammu and Kashmir are areas with temperature gradients in excess of 100 °C/km. However, the majority of hot springs are located between the main central thrust (MCT) and the central Himalayan axis. The hot springs of the Parbati valley, Sutlej valley and Alakananda valley are examples of such occurrences with 20 °C/km temperature gradient. The foothill Himalayan belts show low temperature gradients of 5 °C/km. There are 113 thermal springs in north-western India, of which 112 are in the Himalayan domain (Table 8.5).

(B) Utilisation of Geothermal Resources

The utilisation of geothermal energy has been confined to pilot plants.

(i) **Power Generation:** A 5-kW pilot-scale power plant has been installed at Manikarcan in Himachal Pradesh by the National Aeronautical Laboratory (NAL). This plant operates on a closed-loop Rankine cycle, with Freon-113 as the working fluid. It has been designed and fabricated at NAL.

(ii) **Space Heating:** Space-heating experiments have been carried out at Puga in Ladakh, Jammu and Kashmir. A m^3 hut was heated at temperatures of 20 °C in excess of the ambient.

(iii) **Extraction and Refining of Borax and Sulfur:** Geothermal energy was used for the extraction and refinement of locally occurring borax and sulfur at Puga. The extraction plant could handle 2 tonnes/day of borax ore while the refining plant could handle 500 kg/day of borax. The pilot plant for refining sulfur could handle 100 kg/day of sulfur.

(iv) **Greenhouse Heating:** Geothermal water was utilised at Chhumathang (Jammu & Kashmir) for greenhouse cultivation using discharge from a geothermal drill hole. A temperature of 20–25 °C was maintained inside the greenhouse even when the atmospheric temperature outside went down to –20 °C. A variety of vegetables and flowers could be grown in the greenhouse.

Table 8.5 State-wise distribution of hot springs.

State	*Number of Hot Springs*	*Areas*
Jammu and Kashmir	20	Puga-Chhumathung, Tattapani
Himachal Pradesh	30	Parbati valley, Beas valley, Tattapani
Uttar Pradesh	62	Alaknanda valley, Bhagirathi valley

(v) **Refrigeration:** At Manikaran (Himachal Pradesh), a 7.5-ton capacity cold storage plant was established. It utilises the geothermal water flowing at 90 °C. The cold storage plant is based on an ammonia absorption system.

OBJECTIVE QUESTIONS

8.1 Which of the following are considered to be drawbacks of geothermal energy?
- (a) It is not available everywhere
- (b) It is available only in areas where hot rocks are present near the Earth's surface
- (c) Deep drilling in the Earth to obtain geothermal energy is technically difficult and expensive
- (d) All of them

8.2 Which country was the first to use geothermal energy to generate power?
- (a) Italy
- (b) China
- (c) USA
- (d) UK

8.3 A number of geothermal power plants are running successfully in
- (a) USA and New Zealand
- (b) India and China
- (c) Nepal and Sri Lanka
- (d) UK and Switzerland

8.4 Energy derived from hot spots beneath the Earth is called
- (a) Bioenergy
- (b) Geothermal energy
- (c) Nuclear energy
- (d) Hydrogen energy

8.5 Energy obtained from the heat generated by natural processes within the Earth is known as
- (a) Solar energy
- (b) Geothermal energy
- (c) Ocean energy
- (d) Hydropower energy

8.6 Geothermal energy is extracted from the
- (a) Sun
- (b) Earth
- (c) Sea
- (d) Sky

8.7 Which type of energy does volcanoes possess?
- (a) Mechanical energy
- (b) Geothermal energy
- (c) Electrical energy
- (d) Nuclear energy

8.8 The molten mass of the Earth is called
- (a) Magnous
- (b) Magma
- (c) Hot coke
- (d) Magmus

8.9 The average thickness of the crust above the magma in
   (a) 10–25 km
   (b) 30–40 km
   (c) 100–150 km
   (d) 150–200 km

8.10 Geothermal energy reservoirs are
   (a) Liquid-dominated reservoirs
   (b) Steam-dominated reservoirs
   (c) Hot rocks with no water
   (d) All of them

8.11 While harnessing power from geothermal resources, which of the following materials is produced as a byproduct
   (a) Commercial-grade sulfur
   (b) Commercial-grade coke
   (c) Commercial-grade iron
   (d) Commercial-grade soil

ANSWERS

8.1 **(d)**; 8.2 **(a)**; 8.3 **(a)**; 8.4 **(b)**; 8.5 **(b)**; 8.6 **(b)**; 8.7 **(b)**; 8.8 **(b)**; 8.9 **(b)**; 8.10 **(d)**; 8.11 **(a)**.

REFERENCES

1. Huttrer, G. W. 2001. *Geothermics* **30**, 1–27.
2. Lund, J. W. and Freeston, D. H. 2001. *Geothermics* **30**, 2968.
3. Gupta, H. K. 1980. *Geothermal Resources: an Energy Alternative*. Elsevier Scientific Publishing Company, New York.
4. Uyeda, S. 1988. *Geodynamics; Handbook of Terrestrial Heat Flow Density Determination*. Academic Publishers, Germany. p. 486.
5. Silvestn, M. 1988. *The Energy Future*. Bollati Boringhieri Publication, Turin, p. 209.
6. Haenel, R., Rybach, L. and Stegena, L. 1988. *Handbook of Terrestrial Heat Flow-Density Determination*. Kluwer Academic Publishers. Dordrecht. p. 9–57.
7. Muffler, U. P. and Cataldi, R. 1978. *Geothermics* **7**, 53–89.
8. Hochstein, M. P. 1990. *Classification and Assessment of Geothermal Resources*. UNITAR, New York, pp. 31–57.
9. Benderitter, Y. and Cormy, G. 1990. *Small Geothermal Sources: A Guide to Development and Utilisation*. UNITAR, New York, pp. 59–69.
10. Grant, M., Donaldson, I. G. and Bixley, P. F. 1982. *Geothermal Reservoir Engineering*. Academic press. New York, p. 370.
11. Saito, S., Sakuma, S. and Uchida, T. 1998. *Geothermics* **27**, 573–590.
12. Fridleifsson, L. B. and Freeston, D. H. 1994. *Geothermics* **23**, 175–214.
13. Dickson, M. and Fanelli, M. 2001. *Renew. Energy World*, July–August, pp. 211–217.

The Use of the Earth's Surface Water Power, Namely Ocean Thermal, Tidal and Wave will meet the Challenges on Planet Earth due to Climate Change.

CHAPTER 9

Ocean Thermal, Tidal, Wave and Animal Energy

9.1 OCEAN THERMAL ENERGY CONVERSION (OTEC)

The oceans of the world constitute a natural reservoir for receiving and storing the energy of the Sun. These oceans consist of nearly three times the area of that of land, and they take the solar energy in proportion to their surface area. Water near the surface of tropical and subtropical seas is maintained by this solar radiation at higher temperatures at greater depth or higher altitudes. Due to heat and mass transfer at the surface of the ocean, the maximum temperatures occur just below the surface. It has been estimated that the amount of solar energy absorbed annually by the oceans is equivalent to at least 4000 times the amount of electricity consumed presently by the mankind. The ocean thermal energy conversion efficiency to electricity is 3 per cent. We need less than 1 per cent of this renewable energy to meet all of our energy demands. Ocean thermal energy is therefore a vast solar energy heat collector in the upper layer of the ocean. The ocean is the world's largest solar collector.

In an ocean thermal energy conversion (OTEC) system, the temperature difference between the warm ocean surface water and the cold deep ocean water can be utilised to generate electricity. The temperature difference needs to be about 20 °C. This occurs at a depth of around 1000 m. These conditions exist in tropical coastal areas, between the Tropic of Capricorn and the Tropic of Cancer. Even though OTEC seems technologically sophisticated, its concept dates back to 1881. The French engineer J. A. d'Arsonvil envisioned the idea, a task completed by his student G. Claude in 1930. Research and developments are still continuing to make the concept of OTEC cost effective and a commercially successfully electrical power technology. The theoretical efficiency of OTEC is small ($\sim$2%). There are vast quantities of seawater available for use in power generation. It has been estimated that there could be as much as 10^7 MW power available worldwide from OTEC process. Efforts are continued to extract energy from the oceans through the use of heat engines. This exploits the thermal gradient between the tropical surface water and deeper layers for electrical-power generation. These temperature differences are very steady without energy storage systems persisting over day and night and from season to season. The regions of significant potential for OTEC are observed in many heavily populated coastal areas. For example, the OTEC potential is most promising for Florida, Puerto Rico and Hawaii in the United States. Also, the research and development efforts on OTEC have introduced new applications that include mariculture, fresh-water production and air conditioning. It is an essential element of the system

Advanced Renewable Energy Sources
G. N. Tiwari and R. K. Mishra

Published by the Royal Society of Chemistry, www.rsc.org

that these new applications can be combined with electricity generation in order to expand the economic output of OTEC systems, and thus, improve the prospects for commercialisation.

The design for OTEC apparatus is a straightforward matter of devising a heat engine that operates at the modest temperature differential available. Such an engine can be achieved by using ammonia as the working fluid, because it vaporises and condenses at the available temperatures of the ocean. This is analogous to choosing water as the working fluid matched to the temperature differential between a fossil fuel-fired boiler and a condenser cooled by ambient air or water. Other types of OTEC engines have also been proposed, and are in the research and development stage. The electricity is derived from the cooling of the warmer surface water entering the heat engine. The energy extracted is proportional to the volume of warm water entering the engine in a given time. The unit cost of OTEC-generated electricity is expected to be comparable to that of electricity produced by coal-burning plants. The OTEC technology is suitable for islands where electricity is now generated by oil-burning plants. It has been estimated that more than 80 000 barrels of oil could be saved each day by generating 2000 MW electricity from OTEC. It can be referred to as oil-saving technology.

9.1.1 History

Georges Claude tested his concept at Matanzas Bay in northern Cuba in 1930 for OTEC technologies. His early model generated 22 kW of electric power. But it consumed more power than it generated. Later, Claude launched a second effort for the Brazilian coast in the form of a floating plant. Waves destroyed the cold-water pipe. Claude never achieved his goal of generating net power with an OTEC system.[1]

A French team designed a 3-MWe plant to be built on the west coast of Africa in 1956. They abandoned the project due to various reasons. J. H. Anderson was devoted to OTEC development and invented a submerged-type OTEC power plant using the closed system in USA in 1964. These early research efforts were not very successful.

In the early 1970s, the oil crisis renewed interests in OTEC. The National Science Foundation awarded a grant to the University of Massachusetts to assess the technical and economic feasibility of the OTEC process in 1972. Another grant a year later to Carnegie-Mellon to investigate other elements of the OTEC system was sanctioned. Japan followed the attempts in 1974 with the MITI (Ministry of International Trade and Industry) through the Sunshine Project.

At present, countries such as France, the Netherlands, Sweden, the United Kingdom, Japan. Taiwan, USA, *etc.* are also engaged in OTEC research.

Nearly 20 years of research and development on OTEC systems have produced a wealth of data on the scientific and technical aspects of this technology.

Major stages in OTEC development are listed below:

Year	*Stage*
1981	The OTEC concept was first suggested by J.A. d'Arsonval
1930	Experiment was first carried out by G. Claude
1933	Claude's 1200-kW OTEC ship was constructed
1956	A 3-MWe OTEC plant was built in Africa by a team of French scientists
1964	J. Anderson proposed a submerged OTEC plant
1979	The mini-OTEC (50 kW) was launched by Lockheed Missiles and Space Co. Inc. and the state of Hawaii
1980	OTEC I of the US Department of Energy was deployed
1981	The Nauru plant (120 kW) was established by the Tokyo Electric Power Services Company and the Tokyo Electric Power Company

(Continued).

Year	*Stage*
1981	Tokunoshima plant (50 kW) was established by Kyushu Electric Power Company
1992	An open-cycle OTEC plant generating 210 kW of gross power along with cold water for marine culture was operational at NEPL, Hawaii, USA
1985	Saga University's experimental plant (75 kW) began operating

9.1.2 Working Principle

Solar energy absorption by the water takes place following Lambert's law of absorption. This states that the reduction in the intensity (dI) of a ray of original intensity I, along a path length dx is same for all lengths dx, *i.e.* the fractional absorption dI/I is the same for all lengths dx. Mathematically, this can be expressed as follows:

$$\frac{dI}{I} = -k\,dx \tag{9.1}$$

After integrating with initial condition, *i.e.* $I_{x=0} = I_0$, one gets
or

$$I = I_0 e^{-kx} \tag{9.2}$$

where I_0 and I are the intensities of radiation at the surface ($x = 0$) and at distance x below the surface, k is the extinction coefficient (or absorption coefficient) that has the unit L^{-1}. k has a value of 0.05^{-1} for very clear fresh water; 0.27 for turbid fresh water and 0.50 for very salty water.

EXAMPLE 9.1

Evaluate the water depth in ocean for very salty water for reduction of solar radiation to 90%.

Solution

From Eq. (9.2), we have

$$\frac{I}{I_0} = e^{-kx}$$

Given $\frac{I}{I_0} = 0.10$ and $k = 0.5\,\text{m}^{-1}$, then

$$x = -\frac{1}{k}\log\frac{I}{I_0} = -\frac{1}{0.5}\log(0.10) = 2\,\text{m}$$

Thus, solar intensity decreases exponentially with depth. It depends upon "k" Almost all of the absorption occurs very close to the surface of deep water.

Water density decreases with an increase in temperature. Thus, there occurs no thermal convection current between the warmer and colder regions. Hence, the lighter water is at the top and deep cooler and heavier water is at the bottom. Thermal conduction heat transfer between them across the large depths is very low. Thus, mixing is retarded. There are essentially two infinite heat reservoirs, a heat source at the surface of around 25 °C and a heat sink of about 4 °C at a distance of nearly 1 km below the surface in tropical ocean. Both reservoirs are maintained annually by solar radiation. The concept of ocean thermal energy conversion (OTEC) is based on the utilisation of this temperature difference in a heat engine to generate electrical power.

Figure 9.1 shows the OTEC system for converting the thermal energy to electricity. It is a heat engine. It operates between the "cold" temperature (T_c) at substantial depth and the "hot" temperature, ($T_h = T_c + \Delta T$), of the surface water. A working fluid circulates in a closed-loop cycle. It takes up heat from the warm water through the heat exchanger. The fluid expands to drive a turbine and in turn the turbine drives a generator to produce electricity. The working fluid is cooled by the cold water and the cycle continues. Ocean water may also be considered as the working fluid to run an open cycle.

P_0 is the power given up from the warm water in an ideal system. We assume that a volume flow rate $Q = \{(V/t)\}$ of warm water passes through the system at temperature T_h and leaves at T_c (the cold water temperature of lower depths). Assuming perfect heat exchangers for an idealised system, if $\Delta T = T_h - T_c$.

Then,

$$P_0 = \frac{Q}{t} = \frac{mC\Delta T}{t} = \frac{V}{t}\rho C\Delta T = \rho CQ\Delta T \tag{9.3}$$

where $\rho = 1000\ \text{kg/m}^3$ is the water density, $C = 4200\ \text{J/kg}\ ^\circ\text{C}$ is the specific heat of water. According to the second law of thermodynamics, the maximum mechanical output can be obtained from the heat flow P_0 as

$$P_1 = \eta_{carnot} P_0 \tag{9.4a}$$

where,

$$\eta_{carnot} = \frac{\Delta T}{T_h} = \frac{T_h - T_c}{T_h} \tag{9.4b}$$

η_{carnot} is the efficiency of an ideal Carnot engine operating between T_h (source temperature) and T_c (sink temperature). The output from a real system is substantially less than P_0. Real engines actually do not follow the Carnot cycle. They may operate nearer to the ideal Rankine cycle of

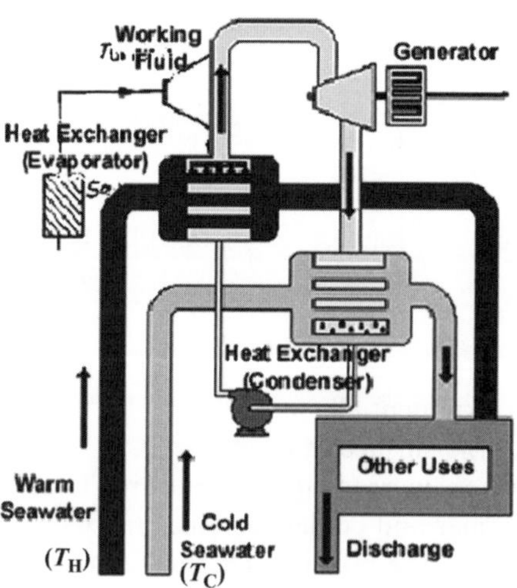

Figure 9.1 A line diagram of OTEC.

vapour turbines. These above equations suffice to illustrate the operation and limitations of OTEC. From Eqs. (9.3) and (9.4), and the ideal mechanical output power is given by

$$P_1 = \left(\frac{\rho CQ}{T_h}\right)(\Delta T)^2 \tag{9.5}$$

It has been evaluated that the flow rate required to yield 1 MW of electricity from an ideal heat engine at a temperature difference of 20 °C becomes 650 t/h. Thus, a substantial flow is required to give a reasonable output at the highest ΔT available. This requires large and expensive machinery. As P_1 depends quadratically on ΔT, experience shows that only sites with $\Delta T > 15\,°C$ have much chance of being economically attractive.

EXAMPLE 9.2 (a)

Calculate the hot temperature in the sea bed to produce 1 MW of electricity from OTEC for a given flow rate of 650 t/h and $\Delta T > 20\,°C$.

Solution

From Eq. (9.5), we have

$$P_1 = \left(\frac{\rho CQ}{T_h}\right)(\Delta T)^2$$

Given, $P_1 = 1$ MW $= 10^6$ W, $\rho Q = 650$ t/h $= (650 \times 1000)/3600 = 18$ kg/s.
C = 4200 J/kg °C (appendix V) and $\Delta T = 20\,°C$.
After substituting the above values, one gets

$$T_h = \frac{\rho CQ}{P_1} \times (\Delta T)^2 = \frac{4200 \times 8 \times (20)^2}{10^6} = 30\,°C$$

EXAMPLE 9.2 (b)

Calculate the Carnot efficiency for Example 9.2(a).

Solution

From Eq. (9.4b), we have

$$\eta_{carnot} = \frac{\Delta T}{T_h} = \frac{20}{30} = 0.67 \approx 67\,\%$$

9.1.3 OTEC Technology

All OTEC technology namely, closed-cycle, open-cycle and hybrid systems work as follows:

(i) **The Closed-Cycle System:** In this case, a working fluid with a low boiling temperature (ammonia or Freon) is used. The fluid is heated and then vaporised in a heat exchanger by the warm seawater. The steam produced drives a steam turbine generator. After passing through the steam turbine, the working fluid vapour is condensed in another heat exchanger. It is cooled by water drawn from the deep ocean. The working fluid is then pumped back through the warm water heat exchanger. The cycle is repeated continuously. The schematic diagram of the closed-cycle OTEC system is shown in Figure 9.2.[2]

The most obvious advantage of the open cycle is that warm seawater is flash evaporated. The need for having a surface heat exchanger is eliminated. The other major advantage is that potable (distilled water) is obtained. The major disadvantages are as follows:

(a) Steam is generated at very low pressure (approximately 0.02 bar). The volume of steam to be handled is high, leading to a very large diameter for the steam turbine. For example, a 1-MW OTEC plant requires a steam turbine of 12 m diameter.
(b) To maintain a vacuum in the flash evaporator, massive vacuum pumps are required.

(ii) **The Open-Cycle System:** In this case, seawater is the working fluid. Warm water from near the surface of the sea is pumped into a flash evaporator. The pressure is lowered by a vacuum pump to the point where the warm seawater boils at ambient temperature. This process produces steam. It drives a low-pressure turbogenerator to generate electricity. After leaving the turbine, the steam is condensed in a heat exchanger cooled by cold and deep ocean water. It also produces desalinated water. The schematic diagram of an open-cycle OTEC process is shown in Figure 9.3. G. Claude tried the first operational OTEC by using warm surface water as the working fluid. Hence, an open-cycle system is also known as a Claude cycle OTEC system.

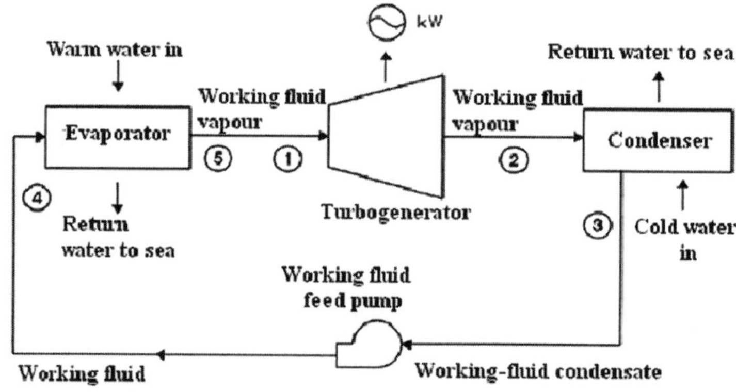

Figure 9.2 Schematic diagram of a closed-cycle OTEC system.

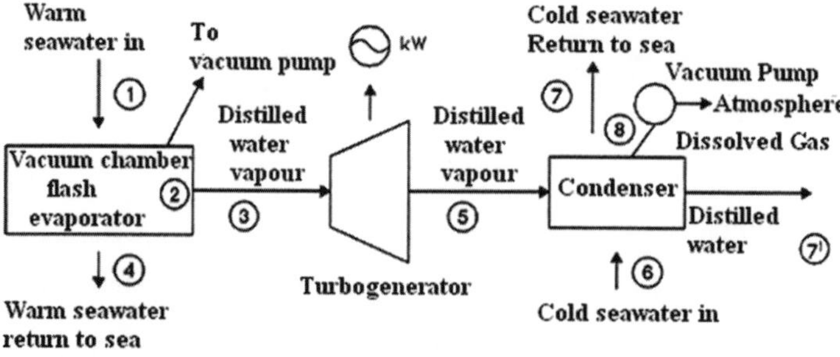

Figure 9.3 Schematic diagram of an open-cycle OTEC system.

On the contrary, the closed-cycle system requires expensive working fluids like Freon or ammonia. But the major advantage of this system is that the fluid evaporates at around 25 °C and does not require vacuum pumps. The pressure at the turbine is of the order of 9 to 6 bar, resulting in compact turbines. For example for a 1-MW OTEC plant, the ammonia turbine has a diameter of about 1.1 m only. The fabrication of such a turbine is technically easier than the fabrication of very large steam turbines for open-cycle system.

(iii) **The Hybrid-Cycle System:** This system combines the features of both the closed- and open-cycles system. In a hybrid system, warm seawater enters a vacuum chamber. It is flash evaporated there into steam. Similar to the open-cycle evaporation process. The steam vaporises a low boiling point fluid in a closed-cycle system. It drives a turbogenerator to produce electricity. The warm surface water after releasing heat to the working fluid gets condensed by the surface condenser to produce distilled water. Thus, the hybrid cycle maximises the use of the thermal resource by producing both electricity and distilled water. In hybrid system, the electricity generation and fresh water production can be maximised.

9.1.4 Location

The OTEC system could be installed in the following forms:

(i) **Floating Plant:** The OTEC plant could be placed on a floating platform at a distance of the deep sea from the coast, *i.e.* 10 to 15 km from the shore line. In this case, a cold water pipe is suspended vertically down. An underwater cable is needed for power transmission to shore. The energy generated may be utilised to produce energy-intensive materials (ammonia or hydrogen) from the seawater. The products may be transported to the main land by ships.

The design of floating plant needs careful considerations to avoid its position under the action of waves, wind and current.

(ii) **Shelf-Mounted Plant:** If the coast is shallow for some distance and if the distance of the 1000 m contour is around 10 km, the OTEC plant can be mounted on a platform at the end of the shallow water area and the generator power can be transmitted to the mainland by underwater cables. A tower can be built in water at a depth of around 50–100 m where the seabed starts sloping down deeply.

(iii) **Land-Based Plant:** The entire OTEC plant could be situated on land with the cold water pipeline running along the ocean bed to a depth of 800–1000 m. The deep water conditions are available within 2–3 km from the coast.

9.1.5 Advantages and Disadvantages

The following are the advantages and disadvantages of OTEC systems

Advantages

(i) OTEC uses clean, abundant environmentally friendly renewable resources. It replaces fossil fuels to produce electricity.

(ii) An OTEC plant produces little or no carbon dioxide. Therefore, power from an OTEC system is clean energy.

(iii) It can produce both fresh water as well as electricity. This is a significant advantage in island areas where fresh water is limited.

(iv) The use of OTEC as a source of electricity helps in reducing the nation's almost complete dependence on imported fossil fuels in coastal regions.
(v) The cold deep seawater is rich in nutrients. It can be utilised for aquaculture.

Disadvantages

(i) Electricity from OTEC costs more than the electricity generated from fossil fuels.
(ii) Seasonal variations and natural calamities affect OTEC performance.
(iii) Ocean depths should be available fairly close to shore-based facilities to avoid transmission/distribution loses.
(iv) Construction of OTEC plants and laying of pipes in coastal water may cause localised damage to reefs and near-shore marine ecosystems.

9.1.6 Applications

OTEC systems can be used to produce distilled (potable) water and support deep water mariculture. It also provides refrigeration and air conditioning as well as aids in crop growth and mineral extraction. The various applications of OTEC are shown in Figure 9.4.

They are briefly described as follows

9.1.6.1 Mariculture

This can be done in deep ocean water. It provides an excellent medium for the cultivation of marine organisms. The cold water at deep ocean is being used to cultivate salmon, trout, opihi (a shellfish

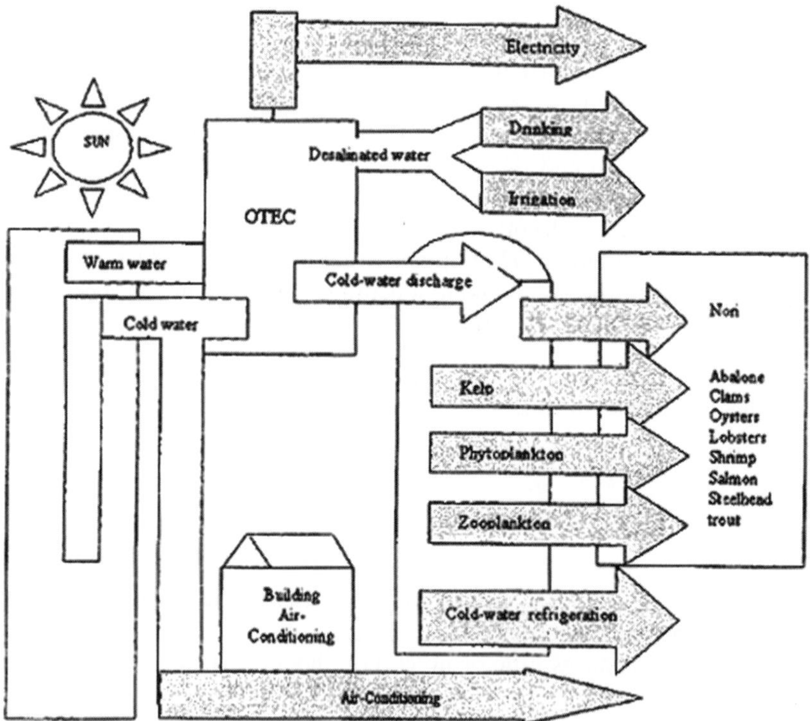

Figure 9.4 Applications of OTEC.

delicacy), oysters, lobsters, sea urchins, abalone, kelp, nori (popular edible seaweed) and other macro- and microalgae.

9.1.6.2 Fresh Water

The condensate of warm water/vapour in the open-cycle and hybrid OTEC systems is distilled water. It is suitable for human consumption as potable water and agricultural purposes. This water is actually purer (less saline) than the water provided by the municipal water system. The market value of distilled water in the islands is higher. It may be even higher in locations with no groundwater resources.

9.1.6.3 Refrigeration and Air-Conditioning

The cold water can be used to create cold storage space and for air conditioning. The laboratory at the Natural Energy Laboratory, Hawaii is air conditioned by OTEC. In this case, the cold seawater passes through a heat exchanger placed inside the room. OTEC technology also supports chilled-soil agriculture. Cold seawater flows through underground pipes. It chills the surrounding soil. The temperature difference between plant roots in the cool soil and plant leaves in the warm air allows many plants to be grown in the subtropics. The natural energy laboratory maintains a demonstration garden near its OTEC plant with more than 100 different fruits and vegetables. Many of which would not normally survive in Hawaii.

9.1.6.4 Agriculture

This involves burying an array of cold water pipes in the ground to create cool weather growing conditions in tropical environments. In addition to cooling the soil, the system also produces drip irrigation. This is created by the atmospheric condensation on the cold water pipes. This was carried out in actual demonstrations in the University of Hawaii, which determined that strawberries and other spring crops and flowers could be grown throughout the year in the tropics using these methods.

9.1.6.5 Aquaculture

Aquaculture is perhaps the most well known byproduct of OTEC. Cold-water delicacies (salmon and lobster) thrive in the nutrient-rich deep seawater from the OTEC process. Microalgae (spirulina, a health food supplement) can also be cultivated in the deep ocean water. The ocean farms of Hawaii are growing salmon, abalone, sea urchins and oysters. They have used a method called polyculture, with four 15-million gallon ponds and its own system of cold water pipelines.

9.1.7 Power Module and Seawater Systems

A Rankine cycle with a working fluid (ammonia with high vaporisation enthalpy) is a first choice for extracting energy from the temperature difference. Closed-cycle systems are also suitable for MW-range OTEC plants. This is due to the compactness of the turbine. It can be easily sealed up for the higher rating. The 1-MW OTEC plant on which National Institute of Ocean Technology (NIOT), Chennai, India currently working is based on a closed Rankine cycle. The power cycle consists of an evaporator, a turbogenerator, a condenser and pumps for seawater and working fluid. The warm seawater as working fluid pumped from sea surface exchanges heat in the evaporator. It undergoes a phase change into vapour and this vapour is allowed to expand in the turbine, producing mechanical energy. The vapour is allowed to condense in the condenser. It is again pumped back to the evaporator using the working fluid pump. The heat is exchanged with the cold seawater pumped from the ocean depth inside the condenser. The thermal design of the power cycle is done by optimising the heat exchanger area for the net power output conditions.[2]

The evaporator and condenser consist of four modules of a plate heat exchanger. It is the largest of its kinds used for such an application. The ammonia side of the evaporator is coated with stainless steel powder to enhance the heat transfer coefficient. The power output may be improved by 20–40% because of the coating. It acts as nucleation points for the vaporisation of ammonia. The ammonia turbine is expected to work with an adiabatic efficiency of 88%. It is a 4-stage axial-flow turbine designed by NREC Massachusetts, USA. The power system flow rates and the net power are very much dependent on the turbine efficiency. The seawater pumps are of the vertical type. It is a mixed-flow type due to the low head and high discharge combination.

The OTEC plant comprises a cold water pipe (CWP), a warm water pipe (WWP), a mixed discharge pipe and ducting on the barge. The CWP is an integral part of the mooring of the barge. There are several configurations for CWP mooring. The CWP is connected to a floating platform using a single-point mooring. The major components of the mooring are a surface buoy, top fitting, bottom weight, anchor chains and the anchor. The platform assumes significance due to the fact that it houses the entire plant. It accommodates the seawater pumps and the cold water pipe or mooring system in a rough marine environment.

9.1.8 Economics

To make the OTEC power cost effective, the unit cost of power produced by OTEC should be comparable with the unit cost of power produced by fossil-fuelled plants. OTEC systems should also have other benefits like enhanced aquaculture, desalination or even air conditioning, which might reduce the cost of electricity generated. The coproduction of fresh water, aquaculture and air conditioning along with power is to be considered for the estimation of unit cost for OTEC plants. It is estimated that the annual revenue from fish culture from a 10-MW OTEC facility is 3.3 *million/year. It can be increased up to* $82.5 million/year with more efficient species. OTEC is economical and the production cost is comparative for higher range of plant.[3] Table 9.1 shows the potential and status in India for OTEC, tidal and wave technologies.[4] It can be noted that although vast potential is available from ocean energy resources, only a little is achieved in harnessing it. This is mainly due to the fact that the technology is not matured or the cost implications are high.

9.1.9 Status and Development of OTEC Systems

The oil crisis in the early 1970s prompted the search for alternative sources of energy. The potential of OTEC was also examined. Research and development of OTEC systems in closed-cycle mode and open-cycle mode were also initiated all over the world to make the concept feasible for power generation.

Year	*Country*	*Net capital power*	*Mode of operation*
1970	Hawaii, USA	15 kWe	Closed cycle
1982	Tokyo, Japan	35 kWe	Closed cycle
1989	Republic of Kiribati	1 MWe	Closed cycle
1989	Hawaii, USA	500 kWe	Closed/hybrid cycle
1992	Hawaii, USA	220 kWe	Open cycle
1997	DOD, India	1 MW	–
2001	DOD, India	10–25 MW	–

Table 9.1 Various ocean-energy technologies.

Ocean-Energy Technology	*Potential (MW)*	*Status (kW)*	*Unit Cost*	
			US \$/kWh	*Rs./kWh*
OTEC	180 000	Nil	0.189	8.12
Tidal	8000	Nil	0.t67	7.17
Wave	40 000	50	0.171	7.5

9.2 TIDAL ENERGY

The tide is the periodic motion of the ocean water. It is caused by celestial bodies, principally, the Moon and the Sun, on the different parts of the rotating Earth.

Tide and tidal currents must be differentiated as the relation between them is not simple. The oceanographer uses the word tide for the vertical rise and fall of the water and the current for the horizontal flow. In its rise and fall, the tide is accompanied by a periodic horizontal movement of the water known as tidal current. These two movements, namely tide and tidal current are intimately related. These form the same phenomenon brought about principally by the tide-producing forces of the Sun and Moon.

Tides are caused by the gravitational force of Moon and Sun acting upon the rotating Earth in an elliptical cycle around its axis and the relative positions of the Earth, Moon and the Sun. Waves have a period of only about six seconds. Tides have a period of 12.5 h. Waves are caused by surface winds, whereas tides are caused by the gravitational forces of Moon and Sun on ocean water. The amplitude of tides covers a wide range from 25 cm to 10 m.

According to Newton's law of gravitation, the gravitational force between the Sun and the Earth can be give as

$$F = G \cdot \frac{m_1 m_2}{r^2} \tag{9.6}$$

The gravitational force decreases as the distance (r) increases. It depends on masses (m_1 and m_2) between which gravitational force exists.

The Moon is much closer to the Earth than is the Sun. Therefore, the Sun's larger mass effect upon the Earth is less than half that of Moon. The force producing the tide is proportional to the quotient of the masses and inversely proportional to the square of the distance separating these masses. Earth, Moon and Sun are also not constantly in the same relative position to one another. Tides on the Earth follow the lunar cycle and their amplitude varies according to the relative position of the Earth, Moon and the Sun. The interaction of the Moon and the Earth results in the oceans bulging out towards the Moon. The gravitational effect is partly shielded by the Earth, resulting in a slightly smaller interaction and the oceans on the side that bulges out away from the Moon due to the centrifugal forces. This is known as the lunar tide. Similarly, the solar tide is due to the gravitational interaction of the Sun. This results in the same effect of bulging towards and away from the Sun on facing and opposing sides of the Earth.

Tidal power is one of the older forms of energy used by human beings, and tide mills were being worked on the coasts of Britain, France and Spain before 1100 AD. They remained in common use for many centuries. They were gradually displaced by the cheaper and more convenient fuels and machines made available by the Industrial Revolution. Many methods of extracting potential or kinetic tidal energy have been tried in the past. Devices used were water wheels, lift platforms, air compressors and water pressurisation, *etc.*

The tide mills were typically operated by filling a storage pond at high tide. Later, the water is allowed to flow from the pond at high altitude to the sea through a water wheel. This is the simplest method of operation. In this case, potential energy stored in the pond is converted into mechanical energy through the water wheel. It is generally referred to as a "single basin, single effect". Also, the harnessing of tidal energy has been used for small mechanical power devices. The best known large-scale electricity generating plant is the 240-MW LA Rance system at an estuary into the Gulf of St Malo in Brittany, France. Other key sites are the Severn Estuary in England and the Bay of Fundy on the eastern boundary between Canada and the United States. There is an important 400-kW capacity prototype tidal power plant at Kislayaguba on the Barents Sea in Russia. In India, there are possible tidal projects in the Gulf of Kutch and Cambay and on a smaller scale, in the Sunderbans regions of the Bay of Bengal.

9.2.1 Working Principle

The daily rise and fall of ocean levels is known as tidal. These are the result of the gravitational force of the Moon and Sun as well as the revolution of the Earth. The Moon and the Sun both exert a gravitational force of attraction on the Earth. The magnitude of the gravitational attraction of an object depend the mass of objects and its distance (Figure 9.5) as given by Eq. (9.6) (Newton's law).

The Moon exerts a larger gravitational force on the Earth due to being very close to Earth compared to the Sun, even with its smaller mass. This force of attraction causes the oceans to bulge along an axis pointing towards the Moon. Tides are produced by the rotation of the Earth beneath this bulge in its watery coating. This results in the rhythmic rise and fall of coastal ocean levels.

The Sun and Moon are not in fixed positions. The positional change with respect to each other influences the difference between low and high tide. This is known as the tidal range. For example, when the Moon and the Sun are in the same plane as the Earth, the tidal range is the superposition of the range due to the lunar and solar tides. This results in the maximum tidal range known as spring tides. Alternatively, when they are at right angles to each other, lower tidal differences are experienced, known as neap tides as shown in Figure 9.5. At half Moon, the Sun and Moon are at right angles, it causes the neap tides. Likewise, at new Moon and full Moon, the Earth, Moon and Sun are positioned in a straight line; it causes the combined effects of gravitational attractions to form very large spring tides. The rise and fall of the water level follows a sinusoidal variation. The periodic raising and lowering of ocean surface follows a number of interacting cycles.

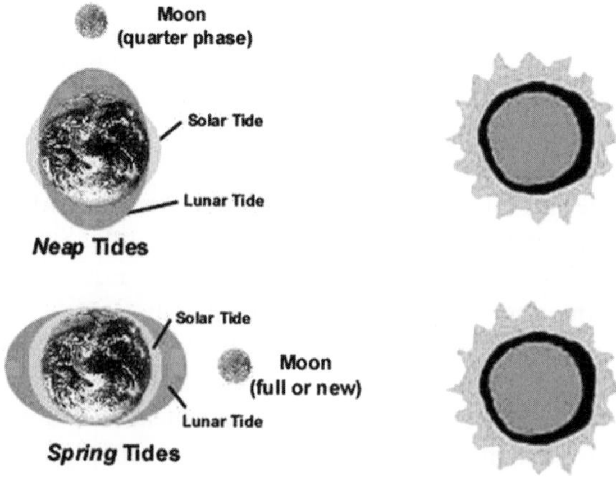

Figure 9.5 Gravitational effect of the Sun and the Moon on tidal energy.

(i) **A Half-Day Cycle:** This is due to the rotation of the Earth within the gravitational fields of the Moon. It results in a period of 12 h 25 min between successive high waters as shown in Figure 9.6. "A" and "B" indicate the high and low tide points, respectively. The average period of time for the water level to fall from A to B and then rise from B to C is each approximately to 6 h 12.5 min. The difference between high and low water levels is known as the range of the tide.

(ii) **A 14-Day Cycle:** This is the result of the superposition of the gravitational fields of the Moon and Sun (Figure 9.7). At new Moon and full Moon, the Sun's gravitational field reinforces the Moon, which results in the maximum tides or spring tides. At quarter phases of the Moon, the Sun's attraction partially cancels. This results in the minimum or neap tides. The range of a spring tide is typically about twice that of a neap tide.

(iii) **Other Cycles:** These are half-year and 19-year cycles. These arise from other interactions between the gravitational fields.

Tidal ranges vary from one ocean location to another.

In midocean: The tidal range is typically about one meter, but frequently this is strongly modified by coastal contours.

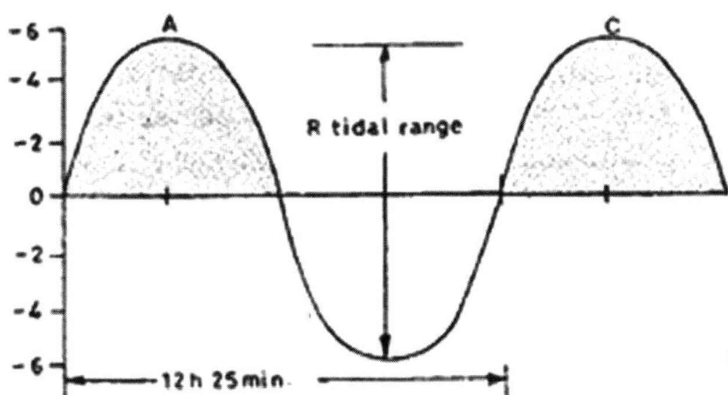

Figure 9.6 The tides of the sea.

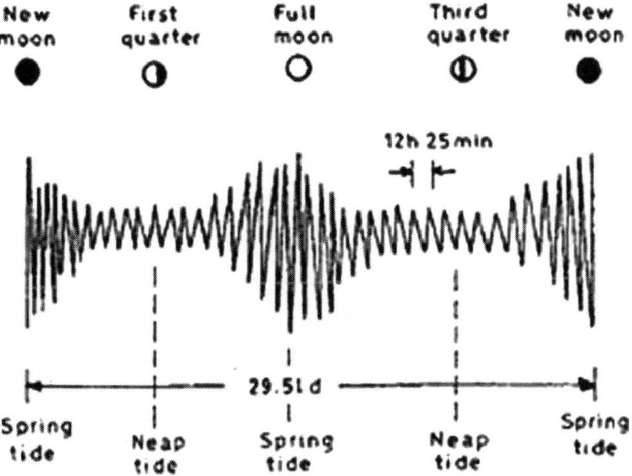

Figure 9.7 Relative high and low tides.

In estuaries: The tidal range may be as great as 10–15 m at the most favorable sites. The amplification arises from proximity to resonance. In a simple model of a channel having uniform cross section, resonance occurs when the length of channel is close to one quarter of the length of the tidal movement. In particular, frictional losses resulting from water moving over the seabed and banks introduce damping. Full understanding of the characteristics of particular prospective sites should be obtained during the construction of a dam across the estuary.

The tide moves a huge amount of water twice each day. Its harnessing provides a great deal of energy. The energy supply is reliable and plentiful. It converts it into useful electrical power, which is not easy. Also, it is renewable in the sense that tides will continue to ebb and flow.

A tidal power plant works like a hydroelectric systems, without having the bigger dam. A dam known as a barrage is built across the estuary. The ebb and flow of the tides can be used to turn a turbine to generate electricity. Electricity can be generated by water flowing both into and out of the estuary. Power generation is more common when water is going out. The major drawback is the mismatch of the principal lunar driven periods of 12 h 25 min and 24 h 50 min, respectively with the human (solar) period of 24 h, so that optimum tidal power generation is not in phase with demand.

The final criterion for the success of a tidal power plant is to have minimum cost per unit kWh of the power produced in comparison with the other systems. This can be minimised (i) if other advantages can be costed as benefits to the project (ii) if interest rates of money are low and (iii) if the output power can be used to save/conserve diesel oil. Smaller schemes for use in outlying regions may prove themselves more economic, but tidal power has the potential to generate significant amounts of electricity at certain sites. It can be a valuable source of renewable energy to an electrical system. The negative environmental impacts of tidal barrages are probably much smaller than those of other sources of electricity.

9.2.2 Tidal Power Calculation

Let the water be trapped and stored at high tide in a basin and allowed to flow through a turbine at low tide, as shown in Figure 9.8.

The basin has a constant surface area A. It remains covered in water at low tide. The trapped water having a mass ρAR at a centre of gravity $R/2$ is assumed to run out at low tide. The maximum potential energy available per tide with water falls through $R/2$ is given by

$$\text{Energypertide} = (\rho AR)g\frac{R}{2} \tag{9.7}$$

where g is the acceleration due to gravity.

The average potential power averaged over the tidal period T becomes

$$\overline{P} = \frac{\rho AR^2 g}{2T}(W) \tag{9.8}$$

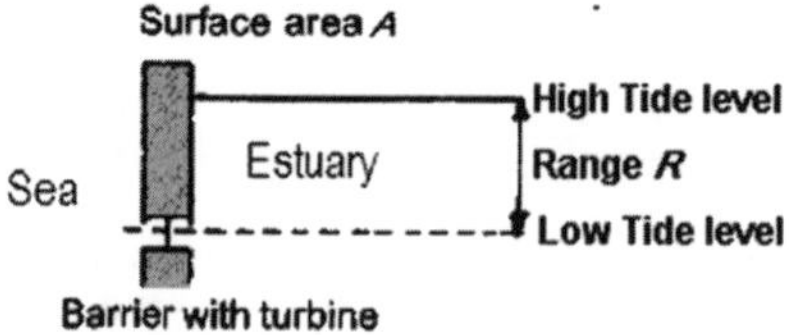

Figure 9.8 Power generation from tides.

From Eq. (9.8), one can see that the amount of energy available from a tide varies approximately with the square of the tidal range.

EXAMPLE 9.3 (a)

Evaluate the range of a tidal wave for a half-day cycle for 600 W/m^2 average potential power.

Solution

From Eq. (9.8), we have the expression for potential power in W/m^2 as

An average potential power per $m^2 = \frac{\bar{p}}{A} = \frac{\rho R^2 g}{2T} (W/m^2)$

Given: $T = 12$ h $= 12 \times 3600$ s, $\rho = 1000$ kg/m^3, $g = 10$ m/s^2, and $\frac{p}{A} = 600\,W/m^2$.
After substituting the values in the above equation one gets

$$R = \sqrt{\frac{\bar{p}}{A} \times \frac{2T}{\rho g}} = \sqrt{\frac{600 \times 2 \times 12 \times 3600}{1000 \times 10}} = 72 \text{ m}$$

EXAMPLE 9.3 (b)

Repeat example for an average period of a half-day cycle, *i.e.* $T = 6$ h.

Solution

In this case, the range of tidal wave will be

$$R = \sqrt{\frac{600 \times 2 \times 6 \times 3600}{1000 \times 10}} = 50 \text{ m}$$

9.2.3 Basic Modes of Tidal Operations

There are two principal ways for power generation from a tidal estuary.

9.2.3.1 Single-Basin Schemes

This involves building a barrage at a suitable point along the estuary, installing turbines and generating electricity.

These operate in the following three basic modes:

(a) **Ebb Generation:** In this case, the operating cycle consists of the following four steps:
   (i) The water is allowed to pass through the sluice gate during the flood tides to fill the basin.
   (ii) The gates are then closed until the receding tide. This creates a suitable head between the basin and the sea.
   (iii) The water is allowed to flow from the basin through the turbines on the ebb tide, which happens until the head is reduced to the minimum operating point. This is a result of the rising tide and the decreasing water level in the basin.

(iv) The gates are closed until the tide rises sufficiently to repeat the first step.

(b) **Flood Generation:** This involves reversing the cycle from sea to basin. The sluices and turbine passage ways are closed against the incoming tide. In this case, water level rises on the seaward side of the barrage. When a sufficient head has been built up, the turbines are brought into operation. This continues until the water in the basin has reached about midtide. Flood generation schemes are generally less favoured due to (i) The estuary above the barrage is kept at low tide for prolonged periods with possible adverse ecological effects and with reduced access for shipping. (ii) The energy produced is less than that in an ebb-generation scheme. This is due to the surface area of the estuary decreasing with depth. It leads to a more rapid reduction in the head of water when generation commences and thus the volume of entrained water that can be used to drive the turbine is reduced.

(c) **Two-Way Generation:** Two-way generation on both ebb and flood is also feasible. This is also known as a double-effect cycle. In this case, the direction of flow through the turbines during the ebb as well as flood tides alternates. The machine acts as a turbine for either direction of flow. In this case, the generation of power is produced during emptying and filling cycles. Both filling and emptying processes take place during short periods of time. The filling occurs when the ocean is at high tide and while water in the basin is at the low-tide level. The emptying occurs when the ocean is at low tide and the basin is at the high-tide level.

Generating electricity on both flood and ebb tide does not lead to a major increase in energy yield for a number of reasons: (i) There is a need to achieve a useful head during the previous flood-generation phase and there is a sizeable reduction in the initial water level for the ebb-generation part of the cycle. (ii) The phase cannot be taken to completion, because it is necessary to open the sluices and reduce the level of water in the basin in preparation for flood generation. The same applies in reverse during the flood-generation phase. (iii) A large number of turbines are required to achieve the same power output. This extra number of turbines may be difficult to accommodate. The turbines are less efficient and more complex if they are required to operate in two directions. The turbine water passage has to be longer. This also leads to a more expensive design. Larger space is required to house the turbines. As in case of flood generation, the reduction in the maximum water levels in the basin means that navigation for large vessels is adversely affected.

There are, however, some merits in two-way generation. In this case, water levels in the basin remain closer to the mean sea level. Some environmental problems are minimised. Power is produced in four blocks rather than two. Thus, it increases the firm power contribution from the tidal scheme. For the same energy output, the peak power generated is less than for the other modes of operation. Transmission costs may be reduced by achieving a higher utilisation of lower capacity lines.

9.2.3.2 Double-Basin Schemes

In this case, the two basins operate as independent ebb- and flood-generation schemes. There is an output characteristic similar to those of two-way generation. There are the provisions of two basins on the landward side. The power house is located in the barrier between the two basins, as shown in Figure 9.9. Power is generated by (a) water flowing from the high basin to the low basin through the turbines and (b) water flowing from the low basin to the sea during ebb tide. Turbogenerators should be capable of efficient generation at low heads along with handling large discharges.

The operation of the double-basin scheme can be controlled with continuous water flow from the upper to the lower basin. Since the water head between the basins varies during each tidal cycle, as well as from day to day, the power generated also varies. The peak power generation does not often correspond in time with the peak demand in the case with a single-basin scheme. One way of improving this situation in the double-basin scheme is to use off-peak power from the tidal power

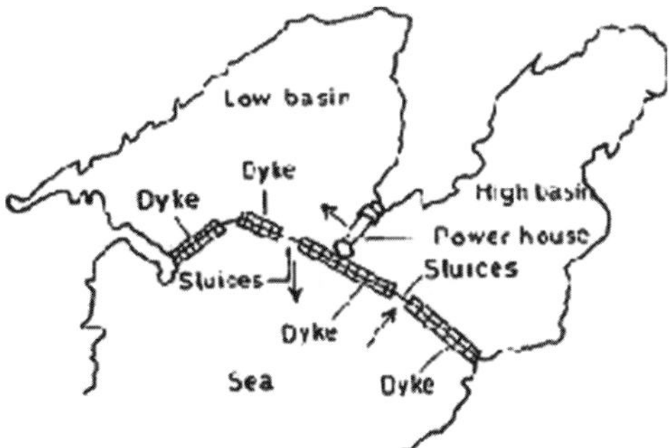

Figure 9.9 Double-basin tidal scheme.

generators. This can also be done from an alternative system, to pump water from the low basin to the high basin. An increased head would be available for tidal power generation during peak demand. This is very similar to pumped water in storage system in hydroelectric power stations.

9.2.4 Advantages/Disadvantages

The following are the advantages/disadvantages of tidal energy:

Advantages

- Once the dam is constructed, tidal power is free.
- It produces no greenhouse gases.
- Fuel is renewable and free.
- Reliable electricity is produced.
- Maintenance is inexpensive.
- Tides are totally predictable.

Disadvantages

- Very expensive to build the tidal power plant.
- The environment is changed for many miles upstream and downstream.
- It provides power for around 10 hours each day.
- There are very few suitable sites for tidal power.
- Seawater is corrosive. The machinery gets corroded. Stainless steel with a high chromium content and a small amount of molybdenum and aluminium bronzes are to be used.

9.2.5 Status of Tidal Power

The first and largest operational tidal barrage plant in the world is the La Rance plant on the Brittany coast of northern France. It was built in the early 1960s. The plant produces 240 MW of electricity due to 2.4 m tidal height at the mouth of the La Rance estuary.

It is estimated that the United Kingdom could generate up to 50.2 TWh/year with tidal power plants. Western Europe as a whole could generate up to 105.4 TWh/year. Total worldwide potential is estimated to be about 500–1000 TWh/year. Only a fraction of this energy is likely to be exploited due to economic constraints.[5] The availability of tidal energy is very site specific.

Major potential sites for barrages include the Bay of Fundy in Canada, with a mean tidal range of 11 m, the highest tides in the world and the Severn Estuary of Britain. The worldwide tidal power sites and estimated potential tidal power are given in Tables 9.2 and 9.3, respectively.

9.3 WAVE ENERGY

Waves in the ocean are caused by the transfer of energy from the wind to the sea. The rate of transfer depends on (i) wind speed and (ii) the distance of interaction of the wind with the water (the fetch). Waves are ultimately a source of solar energy. The Sun heats the Earth's surface. Winds are generated by the differential heating of the Earth. Winds created blow over the surface of the ocean

Table 9.2 World's tidal power sites and estimated potential power.

Location	*Average Range (R) (m)*	*(R^2) (m^2)*	*Basin Area (A) (m^2)*	*(R^2A) (m^4)*	*Average Potential Power(P) (10^3 kW)*	*Potential Annual Energy (10^6 kWh)*
North America						
Bay of Fundy						
Passntaquoddy	5.5	31	262	7930	1800	15 800
Cobscook	5.5	30	106	3210	722	6330
Annapolis	6.4	41	83	3440	765	6710
Minas-Cobequid	10.7	114	777	88 600	19 900	175 000
Amhrerst Point	10.7	114	10	1140	256	2250
Shepody	9.8	96	117	11 200	2520	22 100
Cumberland	10.1	102	73	7450	1680	14 700
Peutcodiac	10.7	114	31	3530	794	6960
Memramcook	10.7	114	23	2620	590	5170
Subtotal					29 027	255 020
South America						
Argentina						
San Jose	5.9	34.8	750.0	26 100	5870	51 500
Europe						
England						
Severn	9.8	96.0	70.0	7460	1680	14 700
France						
Aber-Benoit	5.2	27.0	2.9	78	18	158
Aber-Wrach	5.0	25.0	1.1	28	6	53
Arguenon and Lancieux	8.4	70.6	28.0	1980	446	3910
Frenaye	7.4	54.8	12.0	658	148	1300
La Rance	8.4	70.6	22.0	1550	349*	3060
Rotheneuf	8.0	64.0	1.1	70	16	140
Monttain Michel	8.4	70.6	610.0	43 100	9700	85 100
Somme	6.5	42.3	49.0	2070	466	4090
Subtotal					11 149	97 811
India						
see Table 9.3						
USSR						
Kislaya Inlet*	2.31	5.62	2	11	2*	22
Lumbovskii Bay	4.20	17.60	70	1230	277	2430
White Sea	5.65	31.90	2000	63 800	14 400	126 000
Mezen Estuary	6.60	43.60	140	6100	1370	12 000
Subtotal					16 049	140 452
Grand Total					**63 775**	**559 483**

*Power plants in operation.

Table 9.3 Tidal power potential of India and prospective sites.

	Gujarat State						*West Bengal State*			
	Gulf of Cambay		*Gulf of Kutch*							
Characteristic	*Single Basin*	*Single Basin*	*Single Basin*	*Single Basin*	*Two basin system without pumping*		*Sunderbans Durgdauni*	*Creek Area Sluicing*	*Bella dona creek*	*Pitts Creek*
Mean Tidal Range (m)	6.8	6.8	5.3	5.3	5.3					
Length of Structure (km)	26	32.1	26	31	34	*	*	*	*	*
Deepest Water level above low tide (m)	29	27	13	13	13	*	*	*	*	*
Area of the basin (km^2)	1972	1751	639	538	278	0.76	0.31	0.83	0.45	13.57
Potential capacity	7364	5510	1187	1182	586	1.73	2.29	1.85	2.25	15.4
Annual energy Generation (GWh)	15 394	11 583	3037	2984	1266	3.105	1.544	3.466	1.744	31.193
Estimated ($R \times 10^7$)	19 25.1	*	593.5	*	468	2.8	*	2.2	2.5	15.7

*Source: http://www.tidalpower.com

to form waves. These waves travel for thousands of miles from their point of propagation. They are continuously strengthened by new winds as they pass. As a result, waves retain their energy long after the winds that create them die down. Therefore, waves are one of the renewable energy sources. Wave energy offers the benefits of limitless supply of energy that is environmentally pollution free. Unlike wind and solar energy, power from ocean waves continues to be produced round the clock. Wind tends to die in the morning and at night and solar energy is only available during the daytime. Waves are generally characterised by their height, wavelength and period.

The prospect of extracting energy from ocean waves has attracted some attention from time to time over the past century. Serious efforts to establish an effective technology were started from the mid-1970s after the oil crisis. Major research and developments have been undertaken in different countries. Numerous devices have also been developed. One of the richest nations in terms of potential for wave energy at present is the north of Scotland (UK). In the USA, a reasonable potential for wave energy development may exist off the Pacific Northwest coast. According to the world Energy Council,[6] wave energy could potentially provide up to 2 TW of electricity, approximately 1/5 of the current global energy demand.

The economics of wave energy power are promising and the situation is improving with more advanced technology. Costs have rapidly dropped in the last several years. Now companies are aiming for less than 10 cents/kWh, to 5 cents/kWh. This price would allow wave plants to compete favourably with conventional power plants.[7]

9.3.1 Basics of Wave Motion

When a particle of the medium is displaced slightly from its position and then released, it begins to oscillate. This is due to its inertial and elastic properties. In the course of oscillations, this particle disturbs other particles too. Thus, the displacement of any particle will produce a disturbance from particle to particle. This will continue till all of them have suffered greater or less displacements. This disturbance is termed a progressive wave (wave motion). Thus, wave motion is a kind of disturbance to propagate in the medium with a certain velocity without changing its form.

There are two types of waves: (a) transverse waves (b) longitudinal waves.

(a) **A Transverse Wave:** This is one wave in which the particles of the medium vibrate in a direction at right angles to the direction of propagation of the wave.
The motion of waves on the surface of water in a rectangular vessel partly filled with water is an example of transverse waves. A flat piece of wood W is partly dipped in water. It is moved up and down near one end of the vessel in a periodic manner. This up and down motion generates a series of transverse waves. The topmost position of a wave is called a crest and the lowest position is called a trough. Electromagnetic waves are examples of transverse waves.

(b) **A Longitudinal Wave:** This is one in which particles of the medium vibrate in the direction of propagation of the wave. The particles in longitudinal waves move to and fro about their mean positions along or parallel to the direction of propagation of wave. The waves propagated through air are examples of longitudinal waves.
Any motion repeats itself in equal intervals of time is known as **periodic motion**. Periodic motion of the particle can be easily represented as the projection on any diameter of a point moving in a circle with uniform speed (Figure 9.10).

Let a particle P move around the circle of radius 'a' with uniform speed u. Also, let N is be the foot of the perpendicular drawn from P on diameter yy'. Now consider the movement of N as the point P moves around the circle with angular velocity ω. If P is at x, the foot N is at O. If P reaches y, the foot N also reaches y, if P is at x', N returns to O. As P moves along the circumference to y',

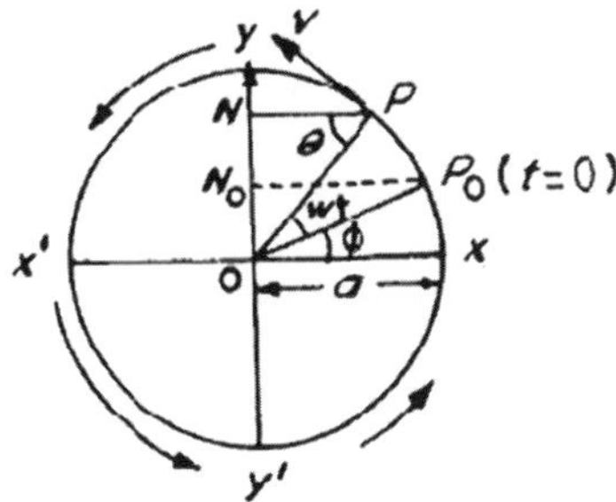

Figure 9.10 Oscillating motion.

foot N reaches y'. When P finally returns to x, the foot N is again at O. Thus, the oscillating motion of N about the mean position O is periodic a transverse wave. The circle $XYX'Y'$ is called the circle of reference and point P is called the reference point.

Let us consider a point P is moving counterclockwise in a circle of radius "a" with angular velocity ω. Let at $t=0$, point P be at P_0, OP_0 makes an angle φ with the x-axis. After time t, it makes an angle θ, Now, $\theta = \omega t + \varphi$.

The displacement of N in time t $ON = OP \sin NPO = OP \sin \theta = a \sin(\omega t + \varphi)$

or, (9.9)

$$y = a \sin(\omega t + \varphi)$$

At $t=0$, if P_0 lies on the other side of $x'ox'$ then $\theta = \omega t - \varphi$ and hence the displacement of N in time "t" is given by,

$$y = a \sin(\omega t - \varphi) \tag{9.10}$$

for

$$\varphi = 0, y = a \sin \omega t \tag{9.11}$$

Equations (9.9) and (9.11) are the basic equations for motion of a particle in a circular motion. The angle φ is called the initial phase of the vibrating particle N. The quantity "a" is the maximum amplitude. This is called the amplitude of vibration. The displacement "y" actually varies between $+a$ and $-a$.

Let O be the position of a point source emitting a progressive wave travelling along the x-axis. The source of waves O performs up and down motion of amplitude "a"and time period T. Then, the displacement y of a particle at O from its mean position at any time t is given by $y = a \sin \omega t$ (Eq. (9.11)). Again, let the wave be travelling with a velocity v in the positive direction of the x-axis. If we consider a particle of the medium at a point P at a distance of x from O, the wave starting from O would reach this point P in x/v s. Hence, this particle will start vibrating x/v s later than the particle at O. The particle at P lags in phase by x/v s that of particle at O. Consequently, the displacement of the particle at O but at a time x/v s earlier, *i.e.* at time $\left(t - \frac{x}{v}\right)$ is given by

$$y = a \sin \omega\left(t - \frac{x}{v}\right) \tag{9.12}$$

Here, $\omega = \frac{2\pi}{T}$, and T is the time period.

Hence, Eq. (9.12) can be written as

$$y = a\sin\frac{2\pi}{T}\left(t - \frac{x}{v}\right) = a\sin\frac{2\pi}{Tv}(vt - x)$$

with $vT = \frac{v}{n} = \lambda$, n is the frequency and λ is the wavelength. The above equation becomes

$$y = a\sin\frac{2\pi}{\lambda}(vt - x) \tag{9.13}$$

Further,

$$v - n\lambda = \lambda/T$$

Then,

$$\begin{aligned} y &= a\sin\frac{2\pi}{\lambda}\left(\frac{\lambda}{T}t - x\right) \\ &= a\sin\frac{2\pi}{\lambda}\left(\frac{\lambda}{T}t - x\right) \\ &= a\sin 2\pi\left(\frac{t}{T} - \frac{x}{\lambda}\right) \end{aligned} \tag{9.14}$$

In the above equation, $2\pi(\frac{t}{T} - \frac{x}{\lambda})$ is called the phase of the periodic motion.
Also,

$$y = -a\sin 2\pi\left(\frac{x}{\lambda} - \frac{t}{T}\right) \tag{9.15}$$

Equation (9.15) is the general equation of wave motion (transverse wave).

Let at any instant "t", phases of two particles whose distances from mean position are x_1 and x_2 are denoted by φ_1 and φ_2. Then, by definition of phase,

$$\varphi_1 = 2\pi\left(\frac{x_1}{\lambda} - \frac{t}{T}\right) \quad \text{and} \quad \varphi_2 = 2\pi\left(\frac{x_2}{\lambda} - \frac{t}{T}\right)$$

The phase difference of two particles is given by

$$\varphi_1 - \varphi_2 = \frac{2\pi}{\lambda}(x_1 - x_2)$$

or,

$$\Delta\varphi = \left(\frac{2\pi}{\lambda}\right) \times \Delta x\,(radian) = \frac{2\pi}{\lambda} \times path\ difference$$

This is the phase difference of those two particles whose path difference is Δx. If $\Delta x = \lambda$, then $\Delta\varphi = 2\pi$. Hence, the phase difference of two particles whose path difference is wavelength.

Let a particle be at distance x from the mean position and its phase at time t_1 be φ_1 and at time t_1 be φ_2, then

$$\varphi_1 = 2\pi\left(\frac{x}{\lambda} - \frac{t_1}{T}\right)$$

and

$$\varphi_2 = 2\pi\left(\frac{x}{\lambda} - \frac{t_2}{T}\right)$$

Hence, the phase difference of two particles at different time is given by

$$\varphi_1 - \varphi_2 = \frac{2\pi}{T}(t_1 - t_2)$$

or,

$$\Delta\varphi = \frac{2\pi}{T} \times \Delta t\,(radian)$$

This is the phase difference in Δt time interval of the particle.

If $\Delta t = T$, then $\Delta\varphi = 2\pi$

Therefore, after a time period, T, the phase of vibration of a particle becomes the same as it was in the initial position. The phase 2π denotes that the particle again starts from the mean position.

9.3.2 Energy Generation

The potential energy arises from the elevation of water above the mean level, *i.e.* $y = 0$. Considering a differential area $y\,dx$ and a mean height, $y/2$, the potential energy is expressed as

$$d(PE) = mg\frac{y}{2} = (\rho y dx L)g\frac{y}{2} = \frac{g\rho y^2 L dx}{2} \tag{9.16}$$

where, m is mass of liquid, kg, g is gravitational acceleration, 9.8 m/s^2, ρ is water density (kg/m^3) and L is an arbitrary width of the two-dimensional wave perpendicular to the direction of wave propagation m.

The equation of a wave as mentioned in Eq. (9.15) is given by,

$$y = a\sin 2\pi\left(\frac{x}{\lambda} - \frac{t}{T}\right)$$

$$= a\sin(\mathrm{kx} - \omega\mathrm{t}) \tag{9.17}$$

where $k = \frac{2\pi}{\lambda}$ m^{-1}, $\omega = \frac{2\pi}{T}$, y is the height above its mean level. m and t is the time in seconds.

Substitute Eq. (9.17) in Eq. (9.16). After integration between 0–T, we get

$$PE = \frac{\rho L a^2 g}{2}\int_0^\lambda \sin^2(kx - \omega t) = \frac{\rho L a^2 g}{2}\int_0^\lambda \frac{1}{2}[1 - \cos 2(kx - \omega t)]dx$$

$$= \frac{\rho L a^2 g}{2}\left(\frac{1}{2}x - \frac{1}{4}\sin 2(kx - \omega t)\right)\Bigg|_0^\lambda$$

$$= \frac{1}{4}g\rho a^2\lambda L$$

$$PE = \frac{1}{4}g\rho a^2\lambda L \tag{9.18}$$

The potential energy (PE) per unit area is $PE/A(J/m^2)$ where $A = \lambda L$ is given by

$$\frac{PE}{A} = \frac{1}{4} g\rho a^2 \tag{9.19}$$

From hydrodynamic theory, the kinetic energy of ocean wave is expressed as

$$KE = \frac{1}{4} g\rho a^2 \lambda L \tag{9.20}$$

Also,

$$\text{Total energy} = \text{Potential energy} + \text{Kinetic energy}$$

Total energy per unit area and power density can be written as follows,

$$\frac{E}{A} = \frac{1}{2} g\rho a^2 \tag{9.21}$$

And power density

$$\frac{P}{A} = \frac{1}{2} g\rho a^2 f \tag{9.22}$$

where f is the frequency and Power $P = \text{energy} \times \text{frequency} = \text{energy/time}$.

EXAMPLE 9.4 (a)

Calculate the frequency of the wave energy for a given power density of 600 W/m^2 and mean wave height of 1 m.

Solution

The mean wave height can be considered as an amplitude of a wave, *i.e.* $a = 1$ m.

From Eq. (9.22),

$$\frac{P}{A} = \frac{1}{2} g\rho a^2 f$$

Given $P/A = 600$ W/m^2, $a = 1$ m, $g = 10$ m/s^2 and $\rho = 1000$ kg/m^3.

Then, the frequency can be obtained as follows,

$$f = \frac{P}{A} \cdot \frac{2}{g\rho a^2} = \frac{600 \times 2}{10 \times 1000 \times 1} = 0.12\,\text{sec}^{-1}$$

EXAMPLE 9.4 (b)

Calculate the time period of Example 9.4(a).

Solution

$$\text{Time period} = \frac{1}{f} = \frac{1}{0.12} = 8.3\,\text{sec}$$

9.3.3 Wave Energy Conversion Devices

The following methods can be used to convert wave energy to electricity:

(i) **The Float or Buoy System:** This uses the rise and fall of ocean swells to drive hydraulic pumps. The object can be mounted to a floating raft on the ocean floor. A series of anchored buoys rise and fall with the waves. The movement of buoys strikes directly an electrical generator and electricity is generated. The electricity is then transmitted ashore by underwater power cables.

Wave motion is primarily horizontal. The motion of the water particles is primarily vertical. Mechanical power is obtained by floats making use of the motion of water up and down.

It is guided by four vertical manifolds. These are part of a platform. There are four large underwater floatation tanks that stabilise the platform. This is supported by buoyancy forces and no vertical or horizontal displacement occurs due to wave action. Thus, the platform is made stationary in space. A piston is attached to a float. It moves up and down inside a cylinder. The cylinder is attached to the platform. It is therefore relatively stationary. The piston and cylinder arrangement is used as a reciprocating compressor. The downward motion of the piston draws air into the cylinder *via* an inlet check valve. This air is compressed by upward motion of the piston. It is supplied to the four underwater floatation tanks through an outlet check valve *via* the four manifolds. In this way, the four floatation tanks serve the dual purpose of buoyancy and air storage. It also serves the four vertical manifolds and float guides. An air turbine is run by the compressed air that is stored in the buoyancy–storage tanks. This drives an electrical generator to produce electricity. The generated electricity is transmitted to the shore through an underwater cable.

(ii) **The Oscillating Water Column Device:** In this case, the in and out motion of waves at the shore enters a column. It forces air to turn a turbine.

When a wave passes on to a partially submerged cavity open under the water as shown in Figure 9.11, a column of water oscillates up and down in the cavity. This can induce an oscillatory motion in the air above the column. It may be connected to the atmosphere through an air turbine. Due to the oscillating water column can increase the air velocity by reducing the cross-sectional area of the channel at the turbine. In this way, the slow wave motions can be coupled to the high-frequency turbine motion.

(iii) **Tapered Channel or Tapchan System:** This consists of a tapered channel (hence the name). It feeds into a reservoir that is constructed on a cliff. The narrowing of the channel causes the waves to increase their amplitude (wave height) as they move towards the cliff face.

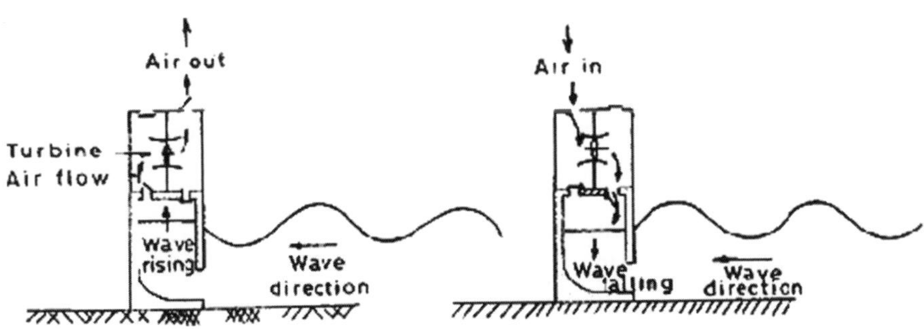

Figure 9.11 Oscillating water columns in wave form.

The cliff face eventually spills over the walls of the channel and into the reservoir that is positioned several meters above mean sea level.

The kinetic energy of the moving wave is converted into potential energy due to storage of water in the reservoir. The stored water is then fed through a water turbine to generate electricity. This is also referred to as a tapchan wave energy device.

9.3.4 Advantages and Disadvantages

The following are the advantages/disadvantages of wave energy systems:

Advantages:

(i) The energy from the wave is naturally concentrated by the accumulation over time and space. It is transported from the point present in the winds.
(ii) A greater extent of power is concentrated in the motion of waves that is due to the movement of air. The power density at the wind energy site may be of the order of few square metres. The power density in a corresponding area of wave motion may be up to 100 times greater.
(iii) It is a renewable energy source.
(iv) Wave power devices do not use large land masses like solar or wind.
(v) These devices are relatively pollution free. They remove energy from the waves. The water is left in a relatively placid (calm) state in their wakes.
(vi) Wave energy systems can shelter the coast. They are therefore useful in harbour areas and coastal erosion zones.

Disadvantages:

(i) The major disadvantage of wave energy is its availability in the ocean as compared to wind. Therefore, the extraction equipment must operate in a marine environment. The cost of its construction, its maintenance, biological growth of marine organisms, lifetime and reliability should be considered in economic analysis.
(ii) Wave energy converters must be capable of withstanding very severe peak stresses in storms.
(iii) Wave energy conversion devices are relatively complicated.
(iv) Economic factors such as capital investment, costs of maintenance, repair and replacement as well as problems of which seem to be very large, are all relatively unknown.

9.3.5 Applications

Wave energy conversion systems can be used for the following applications:

(i) electric power generation;
(ii) hydrogen production by using electrolysis;
(iii) seawater desalination by using reverse osmosis;
(iv) shoreline protection;
(v) commercial mariculture and fish farming;
(vi) navigation aids.

9.3.6 Current Status of Wave Energy

An independent market assessment estimated the worldwide potential of wave energy economic contribution in the electricity market to be in the order of 2000 TWh/year, which is about 12% of world electricity consumption (based on 2009 data) and is comparable to the amount of electricity currently produced worldwide by large-scale hydroelectric projects.

In terms of market value, the potential market for wave energy is worth about $1 trillion worldwide, according to the World Energy Council. In the United States alone, wave technology could supply 6.5 per cent of the nation's energy. The worldwide wave energy installed capacity for various countries is given in Table 9.4.

9.3.6.1 Wave Energy and India

The annual wave energy potential for the India coast is around 175–200 MWh/m. The wave power potential at six locations off the Indian coast is given in Table 9.5.

The wave energy potential along the Indian coast is low compared to the Northern-Latitude countries. There are many coastal areas that are badly eroded by the action of waves. Wave energy could be converted to electricity, if a long barrier of wave energy devices could be arranged parallel to the shore, then. Furthermore, the water between the barrier and the coast becomes calm if the barrier is not far away from the coast. This calm pool of water could be used as a natural harbour

Table 9.4 Worldwide wave energy installed capacity (Source: IEA-OES Annual Report, 2010).

Country	*Installed capacity (kW)*
UK	1300
Canada	1065
Korea	500
Denmark	195
Sweden	149
Norway	4

Table 9.5 Wave power potential along the Indian coastline.

	South-west monsoon			*North-east monsoon*			*Non-monsoon*		
Location	*Mean wave height (m)*	*Mean wave period (s)*	*wave power (kW/h)*	*Mean wave height (m)*	*Mean wave period (s)*	*wave power (kW/h)*	*Mean wave height (m)*	*Mean wave period (s)*	*Wave power (kW/h)*
10–15°N and coast 85°E (off Madras)	1.71	5.80	16.62	1.53	5.86	13.44	1.14	5.50	7.00
15–20°N and coast 85°E (off Vishakapatnam)	2.04	8.25	33.65	1.60	6.28	15.75	1.24	7.10	10.70
20–25°N and 85–95°E (off Calcutta)									
5–10°N and 75–80°E (off Cape Comorin)	1.78	6.29	19.52	1.22	5.35	7.80	1.29	5.46	8.90
10–15°N and 70°E coast (off Cochin)	2.03	6.77	27.34	1.03	5.05	5.25	1.01	5.38	5.37
15–25°N and 70°E coast (off Bombay)	2.63	6.93	46.98	1.00	5.00	4.90	1.01	5.25	5.24

and as a space for aquaculture. This calm area can also be used as a zone of coastal traffic. Thus, the wave energy device becomes multifunctional.

A pilot plant of 150 kW capacity was built at Vizhinjam near Trivandrum by the Ocean Engineering Centre, Indian Institute of Technology, Madras. It was commissioned in October 1991. The power plant consists of a concrete caisson (17 m × 23 m) with an oscillating water column chamber.

9.4 DRAUGHT ANIMAL POWER

Domestic work animals exist in all regions of the world. The use of domestic cattle and buffalo in agriculture dates back to the 14th century B.C. They are significantly contributing to mankind as a useful source of renewable energy in addition to milk, dung, meat, *etc.* The draught animal power (DAP) is complementary to petroleum-based power. It needs full exploitation in view of the current escalation in use as well as rapid depletion of fossil fuels. Other species of animals like horse, camel, donkey, elephant and yak, *etc.* also contribute to the 2/3s of the entire mechanical energy needed for agriculture, forestry, transport, *etc* (Table 9.6).

The world's population depends on animal power as its main energy source. Draught animals are used for agricultural operations for about 50% of the cultivated areas. The draught animal power output was equivalent to the use of 20 million tonnes of petroleum products by motorised equipment valued at that time at US$ 6 billion.[8]

The use of animals for work dwindled in the developed countries as use of mechanical power increased during the 20th century.[9] However, livestock continue to contribute enormously to the energy needs of agriculture into the 21st century. Information on draught animal numbers is scarce. The estimates of 300 million animals are sometimes used. Other estimates are considerably higher with suggestions that 51% of the 921 million cattle in the developing world were used for work in 1994. These were 35% of 135 million buffalo, 65% of 43 million horses, 87% of 43 million donkeys, 70% of 14 million mules and 15% of 19 million camels.[10]

The entire working performance of a draught animal depends on its power output. It is estimated that oxen can reasonably be expected to exert a draught force of about 10% of their body weight and a horse 15% in good condition. This also depends upon their local conditions.

The animals can also generate substantially higher draught force[11] and power output as compared to their body weight, for a short spurt. However, this is not a healthy practise. It leads to quick exhaustion of the animal.

Animal traction continues to increase in many parts of the world, for smallholder farmers. Animal power will continue to be important for food security, self-reliance and poverty alleviation. Animal power can also be used for milling, logging, land excavation and road construction, in addition to threshing crops, hauling carts, operating water-lifting devices, sugarcane crushers, oil ghanis, *etc.*

The progress of civilisation has been closely linked with the service of work animals to mankind. Work done by animals provides the motive power to millions of tillage implements and carts. They also leave behind the bone-meal at the end of their working lives in developing/underdeveloped

Table 9.6 Power developed by various animals.

Animal	*Average weight (kg)*	*Approx. draught (kg)*	*Average speed of work (m/s)*	*Power Developed (kW)*
Bullock	500–900	60–80	0.6–0.85	0.56
He-buffalo	400–900	50–80	0.8–0.9	0.55
Light horse	400–700	60–80	1	0.75
Donkey	200–300	30–40	0.7	0.25
Mule	350–500	50–60	0.9–1.0	0.52

countries. The contribution of work animals to their economy is of the utmost importance. Draught animals are the major source of farm power. They are used for pulling the agricultural implements and equipments.

9.4.1 Renewability

Animal power is particularly suited to family-level farming and to local transport. Animal power is generally affordable and accessible to the smallholder farmers. Agriculture is basically an energy-conversion activity. It is both a producer and consumer of energy. The energy available from the plants is thus normally greater than the energy supplied through photosynthesis. It forms the basis of modern agriculture. Animals consume stored energy of plants to do the tractive work. As biomass production is renewable, the animals dependent on biomass are also renewable sources of energy for farm operations and transport purposes. This is sustainable and environmentally friendly. Close integration of higher crop output, increased income and improved food security and livestock contributes to sustainable production. This energy can be sustained in rural areas with external input. The use of animal power in mixed farming systems encourages sustainable farming practises. Working animals produce their own organic manure in addition to transport for manure of other livestock to the fields. This enhances the fertility and structure of the soil. The transport role of animals is important for carrying farm inputs (seeds, fertilisers and crop protection requisites) and outputs (harvested crops and animal products). Carts facilitate the marketing of produce, stimulating local trade and production. Animal power requires no foreign exchange. Money spent on motors and machinery is exported from rural areas. Money invested in animal power circulates within rural areas. It helps to revitalise rural economics. The duality of animal power for transport and fieldwork proves it to be an important asset for society. It is a viable technology for rural development.

There are roughly 1200 million work animals, comprising 921 million cattle, 135 million buffalo, 43 million horses, 43 million donkeys, 14 million mules and 19 million camels, *etc.* Elephants are also used for transportation in some regions. On an average, if each animal generates roughly half a horse power (hp), then the total work hp generated by animals would be about 600 million hp (approximately 450 GW of electrical power). This is almost equal to the installed capacity of electrical energy in the developing countries at present.

Domestic animals help in reducing the human drudgery and time particularly in farm operations for sustainable food production. In the past, animal power has been a neglected option. However, governments, planners, agencies and the private sector are now taking it more seriously. Animal power should become an integral part of national development strategies. These include those relating to food security, resource conservation, rural transport, employment and women in development. With a favorable policy on environment and developmental, the private sector can sustain and develop animal power technologies. This benefits rural communities and economies. Animal power issues need to be adequately covered in education and training programmes.

9.4.2 Draught Animal Population (DAP) and Farm Power in Asia

The increase in the rural population of Asia has been modest (1.4 times) compared with the much greater growth of urban population (2.6 times).[12] Most of this increase has been due to the introduction of more intensive and efficient agricultural practises. The increase in the area under cultivation has been comparatively small (1.5 times). However, both the improved yields per hectare and the increase in cultivated area require more power input in almost all farming systems. The main source of this extra power input appears to have been met by mechanisation (the 9.3 fold increase in the number of tractors).

It is quite difficult to assess the contribution that draught animals have made and continue to make to the "green revolution" in Asia. Statistics on draught animals are scarce and often unreliable. The numbers of two main species are used almost entirely for draught (donkeys and horses) remained almost constant. The increase in the number of cattle and buffalo, have been relatively modest (1.4 times and 1.5 times, respectively). They are far lower than the increase in agricultural production during the same period. These observations tend to the conclusion that despite large increase in the human population (matched by the increases in overall agricultural production), the use of animal traction as a source of power on farms has generally remained constant.

The draught animal population (DAP) remains an essential feature of many Asian farming systems. The DAP system represents a fundamental component in the social, agricultural and economic fabric of Asian countries. It is stable, productive and underpins the economy of most of the states of Asia. Another most important factor often overlooked is the use of draught animals is a source of employment. This has the potential for diversity at the microeconomic or rural level.[13]

9.4.3 Versatility of Draught Animal Power (DAP)

Draught animals are very versatile, unlike mechanical and electrical power. They are used as a direct source of energy for

(a) **Core Production Operations**
(i) Ploughing, (ii) harrowing, (iii) land levelling, (iv) puddling, (v) planking, (vi) sowing, planting and fertiliser application, (vii) weeding and interculture, (viii) harvesting, (ix) threshing and (x) waterlifting.

(b) **Postharvest Operations**
(i) Chaff cutting. (ii) cane crushing, (iii) oil extraction, and (iv) grinding.

(c) **Oil Extraction**
The crushing of oil seeds by the bullock-driven "Ghani" or wooden "Kolhu" has been practised for a long time. The pestle is turned round at a speed of 5–7 rpm by a pair of bullocks. A "Ghani" with two bullocks takes about 3 h to crush one charge of 16 kg of mustard seed.

(d) **Transport**
Transport accounts for a high proportion of farming activities. Carts driven by a donkey or by two bullocks/buffaloes or by a horse, play an important role in the development. Transport by domestic animal is certainly much older than animal cultivation. The tractive effort of draught animal depends on the weight and characteristics of ground besides on its individual characteristics, *i.e.* breed, sex, age, weight, size and quality of feed, training for the work and health, *etc*. Practical experience has shown that on relatively flat and firm ground, a reasonable useful load for a pair of bullocks is 800 kg and up to 400 kg for a donkey.
Animal power gear as a mechanical output is used to convert the animals' effort to the particular task to be performed and adapt it. It was at the beginning of the 20th century in Europe and North America that the use of animal power gear was at its peak.

It has been reported that the power available from a power gear worked by a single donkey, horse or bullock (400 kg) is 0.16 hp and 0.32 hp, respectively, on a circular tract. However, in practise one may expect with a pair of animals an output of 0.3–0.6 hp for a period of 45 min.

An animal-driven pedal thresher has given an output of 1.5–2.5 quintal/h of paddy grain. A chaff cutter driven by animal power gear is similar to the hand-operated one. The only difference is that it is provided with a smaller diameter flywheel. By use of power gears, the slow speed of bullocks is utilised to operate the machine at about 250 rpm. Its capacity becomes 4 to 6 times that of the hand machine. An agrowaste compaction machine driven by animal power has also met satisfactory success. Animal power gears are more reliable, maintenance is less.

9.4.4 Benefits of Draught Animal Power (DAP)

- Draught animal power is an excellent example of mass application of appropriate technology. It is ideal for small farmers. Tractors and tillers become economical only for the larger farm size, which is above four hectares. There are over 100 million small farms in the developing countries. Animal power has many other benefits, as mentioned below.
- Appropriate for hauling vehicles for small-scale transportation (about one tonne) over short distances (below 20 km). The loading and unloading take time, animal-drawn vehicles are economical.
- Pack animals for carrying goods and people. Donkeys, mules and camels serve as important sources in deserts and hilly regions and in areas where there are no roads or even pathways.
- Elephants and horses are used for logging in India, Burma, Srilanka, Phillipines and a few other countries. Bullocks are used in South America.
- Draught animals leave behind for man after their death, a number of useful products. They yield meat, skin, bone, hoof blood and numerous other byproducts.
- Used for a wide range of land management and erosion control systems. Animal-drawn scoops or levelers can assist with water harvesting or the construction of water ponds. In hilly areas, animals can assist with contour ploughing and terracing.
- Can help to overcome peak labour shortages. It can also assist in the creation of rural employment opportunities in the agricultural (large and small scale), infrastructure support and transport sectors.
- It should be pointed out that draught animals live largely on crop residue and grass. These products are not really useful for man. In fact, there are possibilities of raising leguminous crops that would enrich the soil by fixing nitrogen.

OBJECTIVE QUESTIONS

9.1 For ocean thermal energy conversion (OTEC), there is a need for at least a minimum temperature difference of
(a) 200 °C
(b) 20 °C
(c) 100 °C
(d) all of them

9.2 The annual amount of solar energy absorbed by ocean is
(a) 400 times the amount of electricity used by human beings
(b) 40 times the amount of electricity used by human beings
(c) 4000 times the amount of electricity used by human beings
(d) 40 000 times the amount of electricity used by human beings

9.3 The variation of solar intensity in water with depth
(a) Linear
(b) Nonlinear
(c) Exponential
(d) Constant

9.4 The solar intensity with water depth
(a) Decreases
(b) Increases

(c) Constant
(d) None of them

9.5 The value of the extinction coefficient (absorption coefficient, k) for fresh water
(a) Greater than the value of k for salt water
(b) Less than the value of k for salt water
(c) Equal to the value of k for salt water
(d) None of them

9.6 The value of extinction coefficient (absorption coefficient, k) is
(a) >1
(b) <1
(c) $=1$
(d) $=0$

9.7 The mechanical power available from OTEC for a given T_H is
(a) $\propto (\Delta T)$
(b) $\propto (\Delta T)^2$
(c) $\propto (\Delta T)^3$
(d) none

9.8 The expression for η_{Carnot} of OTEC is
(a) $\Delta T/T_h$
(b) $T_h/\Delta T$
(c) ΔT
(d) $\Delta T/T_c$

9.9 The mechanical power available from OTEC for a given ΔT is
(a) $\propto T_H$
(b) $\propto 1/T_H$
(c) $\propto T_C$
(d) $\propto 1/T_C$

9.10 OTEC technology operates by
(a) The closed cycle
(b) The open cycle
(c) Both closed–open cycle (hybrid cycle)
(d) All of them

9.11 From OTEC, one can have
(a) Clean water
(b) Electrical power
(c) Fossil fuel savings
(d) All of them

9.12 The cost of electricity produced by OTEC is
(a) $<$ cost of electricity produced by fossil fuels
(b) $>$ cost of electricity produced by fossil fuels
(c) $=$ cost of electricity produced by fossil fuels
(d) None of them

9.13 The cost/kWh of electricity from OTEC is
(a) $>$ cost/kWh of electricity from tidal
(b) $>$ cost/kWh of electricity from wave
(c) $<$ cost/kWh of electricity from tidal and wave
(d) $=$ cost/kWh of electricity from tidal and wave

9.14 Tidal energy is due to the Sun and the Moon
(a) Gravitational effect of the Sun and Moon
(b) Gravitational effect of the Earth and the Moon

(c) Gravitational effect of the Earth and the Sun
(d) All of them

9.15 An expression for energy (kWh) per tide is given by
(a) $(\rho AR) \cdot g\frac{R}{2}$
(b) (ρAg)
(c) $(\rho AR)\, g\frac{R^2}{2}$
(d) $(\rho AR)g\frac{R^2}{2}$
where R is the tidal range

9.16 For tidal period T, power generated for one tide is
(a) $(\rho AR)g\frac{R}{2T}$
(b) $(\rho AR)\frac{R^2}{2T}$
(c) $(\rho AR)g\frac{R^2}{2T}$
(d) $(\rho AR)g\left(\frac{R}{2T}\right)^2$

9.17 Energy per tide for a given basin area (A) is
(a) $\propto R$
(b) $\propto R^2$
(c) $\propto R^3$
(d) $\propto 1/R$

9.18 Power from tide for a given basin area (A) and tide range (R) is
(a) $\propto\ 1/T^2$
(b) $\propto\ 1/T$
(c) $\propto\ T$
(d) $\propto\ T^2$

9.19 Tidal energy
(a) Is renewable
(b) Is economical
(c) Depends on fossil fuel
(d) Environmentally friendly

9.20 The wave energy is produced due to
(a) Transfer of energy from the Sun to the ocean (sea)
(b) Transfer of energy from the wind to the sea
(c) Transfer of energy from the Earth to the sea
(d) All of them

9.21 In a transverse wave, the particles of a medium move
(a) In the direction of propagation of the wave
(b) In the opposite direction of propagation of the wave
(c) In the perpendicular direction of propagation of the wave
(d) None of them

9.22 In a longitudinal wave, the particles of medium moves
(a) In the direction of propagation of the wave
(b) In the opposite direction of propagation of the wave
(c) In the perpendicular direction of propagation of the wave
(d) None of them

9.23 The motion of wave is
(a) Steady state
(b) Transient
(c) Periodic
(d) None

9.24 The wave equation is represented by
(a) $y = a\,sin\,(\omega t + \varphi)$
(b) $y = a \sin(\omega t - \varphi)$
(c) $y = a \sin \omega t$
(d) All of them

9.25 The phase of the periodic motion of transverse wave is
(a) $y = 2\pi\left(\frac{x}{\lambda} - \frac{t}{T}\right)$
(b) $y = 2\pi\left(\frac{t}{T} - \frac{x}{\lambda}\right)$
(c) $y = 2\pi\left(\frac{t}{T} - \frac{x}{vT}\right)$
(d) All of them

9.26 The wavelength (λ) of a wave for a given velocity is
(a) $\propto T$
(b) $\propto 1/T$
(c) $= T$
(d) None of them

9.27 An expression for the phase difference of a wave ($\Delta\varphi$) is given by
(a) $\Delta\varphi = \frac{2\pi}{\lambda} \times path\,difference$
(b) $\Delta\varphi = \frac{2\pi}{\lambda} \times path\,difference$
(c) Both (a) and (b)
(d) None of them

9.28 An expression for the PE density of a wave is given by
(a) $\frac{1}{4} g\rho a^2 \lambda L$
(b) $\frac{1}{4} g\rho a^2$
(c) $\frac{1}{2} g\rho a^2$
(d) $\frac{1}{2} g\rho a^2 \lambda L$

9.29 An expression for the KE density of wave is given by
(a) $\frac{1}{4} g\rho a^2 \lambda L$
(b) $\frac{1}{4} g\rho a^2$
(c) $\frac{1}{2} g\rho a^2$
(d) $\frac{1}{2} g\rho a^2 \lambda L$

9.30 The power density of a wave is expressed as
(a) $\frac{1}{2} g\rho a^2 f$
(b) $\frac{1}{2} g\rho a^2$
(c) $\frac{g\rho a^2}{2T}$
(d) None of them

9.31 The power is defined as
(a) *energy* $\times$ *frequency*
(b) $\frac{energy}{time}$
(c) Both (a) and (b)
(d) None of them

9.32 Draught animal energy is
(a) Renewable energy
(b) Nonrenewable
(c) both (a) & (b)
(d) All of them

9.33 Draught animal power is
 (a) Mechanical
 (b) Thermal
 (c) Chemical
 (d) All of them
9.34 Draught animal power emits
 (a) Zero CO_2
 (b) Zero SO_x
 (c) Zero NO_x
 (d) All of them
9.35 Draught animal power is
 (a) More environment friendly
 (b) Economical
 (c) Locally used
 (d) All of them

ANSWERS

9.1 **(b)**; 9.2 **(c)**; 9.3 **(c)**; 9.4 **(c)**; 9.5 **(a)**; 9.6 **(c)**; 9.7 **(b)**; 9.8 **(a)**; 9.9 **(b)**; 9.10 **(d)**; 9.11 **(c)**; 9.12 **(b)**; 9.13 **(a) & (b)**; 9.14 **(d)**; 9.15 **(a) & (b)**; 9.16 **(a) & (b)**; 9.17 **(b)**; 9.18 **(b)**; 9.19 **(a)**; 9.20 **(b)**; 9.21 **(b)**; 9.22 **(a)**; 9.23 **(c)**; 9.24 **(d)**; 9.25 **(d)**; 9.26 **(a)**; 9.27 **(c)**; 9.28 **(b)**); 9.29 **(b)**; 9.30 **(a) & (c)**; 9.31 **(c)**; 9.32 **(a)**; 9.33 **(a)**; 9.34 **(d)**; 9.35 **(d)**

REFERENCES

1. Penney, T. R. and Bharatan, D. January 1987. *Sci. Am.*, **256**, 86.
2. Uhera, H. and Ikegami, Y. 1990. *ASME J. Solar Eng.*, **112**, No. 4.
3. Vega, L. A. 1992. *Ocean Energy Recov.*, **4**, 152–181.
4. Anon. 1990. *International OTEC Association Newsletter*. Republic of China. Vol. 1, No. 3.
5. Pole R. and Fujita, R. M. 2002. *Marine Policy*, **26**, 471–479.
6. Anon. 1993. World Energy Council. Renewable-energy resources: opportunities and constraints 1990–2020. World Energy Council, London.
7. Palaniappan, C., Kolar Ajit Kumar and Haridasan, T. M. *Renewable Energy Technologies*, Narosa Publishing House, New Delhi, 2001.
8. Ramaswamy, N. 1994. *Rev. Sci. Technol. O. I. E.*, **13**, 195–216.
9. Kaushik, S. J., 1998. Animals for work, recreations and sport. In: *Proceedings of RGE Eighth World Conference on Animal Production*. Symposium, 28 June–4 July, 1998. Seoul National University, Seoul, Korea pp. 235–245.
10. Pearson, R. A. 1999. *Work-Animal Power. An Introduction to Animal Husbandry in the Tropics*, 5th edn, Blackwell Scientific Publications, Basingstoke, UK. pp. 782–798.
11. Hopfen, J. J. 1969. Farm implements for arid and tropical regions. F. A. O. Rome, 159p.
12. Lawrence, P. R. and Pearson, R. A. 2002. *Agricult. Syst.*, **71**, 99–110.
13. Campbell, R. S. F. 1993. Draught animals in the AAAP zone and their economic future. The 6th Asian–Australian Association of Animal Production Scientists Congress, 23–28 November, 1992. Bankok, Thailand, pp. 10–16.

Renewable Energy Technology, which is Friendly to the Ecosystem, and Maintains a Good Quality of Environment.

CHAPTER 10

Sustainable Environment

10.1 INTRODUCTION

For a sustainable environment we require long term strategies for energy related issues. In this regard, the use of renewable energy resources is likely to be one of the most efficient and effective solutions. Renewable technologies have been considered to be substantially safer. They can be a solution to many environmental and social problems associated with conventional fossil fuels.

There has been active worldwide research and development in the field of renewable energy resources and systems since the oil crisis in the 1970s. Renewable energy conversion technologies appeared to be most attractive due to the cost effectiveness and easy implementation. Renewable energy technologies produce marketable energy by converting natural phenomena into useful energy forms.[1] These technologies use solar energy directly or indirectly for a continuous stream of energy. It warms us, causes crops to grow *via* photosynthesis, heats the land and sea differentially, which causes winds and consequently waves and, of course, rain leading to hydropower. Tidal rise and fall is the result of the gravitational pull of the Moon and the Sun. Geothermal energy is available from the heat of the Earth's core. These renewable resources represent a massive energy potential equivalent to fossil fuel resources.

The exploitation of renewable energy resources through various technologies is a key component of sustainable environment. They have been considered to offer less environmental degradation compared to other conventional sources (fossil fuels) of energy. In spite of their sustainability and renewability, they are generally diffuse and not fully accessible. Some are intermittent and have distinct regional variability.

Furthermore, in more recent times, it has been realised that application of renewable sources and systems has an adverse impact on the environment. They have become important as environmental concerns due to utility (hydro) costs climbing and labour costs escalating.[2] The sustainability of renewable energy resources is now debatable owing to their adverse environmental impacts. Environmental concerns are an important factor in sustainable developments activities. For a variety of reasons, activities that continually degrade the environment are not sustainable over time from health, ecological and other points of view.

Hydroelectric power was perceived as the cleanest energy production option during the 1950s throughout the world. Water is one of the easily available resources through the hydrologic cycle for the generation of hydroelectric power. It is believed that dams would store water for public water supply, fisheries, recreation for year-round use. After the generation of electricity, the water is

Advanced Renewable Energy Sources
G. N. Tiwari and R. K. Mishra

Published by the Royal Society of Chemistry, www.rsc.org

used for irrigating the agricultural fields in the downstream. In that process, it also recharges the underground aquifers.

It is useful to describe the role of different pollutants and particulate matter during use of various renewable energy sources. These are responsible for producing undesirable effects on human beings and the environment.

10.2 THE ENVIRONMENT

The **environment** supports life and sustains various human activities on planet Earth. It is widely known as the **biosphere**. The biosphere is a shallow layer compared to the total size of the Earth. It extends to about 20 km from the bottom of the ocean to the highest point in the atmosphere at which life can survive without man-made protective devices. The biosphere contains at least 3.3×10^5 species of green plants, 9.3×10^5 species of animal and 8×10^4 species of bacteria and fungi. The total number of existing species on planet Earth is certainly much more and may be close to three million. The essential requisites of life, namely light, heat, water, food and habitats for all these species are supplied by the biosphere.

The biosphere is very complex and large. It is usually divided into smaller units or ecosystems. All ecosystems can be divided into two parts

(a) **The Biotic (Living):** The biotic category can be subdivided into the three functional groups:
   (i) **Producers** (the autotrophic organisms (self-nourishing)): These are largely the green plants and algae.
   (ii) **Consumers** (the heterotrophic organisms (other-nourishing): These are chiefly all animal life, including mammals, fish, insects and birds; these ultimately depend upon consumption of producers to sustain life.
   (iii) **Decomposers** (heterotrophic organisms): These are chiefly the bacteria and fungi that break down complex compounds from waste materials including dead producers and consumers to again make the chemical components available to producers.

(b) **The Abiotic (Nonliving):** Abiotic substances are the basic elements such as phosphorus and nitrogen and the compounds found in the environment. The abiotic substances are the water, the nutrients, oxygen, carbon dioxide, *etc.* The producers in the ecosystem are the large rooted plants and the free-floating minute plants usually algae, called phytoplankton. These store the energy from the Sun and liberate oxygen into the atmosphere. The primary consumers are benthos, or bottom forms and zooplankton with little or no swimming ability. The phytoplankton is consumed by zooplanktons that are in turn eaten by large aquatic life such as fish. Other consumers are the insects, frogs, human beings, *etc.* and it is a category known as detritivores that live on organic wastes. All of these forms produce organic waste and dead organisms. The decomposers, *i.e.* bacteria and fungi utilise the organic carbon and generate CO_2 that is used by the algae. Additional CO_2 is provided from the atmosphere and the respiration of fish. The availability of the nutrients carbon, phosphorus and nitrogen is sufficiently small in a healthy system to limit the production of algae. A dynamic equilibrium is maintained throughout the ecosystem. Chemical elements circulate between the organisms and the environment through pathways comprising the natural cycles. The most important cycles are (a) the hydrologic cycle and (b) the biogeochemical cycles of (i) carbon (ii) nitrogen (iii) phosphorus and (iv) sulfur. These cycles operate in a balanced state with little variation, thereby contributing to the stability of the whole biosphere.

Man has exploited and modified the environment to his advantage in many ways. However, these changes have represented only a small disturbance in a large system. These have not seriously threatened the homostatic mechanisms to maintain the system in a stable condition. This is called the feedback loop. Now unfortunately, man is capable of inducing large enough disturbances in the ecosystem and it can permanently upset the balanced state of the natural cycles.

10.3 NATURAL CYCLES

The natural cycles are as follows:

10.3.1 The Hydrologic Cycle

The hydrologic (water cycle) is the most important natural cycle in the biosphere. More than 97 percent of water is found in the oceans. The remaining 3 percent is found on the continents and in the atmosphere. But more than 70 percent of this latter portion is locked in glaciers and icecaps. Human beings depend on water that is from lakes, streams and ground water. It accounts for less than 1 percent of the total supply. It is this water that is currently being used and reused in many parts of the world.

The hydrologic cycle of the biosphere depends on the reciprocity of evaporation and precipitation. Liquid water from planet Earth goes into the atmosphere by evaporation and transpiration of the plants as vapour. The vapour is returned to planet Earth as rain or snow (precipitation). Figure 10.1 shows the complete hydrologic cycle. Most evaporation occurs over the oceans. But some oceans lose more water by evaporation than they gain by precipitation. The difference is made up by runoff and seepage from the continents. There is more precipitation than evaporation. The continents lose more than 50 percent of the precipitation through evaporation. The remainder is temporarily stored in lakes and rivers or as ground water. It is later discharged into the oceans. The global cycle can be summarised as shown in Table 10.1. Each year the evaporated water is estimated to be 423 000 km^3. The same quantity is precipitated over the whole surface of the Earth. The amount of water is temporarily stored and later discharged into the oceans (37 000 km^3). This is the amount that is potentially available for human needs. Minor local modifications to the hydrologic cycle are usually made by diverting or regulating the runoff water.

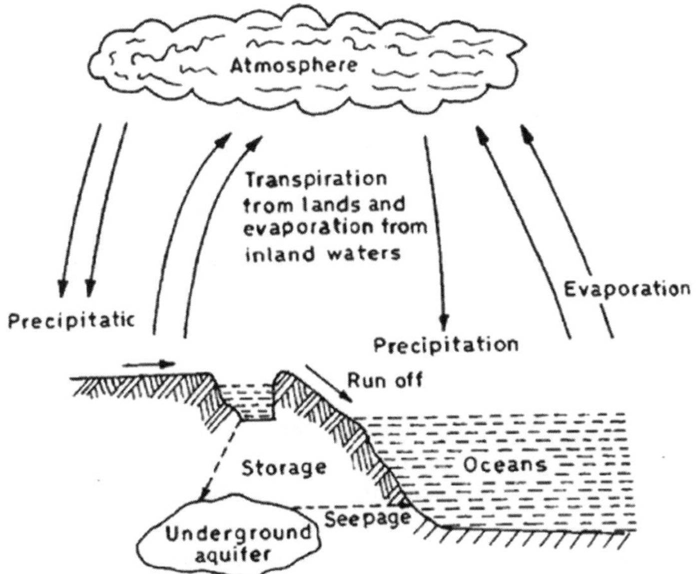

Figure 10.1 View of the hydrologic cycle.

Table 10.1 Water balance on the planet Earth.

	Oceans (km^3/yr)	*Continents* (km^3/yr)	*Whole Earth* (km^3/yr)
Precipitation	324 000	99 000	423 000
Evaporation	361 000	62 000	423 000
Gain by Inflow	37 000	37 000	0

The storage phase of fresh water is used for domestic or industrial purposes and for power generation, flood control, irrigation and recreation.

EXAMPLE 10.1

What is the percentage of water evaporated from the ocean and continents with respect to the whole Earth?

Solution

From Table 10.1, we have

$$\text{Water evaporated (in percent) from the ocean} = \frac{361\,000}{423\,000} \times 100 = 85.3\%$$

$$\text{Water evaporated (in percent) from the continents} = \frac{62\,000}{423\,000} \times 100 = 14.7\%$$

10.3.2 The Carbon Cycle

The biosphere contains a complex mixture of carbon compounds in a dynamic equilibrium of formation, transformation and decomposition. The dynamics of the carbon cycle is shown in Figure 10.2.

The producers reduce carbon dioxide from the atmosphere to organic carbon through photosynthesis. Then, this passes through consumers and decomposers. It usually re-enters the atmosphere through respiration and decomposition. Additional return from producers and consumers occurs through the nonbiological process of combustion. Another major reservoir of carbon is the ocean. It stores more than 50 times as much carbon as the atmosphere. Interchange of carbon dioxide between the atmosphere and the ocean takes place through diffusion. The oceanic reservoir tends to regulate the atmospheric CO_2 concentration.

The Earth has significant reserves of bound carbon in the form of inorganic deposits such as limestone and organic fossil fuel deposits. This consists of mainly coal and petroleum. Some of the bound carbon returns to the atmosphere as carbon dioxide or carbonic acid from aquatic reservoirs. This is because of combustion of fossil fuels, weathering and dissolution of carbonate rocks and volcanic activity.

Combustion of fossil fuels can increase the amount of CO_2 in the atmosphere. Studies have shown that plants tend to grow faster in a CO_2-enriched atmosphere. This benefit is offset by denudation of forests by man, thereby decreasing nature's ability to remove atmospheric CO_2. A detectable increase in the concentration of atmospheric CO_2 from 290 ppm in the nineteenth century to the current level of about 325 ppm, has been observed (Table 10.2). It has been estimated that the atmospheric CO_2 level will be around 380 ppm by the end of the century if the present trend continues.

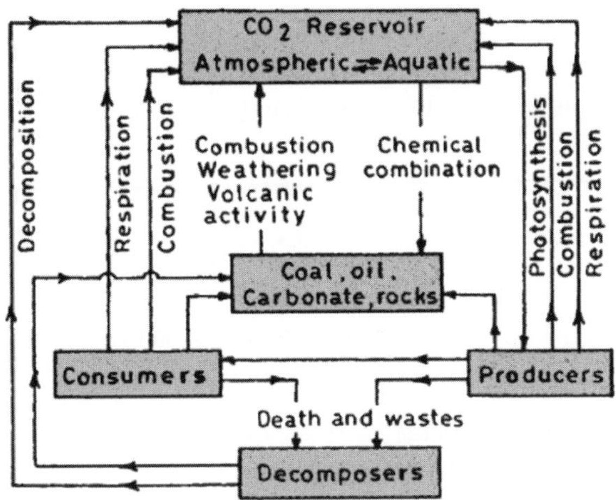

Figure 10.2 View of the carbon cycle.

Table 10.2 Concentration of various atmospheric gases for clean and dry air at ground level on Earth.

Gas	*Concentration, ppm by volume*	*Concentration, % by volume*	*Estimated residence time*
Nitrogen (N_2)	280 000	78.09	Continuous
Oxygen (O_2)	209 500	20.95	do
Argon (Ar)	9300	0.93	do
Carbon dioxide (CO_2)	320	0.032	2–4 years
Neon	18	0.0018	Continuous
Helium (He)	5.2	0.00052	2 million years
Methane (CH_4)	1.5	0.00015	4–7 years
Krypton (Kr)	1	0.0001	Continuous
Hydrogen (H_2)	0.5	0.00005	–
Nitrogen oxide (N_2O)	0.2	0.00002	4 years
Carbon monoxide (CO)	0.1	0.00001	0.5 years
Zenon (Xe)	0.08	0.000008	Continuous
Ozone (O_3)	0.02	0.000002	~60 days
Ammonia (NH_3)	0.006	0.0000006	7 days
Nitrogen dioxide (NO_2)	0.001	0.0000001	–
Nitric oxide (NO)	0.0006	0.00000006	5 days
Sulfur dioxide (SO_2)	0.0002	0.00000002	4 days
Hydrogen sulfide (H_2S)	0.0002	0.00000002	2 days

Atmospheric circulation models have been used to predict the impact of CO_2 concentration on the temperature of the Earth's surface. They have suggested that a doubling of CO_2 concentration from recent recorded levels might raise global temperature by 3 to 4 °C. Such an increase in temperature would cause significant changes in the world's rainfall patterns, some of the ice covers to melt, raise the sea level and lead to the inundation of extensive coastal areas. However, there is no firm evidence to date of an increase in global temperature as a result of increasing CO_2 concentration in the atmosphere. It is certain that the dynamic equilibrium among the major carbon dioxide reservoirs of the biosphere has been disturbed. The effects of such disturbances are a matter of considerable and immediate concern.

EXAMPLE 10.2

Evaluate CO concentration in microgram per m^3 ($\mu g/m^3$) for a given g ppm of CO in air at 0 °C and 1 atm pressure.

Solution

The relation between $\mu g/m^3$ and ppm of a gas is given by

$$\mu g/m^3 = \frac{\text{ppm} \times \text{g mol mass} \times 10^3}{\text{L/mol}}$$

Here, gram molecular mass of CO = 12 + 16 = 28 g/mol
Volume of CO gas per mol at 0 °C and 1 atm pressure = 22.4 L/mol
Volume concentration in ppm = $9 \times 10^{-6} = \text{ppm} \times 10^{-6}$
After substitution in the above equation

$$\mu g/m^3 = \frac{9 \times 10^{-6} \times 28 \frac{g}{\text{mol}} \times 10^3\, L/m^3 \times 10^6\, \mu g/g}{22.4\, L/\text{mol}}$$

g ppm CO in $\mu g/m^3$ = 11 g 250 $\mu g/m^3$ = 11.25 $\mu g/m^3$

10.3.3 The Nitrogen Cycle

Nitrogen constitutes 79 percent of the atmosphere (Table 10.2). However, it cannot be used directly by most forms of life. It must first be "fixed" before it can be utilised by plants and animals. By fixation, nitrogen is converted into its chemical compounds. This is largely nitrates (NO_3) and ammonia (NH_3). The fixation of nitrogen takes place through both physiochemical and biological means. The latter is by far the bigger contributor, while biological fixation is limited to a few examples.

The nitrates are assimilated to form amino acids, urea and other organic residues in the producer, consumer and decomposer cycles. The amino acids and urea are then converted to ammonia through a process called "ammonification". Denitrifying bacteria convert the ammonia into nitrites to complete the cycle, then into nitrates and then back into gaseous nitrogen. In this way, the total amount of nitrogen fixed equals the total amount returned to the atmosphere as gas.

Human beings on planet Earth have interfered with this natural cycle by industrially fixing nitrogen. This includes production of nitrogen fertilisers and oxidation of nitrogen during combustion of fossil fuel. Most of the excess nitrogen (N) is carried off into rivers and lakes and ultimately reaches the ocean. This increases the runoff of nitrogen, which has greatly increased the productivity in many aquatic environments and has contributed to the process of eutrophication.

10.3.4 The Phosphorus Cycle

Phosphorus plays an important role in the growth of living tissue in the metabolic processes of energy transfer. It is quite different from other major elements of the compounds at normal temperatures and pressures in biosphere. It does not form gaseous compounds at normal temperature and pressures and hence, it cannot return to the atmosphere. The main reservoirs of phosphorus are rocks and natural phosphate deposits on Earth. Rain and other natural processes cause phosphorus to be released to the soil. Much of it is fixed in the soil or absorbed onto soil particles. Some of it is lost to the water bodies such as lakes and streams and eventually ends up in the ocean. The phosphorus is deposited partly in shallow sediments and partly in the deep zone.

The land plants take the inorganic phosphate salts from the soil. It converts them into adenosine triphosphate (ATP) and adenosine diphosphate (ADP). These are utilised by the plants as energy carriers for their metabolic reactions. The organic phosphates are transferred to consumers. Subsequently, these are made available as inorganic phosphates for recycling *via* bacterial decomposition.

The inorganic phosphates are leached from the land into freshwater systems. These are taken up rapidly by phytoplankton and are converted to organophosphates. This phosphorus is injected by zooplankton. In turn it is consumed by other organisms. After the death of the organisms, the phosphates are released into the water by bacterial decomposition.

Some of the phosphorus is not deposited in the sediments. It is recirculated by upwelling, which brings the phosphates from unlit depths to the photosynthetic zone. It goes through phytoplankton, zooplankton and animal stages. The zooplankton may excrete as much phosphorus as is stored in their bodies. In doing so, they are instrumental in keeping the cycle running. Of the excreted phosphorus, more than half is in inorganic form. The rest is in the form of organic compounds. It is not certain whether the phytoplankton can directly utilise the organic phosphorus bacterial degradation.

Some of the phosphorus from the ocean is returned to the land through fish harvesting. It is also through the excreta of fish-eating birds. However, it is the fact that more phosphorus is being lost to the depths of the ocean than is being added to land and freshwater systems. This net one-way displacement of phosphorus reserves from land to ocean is of some concern due to the short supply of phosphorus as a nutrient. Hence, it is known as a growth-determining nutrient.

Lack of phosphorus in the soil causes soil infertility. The lost phosphorus, synthetic fertilisers are now being frequently used. The phosphorus cycle is substantially affected due to the accelerated use of fertilisers and synthetic detergents. Phosphorus pollution has contributed to the eutrophication of many water bodies. It may also have an adverse effect on the natural food chains.

10.3.5 The Sulfur Cycle

Sulfur, is a basic constituent of proteins in plants and animals like nitrogen. It is found in a wide variety of forms in the biosphere. Sulfur dioxide (SO_2) and hydrogen sulfide (H_2S) are the important gaseous forms and the sulfate ion (SO_4^-) is the common form found in water and soil.

Some sulfate are reduced directly to sulfides under anaerobic conditions, including H_2S or to elemental sulfur by a class of bacteria. This is known as Desulfovibrio bacteria. It is found largely at the ocean bottom. The hydrogen sulfide thus produced escapes as a gas into the atmosphere. It replenishes the sulfur lost by precipitation. H_2S is rapidly oxidised to sulfates by bacteria of genus Thiobacillus in the presence of oxygen. Bacteria such as Chlorobacteriaceae and Thiorhodaceae oxidise H_2S to elemental sulfur in the absence of oxygen.

Atmosphere (biosphere) receives sulfur (S) through bacterial emission (H_2S), fossil fuel burning (SO_2), wind-blown sea salts (SO_4^{2-}) and volcanic emissions (H_2S, SO_2, SO_4^{2-}) *etc.* Most of the sulfur (S) in the form of SO_2 or H_2S is converted to sulfur trioxide (SO_3). The SO_3 dissolves in water droplets to form sulfuric acid. The sulfates and the acid then precipitates with rain causing adverse ecological effects.

The sulfur cycle is overloaded due to burning of fossil fuels at an increasing rate. As a result, the SO_2 emitted into the atmosphere constitutes a significant fraction of total global sulfur transport.

The balanced state of natural cycles occurring in the environment is very often disturbed in every energy generation and transmission method. Conventional energy (fossil fuel) generating power damages to a great extent, the air, climate, water, land, wildlife, landscape *etc*. However, renewable energy technologies are safer in offering a solution to many environmental problems. Sometimes, their wide scale deployment creates negative environmental implications. These potential problems seem to be a strong barrier for further dissemination of renewable energy technologies in many cases.

In view of these problems, the present chapter presents an overview of the environmental impact assessment of various renewable energy technologies. The analysis provides the potential burdens to the environment during construction, installation and demolition of renewable energy technologies. These include noise and visual intrusion, greenhouse-gas emissions, water and soil pollution, energy consumption, labour accidents, impact on archaeological sites or on sensitive ecosystem, negative and positive socioeconomic effects.

10.4 ASSESSMENT OF IMPACTS OF RENEWABLE ENERGY TECHNOLOGIES ON THE ENVIRONMENT

The global attention has always been focused on the impact of conventional energy sources on the environment. In contrast, the renewable energy sources have enjoyed a "clean" image for sustainable environment. The only major exception to this general trend is the large hydropower projects; experience has taught us that they can be disastrous for the environment. The belief now is that minihydel and microhydel projects are harmless alternatives.

The present section discusses impacts of renewable energy sources on the environment.

10.4.1 Solar Energy

Solar energy is the source of all renewable energy sources. The use of direct solar energy-based systems appears the easiest and cleanest means of tapping renewable energy. The solar systems do not generate air pollution in environment during the operation. The primary environmental, health and safety issues are how they are manufactured, installed and ultimately disposed of.

Direct conversion of solar radiation into utilisable energy can be accomplished in many ways. These include solar thermal systems, photovoltaic systems and solar thermal power plants.

(i) **Solar Thermal Systems**

The solar thermal systems (Chapter 2) operating in medium-temperature ranges has no adverse effect on the environment during operation. It needs only maintenance during operation over its lifetime. The environmental impacts for solar thermal systems have been described in the following aspects. The environment has been affected before use of the system due to embodied energy involved in preparing the materials for development of solar systems (Chapter 2)

(a) **Land Use**

In the case of hot water/air or space heating/cooling, *etc*., no land will be required. The system will usually be installed on the roof of the existing building. The land-use requirements of concentrating collectors providing process heat is also problematic. There occurs the loss of habitat and the change of ecosystem due to land use in the case of large-scale systems.

(b) **Health Hazards**

The solar thermal system has no health hazards and accidental problems due to medium operating temperature range, say up to 100 °C. However, the performance of a solar thermal system is adversely affected due to particulate presence in the environment.

(ii) **Photovoltaic System**

The environmental impacts of the photovoltaic systems are as follows:

(a) **Land Requirement**

The land for PV (photovoltaic) systems should be situated in areas receiving high solar radiation and wind velocity. The land should be inexpensive and unfertilised. The lands are not usable for agriculture or do not have a forest cover. Further, it is required to locate the PV systems not too distant from population centres in order to reduce transmission/distribution losses and expenses on installing transmission lines. It is therefore ideal to install PV system in arid regions.

(b) **Large Materials Requirements**

There are large requirements of the materials used for PV modules and for structural support of PV systems (BOS). It has been estimated that central photovoltaic-based systems require large quantities of exotic inputs, which are toxic and/or explosive such as cadmium sulfide.

(c) **Health Hazards**

Materials used in PV systems (Chapter 2) create health and safety hazards for workers coming in contact with them. The manufacture of solar (photovoltaic) cells often requires hazardous materials such as arsenic and cadmium. Relatively inert silicon is a major material used in solar cells; it is hazardous to workers if it is breathed in as dust. Workers involved in manufacturing photovoltaic modules and components must consequently be protected from exposure to these materials.

(iii) **Solar Thermal Power Plant**

The environmental impacts of solar power plants operating at very high temperature ($>500\,^{\circ}C$) can be classified as follows:

(a) **Land Use**

The large amount of land is required for solar thermal power plants. One square kilometer for every 20–60 MW is required. This poses the problem, especially where wildlife protection is a concern. Most of the sites used for solar thermal power plants are in arid desert areas.

(b) **Construction**

Solar thermal power plant projects have the usual environmental impacts such as landscape, effects on local ecosystems and habitats, noise and visual intrusion, and temporarily pollutant emissions.

(c) **Water Resources**

Parabolic-trough (Chapter 2) and central-tower systems using conventional steam plant to generate electricity require the use of cooling water. This creates a problem for water resources in arid areas. There may be some pollution of water resources through thermal discharges and accidental release of plant chemicals used in heat exchangers.

(d) **Health Hazards**

The accidental release of heat transfer fluids (water and oil) from parabolic-trough and central-receiver systems form health hazards. The hazard may be substantial in some central-tower systems. In this case they use liquid sodium or molten salt as a heat-transfer medium. A fatal accident can occur in a system using liquid sodium. These dangers may be avoided by moving to volumetric systems with the use of air as a heat transfer medium. Central-tower systems have the potential to concentrate light to intensities that could damage eyesight. Under normal operating conditions, this should not pose any danger to operators. But failure of the tracking systems could result in stray beams that might pose an occupational safety risk on site.

The above discussions can be summarised as follows:

(a) It is important to recognise the substantial costs, hazardous wastes and land-use issues associated with solar technologies.
(b) The production of materials for solar energy systems involves hazardous substances. It must be handled cautiously to avoid environmental damage.
(c) Permanent use of large land areas and no reclamation until the plant is decommissioned.
(d) Generation of nonrecyclables during decommissioning fibreglass, glass, coolant, insulations. Additional disposal problems are caused by cadmium and arsenic used in PV-based systems.
(e) Hazard to eyesight from reflectors.
(f) Impact on water resources (for cooling of steam plant) and water pollution due to thermal discharges or accidental discharges of chemicals used by the system.

10.4.2 Hydropower Plant

Major ecological impacts are caused by hydropower projects in all the four habitats associated with the projects. These are the reservoir catchment, the artificially created lake, the downstream reaches of the dammed driver and the estuary into which the river flows.

The environmental stresses are caused by (i) altered timing of river flow, (ii) increased evapotranspiration and seepage water losses, (iii) barriers to aquatic organism movement, (iv) thermal stratification, (v) changes in sediment loading and nutrient levels and (vi) loss of terrestrial habitat to artificial lake habitat. The nesting, mating and other behaviour of riparian organisms are affected as a result of altered river flow and barriers to movement. Impounding and increased human activity in the reservoir catchment leads to deforestation and loss of wildlife. There is often an increase in the incidence of waterborne diseases. Above all, the damming is associated with serious problems of rehabilitation of persons living in the reservoir area. The environmental aspects of hydropower systems have been described as:

(a) **Population Movements**
Large hydropower plants require large reservoir and discharge areas. Many people have to be evacuated to make room for hydropower plants. This leads to a completely new situation for people who have lived in a relatively small and protected environment. The housing, land distribution, working conditions and way of life of affected people change radically. Social consequences are also likely to felt by dislocated people if the population concerned should be pressured into settling down in ecologically vulnerable areas. The sociocultural conditions together with their traditional connection to land, water and other natural resources tend to make them unadaptable to changes and new activities.

(b) **Health Hazards**
Large hydropower plants can increase the extent of water-related diseases to people leaving nearby (like cholera, dysentery and several tapeworm and round worm diseases). The reservoir may improve the living and breeding conditions of disease-causing organisms namely pathogens and their intermediate hosts. Reservoirs with a large volume of stagnant water offer favorable living conditions to pathogens. If the reservoir is employed for irrigation, industrial and drinking water supply, there may be the risk of infection spread by pathogens living in the water. Such infection may spread over large areas.

(c) **Fish Culture**
The composition of fish species gets altered. Since reproduction for some species may be hindered if the operation involves changes in the water level during the spawning period.

Artificial reservoirs tend to contain a less varied composition of species than a natural lake. This is due to the fact that changes in the water flow and water-flow pattern significantly alter nutrient and spawning conditions downstream. The primary production as well as the direct accessibility of nutrient for fish also changes. A surfeit of gas (nitrogen) may occur at dam and turbine outlet, which can cause death to fish.

(d) **Flora and Fauna**
The changes in the fauna and vegetation beyond the watercourse take place due to submerging and water-flow changes. Large reservoirs exert a considerable direct impact on the flora and fauna through submerging the area permanently or periodically.

(e) **Groundwater**
The groundwater level is important for the ecosystem's composition to develop plant and animal species. Groundwater is particularly important as a drinking water source in many developing and underdeveloped countries. The filling of a reservoir of a hydropower plant and the flow of a watercourse reduces the groundwater level in the surrounding area. These areas may influence the quality of the underground water and the sediment transport of the watercourse due to runoff and erosion.

(f) **Excessive Fertilisation**
Due to trapped nutrients in a reservoir, there are possibilities of excessive fertilisation eutrophication in the reservoir. This may cause an increase in the growth of algae or large amounts of higher-order aquatic plants, which requires excess biological oxygen demand. This may threaten the survival of some fish species. Evaporation also causes a concentration of nutrients leading to excessive fertilisation or eutrophication.

(g) **Transport of Nutrients**
A reservoir serves as a trap for nutritious elements and mud flowing in. It possibly leads to a reduction of the total transport of nutrients downstream. Also, the annual variations in supply downstream reduce the biological productions. The marine fishing is also impaired due to a major dam development.

10.4.3 Wind Energy

Wind energy is one of the renewable sources of energy without air or water pollution. It involves no toxic and hazardous substances and poses no threat to public safety. It has concern over the visibility and noise of wind turbines and their impacts on wilderness areas. Some of environmental impacts of wind energy are as follows:

(a) **Land Use**
The wind turbines need to be spread over a wide range area to collect large amounts of energy from the wind. Each is positioned not to interfere with another turbine. Spacing is particularly important to large wind farms. The turbines are typically separated by distances of five to ten rotor diameters. The total land areas used by a wind farm are for foundations, access roads and substations. It is typically around 1% of the dispersed land area of a wind farm. The rest can be used for other purposes. Wind power development is more ideal to farming areas. Wind power development also creates serious land-use conflicts in forested areas. It is required to clear many trees causing a heavy monetary burden. Wind projects often run into stiff opposition from people who regard them as noisy and who fear that their presence may reduce property value near populated areas.

(b) **Visual Impact**
The colour, pattern, shape, rotational speed and reflectance of blade materials can be adjusted to avoid the visual effects of wind turbines including the landscape. Many of the best sites for

wind turbines are in areas of outstanding natural beauty. Well-designed wind turbines can be aesthetically pleasing to the eye like, suspension bridges and some modern buildings. They should be more attractive than microwave transmission towers and television aerials.

(c) **Noise**

There are two principal sources of noise from wind turbines:

(i) **Mechanical sources:** such as from the gearbox, generator and auxiliary motors and

(ii) **Aerodynamic source:** from the blades due to passage of the air. Generally, noise is most audible close to the turbine at low speeds. As the wind speed increases, the background noise from the wind in the trees, grass and bushes tends to dominate. Revolving blades generate noise to be heard in the immediate vicinity of the installation. However, noise does not travel too far.

Some of the noise is of infrasound generated by wind turbines at frequencies below the audible range. This infrasound may cause houses and other structures to vibrate. These low-frequency waves can be eliminated in new buildings after careful design considerations.

(d) **Electromagnetic Interference**

Wind turbines can interfere with the television and microwave signals through reflection. This is caused by the rotor blades of turbine that intercept the television beam. The tower structure also scatters the beam. This can be easily avoided by careful siting and modification of transmitter/receivers.

(e) **Survival of Birds and Wildlife**

The impacts of wind farms on local bird populations have serious concerns. The large numbers of birds might fly into the spinning rotor blades and be killed. Analysis of a 1.5-MW coastal wind farm in the Netherlands concluded that the wind turbines were far less detrimental to birds than a high-voltage electric power line. It is comparable to 1 kilometer of road. Local birds appear to recognise the existence of a wind farm and avoid it. Care should be taken not to site a farm on a sensitive migration route. The effect upon other wildlife is likely to occur during construction. The impact can be minimised with the appropriate level of care. Certainly, the effects of wind turbines on the life of birds are much less in comparison to hunters, overhead electricity cables and road vehicles.

(f) **Human Safety Hazards**

Human safety concerns with wind energy generators are similar to those in the building industry. These include the risk of falling from high buildings during construction and repairs or the failure of parts through fatigue or design. Turbine blades sometimes fail with no serious accidents. Large wind machines are generally located in sparsely populated areas and the risk of human accident is not great. More accidents occur with the large number of small wind machines located within higher population densities.

10.4.4 Biomass Energy

Biomass has both favorable and adverse effects environmentally as a source of renewable energy. It serves as a sink for atmospheric CO_2 and contributes to soil fertilisation and soil stabilisation. It also improves the physical properties of the soil as well as reduction of water runoff and desertification. The deleterious effect of SO_2 in the atmosphere due to low-sulfide fuel is avoided. It is economically an inexpensive source of energy for the rural poor in developing countries. It minimises acid rain.

The conversion of organic wastes into fuel reduces the environmental hazards associated with the disposal of these wastes. If biomass is not properly managed through overgrazing of pastures, over-cutting of trees, slash and burn agricultural practices *etc.*, then soil erosion, desertification and other related detrimental environmental effects occur. Biomass power derived from biomass raises more serious environmental issues than any other renewable resource of energy. Combustion of biomass and biomass-derived fuels produce significant air pollution. There are also concerns about the impacts of using land to grow energy crops. The environmental impacts of biomass energy are as follows:

(a) **Land and Water Resources**
Implementation of biomass energy production programmes requires large amounts of water resources and land. Horticulture is a massive water-consuming activity. It requires more water by several orders of magnitude for every hectare of land. It is needed more in comparison with domestic and industrial uses. It also contributes to underground water pollution through the pesticides and fertilisers. Biomass is inevitably needed in sustaining any intensive cultivation. The land used for energy would have to compete with crops, forests and urbanisation.[3] The competition requires comparison between the crop land needed to feed one person and fuel required per automobile for one year. If an average automobile travels 16 000 km per year and gets 6.2 km/l then 2581 l of petrol is required. Using ethanol, the total in equivalent kcal would be 3875 l. Assuming a zero energy charge means no high-grade fuel used, for the fermentation/distillation processes, then 4.2 ha of land would be required. This is known that about 0.5 ha of land is used to feed each person, which means more than eight times land is required to fuel an automobile than to feed one person.
The removal of biomass from land and underground water may increase soil and water degradation, flooding and removal of nutrients.[4] This might also affect wildlife and the natural ecosystem.

(b) **Soil Erosion and Water Runoff**
It creates soil erosion problems. Although the use of available technologies can minimise erosion they are costly to implement. Producing energy crops such as corn for ethanol requires additional agricultural land. Marginal cropland highly susceptibility to soil erosion would have to be brought under corn cultivation for the purpose. Soil erosion contributes significantly in hastening water runoff. This implies retarding groundwater recharge. The nutrient-rich runoff may harm the quality of receiving rivers, lakes or estuaries due to excessive fertilisation or eutrophication.

(c) **Nutrient Removal and Losses**
Significant nutrient loss is incurred by the harvesting of energy crops. With a corn yield of 8000 kg/ha, 224 kg (N). 37 kg (P), 140 kg (K) and 6 kg (Ca) are the nutrients contained in both grain and residues.[5] Thus, nitrogen as well as other nutrients must be replaced for each subsequent crop. The amount of energy needed to replace the nutrients lost would be the equivalent of at least 46 l of oil per hectare.

(d) **Loss of Natural Biota, Habitat and Wild Life**
The habitat and food sources of wildlife and other biota are changed due to conversion of natural ecosystem into energy-crop plantations.[3] Alteration of forests and wetlands for biopower reduces many habitats and mating areas of some mammals, birds and other biota.
Monoculture plantations of fast-growing trees reduce the diversity of vegetation. It also reduces the value of the areas as habitats for many wildlife species. These monocultures are less stable due to increased energy inputs in the form of pesticides and fertilisers to maintain productivity. Trees in profitable plantations are 2–3 times as dense as those of natural forests. The high stand density may result in greater pest problems.

(e) **Problems in Biomass-Conversion Technologies**
Production of biomass is only one dimension of biopower. Several technologies are available for biomass conversion (Chapter 4). The most widely used are direct combustion and pyrolysis. The impacts of conversion technologies are as follows:

a. air pollution-emissions of particulates (carbon oxides (CO_2), sulfur oxides (SO_2) and nitrogen oxides (NO_x));

b. organic emission of dioxin hydrocarbons, toxic irritants (acid, aldehyde, phenol and benzopyrene, *etc.*);
c. generation of ash and fly ash containing toxic substances with accompanying pollution problems;
d. water pollution (biological oxygen demand, chemical oxygen demand, suspended solids and trace metals, *etc.*);
e. household hazards (accidental fires);
f. occupational hazards (prolonged exposure to toxic and corrosive chemicals).

All the problems of air pollution associated with conversion of biomass to energy are no less important in the conversion of coal and oil. Even these are significant at the very small scale of residential wood burning. The smoke has harmful levels of toxicants. Also in terms of a million kcal output, forest biomass causes only several injuries and illnesses, in comparison to the use of coal and oil mining.

(f) **Social and Economic Impacts**

The major social impacts are (i) the shift in employment, (ii) increase in occupational health and (iii) safety problems. The employment is expected to increase due to meeting the nation's energy demands by biomass resources. The labour forces are required in cutting and harvesting agricultural as well as forest products. They are also required for transport of biomass resources and in the operation of conversion facilities. The requirements of direct labor inputs are 2–3 times greater per million kcal for wood biomass resources than coal. A wood-fired steam power plant requires 4 times more construction workers than a coal-fired plant. They also require 3–7 times more plant maintenance and operation workers. The labour requirement is 18 times more in million kilocuries of ethanol than an equivalent amount of petrol. The possibilities of an increased employment are greater occupational hazards. Significantly, there are more occupational injuries and illnesses with biomass production in comparison with either/coal, oil or gas recovery operations. Agriculture reports 25% more injuries per man-day than all other private industries.

(g) **Price of Biomass**

Food and forest products have a higher economic value (per kcal in their original form) than is converted into either heat, liquid or gaseous form. For example, when one million kcal of corn grain is converted to heat energy then its market value is reduced eight times. Producing liquid fuels (*e.g.* ethanol) is also expensive. For the grain, one gets 20% less when the grain is converted to ethanol.

(h) **As Fuel**

In resource-abundant regions, firewood is obtained from local sources where live trees are felled and allowed to dry for combustion. This puts pressure on trees, bushes and shrubs. This effect is much more serious in resource-scarce areas. The removal of the more easily felled younger trees reduces the regenerative ability of the forest. Excessive removal of too many trees renders the forest susceptible to damage from wind and the Sun. It affects wildlife generation. The removal of residues removes the nutrients that should return to the soil to maintain its fertility. Similarly, the removal of stumps, bushes and shrub destroys the soil's protective cover and binding structure. Finally, whole forests may disappear.

The use of fuel wood directly in the homes is a very serious source of air pollution. It is also a major health hazard for women and children exposed to this pollution for significant lengths of time. It has been reported that the emissions of air pollutants such as carbon monoxide, sulfur dioxide, nitrogen oxides, organics and particulates are much larger from the burning of biomass.

10.4.5 Geothermal Energy

The extraction of large volumes of steam and water are from geothermal heat. The only type of geothermal energy developed is hydrothermal energy. It consists of trapped hot water or steam. However, new technologies are being developed (i) to exploit hot dry rock by drilling deep into rock, (ii) geopressured resources (pressurised brine mixed with methane) and magma (Chapter 8).

The various types of geothermal resource differ in many respects and they raise a common set of environmental issues. Air and water pollution are concerns along with the safe disposal of hazardous waste, siting and land subsidence. Geothermal fluids have a chemical content that is site specific and depends on the rocks of each reservoir. The major environmental issues of geothermal energy are as follows:

(a) **Air Pollution**

Water/steam from major geothermal fields has a content of noncondensable gases (CO_2, H_2S, NH_3, CH_4, N_2 and H_2). It ranges from 1.0 to 50 g/kg of steam. Carbon dioxide (CO_2) is the major component. Its emission into the atmosphere is well below the figures for natural gas, oil or coal-fired power stations per kWh generated.

Hydrogen sulfide (H_2S) is the air pollutant of major concern. Its emissions generally range between 0.5 and 6.8 g/kWh. The H_2S is oxidised to sulfur dioxide and then to sulfuric acid, which may cause acid rain.

Geothermal gases in steam also contain ammonia (NH_3), traces of mercury (Hg), boron vapours (B), methane (CH_4) and radon (Rn). Boron, ammonia and mercury are leached from the atmosphere by rain leading to soil and vegetation contamination. In particular, boron may have a serious impact on vegetation. These contaminants also affect surface water and aquatic life. Mercury (H_g) emissions from geothermal power plants range between 45 and 900 μm/kWh. This is comparable with mercury emissions from coal-fired power plants. Ammonia is discharged into the atmosphere in concentrations between 57 and 1938 mg/kWh due to atmospheric processes, but it is dispersed rapidly.

Radon (^{222}Rn), a gaseous radioactive isotope naturally present in the Earth's crust is contained in the steam. It is discharged into the atmosphere in concentrations of 3700–78 000 Becquerel/kWh.[6]

(b) **Water Pollution**

Water pollution in power production is a potential hazard for rivers and lakes. Most of the pollutants are found in the vapour state in vapour-dominated reservoirs. The pollution of water bodies is more easily controlled than in water-dominated reservoirs. Waste-steam condensate (20% of the steam supply) must be added to the waste water. The water and the condensate generally carry a variety of toxic chemicals (arsenic, mercury, lead, zinc, boron and sulfur, together with significant amounts of carbonates, silica, sulfates and chlorides).

Water and steam are separated at the surface in water-dominated and in hot-water reservoirs. The steam is used for the generation of electricity. The volume of water is disposed of. It may contain large quantities of salts (above 300 g/kg to 70 kg/kWh), which is more than four times the steam supply. It is up to 400 kg/kWh in binary cycle plants.

(c) **Land Subsidence**

The weight of the rocks above a reservoir of groundwater, oil or geothermal fluids is borne by the mineral skeleton of the reservoir rock and by fluids in the rock pores. As fluids are removed, the pore pressure is reduced and then the ground tends to subside. Less subsidence is expected with harder reservoir rock.

Water-dominated fields subside more in comparison with vapour-dominated fields. The withdrawal of huge quantities of underground fields causes substantial ground subsidence from a wet field. Subsidence can be controlled or prevented by the reinjection of spent fluids. However, reinjection may induce microseismicity.

(d) **Induced Seismicity**

Many geothermal reservoirs are located in geologically unstable zones of the Earth's crust at high temperature. These are the zones characterised by volcanic activity, deep Earthquakes and a heat flow. They are also zones with a higher frequency of naturally occurring seismic events. Seismic activity may be induced by water reinjection into the reservoir by reducing rock stress, loosening vertical faults and triggering the release of accumulated tectonic stress.

(e) **Noise**

A noise level of 90–122 dB at free discharge and 75–90 dB through silencers during maintenance are observed. Exhausts, blowdowns and centrifugal separation are some of the sources of noise. This necessitates the installation of silencers on some equipment. The noise causes serious health hazards. Workers on well sites have to wear earplugs and muffs to guard against hearing problems.

(f) **Escaping Steam**

Huge volumes of flash steam escaping into the air could cause dense fog to occur. This may drift across to near by roads causing traffic hazards. The proportion of escaping steam should be maintained.

10.4.6 Ocean Thermal Energy Conversion (OTEC)

OTEC poses some potential environmental threats on a large scale compared to traditional power plants. OTEC power plants have the potential for major adverse impacts on the ocean water quality. These plants affect the surrounding marine environment due to heating of water, the release of toxic chemicals, impingement of organisms on intake screens, entrainment of small organisms by intake pipes. These are concerns for OTEC systems (Chapter 9).

The OTEC plants displace about 4 cubic meter of water per second per MW electricity output. This is from the surface layer as well as from the deep ocean. It discharges them at some intermediate depth between 100 m and 200 m. This massive flow may disturb the thermal structure of the ocean near the plant. It changes salinity gradients and change the amounts of dissolved gases, nutrients, carbonate sand turbidity. These changes may have adverse impacts on the marine environment.

A large discharge (mixed warm and cold water) near the surface creates a plume of sinking cool water. The continual use of warm surface water and cool deep water may over long periods of time lead to slight warming at depth and cooling at the surface. Thermal effects may be significant, as local temperature changes of only 3–4 °C are known to cause high mortality among corals and fishes. Other effects such as reduced hatching success of eggs and developmental inhibition of larvae may result from thermal changes that lower reproductive success.

Cold-water effluent is used for the cultivation of algae, crustaceans and shellfish. In the nutrient-rich water, unicellular algae grow to a density 27 times greater than the density in surface water. They are consumed by filter-feeding shellfish such as clams, oysters and scallops. However, the abundance of nutrients in aquatic ecosystems can spell serious trouble that leads to eutrophication and all the adverse consequences associated with eutrophication.

Toxic chemicals (ammonia and chlorine) from OTEC plants may enter the environment, which will kill local marine organisms. If the working fluid is ammonia and it leaks out, there could be serious consequences to the ocean ecosystem nearby. The use of ammonia in closed-cycle systems would be designed to avoid the environmental contact. A dangerous release would be expected to result only from serious malfunctions such as major breakdown, collision with a ship, major human error, *etc.*

Marine biota may be impinged on the screens covering the warm and cold water intakes of an OTEC plant. Small fishes and crustaceans may be entrained through the system. They experience rapid changes of temperature, salinity, pressure, turbidity and dissolved oxygen. A major change occurring in the cold water pipe is the depressurisation of water coming from a depth of 1000 m to the surface (up to 10^7 Pa).

Sea-surface temperatures could be lowered by the discharge of cold water from the plant in the vicinity of an OTEC plant. This produces impacts on organisms and the microclimate. The pumping of large volumes of cold water from depth of the ocean to the surface releases dissolved gases such as carbon dioxide, oxygen and nitrogen to the atmosphere. This affects water pH and DO (dissolved oxygen) status, causing stress to marine life.

Biocides (chlorine) is used to prevent biofouling of the pipes. Heat-exchanger surfaces may be irritating or toxic to organisms in the ocean ecosystem.

Appropriate measures need to be taken to control the environmental impact before the viability of OTEC can be assessed.

10.4.7 Tidal Energy

The effects of a tidal power system on the environment depend upon the location of the scheme. These effects need to be examined closely before installation of tidal schemes is undertaken. This may result from changes in water levels, changes in water flows and velocities, changes in sedimentation, the physical presence of the barrage and construction activities close to the site. The impacts of tidal plant on environment include such as on ports and navigation; facilities for recreation; water quality; agriculture and land drainage; visual amenity; bird life; migratory fish; the ecosystem in the estuary, *etc.*

Tidal fences and tidal turbines are likely to be more environmental friendly. Tidal fences may have some negative environmental impacts due to blockage of channels that makes it difficult for fish and wildlife to migrate through those channels.

It is also feared that tidal power plant hampers the other natural uses of estuaries such as fishing or navigation.

10.4.8 Wave Energy

Small-scale wave energy plants are likely to have minimal environmental disturbances. However, the very large-scale projects have the potential for harming ocean ecosystems. Covering very large areas of the surface of the ocean with wave energy devices harms the marine life. It produces more widespread effects by altering the way the ocean interacts with the atmosphere. Some of the environmental impacts due to wave energy plants are described as follows:

(a) **Disturbances of Marine Life**
Wave energy plants may have a variety of effects on the climate. This also influences the shore and shallow subtidal areas and the communities of plants and animals. Potential impacts include disturbance or destruction of marine life. This includes changes in the distribution and types of marine life near the shore. Installation of support structure and cable laying for wave energy plants may temporarily interrupt marine life. However, the ecology is likely to recover. Installation of the wave energy plant itself causes a disturbance to local mammal's seals and dolphins. The timing of the installation needs to be chosen carefully.

(b) **Visual Impact**
The visual impact of a wave energy conversion system depends on the type of device and its distance from the shore. In general, a floating buoy system or an offshore platform has much visual impact. Shoreline and near-shore devices have a visual impact. Onshore facilities

and offshore platforms in shallow water change the visual landscape from the natural scenery to an industrial landscape.

(c) **Noise**
Wave energy conversion systems make some noise. The levels are expected to be below the levels of a normal ship. They are expected to be noisier than the surrounding wind and waves with full operation of plant. Any noise that is generated can travel long distances underwater, which can have an impact on certain animals such as seals or whales.

(d) **Threat to Navigation**
The wave energy conversion devices may be a dangerous obstacle to any navigational craft. It cannot see or detect them by radar or by direct sighting. The use of adequate navigational aids on offshore wave energy conversion devices is required.

(e) **Coastal Erosion**
Some energy conversion devices concentrate wave energy into a tapered area. This concentrating surge device for large channel waves to increase wave height for redirection into elevated reservoirs. The water from elevated reservoir then passes through hydroelectric turbines on the way back to sea level and hence generates electricity. Continuous arrays of such onshore or shore-based wave energy devices physically alter coastlines.

10.5 REMEDIAL MEASURES

The renewable energy sources are to some extent safer than the pollution-generating nonrenewable ones. It is required to generate awareness and to take preventive measures towards the adverse environmental impacts of various renewable energy options. A balanced view of their virtues (already well known) as well as shortcomings (not so well known) is needed. The sad euphoria-turned-despair history of hydel power projects may not be repeated.

We have to achieve the following objectives:

(a) help us to do elaborate environmental planning based on "preventive adverse impact assessment" before installation of any renewable energy system;
(b) help us in rational site selection for a renewable energy project to ensure the sustainable environment and minimum adverse impacts;
(c) help us in generating awareness towards the niches of various renewable energy systems.

In the past, major natural resource development activities were taken with only the benefit in mind. For example, in the 1960s, India and some other countries were swept by the "green revolution". During which high-yielding dwarf varieties of plants and intensive agricultural practices were used on a very large scale to produce massive stocks of food grains. This enabled food-deficient countries particularly India to have a food surplus during a period of 3–4 years. During these periods the scientists who worked for the green revolution were heralded as "messiahs". Then, as years passed and the adverse impacts of intensive agriculture began to surface in the form of waterlogging, salinisation, depleted soil productivity and pollution. The very same scientists who were lionised earlier were made targets of public ire and ridicule.

Similarly, Eucalyptus initially, was considered as the "wonder tree" in India. It grew easily and rapidly and each cell of it was utilisable. Some even saw it as embodiment of the mythical "Kalp Vriksha"; the tree capable of giving everything one wants. After a few decades, eucalyptus has become the favourite object of attack for environmental activists. It has been called the enemy of water, soil, wildlife and all other constituents of the environment.

In both instances, the fault did not lie with the instruments of change, *i.e.* green revolution/eucalyptus, there was a lack environmental impact forecasting before large-scale use. A similar fate should not befall renewable energy sources, which can contribute towards true environment-friendly use.

OBJECTIVE QUESTIONS

10.1 The environment is also known as the
 (a) Atmosphere
 (b) Biosphere
 (c) Troposphere
 (d) All of them

10.2 The hydrological cycle is
 (a) Directly related to the carbon cycle
 (b) Directly related to the nitrogen cycle
 (c) Directly related to the sulfur cycle
 (d) Indirectly related to the carbon cycle

10.3 The following gas has a maximum concentration in the atmosphere
 (a) CO_2
 (b) O_2
 (c) N_2
 (d) SO_2

10.4 The following gases have the minimum concentration in the atmosphere
 (a) CO_2
 (b) O_2
 (c) H_2S
 (d) SO_2

10.5 The biosphere exists due to presence of
 (a) Wind energy
 (b) Solar energy
 (c) Tidal and wave
 (d) None

10.6 Reason for acid rain is the pressure of the following gas in atmosphere
 (a) CH_4
 (b) H_2
 (c) NH_3
 (d) H_2S

10.7 The noncondensible gases in the stream of a geothermal field is
 (a) CH_4
 (b) H_2
 (c) NH_3
 (d) H_2S

10.8 The CO_2 gas in the stream of a geothermal field is a
 (a) Major contributor for polluting the environment
 (b) Minor contributor for polluting the environment
 (c) No effect for polluting the environment
 (d) A noncondensible gases

10.9 The OTEC is more responsible for polluting
(a) Biosphere ecosystem
(b) Ocean ecosystem
(c) Both (a) & (b)
(d) None of them

10.10 The tidal power plant hampers
(a) Fishing
(b) Navigation
(c) Both (a) & (b)
(d) None of them

ANSWERS

10.1 **(b)**; 10.2 **(a)**; 10.3 **(a)**; 10.4 **(c)**; 10.5 **(c)**; 10.6 **(b)**; 10.7 **(a)**; 10.8 **(a)**; 10.9 **(b)**; 10.10 **(c)**.

REFERENCES

1. Hartley, D. L. 1990. Perspectives on renewable energy and the environment. In: Tester, J. W., Wood, D. O., Ferrari, N. A. eds. *Energy and the Environment in the 21st Century*. MIT Press, Massachusetts.
2. Dincer, I. 1998. Renewable energy, environment and sustainable development. In *Proceedings of the world renewable Energy Congress*, 20–55 September, Florence, Italy, pp. 2559–2562.
3. Gadgil, M. 1993. *Ambio*, 22 (2–3), 167–172.
4. Tandon, H. L. S. 1995. *Recycling of Crop, Animal and Industrial Wastes in Agriculture*, Fertiliser Development And Consultation Organisation, New Delhi.
5. Pimental, D., Frqiend, C. Olson, L., Schmidt, S., Wagner, W. B., Johnson, K., Westman, A, Whelan, A. M. Fegalia, K., Poole, P., Klein, T., Sobin, R., and Bochner, A. 1983. Biomass energy: environmental and social costs. *Environmental Biology Report*, Cornell University Ithaca, New York. pp. 2–83.
6. Layton, D. W., Anspaugh, L. R. and O'Banion, K. D. 1981. Health and environmental effects, documents on geothermal energy – 1981. Lawrence Livermore Laboratory, California, Report No. UCRL-53232, p. 61.

The Carbon Credit Earned by Using Second Law Thermodynamics (Exergy Analysis) from Renewable Energy Sources Makes the System More Economical.

CHAPTER 11

Energy and Exergy Analysis

11.1 INTRODUCTION

Energy drives human life. It is crucial for continued human development, and for human life. There is a need for a secure and accessible supply of energy for the sustainability of modern societies. Now, energy has become an integral part of human life for almost every activity, *e.g.* domestic, transport, industrial, medical, *etc*. So, there is a need for energy security for sustainability of the growing world population. Continuation of the use of fossil fuels (conventional energy sources) is set to face multiple challenges namely (i) depletion of fossil fuel reserves, (ii) global warming and other environmental concerns, (iii) geopolitical and military conflicts, and (iv) continuing fuel price rise. These problems will create an unsustainable situation. Renewable energy is the only solution to the growing energy challenges/crisis. Renewable energy resources such as solar, wind, biomass and wave and tidal energy as discussed in various chapters, are abundant, inexhaustible and environmentally friendly. Bentley[1] has overviewed the global oil and gas depletion. It is reported that fossil fuels (conventional energy resources) are being exhausted through their uncontrolled harnessing of limited resources. The world relies heavily on fossil fuels to meet its energy requirements. Fossil fuels such as oil, gas and coal are providing almost 80% of the global energy demands.[2] On the other hand, presently renewable energy and nuclear power are, respectively, only contributing 13.5% and 6.5% of the total energy needs in the world. The enormous amount of energy being consumed across the world is having adverse implications on the ecosystem of the planet.

Fossil fuels are inflicting enormous impacts on the environment. Climatic changes driven by human activities cause the production of greenhouse gas emissions (GHG). This is directly impacting on the environmental condition. According to the World Health Organisation (WHO) as many as 160 000 people die each year from the side effects of climate change. This number could almost double by 2020. These side effects range from malaria to malnutrition and diarrhoea that follow in the wake of floods, droughts and warmer temperatures.[3]

With the exception of humans every organism's total energy demand is its supply of energy in the form of food derived directly or indirectly from the Sun's energy. For humans the energy requirements are not just for heating, cooling, transport and manufacture of goods but also those related to agriculture. Solar energy is a renewable, environmentally friendly, pollution free and freely available energy source on planet Earth. **Basically, solar energy is the direct source of all renewable and indirectly source of all nonrenewable sources.** In this perspective, over the last two decades solar energy systems have experienced rapid growth globally. However, energy and exergy analysis of

Advanced Renewable Energy Sources
G. N. Tiwari and R. K. Mishra

Published by the Royal Society of Chemistry, www.rsc.org

renewable energy systems can be used to estimate the environmental impact of different activities for producing materials, *i.e.* the more energy is required, the greater the environmental impact.[4]

From renewable energy systems (see previous chapters), broadly the following forms of energy can be obtained:

(i) thermal energy;
(ii) mechanical energy; and
(iii) electrical energy.

From solar energy, both thermal and electrical energy are achieved (Chapters 2 and 3). The mechanical and electrical energy can be obtained from biofuels (Chapters 4 and 5), hydropower systems (Chapter 6), wind energy conversion systems (Chapter 7), geothermal energy systems (Chapter 8) and ocean, tidal, waves and animal systems (Chapter 9). Energy analysis (first law of thermodynamics) can be applied for thermal and mechanical energy. In the case of electrical, exergy analysis is mandatory (second law of thermodynamics).

In view of the above, renewable energy technology has to meet the following two main criteria:

(i) cost effectiveness;
(ii) the maximum net annual energy and exergy yield.

The net maximum annual thermal energy and exergy yield for renewable energy systems means the net sum of either

(a) annual electrical energy output and equivalent annual exergy of thermal energy;
(b) annual equivalent thermal output of electrical and annual thermal energy; or
(c) annual direct sum of thermal and electrical energy if available thermal energy replaces the thermal energy generated by electrical energy.

The total energy requirement for manufacturing of renewable energy technology, energy matrices namely energy payback time (EPBT), energy production factor (EPF), lifecycle conversion efficiency (LCCE) and also CO_2 emissions will be evaluated in this chapter.

11.2 ENERGY MATRICES

In the recent past, developments in the design and manufacture of renewable energy technology have been very rapid. They are now predicted to become a major player to meet the energy needs of human beings for a good standard of living, particularly in developing and underdeveloped countries. The energy matrices are important for renewable technologies as their use makes no sense if the energy used in their manufacture is more than they can produce in their lifetime. The energy payback time is always be one of the criteria used for comparing the viability of one renewable technology against another. For example, the energy analysis of a PV module was conducted by Hunt[5] and it was reported that the energy payback time (EPBT) of a PV module is 12 years. The results reported by Hunt[5] are also in general agreement with those of Kato et al.[6] for crystalline silicon (c-Si) solar-cell modules. Aulich et al.[7] evaluated the EPBT for a crystalline silicon module and it was concluded that the EPBT is 8 years; in this case plastic materials were used for encapsulation for the Siemens C process. The energy payback time (EPBT) for a crystalline silicon (c-Si) solar-cell module under Indian climatic condition for annual peak load duration is about four years.[8] Lewis and Keoleian[9] predicted the energy payback time (EPBT) for an amorphous silicon (a-Si) solar-cell module with efficiency of 5% as 7.4 years for the climatic conditions of Detroit, USA; the EPBT gets reduced to 4.1 years with the increase in the efficiency of the module to 9%. Srinivas et al.[10] reported that the energy payback time for an amorphous silicon (a-Si) solar-cell module reduces to 2.6 years after considering gross energy requirement (GER) and the hidden energy. Battisti and Corrado[11] investigated the energy payback time for a conventional multicrystalline building integrated system, retrofitted on a tilted roof, located in Rome (Italy); with the yearly global insolation on a horizontal plane was taken as 1530 kWh/m^2 y. They concluded that the energy payback time gets reduced from 3.3 year to 2.8 year.

To reduce the energy payback time as mentioned in the case of PV module, the following considerations for other renewable energy technologies should be adopted:

(i) increase of efficiency;
(ii) use the cost effective materials with low energy densities for longer life;
(iii) with minimum annual maintenance.

11.2.1 Energy Payback Time (EPBT)

The EPBT depends on (i) the energy spent to prepare the materials used for fabrication of the system and its components. **This is referred to as embodied energy.** and (ii) the annual energy yield (output) obtained from such a system. To evaluate the embodied energy of various components of system, the energy densities of different materials are required. This will be discussed in detail in the next section. **The energy payback time (EPBT) is the total time period required to recover the total energy spent to prepare the materials (embodied energy) used for fabrication of renewable energy technologies.** It is the ratio of the embodied energy and the annual energy output from the system. It can be mathematically can be expressed as

$$\text{EPBT} = \frac{\text{Embodied Energy}\,(E_{\text{in}})}{\text{Annual Energy Output}\ (E_{\text{out}})} \tag{11.1}$$

The numerical values of EPBT should be as low as possible to make the renewable energy technologies cost effective to users.

11.2.2 Energy Production Factor (EPF)

This is used to predict the overall performance of the system. It is defined as the ratio of the output energy and the input energy or it can also be expressed as the inverse of EPBT.

(a) On Annual Basis

$$\chi_a = \frac{E_{\text{out}}}{E_{\text{in}}} \tag{11.2a}$$

or,

$$\chi_a = \frac{1}{T_{\text{epb}}} \tag{11.2b}$$

If $\chi_a \rightarrow 1$, for $T_{\text{epb}} \rightarrow 1$ the system is worthwhile, otherwise it is not worthwhile from an energy point of view.

(b) On Basis of Lifetime

$$\chi_a = T\frac{E_{\text{out}}}{E_{\text{in}}} \tag{11.2c}$$

In this case χ_a should be more than one. The value of χ_a should be a maximum for the cost effectiveness of the system.

11.2.3 Lifecycle Conversion Efficiency (LCCE)

The LCCE is the net energy productivity of the system with respect to the solar input (radiation) over the lifetime of the system, (T years). The expression for LCCE is given by

$$\phi(t) = \frac{E_{out} \times T - E_{in}}{E_{sol} \times T} \tag{11.3}$$

The numerical value of LCCE is always less then one. However, if the value of LCCE approaches one, it is the best technology from the energy point of view.

One can observe that embodied energy (E_i), annual electrical output (E_{out}) and life of renewable energy technology have very important roles in the evaluation of the energy matrices (Eqs. (11.1)–(11.3)).

11.3 EMBODIED ENERGY

The concept of embodied energy is a relatively new area of environmental assessment. It has started to be included in lifecycle energy calculations of any technology producing energy. **Embodied energy is defined as: "the quantity of energy required by all of the activities associated with a production process, including the relative proportions consumed in all activities upstream to the acquisition of natural resources and the share of energy used in making energy equipment and in other supporting functions, *i.e.* direct energy plus indirect energy".**[12] Thus, the aim of any embodied energy analysis is to quantify the amount of energy used to manufacture a material, product, component, and element. This involves the assessment of the overall expenditure of energy required to extract the raw material, manufacture products and components, build, and maintain the component element whichever is being assessed. A secondary aim is to establish the embodied energy required to construct and maintain the item, component or building over the whole lifecycle.

Like operational energy, embodied energy is an indicator of the level of energy consumption. Reducing energy consumption through better design with higher efficiency has been a goal of designers for many years. But the embodied energy portion of this consumption has largely been ignored. There are several reasons for this omission, including no clear assessment methodology, lack of data, lack of understanding and a common belief that the embodied energy portion of assets energy consumption is insignificant. However, over recent years, the methodologies for assessment have improved, data reliability and access has increased. Recent reports have indicated that the embodied energy portion may be as high as 20 times the annual operational energy of an office building.[12]

11.3.1 Embodied Energy Analysis

Embodied energy analysis involves identifying energy-consuming processes and calculating their contribution within the total product creation process. This usually involves several individual actions.

To be able to quantify the energy embodied in the construction of an asset, the quantities of materials must first be estimated through a process of desegregation and decomposition to a level of detail that allows for the separation of components into their principal materials. The energy intensities of each material can then be multiplied by the quantities of individual materials and the products aggregated to obtain the total for each material, element. In addition to the embodied energy value, other environmental indicators can also be calculated, such as CO_2 emissions. This is the basis of lifecycle cost analysis (LCA) work (Chapter 12).

11.3.2 Energy Density

Energy densities (intensities) are derived from energy analysis studies from various national and international sources. Among the difficulties encountered in using a wide variety of sources to verify values is the need to clarify definitions of system boundaries or whether the values are in terms of primary energy or delivered energy. To obtain an accurate and reliable database of energy intensities for all materials used in water assets is an enormous task in itself. It is a necessity for detailed comparisons of materials. The main requirement of embodied energy calculations at the design stage is obtaining accurate and useable material quantities and then combining them with currently available energy-intensity values.[13]

There are several methods used to carry out an energy analysis including:

- **Process analysis** – a commonly used procedure that involves identifying a system boundary around a particular process and determining the requirements for direct energy and indirect energy (through the provision of other goods and services crossing the system boundary and capital equipment, including buildings). The critical aspect of a process analysis is the definition of the system boundary. Considerable ranges of results are possible by the selection of different system boundaries. For a particular manufacturing process the system boundary may be the factory fence, or may include the requirements "upstream" for the provision of natural resources within the system boundary.
- **Input–output analysis** – developed for economic analysis, used by government economists who have collected data for the compilation of input–output matrices, which trace economic flows of goods and services between sectors of an economy. In Australia, the Australian Bureau of Statistics publishes input–output matrices for the 109 economic sectors every 5 years. A row in the matrix lists all the sales of a sector and a column lists all the purchases (in dollars of input per \$100 of output). Thus, the energy intensity of a sector, expressed in gigajoules (10^9 J) of energy per \$100 of sector output (GJ/\$100), can be derived by dividing purchases from individual energy supply sectors by the appropriate tariffs.
- **Hybrid analysis** – direct energy and quantities of goods and services are obtained for critical aspects of the process under consideration by process analysis. This could, for example, mean that for materials where the manufacture represents the main bulk of the overall environmental impact, the production processes are examined and quantified in detail by the process analysis method. The energy intensities of goods and services further upstream are then obtained using input–output analysis. With this approach the errors associated with input–output analysis are thus removed from a large proportion of the result, but the energy intensities derived only apply to materials and products manufactured by the specific process(es) audited and can not be applied globally.

Traditionally, input–output analyses have been used to derive the embodied energy intensities, as the resultant energy intensities were more complete than those derived from process analysis. Nevertheless, the accuracy of input–output analyses are inherently unreliable, but provided a common basis for comparison purposes. This method greatly reduces the errors associated with input–output analysis and is now considered the preferred method for embodied energy studies.

The energy density of different materials generally used to manufacture renewable energy technology is given in Appendix IV.

11.4 EMBODIED ENERGY AND ANNUAL OUTPUT OF RENEWABLE ENERGY TECHNOLOGIES

The calculation of embodied energy of a given renewable energy technology is straightforward except for a PV module that goes through various high-technology processes unlike other renewable energy technology. The calculation of embodied energy of a given renewable energy technology except PV module can be achieved as follows:

(i) Multiply the mass of different materials (m_i) used for manufacturing renewable energy technology by corresponding energy density (e_i) given in Appendix IV which gives $m_i\ e_i$.
(ii) Sum each of the numerical products to give a total embodied energy, *i.e.* summation of each $m_i\ e_i$.

11.4.1 PV Module

The total embodied energy required for making individual components of the PV module is difficult and their manufacturing energy needs to be evaluated. Energy requirement in different processes for production of PV modules are given in Table 11.1.

The embodied energy of PV modules for 1 m^2 for the following specifications is given as:

Processes	*MG-Si (kWh)*	*EG-Si (kWh)*	*Eg-Si for Cz-Si (kWh)*	*Fabrication Cell (kWh)*	*(kWh)*	*Total Module (kWh)*
Case (i)	48	230	483	120	190	1071
Case (ii)	26.54	127.30	267.33	60.3	125.4	607
Case (iii)	4.80	23.00	48.30	90	95	261

The above data has been generated by excluding the embodied energy of BOS of PV system.

The basis of the above data generated is due to the following reasons:

(i) Reduction of mass of solar cell: This may be due to development of new materials. Case (iii) has been considered as 10% of case (i).
(ii) Cell processing energy: This is reduced by 75% of case (i) to case (iii).
(iii) Cell efficiency: This is increased by 4%.
(iv) Elimination of wafer trimming and packaging.
(v) Reduction of embodied energy of module assembly by 50%.

Case (ii) is based on the average value of case (i) and case (ii), respectively.[14] The above results show that an embodied energy of a PV module is reduced significantly from 1071 to 261 kWh/m^2 due to the various reason described above.

Table 11.1 Energy requirement (energy density) in different processes for production of PV module.

Process	*Energy Requirement*
Silicon purification and processing	
(a) Metallurgical grade silicon (MG-Si) production from silicon dioxide (quartz, sand)	20 kWh per kg of MG-Si
(b) Electronic grade silicon (EG-Si) production from MG-Si	100 kWh per kg of EG-Si
(c) Czochralski silicon (Cz-Si) production from EG-Si	290 kWh per kg of EG-Si
Solar-cell fabrication	120 kWh per m^2 of Silicon cell
PV-module assembly	190 kWh per m^2 of PV module
Roof-top integrated PV system	200 kWh per m^2 of PV module

The PV module itself is called the system. Other components are called the balance of the system (BOS). This comprises wiring, electronic components, foundation, support structure, battery, installation, *etc*. For open-field installation, the concrete, cement and steel are the main components used for the foundation and frame, which requires maximum energy. The energy requirement (Energy density) for open-field installation is 500 kWh/m^2 of panel. For rooftop integrated PV system, the energy requirement is reduced to 200 kWh/m^2 of panel due to the absence of a foundation and structure for frame. Further, the present embodied energy can have a lower value (say about 75% of its present value) in future due to development of new materials used for BOS.

Hence, the total embodied energy of a PV module with installation for open-field and roof-top integration to a building will be 1571 and 1271 kWh/m^2, respectively for case (i).

$$\text{Annual output of PV module} = \eta_m \times \bar{I} \times A_m \times N \times n_0$$

where η_m and A_m are electrical efficiency and area of PV module. N and n_0 are the numbers of sunshine hours and clear days in a year. The $\bar{I}$ is the annual average values of solar intensity, which varies from place to place, *e.g.* its values for Port Hedland (NW Australia, Sydney and India are 2494, 1926 and 1800 kWh/m^2/year, respectively. This indicates that the annual electrical output for Port Hedland (NW Australia) will be maximum.

For the following parameters namely $\eta_m = 0.12$, $A_m = 1$ m^2 and for **Port Hedland (NW Australia)** climatic condition

Annual output of PV module for 1 m^2 $= 0.12 \times 2494 \times 1 = 299.28$ kWh

Similarly, for Sydney and Indian climatic condition for the same design parameters,

Annual output of PV module for 1 m^2 $= 0.12 \times 1926 \times 1 = 231.12$ kWh

and,

Annual output of PV module for 1 m^2 $= 0.12 \times 1800 \times 1 = 216$ kWh

If the life of a PV module is considered as $T = 30$ yr, then the matrices of a PV module can be obtained from Eqs. (11.1)–(11.3). The results are given below:

	Port Hedland			*Sydney*			*India*		
City	*EPBT*	*EPF*	*LCCE*	*EPBT*	*EPF*	*LCCE*	*EPBT*	*EPF*	*LCCE*
Case (i)	3.60	8.30	0.105	4.64	6.47	0.101	4.96	6.05	0.100
Case (ii)	2.00	15.0	0.112	2.62	11.45	0.109	2.81	10.68	0.109
Case (iii)	0.87	34.5	0.116	1.13	26.55	0.115	1.21	24.79	0.115

From the above table one can conclude that the importance of an embodied energy (E_i), annual electrical output (E_{out}) and the life of the renewable energy technology (T). This shows that the PV module is best suited for Port Hedland (NW Australia) due to the lowest value of EPB and the highest value of EPF and LCCE.

11.4.2 Flat Plate Collector

The embodied energy of a solar still, flat plate collectors (FPC), a PV module and a pump is summarised in Table 11.2. The embodied energy for two flat plate collectors each having an area of 2 m^2 is given in Table 11.2. The value of the embodied energy for two flat plate collectors is 2315.10 kWh.

Embodied energy for flat plate collector (FPC) of 2 m^2 = 1157.11 kWh

Annual solar radiation incident on FPC of 2 m^2 $= 400 \times 8 \times 2 \times 268$ Wh = 1715 kWh

For the following parameters namely $\eta_c = 0.70$ (efficiency of FPC), $A_m = 2$ m^2 and 1715 kWh for New Delhi clear sky (Blue sky) condition.

An overall annual thermal energy for one FPC $= 0.7 \times 1715$ kWh = 1205.5 kWh

Table 11.2 Breakdown of embodied energy of different components of hybrid (PVT) active solar still.

Components	*Items*	*Quantity*	*Total weight (kg)*	*Embodied energy (MJ/kg)*	*Total Embodied energy*	
					MJ	*kWh*
Solar Still (1 m^2)	GRP body	1	21.17	92.3	1954.0	542.8
	Glass cover 4 mm	1	1.16	40 060 MJ/m^3	185.9	51.6
	M S clamping frame	1	5	34.2	171.1	47.5
	M S clamp	8	2	34.2	68.4	19.0
	Mild steel stand	1	14/20	34.2	478	133
	Inlet/outlet nozzle	2	0.100	44.1	4.4	1.2
	Gaskets 8.9 m	1	2.1	11.83	24.8	6.9
Subtotal						**802**
Flat plate Collector Quantity 2	Copper riser 1/2″	20×1.8 = 36 m	8.2	81.0	664.2	184.5
	Header 1″	4×1.15 = 4.6 m	3.8	81.0	307.8	85.5
	Al box	2	10	199.0	1990.0	552.0
	Cu sheet	2	11	132.7	1460	405.6
	Glass cover toughened 4 mm	2 (3.75 m^2)	0.01464 m^3	66 020 MJ/m^3	966.5	268.3
	Glass wool	13 m^2	0.064 m^3	139 MJ/m^3	8.89	2.5
	Nuts/bolts/screws	32	1	31.06	31.06	8.6
	Union/elbow	8	1.5	46.8	70.2	19.5
	Nozzle/flange	8	1	62.1	62.1	17.3
	Mild steel stand	1	40	34.2	1368	380
	Paint	1L	1L	90.4	90.4	25.1
	Rubber gasket	18 m	4.2	11.83	49.7	13.8
	G I pipes 1/2″		9.5	44.1	418.9	116.4
	Al frame 1″	12 m	2.5	170	425	118
	Al sheet 24 gauge		2.5	170	425	118
Subtotal						**2315.1**
PV module	Glass to glass	1	0.605 m^2	3612/ m^2	2185.2	607
	BOS			475.2	475.2	132
Subtotal						**739**
Water pump	Copper wire		0.150	110.19	16.5	4.6
	Copper commuter	2	0.04	70.6	2.8	0.78
	Si-steel armature	1	0.05	*	*	*
	Wire insulation	2	0.01	*	*	*
	Motor body (SS)	1	0.100	36.1	3.61	1.0
	Casing (brass)	1	0.300	62.0	18.6	5.2
	Bearings	2	0.030	*	*	*
	Steel shaft	1	0.050	12.5	0.625	0.17
	Impellers(plastic)	1	*	*	*	*
	Nuts/screws/flange		0.100	31.06	3.1	0.86
Subtotal						**12.61**
Total embodied energy of hybrid active still						**3868.6**

The energy matrices of one flat plate collector have been evaluated for $T = 15$ yr by using Eqs. (11.1)–(11.3) as follows:

$$\text{Energy payback time (EPBT)} = \frac{1157.11}{1205.5} = 0.96\,\text{yr}$$

$$\text{Energy production factor (EPF)} = \frac{1205.5 \times 15}{1157.11} = 15.63$$

$$\text{Lifecycle conversion efficiency (LCCE)} = \frac{1205.5 \times 15 - 1157.11}{1715 \times 15} = 0.66$$

The above-calculated energy matrices are based on the annual thermal energy of FPC and it satisfies all the conditions mentioned in Section 11.2. Hence, the use of FPC is economical.

11.4.3 Hybrid Flat Plate Collector

The embodied energy for one hybrid flat plate collector of $2\,\text{m}^2$ will be approximately the sum of the embodied energy for one flat plate collector (FPC) of 2 m^2 (1157.11 kWh) and the embodied energy for one PV module of 0.605 m^2. From Table 11.2, this value will be about 1764.11 kWh.

11.4.4 Hybrid Air Collector

The calculation of total embodied energy for two hybrid air collectors connected in series is given in Table 11.3. The embodied energy for two hybrid air collectors of 2.64 m^2 is 3297.7 kWh. In this case, the embodied energy of PV module 2587.2 kWh and the remaining (710.5 kWh) is the embodied energy for BOS. The embodied energy for two hybrid air collectors connected in series as done in the case of the PV module will be as follows:

Case (i) **Embodied energy** = (2.64 × 980 + 710.5) = 3298 kWh
Case (ii) **Embodied energy** = (2.64 × 607 + 710.5) = 2313 kWh
Case (iii) **Embodied energy** = (2.64 × 261 + 710.5) = 1399 kWh

The annual thermal and electrical energy of hybrid air collectors (2.64 m^2) has been evaluated as 987 and 292 kWh, respectively. By using the conversion factor,

(a) The overall annual thermal energy for two hybrid air collectors = 1753 kWh.
(b) The overall annual exergy for two hybrid air collectors = 334 kWh.
(c) The direct sum of both types of annual energy for two hybrid air collectors = 1279 kWh.

Table 11.3 Breakdown of embodied energy of different components of PVT air collectors.

Sr. No.	*Component*	*Quantity*	*Energy density (kWh/kg)*	*Total embodied energy (kWh)*
1.	M.S. support structure	60 kg	8.89	533.4
2.	Wooden duct	10 kg	2.89	28.9
3.	PV module (glass-tedlar type)	2.64 m^2	980 kWh/m^2	2587.2
4.	Battery	one	-	121.4
5.	D C fan	one		
	(i) Aluminium	0.390 kg	55.28	21.56
	(ii) Iron	0.220 kg	8.89	1.96
	(iii) Plastic	0.120 kg	19.44	2.33
	(iv) Copper wire	0.050 kg	19.61	0.98
			Total	**3297.7**

For a 268 clear-sky condition, 8-h sunshine days and an annual average solar radiation of 400 W/m^2 the annual solar radiation on two hybrid air collectors of area 2.64 m^2 is given by

Annual solar radiation $= 400 \times 8 \times 2.64 \times 268 = 2\,264\,064$ Wh $= 2264$ kWh

	Overall thermal			*Overall exergy*			*Direct sum of thermal and electrical energy*		
Condition	*EPBT*	*EPF*	*LCCE*	*EPBT*	*EPF*	*LCCE*	*EPBT*	*EPF*	*LCCE*
Case (i)	1.88	15.96	0.726	9.87	3.04	0.099	2.58	11.62	0.516
Case (ii)	1.32	22.72	0.740	6.93	4.33	0.114	1.81	16.57	0.531
Case (iii)	0.80	37.50	0.754	4.19	7.15	0.127	1.09	27.52	0.544

From the above results one can infer that case (iii) gives the most favorable energy results from the user's point of view as per expectation for all condition mentioned above. This shows that there is a strong need to produce less high-grade energy consumed materials to produced PV module or any renewable energy systems.

11.5 EXERGY ANALYSIS

Exergy analysis is based upon the second law of thermodynamics, which stipulates that all macroscopic processes are irreversible. Every such irreversible process entails a nonrecoverable loss of exergy. This is expressed as the product of the ambient temperature and the entropy generated (the sum of the values of the entropy increase for all the bodies taking part in the process). Some of the components of entropy generation can be negative, but the sum is always positive.

Energy *vs.* exergy: As water drops over the falls, its potential energy is converted *via* kinetic energy to thermal energy, but on the whole it is conserved. Still, we can see that something – its ease of use in performing work – is being lost here. This lost quantity is called exergy.

The fact is that the quality of energy is more important than the quantity. The exergy analysis of a solar thermal system enables us to identify the sources of irreversibility and inefficiencies with the aim of reducing the losses and achieving the maximum resource and capital savings. This can be achieved by careful selection of the technology and optimisation of design of the system and components. The alternative means of comparing the thermal system meaningfully is exergy analysis.

"Exergy is the property of system that gives the maximum amount of useful work obtained from the system when it comes in equilibrium with a reference to the environment".

The maximum part of the input thermal energy that can be converted to work is called the available energy and that rejected to the surroundings is called the unavailable energy. Therefore;

Heat supplied (energy) = Available energy (exergy) + unavailable energy (anergy)

11.5.1 Exergy of a Process

The maximum work available (W_{max}) from the heat source at T_1 (in K) and the sink at (ambient) temperature T_0 (K), is expressed as;

$$W_{max} = exergy = \left(1 - \frac{T_0}{T_1}\right) \times Q_1 \tag{11.4a}$$

where Q_1 is the heat energy supplied at T_1.

For a given ambient temperature T_0, an increase in source temperature T_1 gives more exergy and less energy for the same heat transfer/energy input. The exergy of a system decreases as the process loses its quality.

$$\text{The unavailable part of the energy} = T_0 \Delta s \tag{11.4b}$$

where Δs is the change in the entropy of system during the change in process.

EXAMPLE 11.1

Calculate the maximum work available (W_{max}) from the heat source at $T_1 = 40$ °C, 60 °C, 80 °C and ambient temperature = 20 °C when $Q_1 = 150$ kWh.

Solution

Use Equation 11.4a for 40 °C, we have

$$W_{max} = \left(1 - \frac{20 + 273}{40 + 273}\right) \times 150 = 9.58\,\text{kWh}$$

Similarly, for 60 °C and 80 °C

$$W_{max} = 18\,\text{kWh and}$$
$$W_{max} = 25.5\,\text{kWh}$$

It is concluded that the maximum work is available at higher source temperature when the sink temperature is constant.

11.5.2 Exergy Efficiency

The exergy efficiency is a very useful performance parameter for the evaluation of thermodynamic systems. It is being recognised by various researchers. The thermal efficiency of the system is defined on the basis of the first law of thermodynamics, which includes the energy balance equation for the system to account for energy input, desired energy output, and energy losses. The exergy efficiency of the system is based on the second law of thermodynamics, which accounts for total exergy inflow, exergy outflow, and exergy destruction for the process.

The general exergy balance for the system can be written as:

$$\sum \dot{E}x_{in} - \sum \dot{E}x_{out} = \sum \dot{E}x_{dest} \quad (11.5a)$$

or,

$$\left(\sum \dot{E}x_{\text{heat}} + \sum \dot{E}x_{\text{mass,in}}\right) - \left(\sum \dot{E}x_{\text{work}} + \sum \dot{E}x_{\text{mass,out}}\right) = \sum \dot{E}x_{\text{dest}} \quad (11.5b)$$

or,

$$\left(\sum \left(1 - \frac{T_0}{T_k}\right)\dot{Q}_k + \sum \dot{m}_{in}\psi_{in}\right) - \left(W + \sum \dot{m}_{out}\psi_{out}\right) = \sum \dot{E}x_{dest} \quad (11.5c)$$

where $\dot{Q}_k$ is the rate of heat transfer through the boundary at location, k, at temperature T_k (in K).

The ψ is the specific flow exergy, which is defined as:

$$\psi = (h - h_0) - T_0(s - s_0) \quad (11.6)$$

where h and s are the specific enthalpy and entropy, respectively and subscript 0 refers to these properties at the restricted dead state.

Now, the exergy destruction or the irreversibility may be written as:

$$\dot{E}x_{dest} = \dot{I} = \dot{T}_0\dot{S}_{gen} \tag{11.7}$$

where the rate of entropy generation $\dot{S}_{\text{gen}} = \sum \dot{m}_{\text{out}}s_{\text{out}} - \sum \dot{m}_{\text{in}}s_{\text{in}} - \sum \frac{\dot{Q}_k}{T_k}$.

It is proposed that when the exergy destruction or irreversibility $\sum \dot{E}x_{\text{in}} - \sum \dot{E}x_{\text{out}}$ is minimised, there will be maximum improvement in the exergy efficiency for a process or system. It is also useful to employ the concept of an exergetic "improvement potential" while analysing different processes or sectors of the economy. The rate of "improvement potential" can be expressed as:

$$IP = (1 - \varepsilon)\left(\sum \dot{E}x_{\text{in}} - \sum \dot{E}x_{\text{out}}\right) \tag{11.8}$$

The exergy efficiency or second law efficiency is the ratio of actual performance of system to the ideal performance of the system or it is defined as the ratio of exergy output (product exergy) to exergy input and expressed as:[15]

$$\varepsilon = \frac{\text{Rate of useful product energy}}{\text{Rate of exergy input}} = \frac{\dot{E}x_{\text{out}}}{\dot{E}x_{\text{in}}} = 1 - \frac{\dot{E}x_{\text{dest}}}{\dot{E}x_{\text{in}}} \tag{11.9}$$

where $\dot{E}x_{\text{dest}}$ is the rate of exergy destruction.

11.5.3 Solar Radiation Exergy

Exergy is the property of a matter and not of any phenomenon. The matter may be either a substance (which has a rest mass larger than zero) or a field matter, for which the rest mass is zero, *e.g.* the matter of the considered heat radiation, a field of surface tension, magnetic field, acoustic field, or gravitational field. The term "radiation" or "emission" means either the radiation phenomenon or the radiation product (the matter of the electromagnetic field). Therefore, "the exergy of a phenomenon" is an inadequate scientific jargon often used by various researchers. It should be the "change in exergy of the heat source" instead of the "**exergy of heat**".[16] The term solar radiation exergy is generally referred to as the exergy of the Sun and it is the exergy input from the Sun to any solar system or device.

The conversion of thermal radiation can be through various processes, *e.g.* work, heat and some other various processes (*e.g.* growth of natural plants or plant vegetation, *etc.*). The energetic and exergetic conversion efficiency of thermal radiation into work or heat is given in Table 11.4.

Table 11.4 Conversion efficiency of thermal radiation.[16]

S.No.	*Efficiency*	*Radiation to work conversion*	*Radiation to heat conversion*
1.	Energetic, η_e	$\eta_e = \frac{W}{e}$; $\eta_{e\,\max} = \frac{b}{e} = U_{ee}$ or Ψ	$\eta_e = \frac{e - e_a}{e} = 1 - \left(\frac{T_a}{T}\right)^4$
2.	Exergetic, η_{ex}*	$\eta_{ex} = \frac{W}{b} = \frac{W}{e \times U_{ee}}$	$\eta_{ex} = \frac{b_q}{b}$

* The exergetic efficiency, η_{ex} is also denoted by 'ε' by some researchers.
where W is the work performed due to utilisation of the radiation and b ($= W_{\text{max.}}$) is the exergy of radiation and U_{ee} (= Unified efficiency expression).

Thus, from Table 11.4, solar radiation exergy (radiation to work conversion) can be expressed as:

$$\dot{E}x_{\text{sun}} = b = e \times U_{\text{ee}} \tag{11.10}$$

If $I(t)$ is incident solar radiation (*i.e.* solar intensity /energy from the Sun) on surface area A of the solar device/system at the Earth, the energy of thermal radiation (e) can been expressed as $\{I(t) \times A\}$ and thus the exergy input, *i.e.* radiation exergy (radiation to work conversion) can be written as:[17]

$$\dot{E}x_{\text{sun}} = \{A \times I(t)\} \times U_{\text{ee}} = \{A \times I(t)\} \times \left[1 - \frac{4}{3} \times \left(\frac{T_0}{T_s}\right) + \frac{1}{3} \times \left(\frac{T_0}{T_s}\right)^4\right] \tag{11.11}$$

T_0 = Surrounding or environment temperature (K) = T_a;
T_s = Sun surface temperature = T_{Sun} = 6000 K;
T_a = Ambient air tempèrature (K)

The input, output and unified efficiency expression (U_{ee}) of utilisation of thermal radiation given by three researchers is shown in Table 11.5.

EXAMPLE 11.2

Calculate the unified efficiency (U_{ee}) using the expression of the Petela model and radiation exergy when surrounding temperature = 20 °C, $A = 2\ \text{m}^2$, $I(t) = 750\ \text{W/m}^2$.

Solution

Using Table 11.5 and Eq. (11.11), we have

$$\dot{E}x_{\text{Sun}} = 2 \times 750 \times \left(1 - \frac{4}{3} \times \left(\frac{20 + 273}{6000}\right) + \frac{1}{3} \times \left(\frac{20 + 273}{6000}\right)^4\right) = 1.4\ \text{kW}$$

Table 11.5 The input, output and unified efficiency expression (U_{ee}) of utilisation of thermal radiation.

S.No.	*Researcher*	*Input*	*Output*	U_{ee}
1.	Petela[16]	Radiation Energy	Absolute work	$1 - \frac{4}{3} \times \left(\frac{T_0}{T_s}\right) + \frac{1}{3} \times \left(\frac{T_0}{T_s}\right)^4$
2.	Battisti[11]	Radiation Energy	Useful work Radiation Exergy	$1 - \frac{4}{3} \times \left(\frac{T_0}{T_s}\right)$
3.	Lewis[9]	Heat	Net work of a heat engine	$1 - \left(\frac{T_0}{T_s}\right)$

where T_s and T_0 are surface temperature of the Sun and environment temperature at Earth, respectively.

11.5.4 Exergetic Analysis of Flat Plate Collector

An exergy analysis of a flat plate collector (FPC) can be carried out with the aim of providing some ways to save cost and keep the efficiency of the integrated system to the desired extent and at the same time determine related exergy losses. The change in kinetic exergy in the utilisation procedure is negligible since most domestic-scale solar water heaters are driven by the difference of the density of water. Exergetic analysis of the collector–integrated system involves analysis of the collector and analysis of the integrated system.

The following equation can be used to calculate exergy input ($\dot{E}xc$) from the collector to the integrated system.

$$\dot{E}xc = \dot{m}\,C(T_{\text{fo}} - T_{\text{a}}) = \dot{m}\,C\,T_a \ln\frac{T_{\text{fo}}}{T_{\text{a}}} \tag{11.12}$$

where,

$\dot{E}x_{\text{c}}$ = exergy output from collector (W)
$\dot{m}$ = mass flow rate of collector fluid (kg/s)
T_{fo} = outlet temperature of fluid from collector (K)
T_{fi} = inlet temperature of fluid from collector (K)

The **exergy efficiency** of the collector can be expressed as:

$$\varepsilon_c = \frac{\dot{E}xc}{\dot{E}x_{sun}} \tag{11.13}$$

The exergy efficiency of a solar collector will increase with increase in collector efficiency. The exergy efficiency of FPC is low as it transfers low-entropy (high-temperature) solar radiations to high-entropy (low-temperature) energy of working fluid. However, the concentrating collectors have high exergy efficiency because they produce low-entropy (high-temperature) fluids.

EXAMPLE 11.3

Calculate the exergy output from collector when $T_{\text{fo}} = T_{\text{a}}$ and $T_{\text{fo}} - T_{\text{a}} = 35$ °C. When the mass flow rate 0.06 kg/s, $C = 4190$ J/kg K.

Solution:

Use Eq. (11.12), we get

$$\dot{E}x_{\text{c}} = 0.06 \times 4190 \times 35 - 0.06 \times 4190 \times 25 \ln\frac{60}{35} = 5.41\ \text{kW}$$

11.5.5 Exergetic Analysis of PVT Systems

The exergy analysis is a powerful tool for the evaluation of the thermodynamic systems. The energy efficiency of the thermal system is the ratio of energy recovered from the product to the original energy input. The exergetic efficiency can be defined as the ratio of the product exergy to exergy inflow.

The exergy analysis is based on the second law of thermodynamics, which includes accounting for the total exergy inflow, exergy outflow and exergy removed from the system.

$$\sum \dot{E}x_{in} - \sum (\dot{E}x_{thermal} + \dot{E}x_{electrical}) = \sum \dot{E}x_{dest} \tag{11.14}$$

where, Exergy of radiation,

$$\dot{E}x_{in} = A_c \times N_c \times I(t) \times \left[1 - \frac{4}{3} \times \left(\frac{T_a}{T_s}\right) + \frac{1}{3} \times \left(\frac{T_a}{T_s}\right)^4\right],^{24} \tag{11.15a}$$

$$\text{Thermal exergy} = \dot{E}x_{thermal} = \dot{Q}_u\left[1 - \frac{T_a + 273}{T_{fo} + 273}\right] \tag{11.15b}$$

$$\text{Electrical exergy} = \dot{E}x_{electrical} = \eta_c \times A_c \times N_c \times I(t) \tag{11.15c}$$

and,

$$\text{overall exergy} = \dot{E}x_{thermal} + \dot{E}x_{electrical} \tag{11.16}$$

where, A_c is area of collector and T_s is the Sun temperature in Kelvin.

11.5.6 Exergetic Analysis of Wind Energy System

Wind energy is in the form of (kinetic) when wind goes through a wind turbine's (W/T) rotor's surface (turning part) it is given by:

The exergy of wind energy is the useful amount of energy derived from wind. The kinetic energy available from the wind is given by

The kinetic energy associated with the wind $= \frac{1}{2}\dot{m}v^2 = \frac{1}{2}\rho a v^3$

All the kinetic energy of the wind is converted into electrical energy for maximum energy output. A wind turbine (WT) system does not exploit all the kinetic energy of wind that passes through its blades.

The energy efficiency of a wind turbine (WT) is governed by the following efficiency factors:

(i) Power coefficient (C_p) : According to the Betz criteria, a wind turbine (WT) can only exploit up to 59.3% (16/27) of incoming wind energy to the turbine.
(ii) Electric generator efficiency (η_g): This can approach the value of 90–95%.
(iii) Subsystems efficiency factor (η_b): This can approach 95% for modern technologically developed wind turbines (WT). Exergy losses also appear in electronic devices that consume 1–2% of the energy for smooth operation.

An expression for wind turbine (WT) energy efficiency is written as:

$$\eta_{WT} = \frac{P_{electrical}}{P_{KE}} = \frac{\dot{P}_{out}}{\frac{1}{2}\rho a v^3} \tag{11.17}$$

For one wind turbine (WT), the exergy efficiency is given by

$$\eta_{WT,Ex} = \frac{E_{electrical}}{E_{KE}} = \frac{W_{electrical}(\text{Wh})}{P_{Wind} \times 8760(\text{Wh})}$$

There is 8760 h in one year. In the above equation it is assumed that the wind is operating 24 h/day.

It is also well known that a wind turbine (WT) cannot exploit the complete power of wind in the form of electrical energy.

11.5.7 Exergetic Analysis of Hydropower System

Our second example is that of hydroelectric power generation due to potential energy. Unlike wind power as described above, all of the available potential energy can be converted directly into work. Our favourite example is that of the Shoshone Hydropower plant in Glenwood Canyon, Colorado. The unique aspect of this plant is that unlike traditional plants that have the dam located at the same location, the Shoshone dam is located two miles upstream, and the water flows through a tunnel in the wall of the canyon to the power plant. At the power plant the water exits the canyon wall and drops to the hydroelectric turbines to generate power.

$$\text{Power (W)} = \dot{m}gz \tag{11.18}$$

where,

$\dot{m}$ (kg/s) is the mass flow rate of the water through the turbine rotor
g (m/s^2) is the acceletarion due to gravity (=9.81 m/s^2)
z (m) is the height of the water source above the turbine inlet

The Shoshone plant can provide up to 15 MW power, which is enough power for about 15 000 households.

EXAMPLE 11.4

Find the exergy analysis of a control volume of a hydropower system

Solution

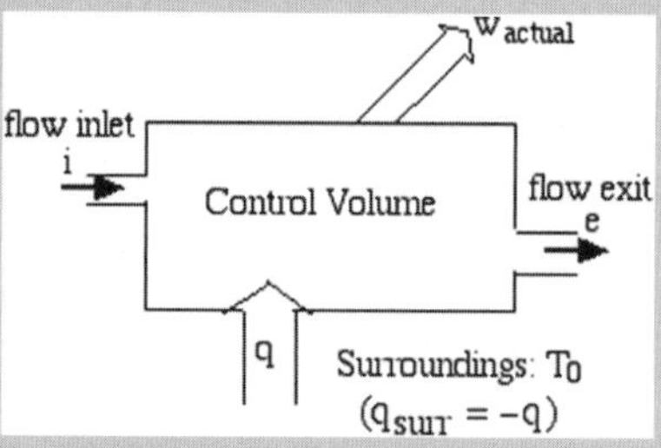

Referring to the above figure, according to the first law of thermodynamics (energy conservation) we have :

$$q - w_{\text{actual}} = \Delta h = (h_e - h_i) \tag{i}$$

According to the second law of thermodynamics (entropy generation) we have

$$s_{\text{gen}} = \Delta s + \frac{q_{\text{sur}}}{T_0} = \Delta s - \frac{q}{T_0}$$

$$q = T_0 \Delta s - T_0 \cdot s_{\text{gen}} \tag{ii}$$

Exergy Analysis

First eliminate q from Eqs. (i) and (ii) to get the following equation

$$T_0 \cdot \Delta s - T_0 \cdot s_{gen} - w_{actual} = \Delta h$$

$$w_{actual} = -\Delta h + T_0 \cdot \Delta s - T_0 \cdot s_{gen}$$

$$w_{actual} = (h_i - h_e) - T_0(s_i - s_e) - T_0 \cdot s_{gen} \quad \text{(iii)}$$

The maximum work for a reversible system will occur with no entropy generation. We can define the **irreversibility (irrev)** as follows:

$$\text{irrev} = T_0 \cdot s_{gen} \quad \text{(iv)}$$

Thus, for irrev = 0, the **reversible work** becomes:

$$w_{rev} = (h_i - h_e) - T_0(s_i - s_e) \quad \text{(v)}$$

Now, define the **second law efficiency** (η_{II}) for either a work-producing or a work-absorbing device as follows:

$$\eta_{II} = \frac{w_{actual}}{w_{rev}} \quad \text{(for a work producing device)}$$
$$\eta_{II} = \frac{w_{rev}}{w_{actual}} \quad \text{(for a work absorbing device)} \quad \text{(vi)}$$

The **exergy** (availability), ϕ, of the working fluid at either the inlet or the outlet port is defined as the maximum available work when the state of that working fluid is reduced to the dead state 0, thus:

$$\psi = (h - h_0) - T_0(s - s_0) \quad \text{(vii)}$$

Thus, the **reversible work** of the control volume can also be defined in terms of the difference in exergy between the inlet and exit ports, thus:

$$w_{rev} = (\psi_i - \psi_e) \quad \text{(viii)}$$

11.5.8 Exergetic Analysis of Geothermal Power System

Assume that a village has discovered underground geothermal well containing water at 95 °C, 100 m below the surface. They are evaluating the option of using the water from this well to provide power to the village. Determine the maximum available work [kJ/kg] that could be obtained from a power plant that will pump water from the well, extract energy from the water and discharge the water to a lake at the surrounding temperature $T_0 = 25$ °C. Assume that the water temperature can be reduced to 35 °C at the outlet of the power plant.

We now do an exergy analysis to determine the maximum available (reversible) work by eliminating q from the above energy and entropy equations:

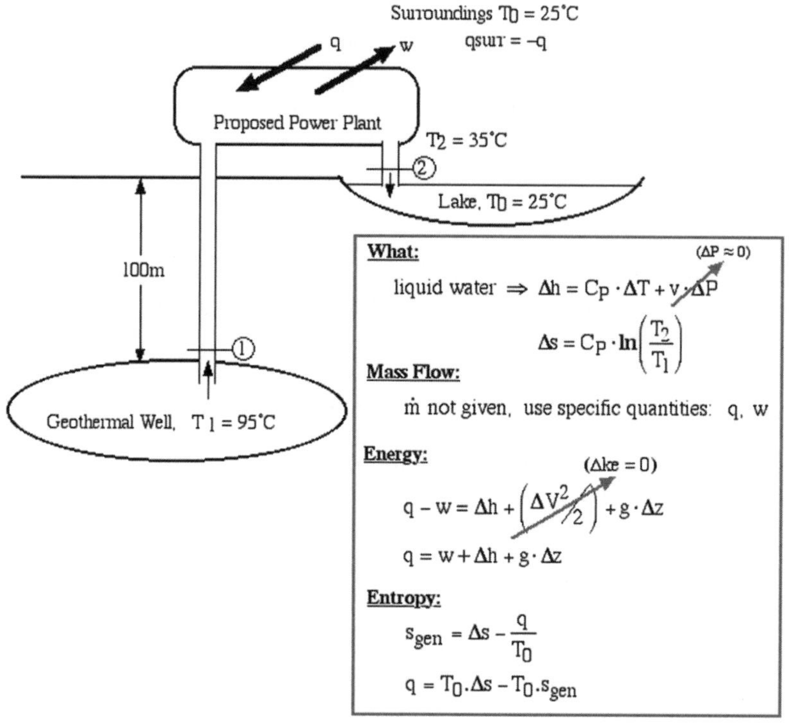

Energy:
Equating Equations (1) and (2) above:

$$q = w + \Delta h + g.\Delta z = T_0.\Delta s - T_0.s_{gen} \quad \text{(reversible)}$$

$$w_{rev} = -\Delta h + T_0.\Delta s - g.\Delta z$$

$$\boxed{w_{rev} = (h_1 - h_2) - T_0(s_1 - s_2) + g(z_1 - z_2)}$$

This equation could have been derived directly from the definition of exergy:

$$\psi = (h - h_0) - T_{01}(s - s_0) + g(z - z_0) + \left({}^{V^2}\!/_2\right)$$

$$w_{rev} = (\psi_1 - \psi_2)$$

Finally, we obtain our numerical answer by considering the question "**What**" above, in which we substitute the enthalpy (h) and entropy (s) values for liquid water, an incompressible liquid:

$$w_{rev} = C_P(T_1 - T_2) - T_0 \cdot C_P \ln\left(\frac{T_1}{T_2}\right) + g(z_1 - z_2)$$

$$w_{rev} = 4.18\left[\frac{\text{kJ}}{\text{kg} \cdot \text{K}}\right](95 - 35)[\text{K}] - 298[\text{K}] \cdot 4.18\left[\frac{\text{kJ}}{\text{kg} \cdot \text{K}}\right] \cdot \ln\left(\frac{273 + 95}{273 + 35}\right)$$
$$+ 9.81\left[\frac{\text{m}}{\text{s}^2}\right](-100 - 0)[\text{m}]\left[\frac{\text{N}}{\text{kg} \cdot {}^{\text{m}}\!/_{\text{s}^2}}\right]\left[\frac{\text{kJ}}{1000\,\text{Nm}}\right]$$

$$w_{rev} = 250.8 - 221.7 - 0.98 = 28.1 \left[\frac{kJ}{kg}\right]$$

11.6 CO_2 EMISSIONS

The energy consumption of a country is one of the indicators of its socioeconomic development. The per capita electricity consumption of India is one of the lowest in the world. The per capita electricity consumption is about 30% of that in China, about 22% of that in Brazil and about 3.18% of that in USA. With development, the per capita energy consumption is likely to increase. To achieve a per capita energy consumption equal to that of Brazil (which is still a developing country like India) our energy production and consumption must be quadrupled and to achieve the European average (about 6500 kWh/capita), we must increase our energy production and consumption by 15.5 times. At present, our annual economic growth rate is 8–10% per annum. To sustain this growth rate we desperately need additional secured and reliable energy sources. For energy India depends on oil and gas imports, which account for over 65% of its consumption; this is likely to increase further considering the economic development, rise in living condition of people and rising prices. Coal, which currently accounts for over 60% of India's electricity production, is the major source of emission of greenhouse gases and that of acid rain. India will become the third biggest polluter in the world after USA and China if we keep depending on coal as the main source of electricity in the years to come. In the business-as-usual scenario, India will exhaust its oil reserves in 22 years, its gas reserves in 30 years and its coal reserves in 80 years.[18] More alarmingly, the coal reserves might disappear in less than 40 years if India continues to grow at 8% a year.[19]

Greenhouse gases are the gases present in the Earth's atmosphere that reduce the loss of heat into space and therefore contribute to global temperatures through the greenhouse effect. Greenhouse gases are essential to maintaining the temperature of the Earth; without them the planet would be so cold as to be uninhabitable. However, an excess of greenhouse gases can raise the temperature of a planet to lethal levels, as on Venus where the 90 bar partial pressure of carbon dioxide (CO_2) contributes to a surface temperatures of about 467 °C (872 °F). Greenhouse gases are produced by many natural and industrial processes, which currently result in CO_2 levels of 380 ppmv in the atmosphere. Based on ice-core samples and records current levels of CO_2 are approximately 100 ppmv higher than during immediately preindustrial times, when direct human influence was negligible. Carbon emissions from various global regions during the period 1800–2000 AD are shown in Figure 11.1a.

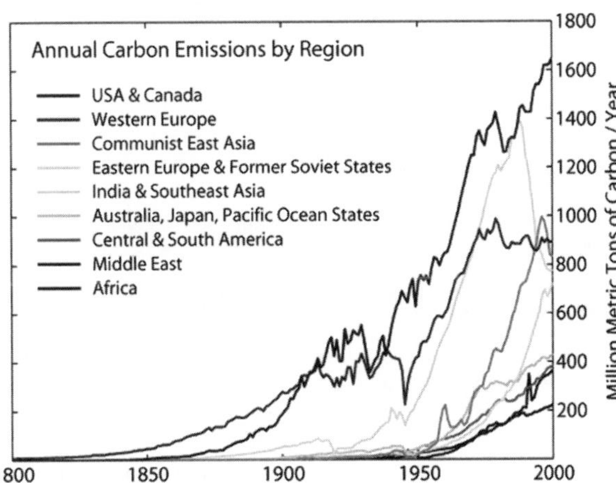

Figure 11.1a Carbon emissions from various global regions during the period 1800–2000 AD.[19]

The average carbon dioxide (CO_2) equivalent intensity for electricity generation from coal is approximately 0.98 kg of CO_2/kWh.[20] If the PV system has a lifetime of 35 years, the CO_2 emissions per year by each component can be calculated as

$$CO_2 \text{ emissions per year} = \frac{\text{Embodied energy} \times 0.98}{\text{Lifetime}} \tag{11.19}$$

The CO_2 emissions per year for a PV module (glass to glass) (effective area = 0.60534 m^2 and size = 1.20 m × 0.55 m × 0.01 m) in present conditions is given in Table 11.6

The CO_2 emissions for different PVT systems are shown in Figure 11.1b

EXAMPLE 11.4

Calculate the carbon dioxide emission per year from a solar water heater in a lifetime of 10, 20 and 30 years, when the total embodied energy required for manufacturing the system is 3550 kWh.

Solution:

Using Eq. (11.19(, we have

For lifetime = 10 yr

$$CO_2 \text{ emissions per year} = \frac{3550 \times 0.98}{10} = 347.9 \text{ kg of } CO_2$$

Similarly, for lifetimes = 20 and 30 yr CO_2 emissions per year is 173.9 and 115.9 kg of CO_2, respectively.

Table 11.6 CO_2 emissions per year from a PV module (glass to glass) (effective area = 0.60534 m^2).

S.No.	*Components*	*Embodied energy (kWh)*	*CO_2 emissions (kg)*
1	MG–Si	26.54	0.74
2	EG–Si	127.30	3.56
3	Cz–Si	267.33	7.49
4	Solar cell fabrication	60.29	1.69
4	Solar cell fabrication	60.29	1.69
5	PV module assembly	125.40	3.51
Total		**606.86**	**16.99**

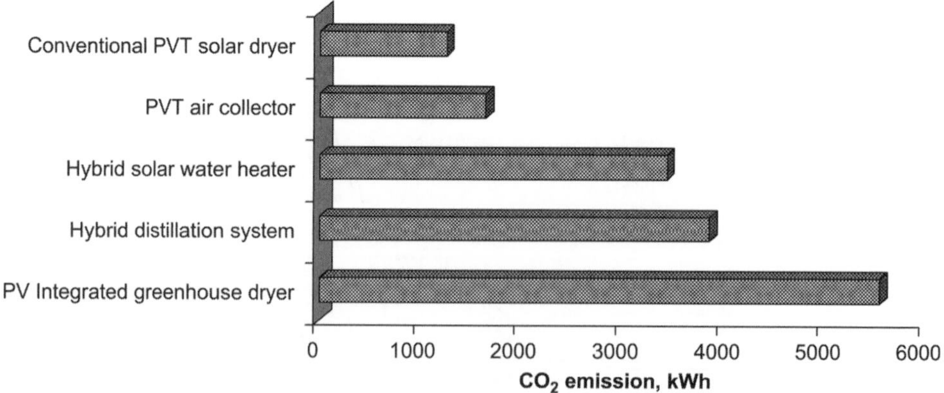

Figure 11.1b CO_2 emissions for different PVT systems.

11.7 EARNED CARBON CREDIT[21]

The trading of carbon credits was therefore created to curb the effect of greenhouse gases by reducing the carbon footprint. Carbon credits are defined as "a key component of national and international emissions trading schemes that have been implemented to mitigate global warming". They provide a way to reduce greenhouse-effect emissions on an industrial scale by capping the total annual emissions and letting the market assign a monetary value to any shortfall through trading. Credits can be exchanged between businesses or bought and sold in international markets at the prevailing market price. Credits can be used to finance carbon-reduction schemes between trading partners around the world. There are also many companies that sell carbon credits to commercial and individual customers who are interested in lowering their carbon footprint on a voluntary basis. These carbon offsetters purchase the credits from an investment fund or a carbon development company that has aggregated the credits from individual projects. The quality of the credits is based in part on the validation process and sophistication of the fund or development company that acted as the sponsor to the carbon project.

11.7.1 Formulation

If unit power is used by a consumer and the losses due to poor domestic appliances is L_a, then the transmitted power should be $\frac{1}{1-La}$ units. If the transmission and distribution losses are L_{td}, then the power that has to be generated in the power plant is $\frac{1}{1-L_a} \times \frac{1}{1-L_{td}}$ units.

The average CO_2 equivalent intensity for electricity generation from coal is approximately 0.98 kg of CO_2 per kWh at the source. Thus, for unit power consumption by the consumer, the amount of CO_2 emission is $\frac{1}{1-L_a} \times \frac{1}{1-L_{td}} \times 0.98$.

The annual CO_2 emission can be expressed as:

$$\mathrm{CO_2 emission\,per\,year} = \frac{E_{\mathrm{in}}}{n_{\mathrm{sys}}} \times \frac{1}{1 - L_a} \times \frac{1}{1 - L_{td}} \times 0.98\,\mathrm{kg} \tag{11.20}$$

where E_{in} is the embodied energy input and n sys is the lifetime of the system.

The CO_2 emission over the lifetime of the system is:

$$E_{\mathrm{in}} \times \frac{1}{1 - L_a} \times \frac{1}{1 - L_{td}} \times 0.98\,\mathrm{kg} \tag{11.21}$$

The net CO_2 mitigation over the lifetime of the system is:
Total CO_2 mitigation – total CO_2 emission

$$= (E_{\mathrm{aout}} \times T_{\mathrm{LS}} - E_{\mathrm{in}}) \times \frac{1}{1 - L_a} \times \frac{1}{1 - L_{td}} \times 0.98\,\mathrm{kg} \tag{11.22}$$

where E_{aout} is the overall exergy gain, which is the sum of the annual electrical exergy ($\dot{E}_{x_{el}}$)and the annual thermal exergy equivalent ($\dot{E}_{x_{th}}$). Substituting its value wc have: the annual exergy output as

$$E_{\mathrm{aout}} = [\eta_{ca} \times I(t) \times bL \times n_s n_p] + \left[\dot{Q}_u \times \left(1 - \frac{T_a}{T_{\mathrm{air\,out}}}\right)\right] \tag{11.23}$$

The net CO_2 mitigation over the lifetime in tonnes of CO_2 is given by:

$$(E_{\mathrm{aout}} \times n_{\mathrm{sys}} - E_{\mathrm{in}}) \times \frac{1}{1 - L_a} \times \frac{1}{1 - L_{td}} \times 0.98 \times 10^{-3} \tag{11.24}$$

If CO_2 emission is being traded at US\$ C per tonnes of CO_2 mitigation, then the carbon credit earned by the system is:

$$= US\$\ C \times \left[(E_{\text{aout}} \times n_{\text{sys}}) \times \frac{1}{1 - L_{\text{a}}} \times \frac{1}{1 - L_{\text{td}}} \times 0.98 - E_{\text{in}}\right] \times 10^{-3} \tag{11.24a}$$

11.7.2 A Case Study with the BIPVT System

If unit power is used by a consumer and the losses due to poor domestic appliances are around 20%, then the transmitted power should be $\frac{1}{1-0.2} = 1.25$ units. If the transmission and distribution losses are 40%, common in Indian conditions, then the power that has to be generated in the power plant is $\frac{1.25}{1-0.2} = 2.08$ units. The average CO_2 equivalent intensity for electricity generation from coal is approximately 0.98 kg of CO_2 per kWh at the source. Thus, for unit power consumption by the consumer the amount of CO_2 emission is $2.08 \times 0.98 = 2.04$ kg. For the BIPVT system, the annual CO_2 mitigation in tonnes of CO_2 is given by:

$$\left(E_{\text{aout}} - \frac{E_{\text{in}}}{n_{\text{sys}}}\right) \times 2.04 \times 10^{-3} \tag{11.25}$$

Assume that the overall embodied energy for monocrystalline silicon (c-Si, $n_{\text{sys}} = 30$ yr), multi-crystalline silicon (p-Si, $n_{\text{sys}} = 30$ years), ribbon silicon (r-Si, $n_{\text{sys}} = 25$ yr), amorphous silicon (a-Si, $n_{\text{sys}} = 20$ yr), cadmium telluride (CdTe, $n_{\text{sys}} = 15$ yr) and copper indium gallium selenide (CIGS, $n_{\text{sys}} = 5$ yr) technological BIPVT systems are 607 613, 540 628, 409 716, 272 324, 211 984 and 63 937 MJ, respectively. The overall exergy calculations for the climatic conditions of New Delhi show that c-Si, p-Si, r-Si, a-Si, CdTe and CIGS BIPVT systems covering 45 m^2 of roof area generate 16 224, 14 352, 12 512, 7790, 9547 and 11 037 kW of overall exergy output, respectively. Thus, the annual CO_2 mitigation for c-Si, p-Si, r-Si, a-Si, CdTe and CIGS systems are 77.83, 68.64, 58.45, 29.44, 41.29 and 54.97 tonnes, respectively. If CO_2 emissions are being traded at US\$ 20 per tonnes of CO_2 mitigation, then the carbon credit earned by the BIPVT system with the c-Si, p-Si, r-Si, a-Si, CdTe and CIGS technologies are US\$ 1557, 1373, 1169, 589, 826 and 1099, respectively. This shows that the monocrystalline silicon BIPVT system gives the highest earnings through carbon credit trading.

OBJECTIVE QUESTIONS

11.1 1 kWh is equal to
- (a) 36 MJ
- (b) 3.6 MJ
- (c) 0.36 MJ
- (d) 360 MJ

11.2 The energy density is maximum for
- (a) Nonmetal
- (b) Metal
- (c) Glass
- (d) None of them

11.3 The embodied energy of most of the renewable energy technologies is always
- (a) Less than energy density
- (b) Higher than energy density
- (c) Equal to energy density
- (d) None of them

11.4 The energy payback time (EPBT) should be preferably
  (a) < 1
  (b) > 1
  (c) $= 1$
  (d) $= 0$

11.5 For zero embodied energy, CO_2 emission is
  (a) < 1
  (b) > 1
  (c) $= 1$
  (d) $= 0$

11.6 At the source of a coal-based power generation, 1 kWh is equivalent to
  (a) 0.98 kg CO_2 emission
  (b) 9.8 kg CO_2 emission
  (c) 0.098 kg CO_2 emission
  (d) none of them

11.7 For wind energy mechanical power, EPBT is
  (a) < 1
  (b) > 1
  (c) $= 1$
  (d) $= 0$

11.8 The energy payback time (EPBT) for FPC mainly depends on
  (a) Ambient air
  (b) Wind velocity
  (c) Solar intensity
  (d) None of them

11.9 The energy payback time (EPBT) for a PV module with BOS is
  (a) Reduced
  (b) Increased
  (c) Unaffected
  (d) None of them

11.10 The energy payback time (EPBT) for a PV module at present is
  (a) < 1
  (b) > 1
  (c) $= 1$
  (d) $= 0$

11.11 The energy payback time (EPBT) is
  (a) Ratio of embodied energy to annual energy output
  (b) Ratio of annual energy output to embodied energy
  (c) Ratio of annual energy output to annual input energy
  (d) None of them

ANSWERS

11.1 **(b)**; 11.2 **(b)**; 11.3 **(a)**; 11.4 **(d)**; 11.5 **(d)**; 11.6 **(a)**; 11.7 **(a)**; 11.8 **(c)**; 11.9 **(b)**; 11.10 **(a)**; 11.11 **(a)**.

REFERENCES

1. Bentley, R. W. *Energy Policy*, 2002, 30(3), 189–205.
2. Alsema, E. A. and Niluwlaar, E. *Energy Policy*, 2000, 28, 999–1010.

3. Krauter, S. and Ruther, R. *Renew. Energy*, 2004, 29, 345–355.
4. Frankl, P., Masini, A., Gamberale, M., and Toccaceli, D. *Prog. Photovolt.: Res. Appl.*, 1998, 6(2), 137–146.
5. Hunt, L. P. *IEEE PV Specialists Conference*, Piscataway, NJ, 1986, p 347–352.
6. Kato, K., Murata, A. and Sakuta, K. *Prog. Photovolt.: Res. Appl.*, 1998, 6(2), 105–115.
7. Aulich, H. A., Schulz, F. W. and Strake, B. *IEEE PV Specialist Conference*, Piscataway, NJ, 1986, pp. 1213–1218.
8. Prakash, R. and Bansal, N. K. *Energy Sources*, 1995, 17, 605–613.
9. Lewis, G. and Keoleian, G. 1996. National Pollution Prevention Centre, School of Natural Resources and Environment, University of Michigan.
10. Srinivas, K. S., Vuknic, M., Shah, A. V. and Tscharner, R. 1992. *6th International Photovoltaic Science and Engineering Conference (PVSEC-6)*, New Delhi, India, pp. 403–413.
11. Battisti, R. and Corrado, A. *Energy*, 2005, 30, 952–967.
12. Treloar, G. J. 1994. *Energy analysis of the construction of office buildings*, Master of Architecture Thesis, Deakin University, Geelong.
13. Boustead, I. and Hancock, G. F. 1979. *Handbook of Industrial Energy Analysis*, Ellis Horwood Publishers, Chichester, UK, 309–410.
14. Tiwari, G. N. and Ghosal, M. K. 2005. *Renewable Energy Resources: Basic Principles and Applications*, Narosa Publishing House, New Delhi, India.
15. Akpinar, E. K., Midilli, A. and Bicer, Y. *J. Food Eng.*, 2006, 72, 320–331.
16. Petela, R. *Solar Energy*, 2003, 74(6), 469–488.
17. Szargut, J. *Energy*, 2003, 28(11), 1047–1054.
18. Kalshian, R. 2008. Energy versus emissions: The big challenge of the new millennium, By Info Change News & Features, www.infochangeindia.org/agenda5_01.jsp, accessed 21 March 2008.
19. Per capita greenhouse gas emissions on world map, http://en.wikipedia.org, accessed 20 July 2008.
20. Watt, M., Johnson, A., Ellis, M. and Quthred, N. *Prog. Photovolt. Res. Appl.*, 1998, 6(2), 127–136.
21. Agrawal, B. and Tiwari, G. N. 2010. *Building Integrated Photovoltaic Thermal Systems*, RSC Publishing, Cambridge.

The Cost per kWh of Power Available from Renewable Energy Sources and Fossil Fuel should be Compared without any Subsidy to Both Sources of Energy.

CHAPTER 12

Economics of Renewable Energy

12.1 INTRODUCTION

Interest in the development of and dissemination of renewable energy technologies has again been rekindled in view of increasing global climate change concerns. In addition to the development of new and appropriate technology, issues related to their financial and economic viability and financing of renewable energy system are being given considerable importance. Technoeconomic analysis is the area of engineering where engineering judgment and experience are utilised. Analysis is used for project cost control, profitability analysis, planning, scheduling and optimisation of operational research, *etc*. In the case of renewable energy systems, it is necessary to work out its economic viability so that the users of the technology may know the importance and they can utilise the area under their command to their best advantage.

An effective economic analysis can be made by the knowledge of cost analysis, using cash-flow diagrams and some other methods.

Techno-economic analysis of renewable energy systems mainly depends on the following factors:

- initial investment/present value/first cost for construction of renewable energy system;
- annual operating cost of renewable energy system;
- annual maintenance cost;
- annual energy output either in terms of thermal energy or electrical energy;
- the rate of interest;
- overhauling cost of renewable energy system, if any, during life of the system;
- life of the system and its salvage value.

In addition to the above points, it is also necessary to consider the following points for the economic analysis:

(i) an impact on the environment due to CO_2 emission by the embodied energy (one time) of renewable energy systems;
(ii) the energy used to operate it (annually) and pretreatments, *etc*.; and
(iii) CO_2 credit (CC) earned due to use of the renewable energy system, *etc*.

Advanced Renewable Energy Sources
G. N. Tiwari and R. K. Mishra

Published by the Royal Society of Chemistry, www.rsc.org

For effective economic analysis of renewable energy systems, the subsequent sections deal with the knowledge of cost analysis, cash-flow diagram, payback time and benefit/cost (B/C) analysis *etc*.

12.2 COST ANALYSIS

Financial evaluation of renewable energy technologies necessitates that various energy-resource technology combinations for a given end use are compared with each other. For such comparisons it is necessary that monetary values at different points in time be reduced to an equivalent basis.

12.2.1 Capital Recovery Factor

Let P be the present amount invested at zero ($n=0$) time at the interest rate of i per year. If S_n be its future value at the end of n years, then the cash flow can be diagrammatically shown as follows:

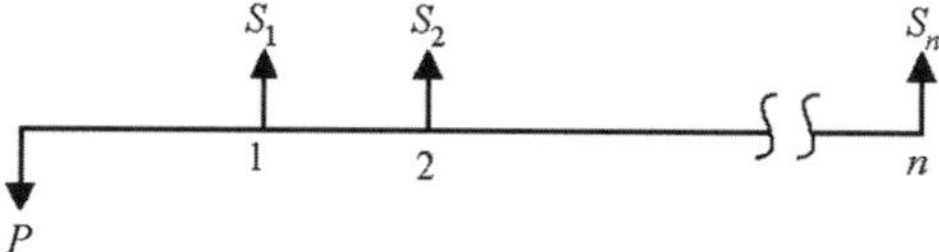

At the end of one year, the time value of investment P is given by

$$S_1 = P + iP = P(1+i)$$

At the end of the second year the value becomes

$$S_2 = S_1 + i\,S_1 = P(1+i) + iP(1+i) = P(1+i)(1+i) = P(1+i)^2$$

Similarly, at the end of the 3rd and nth years, respectively, the value becomes

$$S_3 = P(1+i)^3 \text{ and } S_n = P(1+i)^n$$

For simplicity, assuming S_n to be S, the above equation can be written as

$$S = P(1+i)^n \tag{12.1a}$$

Here, $S > P$ and for $i > 0$, considering compound interest law, the above equation can be simplified as

$$S = P\,F_{PS}$$

i.e.

Future value = (Present value) × (Future value factor) (12.1b)

The F_{PS} is more completely designated as $F_{PS,i,n}$, hence

$$F_{PS,i,n} = (1+i)^n \tag{12.1c}$$

where $F_{PS,i,n}$ is known as the **compound interest factor or future value factor**, which evaluates the future amount if the present amount is known, i.e. conversion of P into S. Thus, the compound interest factor when multiplied with the present value gives the future value.

If one year is divided into p equal units of period, then n becomes np and i become i/p, which is the rate of return per unit period. Substitution of these values in Eqs. (12.1b) and (12.1c), one gets

$$S = P\left(1+\frac{i}{p}\right)^{np}$$

The above equation can be further written as,

$$S = P\left[\left(1+\frac{i}{p}\right)^{p}\right]^{n}$$

where the expression $(1+i/p)^p$ can be expressed as follows

$$\left(1+\frac{i}{p}\right)^{p} = 1+\text{effective rate of return}$$

or

$$\text{effective rate of return} = \left(1+\frac{i}{p}\right)^{p} - 1 = i \text{ for } p = 1$$

$$> i \text{ for } p > 1 \quad (12.2)$$

For simple interest,

$$S = P(1+ni) = P + iPn \quad (12.3)$$

Equation (12.1a) can be rewritten as,

$$P = S/(1+i)^n$$

or,

$$P = S(1+i)^{-n} \quad (12.4a)$$

This shows that the future amount (at the nth year) is reduced when converted against the calendar to the present value (at zeroth time), assuming i to be positive.

Further, Eq. (12.4a) can be rewritten as

$$P = SF_{SP} \quad (12.4b)$$

or,

Present value = future value × (present value factor)

The numerical value of F_{SP} will always be less than unity due to the denominator always being higher than one. For this reason, present-worth calculations are generally referred to as **discounted cash flow (DCF)** method. Other terms generally used in reference to **present-worth (PW)** calculations are **present value (PV)** and **net present value (NPV)**.

From Eqs. (12.1b) and (12.4b), F_{PS} and F_{SP} can be related as,

$$F_{PS} = \frac{1}{F_{SP}}$$

or

$$F_{PS} \cdot F_{SP} = 1 \quad (12.5)$$

EXAMPLE 12.1

A low-interest loan of USD \$ 2000 is provided for the purchase of a low-capacity hybrid solar dryer for a period of 18 months at a simple interest rate of 5 percent. What is the future amount due at the end of the loan period?

Solution

Simple interest $(I) = P\ n\ I$

$$= 2000 \times 18/12 \times 5/100 \text{ (18 months} = 12/18 \text{ years, } 5\% = 5/100)$$

or,

$$I = \text{USD \$ } 150$$

Thus, the total amount due at the end of the loan period

$$= 2000 + 150 = \text{USD \$ } 2{,}150$$

EXAMPLE 12.2

If USD \$ 20 000 compounds to USD \$ 28 240 in 4 yr of a given solar system, what will be the rate of return?

Solution

Using Eq. (12.1a), $S = P\,(1 + i)^n$ and substituting $S =$ USD \$ 28 240; $P =$ USD \$ 20 000 and $n = 4$, we get

$$28\,240 = 20\,000\ (1 + i)^4 \text{ or } (1 + i)^4 = 1.412$$

Solving the above equation, we get

$$i = 0.09 \text{ or } 9\% \text{ per year.}$$

EXAMPLE 12.3

How long will it take for money to double if compounded annually at 10% per year?

Solution

Let us assume that the money doubles in n years Then $S = 2\ P$.
Using Eq. (12.1a) and substituting $S = 2\ P$, we get

$$2\ P = P\ (1 + 0.10)^n$$
$$2 = (1 + 0.10)^n$$

Solving the above equation, we get

$$\log 2 = n \log 1.1, \text{ i.e. } n = 7.3 \text{ yr}$$

The money will be doubled in 7.3 yr

EXAMPLE 12.4

Calculate the effective rate of return for 10% interest for $p=5$ and $p=12$.

Solution

From Eq. (12.3), we have

$$\text{Effective rate of return} = \left(1+\frac{i}{p}\right)^p - 1$$

For $p=5$; the

$$\text{Effective rate of return} = \left(1+\frac{0.10}{5}\right)^5 - 1 = (1.02)^5 - 1 = 0.104$$

For $p=12$; the

$$\text{Effective rate of return} = \left(1+\frac{0.10}{12}\right)^{12} - 1 = 0.1047$$

EXAMPLE 12.5

It is estimated that about 120 million households in the country can benefit from the use of improved PVT drying techniques. What is the required growth rate to achieve the potential in the next 20 years if the number of improved drying techniques disseminated so far is 30 million?

Solution

For $S=120$ million; $P=30$ million and $n=20$ years

From Eq. (12.1a), we have after taking log of both sides,

$$\log(1+i) = 1/n \log(S/P)$$
$$\text{or } \log(1+i) = 1/20 \log(120/30)$$
$$\text{or } \log(1+i) = 0.05 \log 4$$
$$\text{or} = 0.030103$$

which may be solved to give

$$i \approx 0.07177 \text{ (or } \approx 7.18\%)$$

Thus, a compound rate growth rate of more than 7 percent could be required to achieve the estimated potential of improved drying techniques utilisation in the country in the next 20 years.

EXAMPLE 12.6

A farmer borrows USD \$ 2000 to buy a PV/T hybrid solar dryer and returns USD \$ 2100 at the end of six months. What was the rate of interest paid by the farmer?

Solution

We have $S=2100$, $P=2000$ and $n=6/12$. Thus, using Eq. (12.3) we can write:

$$2100 = 2000\left(1+\frac{6}{12}i\right)$$

Simplifying, we can write:

$$1.05 = 1 + 0.5\,i,$$

or,

$$i = 0.10 \text{ or } 10\%$$

EXAMPLE 12.7

The owner of a small restaurant borrows USD $ 10 000 for a hybrid PVT solar water heater at 10% for 4 yr and 4 months. Considering compound interest, calculate the money paid.

Solution

Using Eq. (12.3), we have

$$S = 10\,000\left[1 + \frac{4}{12} \times 0.1\right] = \text{Rs. } 10\,333$$

The future amount after 4 months is evaluated as USD $ 10 333, which becomes P for another 4 yr. For compound interest, using Eqs. (12.1b) and (12.1c), we have:

$$S = PF_{PS,10\%,4} = 10\,333(1 + 0.1)^4$$

Thus, substituting the numerical values in above equation

$$S = 10\,333(1.4641) = \text{USD } \$\,15\,129.$$

12.2.2 Unacost

In solving engineering economic problems it is convenient to diagram expenditures (debits) and receipts (credits) as vertical lines positioned along a horizontal line representing time. Expenditures and receipts can point in opposite directions. By using this concept, a uniform annual amount will be discussed.

The smallest unit of time considered normally is a year. Consider a uniform end-of-year annual amount R (**unacost**) for every year for a period of n years. The diagram for this is as shown below:

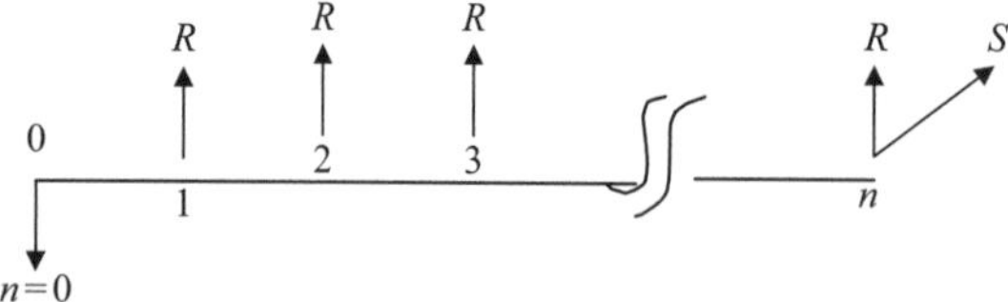

Let P_R represents single present value at initial time (*i.e.* at $n = 0$), then by Eq. (12.4a), we get

$$P_R = R\left[\frac{1}{1+i} + \frac{1}{(1+i)^2} + \cdots\cdots + \frac{1}{(1+i)^n}\right] \tag{12.6a}$$

This can be written as

$$P = R\sum_{1}^{n} \frac{1}{(1+i)^n}$$

$$\textbf{Present-worth factor (PWF)} = \frac{1}{(1+i)^n}$$

Equation (12.6a) is a geometric series, which has $1/(1+i)$, as the first term and $1/(1+i)$ as the ratio of n successive terms. The term summation of geometric series in Eq. (12.6a) can be further evaluated as:

$$\sum_{1}^{n}\frac{1}{(1+i)^n}=\frac{1}{(1+i)}\left[\frac{1-\left\{\frac{1}{(1+i)}\right\}^{n}}{1-\frac{1}{(1+i)}}\right]=\frac{(1+i)^n-1}{i(1+i)^n}$$

Substituting in Eq. (12.6a), we get

$$PR=R\left[\frac{(1+i)^n-1}{i(1+i)^n}\right]=R\,F_{RP,i,n} \tag{12.6b}$$

or,

Present value = (Unacost) × (Unacost present value factor)

where

$$F_{RP,i,n}=\left[\frac{(1+i)^n-1}{i(1+i)^n}\right] \tag{12.6c}$$

$F_{RP,i,n}$ is the **equal-payment series present value factor or annuity present value factor**.

If R is considered as cost of annual energy saving and CC is the carbon credit due to use of renewable energy technology, then Eq. (12.6b), which is present worth for net income from renewable energy technology, becomes

$$P_R=[R+\mathrm{CC}]\,F_{RP,i,n} \tag{12.6d}$$

Equation (12.6b) can also be rewritten as

$$R=P_R\left[\frac{i(1+i)^n}{(1+i)^n-1}\right]=P_R\,F_{PR,i,n}$$

or,

Unacost = (Present value) × (Capital recovery factor)

where, $F_{PR,i,n}$

$$F_{PR,i,n}=\left[\frac{i(1+i)^n}{(1+i)^n-1}\right]=CRF \tag{12.7b}$$

This is also known as the **capital recovery factor** (CRF). The relation between the **equal-payment series present value factor and** the **capital recovery factor** can be obtained by Eqs. (12.6c) and (12.7b) as

$$F_{RP,i,n}=\frac{1}{F_{PR,i,n}} \tag{12.7c}$$

EXAMPLE 12.8

A large-capacity water heating system is expected to save USD $ 4000 every year in terms of fuel savings. If the effect of escalation in the price of fuel saved is neglected, what is the present worth of fuel saving in the 5th, 10th, 15th, 20th, 25th and 30th years for a discount rate of 12 percent?

Solution

Given that the amount of fuel saving is USD $ 4000 per year and $i = 0.12$. The values of the present-worth factors and the corresponding present worth of annual fuel saving for the desired years are tabulated below:

Year (n) savings	*Present-worth factor* $(PWF) = \left[\frac{1}{(1+i)^n}\right]$	*Present worth of fuel savings (4000 × PWF)*
5	$1/(1.2)^5 = 0.5670$	2269.7
10	$1/(1.2)^{10} = 0.3220$	1287.8
15	$1/(1.2)^{15} = 0.1827$	730.7
20	$1/(1.2)^{20} = 0.1037$	414.6
25	$1/(1.2)^{25} = 0.0588$	235.2
30	$1/(1.2)^{30} = 0.0334$	133.5

It may be noted that the present worth of fuel savings in later years of the useful life of the domestic solar water heating system is rather small. Thus, the present value analysis of a renewable energy system with longer useful life may not be representative of its actual usefulness to the user.

12.2.3 Sinking Fund Factor (SFF)

The future value S at the end of n years can be distributed into an equal uniform end-of-year annual amount R as discussed above. It will also be know as a *uniform end-of-year annual amount* but corresponding to the future value S.

Equation (12.7a) can be expressed in terms of S by using Eq. (12.4a) as

$$R = S(1+i)^{-n} \cdot \left[\frac{i(1+i)^n}{(1+i)^n - 1}\right] = S \cdot \left[\frac{i}{(1+i)^n - 1}\right] = S\,F_{SR,i,n} \qquad (12.8a)$$

or,

Unacost = (Future amount) × (Sinking-fund factor)

where,

$$F_{SR,i,n} = \left[\frac{i}{(1+i)^n - 1}\right] = \text{SFF} \qquad (12.8b)$$

This is referred to as sinking-fund factor (SFF). This is mostly used to calculate the **uniform end-of-year annual amount** corresponding to the salvage value of any system in future after completion of the system life. Equation (12.8a) can be rewritten as:

$$S = R\left[\frac{(1+i)^n - 1}{i}\right] = R\,F_{RS,i,n} \qquad (12.9a)$$

or,

Future amount = (Unacost) × (Equal payment series future value factor)

where,

$$F_{RS,i,n} = \left[\frac{(1+i)^n - 1}{i}\right] \qquad (12.9b)$$

This is known as the **equal payment series future value factor.** The reciprocal relation between the **sinking-fund factor** and the **equal payment series future value factor** can be obtained by Eqs. (12.8b) and (12.9b) as

$$F_{SR,i,n} = \frac{1}{F_{RS,i,n}} \tag{12.9c}$$

A uniform beginning of year annual amount, (say) R_b, can be derived in terms of P and S as R and R_b have the following relationship

$$R = R_b(1 + i) \tag{12.9d}$$

The values of various conversion factors with number of years for a given rate of interest are given in the Table 12.1.

Table 12.1 The values of conversion factors.[5]

i=0.04							
N	F_{PS}	F_{SP}	F_{RP}	F_{PR}	F_{RS}	F_{SR}	F_{PK}
1	1.04	0.962	0.962	1.04	1	1	26
2	1.082	0.925	1.886	0.53	2.04	0.49	13.255
3	1.125	0.889	2.775	0.36	3.122	0.32	9.009
4	1.17	0.855	3.63	0.275	4.246	0.235	6.887
5	1.217	0.822	4.452	0.225	5.416	0.185	5.616
6	1.265	0.79	5.242	0.191	6.633	0.151	4.769
7	1.316	0.76	6.002	0.167	7.898	0.127	4.165
8	1.369	0.731	6.733	0.149	9.214	0.109	3.713
9	1.423	0.703	7.435	0.134	10.583	0.094	3.362
10	1.48	0.676	8.111	0.123	12.006	0.083	3.082
11	1.539	0.65	8.76	0.114	13.486	0.074	2.854
12	1.601	0.625	9.385	0.107	15.026	0.067	2.664
13	1.665	0.601	9.986	0.1	16.627	0.06	2.504
14	1.732	0.577	10.563	0.095	18.292	0.055	2.367
15	1.801	0.555	11.118	0.09	20.024	0.05	2.249
16	1.873	0.534	11.652	0.086	21.825	0.046	2.146
17	1.948	0.513	12.166	0.082	23.697	0.042	2.055
18	2.026	0.494	12.659	0.079	25.645	0.039	1.975
19	2.107	0.475	13.134	0.076	27.671	0.036	1.903
20	2.191	0.456	13.59	0.074	29.778	0.034	1.84
i =0.06							
N	F_{PS}	F_{SP}	F_{RP}	F_{PR}	F_{RS}	F_{SR}	F_{PK}
1	1.06	0.943	0.943	1.06	1	1	17.667
2	1.124	0.89	1.833	0.545	2.06	0.485	9.091
3	1.191	0.84	2.673	0.374	3.184	0.314	6.235
4	1.262	0.792	3.465	0.289	4.375	0.229	4.81
5	1.338	0.747	4.212	0.237	5.637	0.177	3.957
6	1.419	0.705	4.917	0.203	6.975	0.143	3.389
7	1.504	0.665	5.582	0.179	8.394	0.119	2.986
8	1.594	0.627	6.21	0.161	9.897	0.101	2.684
9	1.689	0.592	6.802	0.147	11.491	0.087	2.45
10	1.791	0.558	7.36	0.136	13.181	0.076	2.264
11	1.898	0.527	7.887	0.127	14.972	0.067	2.113
12	2.012	0.497	8.384	0.119	16.87	0.059	1.988
13	2.133	0.469	8.853	0.113	18.882	0.053	1.883

Table 12.1 (*Continued*).

$i = 0.06$							
N	F_{PS}	F_{SP}	F_{RP}	F_{PR}	F_{RS}	F_{SR}	F_{PK}
14	2.261	0.442	9.295	0.108	21.015	0.048	1.793
15	2.397	0.417	9.712	0.103	23.276	0.043	1.716
16	2.54	0.394	10.106	0.099	25.672	0.039	1.649
17	2.693	0.371	10.477	0.095	28.213	0.035	1.591
18	2.854	0.35	10.828	0.092	30.906	0.032	1.539
19	3.026	0.331	11.158	0.09	33.76	0.03	1.494
20	3.207	0.312	11.47	0.087	36.786	0.027	1.453

$i = 0.08$							
N	F_{PS}	F_{SP}	F_{RP}	F_{PR}	F_{RS}	F_{SR}	F_{PK}
1	1.08	0.926	0.926	1.08	1	1	13.5
2	1.166	0.857	1.783	0.561	2.08	0.481	7.01
3	1.26	0.794	2.577	0.388	3.246	0.308	4.85
4	1.36	0.735	3.312	0.302	4.506	0.222	3.774
5	1.469	0.681	3.993	0.25	5.867	0.17	3.131
6	1.587	0.63	4.623	0.216	7.336	0.136	2.704
7	1.714	0.583	5.206	0.192	8.923	0.112	2.401
8	1.851	0.54	5.747	0.174	10.637	0.094	2.175
9	1.999	0.5	6.247	0.16	12.488	0.08	2.001
10	2.159	0.463	6.71	0.149	14.487	0.069	1.863
11	2.332	0.429	7.139	0.14	16.646	0.06	1.751
12	2.518	0.397	7.536	0.133	18.977	0.053	1.659
13	2.72	0.368	7.904	0.127	21.495	0.047	1.582
14	2.937	0.34	8.244	0.121	24.215	0.041	1.516
15	3.172	0.315	8.559	0.117	27.152	0.037	1.46
16	3.426	0.292	8.851	0.113	30.324	0.033	1.412
17	3.7	0.27	9.122	0.11	33.75	0.03	1.37
18	3.996	0.25	9.372	0.107	37.45	0.027	1.334
19	4.316	0.232	9.604	0.104	41.446	0.024	1.302
20	4.661	0.215	9.818	0.102	45.762	0.022	1.273

$i = 0.10$							
N	F_{PS}	F_{SP}	F_{RP}	F_{PR}	F_{RS}	F_{SR}	F_{PK}
1	1.1	0.909	0.909	1.1	1	1	11
2	1.21	0.826	1.736	0.576	2.1	0.476	5.762
3	1.331	0.751	2.487	0.402	3.31	0.302	4.021
4	1.464	0.683	3.17	0.315	4.641	0.215	3.155
5	1.611	0.621	3.791	0.264	6.105	0.164	2.638
6	1.772	0.564	4.355	0.23	7.716	0.13	2.296
7	1.949	0.513	4.868	0.205	9.487	0.105	2.054
8	2.144	0.467	5.335	0.187	11.436	0.087	1.874
9	2.358	0.424	5.759	0.174	13.579	0.074	1.736
10	2.594	0.386	6.145	0.163	15.937	0.063	1.627
11	2.853	0.35	6.495	0.154	18.531	0.054	1.54
12	3.138	0.319	6.814	0.147	21.384	0.047	1.468
13	3.452	0.29	7.103	0.141	24.523	0.041	1.408
14	3.797	0.263	7.367	0.136	27.975	0.036	1.357
15	4.177	0.239	7.606	0.131	31.772	0.031	1.315
16	4.595	0.218	7.824	0.128	35.95	0.028	1.278
17	5.054	0.198	8.022	0.125	40.545	0.025	1.247
18	5.56	0.18	8.201	0.122	45.599	0.022	1.219
19	6.116	0.164	8.365	0.12	51.159	0.02	1.195
20	6.728	0.149	8.514	0.117	57.275	0.017	1.175

EXAMPLE 12.9

Derive an expression for R_b in terms of P and S.

Solution

Substitute the expression of R from Eq. (12.6b) into Eq. (12.9a)

$$R_b(1+i) = PF_{PR,i,n}$$

so,

$$R_b = \frac{P}{(1+i)} \cdot F_{PR,i,n}$$

Similarly, from Eq. (10.8b),

$$R_b = \frac{S}{(1+i)} \cdot F_{SR,i,n}$$

EXAMPLE 12.10

The estimated salvage value of a large capacity PVT solar water heater at the end of its useful lifetime of 20 years is 5000. Determine its present worth for a discount rate of 10 percent.

Solution

From Eq. (12.6a)

$$P = S\left[\frac{1}{(1+i)^n}\right]$$

In the present example,

$$S = \text{USD } \$\, 5000;\ n = 20\,\text{years};\ i = 10\%$$

Thus, the present worth

$$P = 5000\ [1/(1+0.1)^{20} = \text{USD } \$\, 743.22$$

12.3 CASH FLOW

Cash flow is generally known as the single most pressing concern of any economic analysis. In its simplest form, cash flow is the movement of money in and out of any business and is the life-blood of all growing businesses and the primary indicator of business health. The cash flow is understood graphically on a timescale with the help of a line diagram known as **cash-flow diagram**. The net cash flow is calculated as:

$$\textbf{Net cash flow} = \textbf{Receipts (Credits)} - \textbf{Expenses (Debits)} \tag{12.10}$$

As discussed above this net cash flow can be represented by a cash-flow diagram

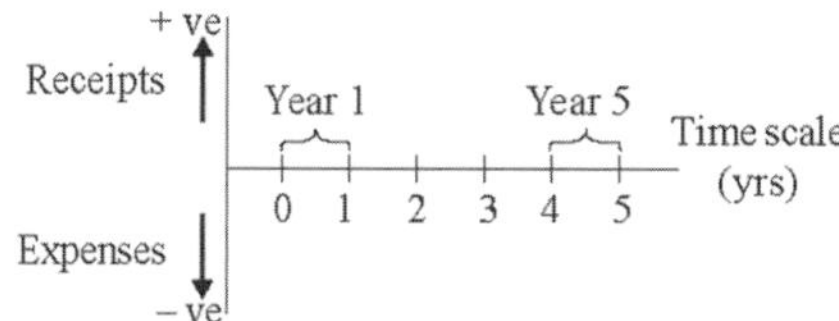

In the above cash-flow diagram, a **uniform end-of-year annual amount (R)** can be considered at the end of each year on the timescale. This cash-flow diagram will be used in the following examples.

EXAMPLE 12.11

A person plans to create a forborne annuity by depositing USD \$ 1000 at the end of the year, for 8 yr. He wants to withdraw the money at the end of 14 yr from now to buy a hybrid solar water heater. Find the accumulated value at the end of the fourteenth year, if money is worth 10 percent per year.

Solution

Let X be the amount available at the 14th year, which can be considered as a receipt. The cash-flow diagram for the payment is,

0 1 2 8 9 14 X

1000 1000 1000

The present value (zero time) can be calculated by using Eq. (12.6a), as

$$P = 1000\, F_{RP,\,10\%,8} = 1000 \times 5.3349 = \text{USD \$ } 5334.90$$

If this amount is deposited for 14 years, then the future value at the end of 14 years (Eq. (12.1a)) will be

$$S = 5334.9 \times F_{PS,\,10\%,14} = 5334.9\,(3.79) = \text{USD \$ } 20\,259$$

The above cash-flow diagram can also be drawn by considering USD \$ 1000 paid for 14 years less USD \$ 1000 paid as annuity for the last 6 years.

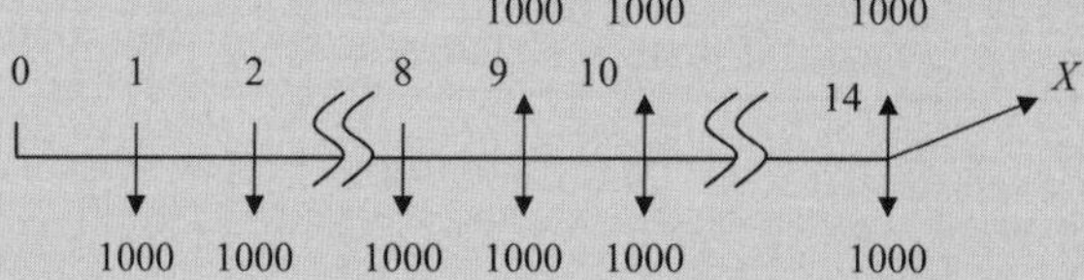

By using Eq. (10.9a) we get

$$S = 1000\left[\frac{(1.10)^{14} - 1}{0.10} - \frac{(1.10)^{6} - 1}{0.10}\right] = 1000[27.975 - 7.7156]$$

or,

$$S = \text{USD \$ } 20\,259$$

EXAMPLE 12.12

A person wants a down payment of USD \$ 2000 on a hybrid solar systems of an amount of USD \$ 10 000. An annual end-of-year payment (R) of USD \$ 1174.11 is required for 12 years. However, the person elects to pay USD \$ 1000 yearly and a balance payment at the end. Find the balance payment if money is worth 10 percent interest.

Solution

Let X is the balance payment. The cash-flow diagram is

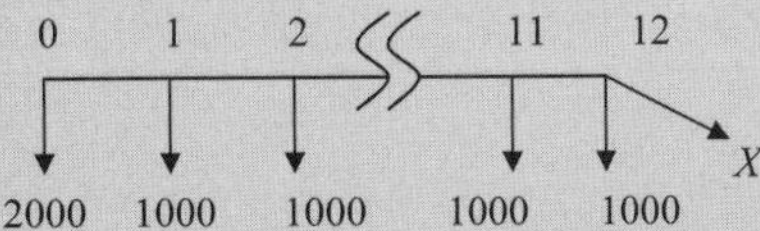

After using the cash-flow diagram, Eqs. (12.4b) and (12.6a), we can write,

$$10\,000 = 2000 + 1000\,F_{RP,10\%,12} + X\ F_{SP,10\%,12}$$
$$= 2000 + 1000(6.8137) + X(0.31863)$$
$$X = 3723.10$$

The balance payment is USD \$ 3723.10

EXAMPLE 12.13

A person decides to spend USD \$ 3000 in the first, second, third and fourth year on energy-efficient equipment and agrees to set aside a certain amount now and each year thereafter until the fourth year. If the contribution forms an arithmetical progression for all years increasing by 20 percent after the first year; calculate his first contribution if money is worth 10 percent.

Solution

Let us assume that his first contribution is x. The cash-flow diagram can be shown as:

Consider two years from now as the focal point. Now, using the time-value conversion relation in the above cash-flow diagram, we get,

x $1.2x$ $1.4x$ $1.6x$ $1.8x$

0 1 2 3 4

3000 3000 3000 3000

$$x\,(1.10)^{-2} + 1.2x\,(1.10)^{-1} + 1.4x\,(1.10)^{0} + 1.6x\,(1.10)^{1} + 1.8x\,(1.10)^{2}$$

$$= 3000\,(1.10)^{-1} + 3000\,(1.10)^{0} + 3000\,(1.10)^{1} + 3000\,(1.10)^{2}$$

$$7.2553\,x = 3000 \times 4.2191$$

$$x = 1744.56$$

The first contribution would be USD \$ 1744.56.

12.4 COST COMPARISONS WITH EQUAL DURATION

In this section a uniform expense is referred to as a uniform end-of-year cost.

EXAMPLE 12.14

Two hybrid PVT solar systems have the following cost comparison. Which system is more economical if the money is worth 10 percent per year?

Economic Components	*System (A)*	*System (B)*
First cost (USD $)	30 000	15 000
Uniform end-of-year maintenance per year (USD $)	000	5000
Overhaul, end of the third year (USD $)	–	3500
Salvage value (USD $)	4000	1000
Life of the system (years)	5	5
Benefit from quality control as a uniform end-of-year amount per year (USD $)	1000	–

Solution

Cash-flow diagram for each system have been shown as follows:

System A

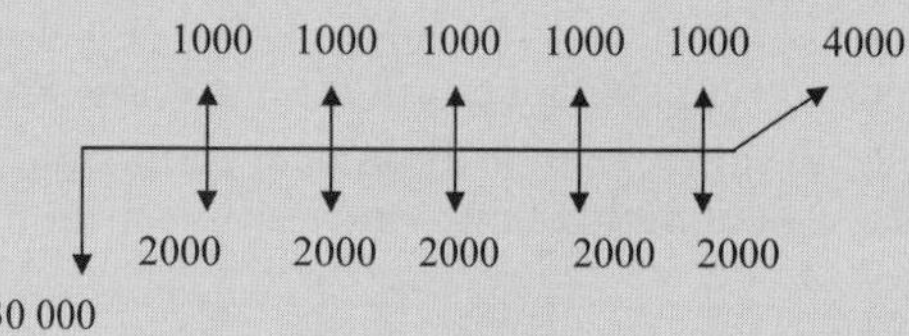

System B

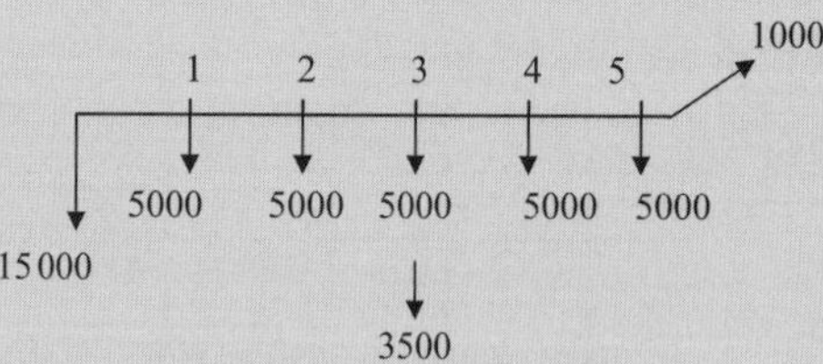

The present value of the costs for system A can be obtained by using Eqs. (12.4b) and (12.6a) as:

$$P_{AS} = 30\ 000 + (2000 - 1000)\ F_{RP,\,10\%,\,5} - 4000\ F_{SP,\,10\%,\,5}$$
$$= 30\,000 + 1000\,(3.7908) - 4000\,(0.62092) = \text{USD}\ \$\,31,307.12$$

The present value of the costs for system B can be obtained by using Eqs. (12.4b) and (12.6a) as follows:

$$P_{BS} = 15000 + 5000\ F_{RP,10\%,5} + 3500\ F_{SP,10\%,3} - 1000\ F_{SP,10\%,5}$$
$$= 15\ 000 + 5000 \times 3.7908 + 3500 \times 0.75131 - 1000 \times 0.62092$$
$$= 15\ 000 + 18\ 954 + 2629.55 - 620.92 = \text{USD}\ \$\ 35\ 962.63$$

From the above calculations, it is clear that system A is more economical than system B.

12.5 COST COMPARISONS WITH UNEQUAL DURATION

If two energy-efficient systems have different duration of lives, a fair comparison can be made only on the basis of equal duration. One of the methods for comparison is to compare single present value of costs on the basis of a common denominator of their service lives.

12.5.1 Single Present Value Method

EXAMPLE 12.15

Two energy-efficient systems have the following cost comparison. Which system is more economical if the money is worth 10 percent per year?

Cost Components (USD $)	*System (A)*	*System (B)*
First cost	20 000	30 000
Uniform end-of-year maintenance	4000	3000
Salvage value	500	1500
Service life, years	2	3

Solution

The cash-flow diagrams for both systems are first reduced to single present value of the cost.

System A

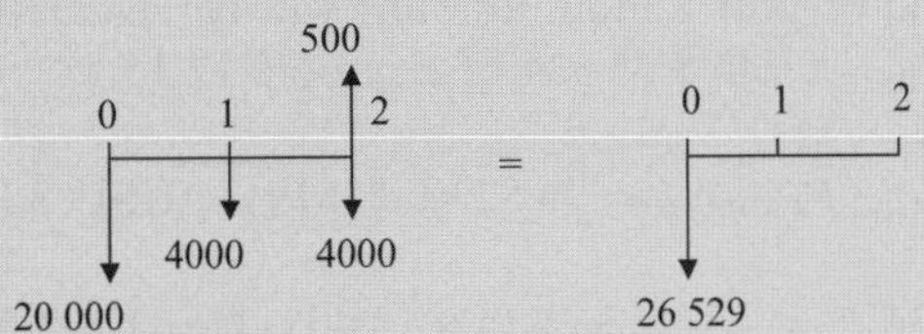

System B

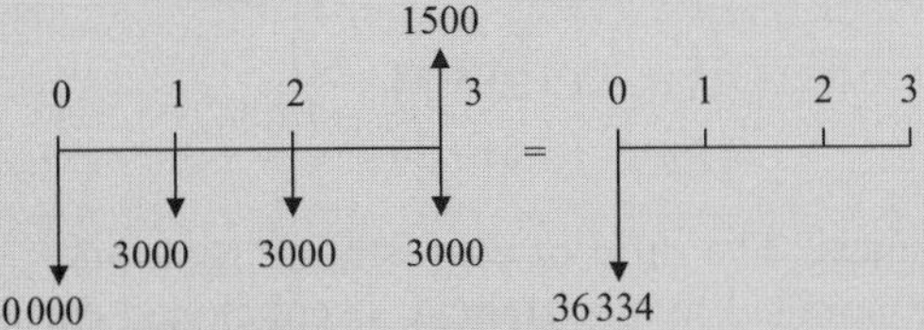

The simplified diagrams are now repeated to obtain the 6-yr duration. Note that the present value of system A is 26 529 at its time of installation.

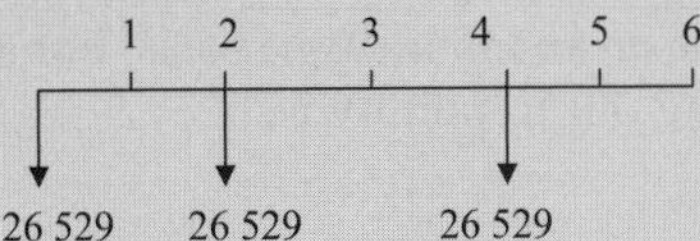

Similarly, the present value of system B is USD $ 36 334 at the time of installation. The cash-flow diagram for the six-year duration is

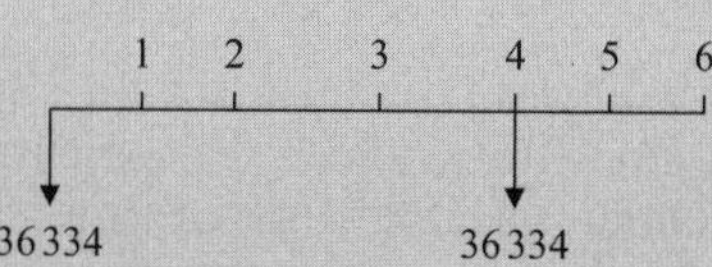

The present value of each of the preceding diagrams at 10 percent per year is

$$\begin{aligned} P_{A6} &= 26\,529 + 26\,529\ F_{SP,10\%,2} + 26\,529\ F_{SP,10\%,4} \\ &= 26\ 529 + 26\ 529\ (1.10)^{-2} + 26\ 529\ (1.10)^{-4} \\ &= 26\ 529 + 21\ 924.7 + 18\ 119.66 = \text{USD}\ \$\ 66\ 573.4 \end{aligned}$$

Similarly,

$$P_{B6} = \text{USD}\ \$\ 63\ 632.2$$

The ratio of cost is

$$\frac{P_{A6}}{P_{B6}} = \frac{66\ 573.45}{63\ 632.27} = 1.0462$$

Thus, system B is more economical than system A.

12.5.2 Cost Comparison by Annual Cost Method

In this case, the uniform end-of-year annual amount will be calculated by using Eq. (10.7b) for $P_{A2} = \text{USD}\ \$\ 26\,529$ and, $P_{B3} = \text{USD}\ \$\ 36\,334$ of Example 12.15.

$$R_A = P_{A2}\, F_{PR,10\%,2} = 26\,529(0.57619) = \text{USD}\ \$\,15\,285.74$$

$$R_B = 36\ 334\ F_{PR,10\%,3} = 36\ 334 \times (0.40211) = \text{USD}\ \$\ 14\,610.26$$

The unacost for two systems are,

$$\frac{R_A}{R_B} = \frac{15\ 285.74}{14\ 610.26} = 1.0462$$

System B is more economical. The ratio of cost is more than one.
The system B is more economical than system A as concluded earlier.

12.5.3 Cost Comparison by Capitalised Cost

Capitalised cost is the present value on an infinite time basis. For a system costing P_n and lasting for n years, the present value replacing out to infinity is,

$$K = P_n \sum_{x=0}^{\text{infinity}} \frac{1}{(1+i)^{xn}} = P_n \left[1 + \frac{1}{(1+i)^n} + \frac{1}{(1+i)^{2n}} + \cdots\cdots \right] \qquad (12.11)$$

This is a geometric series with the first term as 1 and the ratio of the consecutive terms as $1/(1+i)^n$. Its summation is given by,

$$\sum \frac{1}{(1+i)^{xn}} = 1\frac{1-\left(\frac{1}{(1+i)^n}\right)^{\text{infinity}}}{1-\frac{1}{(1+i)^n}} = \frac{(1+i)^n}{(1+i)^n-1} \tag{12.12a}$$

Equation (12.12) becomes,

$$K = P_n F_{PK,i,n} \tag{12.12b}$$

or,

Capitalised cost = Present value basis n years duration × (capitalised cost factor)

where K is the capitalised cost and $F_{PK,i,n}$ the factor that converts a present value to capitalised cost (Table 12.1), also known as capitalised cost factor and given as,

$$F_{PK,i,n} = \frac{i(1+i)^n}{(1+i)^n-1} \tag{12.13a}$$

From Eq. (10.7b), $F_{PR,i,n}$ is given as,

$$F_{PR,i,n} = \frac{i(1+i)^n}{(1+i)^n-1} \tag{12.13b}$$

After comparison of Eqs. (12.13a) and (12.13b), $F_{PR,i,n,}$ and $F_{PK,i,n,}$ gives the relationship.

$$F_{PR,i,n} = i F_{PK,i,n} \tag{12.13c}$$

or,

Capital recovery factor = rate of return × (capitalised cost factor)

Similarly, from Eqs. (12.7a) and (12.12b) R and K are related as

$$R = iK \tag{12.14}$$

or,

Unacost = rate of return × (capitalised cost factor)

In this case, we solve Example 12.11 by using the capitalised cost method.
From Example 12.15, we have, P_{A2} = USD \$ 26 529 and P_{B3} = USD \$36 334.
By using Eq. (12.12b), we get

$$K_A = P_{A2}, F_{PK,10\%,2} = 26\,529 \times (5.7619) = \text{USD }\$\,152\,857.45$$

$$K_B = 36\,334\,F_{PK,10\%,3} = 36\,334 \times (4.0211) = \text{USD }\$\,146\,102.65$$

The ratio of cost is,

$$\frac{K_A}{K_B} = \frac{152\,857.45}{146\,102.65} = 1.0462$$

It is clear from the above calculation that the results obtained are the same as in the earlier solution. From this calculation, we can conclude that the system B is more economical than system A.

As a matter of fact, it is possible to convert a present value P_{n1}, of n_1 year's duration to an equivalent present value P_{n2} of n_2 year's duration.

$$P_{n1}\,F_{PR,i,n} = P_{n2}\,F_{PR,i,n2}$$

$$P_{n2} = P_{n1}\frac{F_{PR,i,n}}{F_{PR,i,n2}} \tag{12.15}$$

Hence, applying Eq. (12.12b) gives,
As reported earlier,

$$P_{A2} = \text{2-year duration} = \text{USD \$ 26 529}$$
$$P_{B3} = \text{3-year duration} = \text{USD \$ 36 334.}$$

Convert the present value of system B to an equivalent value for 2 years duration using Eq. (12.15)

$$P_{B2} = P_{B3}\frac{F_{PR,10\%,3}}{F_{PR,10\%,2}} = 36\,334\frac{0.40211}{0.57619} = \text{USD \$25 356.68}$$

$$\frac{P_{A2}}{P_{B2}} = \frac{26\ 529}{25\ 356} = 1.0462$$

Further, the result of cost ratio is the same as that obtained by various methods discussed earlier. Hence, system B is more economical.

12.6 ANALYTICAL EXPRESSION FOR PAYOUT TIME

The payback period (n), the number of years necessary to exactly recover the initial investment P, is computed by summing the annual cash-flow values and estimating n through the relation:

0 = –initial investment + sum of annual cash flows

For equal annual savings,

$$\text{Payback period (yrs)} = \frac{\text{Initial capital cost}}{\text{Annual operating cash flow}} \tag{12.16a}$$

For unequal annual savings,

$$0 = -P + \sum_{t=1}^{n} \mathrm{CF}_t(F_{SP,i\%,t}) \tag{12.16b}$$

where CF_t is the net cash flow at the end of year t. If cash flow is the same each year, the F_{RP} factor may be used in the above relation:

$$0 = -P + \mathrm{CF}_t(F_{RP,i\%,n}) \tag{12.17}$$

i.e. after n years, the cash flow will recover the investment and a return of i percent. If the expected retention period (life) of the asset/project is less than n years, then the investment is not advisable.

Considering i to be zero, Eq. (12.16b) becomes,

$$0 = -P + \sum_{t=1}^{n} \mathrm{CF}_t \tag{12.18}$$

and if CF_t values are assumed equal, then

$$n = \frac{P}{\mathrm{CF}} \qquad (i.e\ P = n \times \mathrm{CF}) \tag{12.19a}$$

There is little value in technoeconomic study for n computed from Eqs. (12.18) and (12.19). When $i\% > 0$ is used to estimate n, the results incorporate the risk considered in the project undertaken.

Using Eqs. (12.7c) and (12.17), the expression for the payback period for unequal annual savings can be written as,

$$n = \frac{\ln\left[\dfrac{\mathrm{CF}}{\mathrm{CF} - P \times i}\right]}{\ln[1 + i]} \tag{12.19b}$$

EXAMPLE 12.16

Energy-efficient systems purchased for USD \$ 18 000 is expected to generate annual revenues of USD \$ 3000, and have salvage value of USD \$ 3000 at any time during 10 years of anticipated ownership. If a 15 percent per year required return is imposed on the purchase, compute the payback period.

Solution

The cash flow for each year is USD \$ 3000 (P) with an additional revenue of USD \$ 3000 in year n. The cash-flow diagram is shown below:

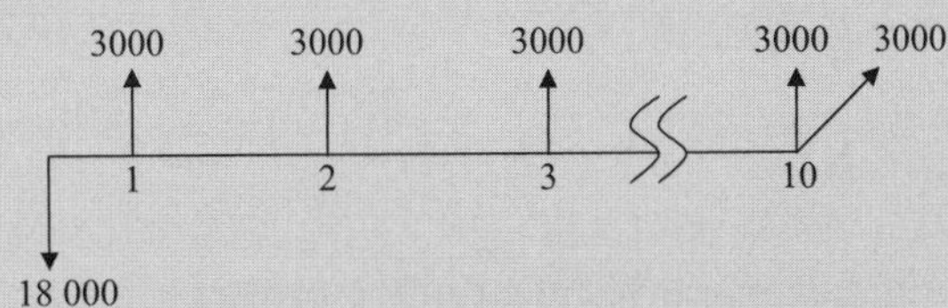

After using Eqs. (12.16b) and (12.17), respectively, for the above cash flow, one gets

$$0 = -P + \mathrm{CF_t}\left(F_{RP,15\%,n}\right) + S.V\left(F_{SP,15\%,n}\right)$$

or

$$0 = -18\,000 + 3000\,F_{RP,15\%,n} + 3000\,F_{SP,15\%,n}$$

The resulting payout time can be evaluated after further using Eqs. (12.4b) and (12.6c) for F_{SP} and F_{RP}, respectively, and we get $n = 15.3$ yr, which is not economical with such a high interest rate.

For $i = 0$, Eq. (12.18) can be used and we get,

$$0 = -18\,000 + n(3000) + 3000$$

The resulting payout time is 5 yr, which is most economical, without the interest rate.

12.7 NET PRESENT VALUE

The difference between the present value of the benefits and the costs resulting from an investment is the net present value (NPV) of the investment. A positive NPV means a positive surplus, indicating that the financial position of the investor will be improved by undertaking the project. Obviously, a negative NPV would indicate a financial loss. An NPV of zero would mean that the present value of all benefits over the useful lifetime is equal to the present value of all the costs. In mathematical terms,

$$\mathrm{NPV} = \sum_{j=0}^{n} \frac{B_j - C_j}{(1+i)^j} \tag{12.20}$$

where B_j stands for benefits at the end of the period j, C_j for costs at the end of period j, n is the useful life of the project and i the interest rate. Equation (12.20) involves subtracting of the cost from the benefits at any period and then multiplying the result by a single payment of the present-worth factor for that period. Finally, the NPV is determined by algebraically adding the results for all the periods under consideration.

It often happens that $(B_j - C_j)$ is constant for all j except for $j=0$. In such a case, Eq. (12.20) can be modified as

$$\mathrm{NPV} = (B_0 - C_0) + \sum_{j=1}^{n} \frac{B_j - C_j}{(1+i)^j}$$

Since B_0, the benefits in the 0th year, are invariably zero and $(B_j - C_j)$ is constant $(= B - C)$ for $j=1$ to n,

$$\mathrm{NPV} = -C_0 + (B - C)\sum_{j=1}^{n} \frac{1}{(1+i)^n}$$

or,

$$\mathrm{NPV} = -C_0 + (B - C)\left[\frac{(1+i)^n - 1}{i(1+i)^n}\right] \tag{12.21}$$

with C_0 representing the initial capital investment in the project.

EXAMPLE 12.17

A PV system for water pumping costs USD $ 10 000 to purchase and install on the field of a farmer. It is expected to save USD $ 1200 worth of diesel annually for the farmer and its annual maintenance cost is estimated at USD $ 100. Calculate the NPV of the investment on the PV system if the useful life of the system is 30 years and the interest rate is 8 percent.

Solution

Net annual benefits of using a PV system = 1200–100 = USD $ 1100

Since the amount of net annual benefits is constant over the useful life of the system, Eq. (12.21) can be used for determining the NPV.

$$i.e.\ \mathrm{NPV} = -C_0 + (B - C)\left[\frac{(1+i)^n - 1}{i(1+i)^n}\right]$$

$$= -10000 + (1100)\left[\frac{(1+0.08)^30 - 1}{0.08(1+0.08)^{30}}\right]$$

$$= -10\,000 + 12384$$

$$= 2384$$

Therefore, the investment in the PV system is a financially viable investment for the farmer.

Equations (12.20) and (12.21) are based on the assumption that the interest rate i remains constant over time. The NPV could also be calculated with different rates of interest rate over the jth period, Eq. (10.20) for NPV can be modified as

$$\mathrm{NPV} = (B_0 - C_0) + \frac{B_1 - C_1}{(1+i_1)} + \frac{B_2 - C_2}{(1+i_1)(1+i_2)} + \cdots\cdots + \frac{B_j - C_j}{(1+i_1)(1+i_2)\ldots\ldots(1+i_j)} + \cdots\cdots$$
$$\cdots\cdots + \frac{B_n - C_n}{(1+i_1)(1+i_2)\ldots\ldots(1+i_n)} \tag{12.22}$$

The acceptance criteria of an investment project, as evaluated from the NPV method are:

(a) **NPV > 0, accept the project**
(b) **NPV = 0, remain indifferent**
(c) **NPV < 0, reject the project**

As mentioned earlier, a positive NPV represents a positive surplus and therefore the project may be accepted subject to the availability of funds. A project with negative NPV should be rejected as the funds may be advantageously invested in the other projects. Thus, unless the project is mandatory only those investments having positive NPV may be accepted. In the case of mutually exclusive alternative investments the project with the highest positive NPV should be chosen.

12.7.1 Limitations of NPV Method

As regards the limitations of the NPV method, the following points are worth mentioning:

(a) The NPV method focuses only benefits and does not distinguish between and investment involving relatively large costs and benefits and one involving much smaller costs and benefits, as long as the two projects results in equal NPV. Thus, it does not give any indication of the scale of effort required to achieve the results.
(b) The results of NPV method are quite sensitive to the interest/discount rate chosen. Thus, failure to select an appropriate value of the interest rate used in the computation of NPV may alter or even reverse the relative ranking of different alternatives being compared using this method. For example, with a very low value of interest rate, an alternative with benefits spread far into the future may unjustifiably appear more profitable than an

alternative whose benefits are more quickly realised but is of a lower amount in undiscounted terms.

From Eq. (12.20) it may be noted that as the interest rate i is increased, each cash flow in the future is discounted to the present by a factor of the general form $1/(1+i)^j$. As i approach infinity, $1/(1+i)^j$ approaches zero. Mathematically, for two extreme values of the interest rate, Eq. (12.20) gives

$$\text{NPV} = \sum_{j=0}^{n} B_j - C_j \quad \text{for } i = 0,$$

$$\text{and NPV} = -C_0 \quad \text{for } i = \infty$$

EXAMPLE 12.18

The cost of a BIPV air-circulating collector is USD \$ 35 000. During its useful life of 20 years, besides other routine maintenance costs of USD \$ 300 each year, replacement of the wooden duct in the 10th year is expected to cost USD \$ 4000. Determine the equivalent annual cost of the system for an interest rate of 10 percent.

Solution

Present values of all the costs associated with the BIPV air collector

$$= 35\,000 + 300\,[1/(1+0.1) + 1/(1+0.1)^2 + \cdots\cdots + 1/(1+10.1)^{10}] + 4000/(1+0.1)^{10}$$

$$= 35\,000 + 300\left[\frac{(1+0.1)^{20}-1}{0.1\,(1+0.1)^{20}}\right] + \left[\frac{4000}{(1+0.1)^{10}}\right]$$

$$= 35\,000 + 300\,(8.51) + 1\,542.2$$

$$= \text{USD } \$\,39\,095.4$$

Hence, the equivalent annual cost of the BIPV system is

$$= 39\,095.4\left[\frac{0.1\,(1+0.1)^{20}}{(1+0.1)^{20}-1}\right]$$

$$= 39\,095.4(0.117)$$

$$= \text{USD } \$\,4574.1$$

12.8 BENEFIT/COST ANALYSIS

The benefit/cost ratio is another method of analysing and making a decision on investments. As its name suggests, the benefit/cost (B/C) ratio method of analysis is based on the ratio of the benefit/costs associated with a particular project. The ratio of benefit/costs as a measure of financial or economic efficiency conceptually simple and quite versatile and it measures cost efficiency. Obviously, the first step in a B/C ratio analysis is to identify the costs and benefits separately. In general, the benefits are advantages (fuel saving in the case of energy projects) expressed in monetary terms and the disadvantages are the associated disbenefits. The costs are the anticipated expenditures for construction, installation, operation, maintenance *etc*. The B/C ratio method has frequently been used by government agencies for projects whose benefits are reaped by the common public and the costs are incurred by the government. Therefore, the determination of whether an item is to be considered as a benefit, disbenefit or cost depends on who is affected by the consequences of the project implementation.

A project is considered to be attractive when the benefits derived from its execution exceed its associated costs.

The conventional B/C ratio is calculated as,

$$B/C = (\text{Benefits} - \text{Disbenefits})/\text{cost} = (B - D)/C$$

The modified B/C ratio, which is gaining support includes operation and maintenance (O & M) costs in the numerator and treats them in a manner similar to disbenefits, and is given by.

$$B\ /C = (\text{Benefits} - \text{Disbenefits} - \text{O \& M cost})/(\text{Initial investment})$$

The salvage value can also be considered in the denominator.

The B/C ratio influences the decision on the project approval.

If

$$\boldsymbol{B/C} > 1, \quad \text{accept the project}$$
$$\boldsymbol{B/C} < 1, \quad \text{reject the project.}$$

Thus, in the case of mutually exclusive projects, the B/C ratio gives a method to compare them against each other.

Benefits (B): Benefits are the advantages to the owner.

Disbenefits (D): The project involves disadvantages to the owner.

Costs: The anticipated expenditures for construction, operation, maintenance, *etc*.

Owner: Public: One who incurs the costs as the government?

Let B and C be the present values of the cash inflows (benefits) and outflows (costs) defined as

$$B = \sum_{j=0}^{n} \frac{B_j}{(1+i)^j} \tag{12.23}$$

$$C = \sum_{j=0}^{n} \frac{C_j}{(1+i)^j} \tag{12.24}$$

where B_j and C_j, respectively, represent the benefits and costs at the end of the jth period and n the useful life of the project.

The equivalent present value cost C (Eq. (12.24)) may be split into two components – (i) the initial capital expenditure, and (ii) the annual costs accrued in each successive period. if it is assumed that the initial investment is required in the first m periods and that the annual costs accrue in each of the following periods till the end of the useful life of n periods, the above two components of the equivalent present value cost C may be expressed as

$$C_0 = \sum_{j=0}^{m} \frac{C_j}{(1+i)^j} \tag{12.25}$$

and

$$C'' = \sum_{j=m+1}^{n} \frac{C_j}{(1+i)^j} \tag{12.26}$$

with

$$C = C_0 + C'' \tag{12.27}$$

Using the above three expressions (Eqs. (12.25) to (12.27)), the following three types of benefit/cost ratios are usually defined.

(i) Aggregate B/C ratio

This is the ratio of the present value of total benefits to total costs.

$$\left(\frac{B}{C}\right)_{\text{aggregate}} = \frac{B}{C} = \frac{B}{C_0 + C''}, \quad C > 0 \text{ (or } C_0 + C'' > 0) \tag{12.28}$$

or,

$$\left(\frac{B}{C}\right)_{\text{aggregate}} = \frac{\sum_{j=0}^{n} \frac{B_j}{(1+j)^n}}{\sum_{j=0}^{n} \frac{C_j}{(1+j)^n} + \sum_{j=m+1}^{n} \frac{C_j}{(1+j)^n}} \tag{12.29}$$

Obviously, to accept a project the ratio $\left(\frac{B}{C}\right)_{\text{aggregate}}$ must be greater than 1.

(ii) Net B/C ratio

In another definition of B/C ratio, only the initial capital expenditure is considered as a cash outlay, and equivalent benefits become net benefits (*i.e.* annual revenues minus annual outlays).

The net benefit/cost ratio is expressed as

$$\left(\frac{B}{C}\right)_{\text{net}} = \frac{B - C'}{C_0}, \quad C_0 > 0 \tag{12.30}$$

Once again, for a project to be viable, the ratio $\left(\frac{B}{C}\right)_{\text{net}}$ must be greater than 1. The benefit/cost ratio defined in this manner essentially provides an index that indicates the benefits expected per unit of capital investment and can hence be used as a profitability index. it may be noted that it is simply a comparison of the present value of net revenues with the present value of capital investment. Thus, $\left(\frac{B}{C}\right)_{\text{net}}$ ensures that there is a surplus at time zero and the project is favorable.

Advantages and limitation of the B/C ratio: The benefit/cost ratio method offers the following advantage over other measures of evaluating different alternatives.

(a) It compares alternatives on a common scale and permits evaluation of different-sized alternatives.
(b) It can be used to rank alternatives projects to determine the most profitable alternative for an investor with a limited budget.
(c) It directly provides an indication of whether a project is worthwhile.
(d) It can also be used to determine the optimal size of a project if it is computed for increment in the investment size.

The shortcomings of the benefit/cost ratio include:

(a) The benefit/cost ratio is influenced by the decision as to whether an item is classified as a cost or a disbenefit, *i.e.* whether it appears in the denominator or numerator of the ratio. Often it may be an arbitrary decision but can lead to inefficient ranking of investment alternatives.
(b) The simple benefit/cost ratio cannot be used to determine the efficient scale of a given project. Incremental analysis is required to be undertaken for this purpose.

EXAMPLE 12.19

The latest building regulation in a city stipulates that all new student hostels must use solar energy for water heating. The manager of a hostel under construction is considering two hybrid solar water heating systems to supplement a natural gas-fired water heating system. One of the day systems (alternative X) is based on double-glazed flat plate collectors and the other (alternative Y) uses evacuated tubular collectors. Both the options have a useful life of 20 years and the associated costs and benefits are tabulated below:

	Amount (USD \$)	
	Alternative X	*Alternative Y*
Capital cost	3 200 000	2 700 000
Annual maintenance cost	50 000	80 000
Annual benefits due to fuel savings	600 000	560 000

Which option should be preferred on the basis of incremental net benefit/cost ratio? Use an interest rate of 10 percent and also assume that the salvage value is negligible for both alternatives. What if the benefit/cost ratio for each alternative is computed and the alternatives with higher benefit/cost ratio are selected?

Solution

As stipulated by the latest building regulations in the city one of the two alternatives should be chosen. Thus, the lower (alternative Y) need not be compared with the do nothing alternative. Instead, alternatives X and Y are compared with each other in terms of their incremental costs and benefits. The incremental capital cost of alternative X over alternative Y is USD \$ 500 000 (USD \$ 3 200 000–USD \$ 2 700 000) similarly the incremental net annual benefits of the alternative X over net annual benefits of alternatives Y are

$$(600\,000 - 50\,000) - (560\,000 - 80\,000) = 550\,000 - 480\,000 = \text{USD } \$\ 70\,000$$

The cumulative present worth of the incremental benefits over 20 yr of useful life of alternative X over alternative Y is

$$= \sum_{j=1}^{n} \frac{70\,000}{(1+i)^j}$$

$$= 70\,000 \left[\frac{(1+0.1)^{20} - 1}{0.1(1+0.1)^{20}} \right]$$

$$= 70\,000\,(8.51)$$

$$= \text{USD } \$\,5\,95\,949.4$$

Thus, the net incremental benefit/cost ratio = 595 949.4/500 000 $\approx$ 1.19

A value greater than one for the ratio of net incremental benefits to incremental capital cost implies that the additional discounted benefits more than justify the extra capital cost of alternative X compared to alternative Y. Therefore, alternative X should be selected for installation on the hostel.

The computation for the net benefit/cost ratio for each alternative independent of each other are given below.

The net benefit/cost ratio for alternative X is

$$= \frac{(60\ 000 - 50\ 000)\left[\dfrac{(1+0.1)^{20}-1}{0.1(1+0.1)^{20}}\right]}{3\ 200\ 000}$$

$$= \frac{50\ 000 \times 8.51}{3\ 200\ 000}$$

$$= 1.463$$

Similarly, the net benefit/cost ratio for alternative Y is

$$= \frac{(5\ 600\ 000 - 80\ 000)\left[\dfrac{(1+0.1)^{20}-1}{0.1\,(1+0.1)^{20}}\right]}{2\ 700\ 000}$$

$$= \frac{480\ 000 \times 8.51}{2\ 700\ 000}$$

$$= 1.513$$

It may be noted that an appraisal of the two alternatives using their net benefit/cost ratios would suggest that the alternative Y is selected.

As the results obtained with the two methods do not match, the net present values of both the alternatives are determined to identify the correct method.

NPV of alternative X is

$$NPV_X = -3\,200\,000 + (6\,00\,000 - 50\,000)\left[\frac{(1+0.1)^{20}-1}{0.1\,(1+0.1)^{20}}\right]$$

$$= -3\ 200\ 000 + 550\ 000\ (8.51)$$

$$= \text{USD } \$\ 1\,482\,460$$

NPV of alternative Y is

$$NPV_Y = -2\,700\,000 + (560\,000 - 80\,000)\left[\frac{(1+0.1)^{20}-1}{0.1\,(1+0.1)^{20}}\right]$$

$$= -2\,700\,000 + 480\,000\,(8.51)$$

$$= \text{USD } \$\,1\,386\,510$$

i.e.

$$NPV_X > NPV_Y$$

Thus, the appraisal based on incremental costs and benefits is correct.

12.9 INTERNAL RATE OF RETURN (IRR)

The internal rate of return (IRR) is a widely accepted discounted measure of investment worth and is used as an index of profitability for the appraisal of projects. The IRR is defined as the rate of interest that equates the present value of a series of cash flows to zero. Mathematically, the internal rate of return is the interest rate i_{IRR} that satisfies the equation

$$\text{NPV}(i_{\text{IRR}}) = \sum_{j=0}^{n} \frac{B_j - C_j}{(1 + i_{\text{IRR}})^j} = 0 \tag{12.31}$$

Alternatively, the internal rate of return is the interest rate that causes the discounted present value of the benefits in a cash flow to be equal to the present value of the costs, *i.e.*

$$\sum_{j=0}^{n} \frac{B_j}{(1 + i_{\text{IRR}})^j} = \sum_{j=0}^{n} \frac{C_j}{(1 + i_{\text{IRR}})^j} \tag{12.32}$$

Multiplying both sides of Eq. (12.31) by $(1 + i_{\text{IRR}})^n$

$$(1 + i_{\text{IRR}})^n \text{NPV}(i_{\text{IRR}}) = \sum_{j=0}^{n} \left\{ \frac{B_j - C_j}{(1 + i_{\text{IRR}})^j} \right\} (1 + i_{\text{IRR}})^n$$

or,

$$\text{NPV}(i_{\text{IRR}})(1 + i_{\text{IRR}})^n = \sum_{j=0}^{n} (B_j - C_j)(1 + i_{\text{IRR}})^{n-j} = 0 \tag{12.33}$$

IRR is widely used in the appraisal of projects because (i) the IRR on a project is its expected rate of return, (ii) it employs a percentage rate of return as the decision variable that suits the banking community, and (iii) for situations in which IRR exceeds the cost of the funds used to finance the project – a surplus would remain after paying for the capital.

12.9.1 Iterative Procedure for Computation of IRR

The following step-by-step procedure is suggested for computation of IRR by the iterative approach.

Step 1: Make a guess at a trial rate of interest.
Step 2: Using the guessed rate of interest, calculated the NPV of all disbursements and receipts.
Step 3: If the calculated value of NPV is positive then the receipts from the investments are worth more than the disbursements of the investments and the actual value of IRR would be more than the trial rate. On the other hand, if NPV is negative the actual value of IRR would be less than the trial rate of interest. Adjust the estimate of the trial rate of return accordingly.
Step 4: Precede with steps 2 and 3 again until one value of $i\ (= i_1)$ is found that results in a positive (+) NPV and next higher value of $i\ (= i_2)$ is found with a negative NPV.
Step 5: Solve for the value of IRR by interpolation using the values of i_1 and i_2 as obtained in step 4 (Figure 12.1a).

$$\text{IRR} = i_1 \left(\frac{i_2 - i_1}{\text{NPV}_1 - \text{NPV}_2} \right) \text{NPV}_1 \tag{12.34}$$

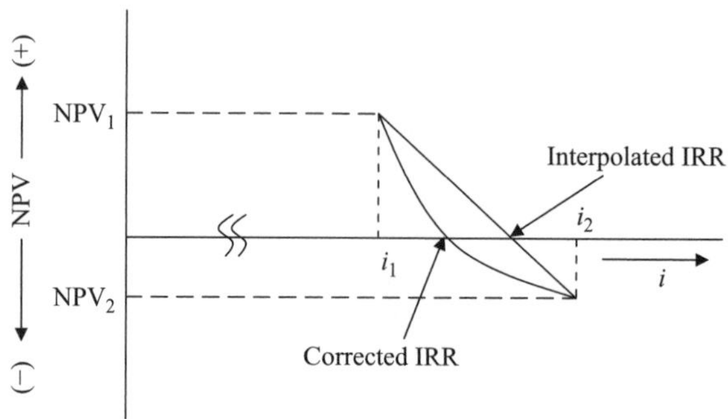

Figure 12.1a Interpolation of IRR.

An important aspect of the iterative method of computing IRR is making the initial estimate. If the initial estimate is too far from the actual value of IRR, a large number of trials will have to be made to obtain the two consecutive values of interest rate (i_1 and i_2) to permit accurate interpolation. It should be noted that the initial estimate of the IRR will always be somewhat in error and several iterations will normally be required to determine i_1 and i_2. A simple approach for making a guess of the first trial rate of return is given below.

The NPV of a capital investment C_0 resulting in uniform net annual cash flows of amount A for an infinite time horizon can be expressed as

$$\text{NPV} = -C_0 + \left[\frac{A}{(1+i)} + \frac{A}{(1+i)^2} + \cdots\cdots\right]$$

where i is the interest rate.

Thus,

$$\text{NPV} = -C_0 + \frac{A}{(1+i)}\left[\frac{1}{1 - \dfrac{1}{(1+i)}}\right]$$

$$= -C_0 + \frac{A}{i}$$

Since NPV = 0 at i = IRR, we have

$$-C_0 + \frac{A}{\text{IRR}} = 0$$

or,

$$\text{IRR} = \frac{A}{C_0}$$

In actual practice, for investment projects with finite life the IRR shall be less than A/C_0. However, to begin with, for cases with uniform periodic cash flows, the figure A/C_0 or a value close to it may be used as the trial rate of return in the iterative procedure used for determining IRR.

The above interpolation between two consecutive values of interest rates that bracket the IRR always overestimates its true value. This is because of the fact that the linear interpolation technique makes an implicit assumption that between two interest rates i_1 and i_2 the IRR changes, following a straight line whereas the true value of IRR follows a concave curvilinear function between the two values. However, the error introduced by interpolation is usually very small. Referring to Figure 12.1a, the true value of IRR is that value of i for which the NPV (i) function intersects the horizontal axis, whereas the interpolated value of IRR is somewhat higher than the true value. Obviously, the interpolation error would become less and less as the incremental change in the trial values of i used in the iteration is made smaller and smaller.

EXAMPLE 12.20

Calculate the internal rate of return for the investment in a heat exchanger that will costs USD \$ 500 000 to purchase and install, will last 10 yr and will result in fuel savings of USD \$ 145 000 per year. Also, assume that the salvage value of the heat exchanger at the end of 10 yr is negligible.

Solution

Let the first guess at the value of IRR be 25%.

$$\text{NPV at } 25\% = 145\,000\left[\frac{(1+0.25)^{10}-1}{0.25\,(1+0.25)^{10}}\right] - 500\,000$$

$$= 145\ 000\ (3.57) - 500\ 000$$

$$= \text{USD \$ } 17\ 722.$$

Since the NPV at 25 percent is positive, the IRR shall be greater than 25 percent. If the next trial value is chosen at 30 percent, then,

$$\text{NPV at } 30\% = 145\,000\left[\frac{(1+0.3)^{10}-1}{0.3\,(1+0.3)^{10}}\right] - 500\,000$$

$$= 145\,000\,(3.09) - 500\,000$$

$$= \text{USD \$ } -51\,724.$$

Obviously, the true IRR lies between 25 percent and 30 percent. By interpolating between the two, the IRR can be estimated as

$$\text{IRR} = 0.25\left(\frac{0.3-0.25}{17\ 722+51\ 724}\right)17\ 722$$

$$=0.26275$$

or, IRR = 26.275%

A better estimate of the true IRR may be obtained by using smaller incremental changes in the interest rate.

EXAMPLE 12.21

Installation of a USD \$ 5 000 000 energy-management system in an industry is expected to result in a 25 percent reduction in electricity use and a 40 percent savings in process heating costs. This translates to net yearly savings of USD \$ 600 000 and USD \$ 750 000, respectively. If the energy-management system has an expected useful life of 20 years, determine the internal rate of return on the investment. The salvage value need not be considered in the analysis.

Solution

$$\text{Total annual benefits} = \text{USD \$ 600 000} + \text{USD \$ 750 000}$$
$$= \text{USD \$ 1 350 000}$$

$$\text{NPV of the investment} = -5\ 000\ 000 + 1\ 350\ 000\left[\frac{(1+i)^{20}-1}{i(1+i)^{20}}\right]$$

NPV at $i = 0.27$

$$= -5\ 000\ 000 + 1\ 350\ 000\left[\frac{(1+0.27)^{20}-1}{0.27(1+0.27)^{20}}\right]$$

$$= -5\ 000\ 000 + 4\ 958\ 034$$
$$= -41\ 965$$

NPV at $i = 0.26$

$$= 5\ 000\ 000 + 1\ 350\ 000\left[\frac{(1+0.26)^{20}-1}{0.27(1+0.26)^{20}}\right]$$

$$= 5\ 000\ 000 + 5\ 141\ 263$$

$$= 141\ 263$$

Thus, the IRR can be obtained by interpolating between $i = 0.26$ and $i = 0.27$ in the following manner:

$$\text{IRR} = 0.26\left(\frac{0.27 - 0.26}{141\ 263 + 41\ 965}\right)141\ 263 = 0.2677$$

i.e. the internal rate of return is 26.77%.

12.9.2 Multiple Values of IRR

The NPV of a set of cash receipts and disbursements can be expressed as an nth-degree polynomial of the form

$$\text{NPV}(i_{\text{IRR}}) = 0 = F_0 + F_1 x + F_2 x^2 + \cdots\cdots + F_n x^n \tag{12.35}$$

where $x = 1/(1+i)$ and the F_i s are coefficients of the n terms in the polynomial.

For the above polynomial, in principle, there may be n different roots or values of x that satisfy Eq. (12.35). Thus, it is possible that the NPV(i) function crosses the i-axis several times, as shown in Figure 12.1b.

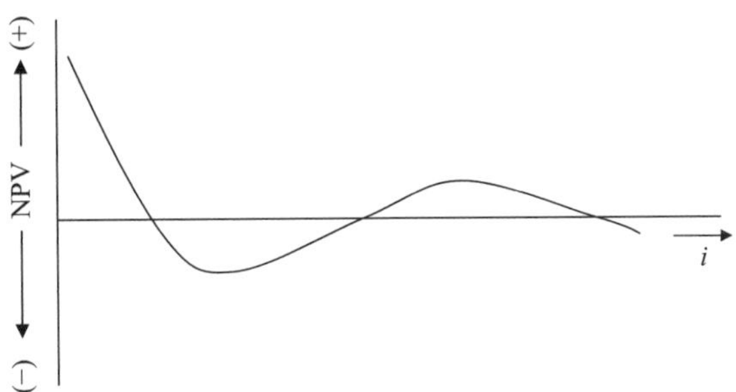

Figure 12.1b Multiple values of IRR.

It may be noted that a unique value of IRR of special interest in applying the IRR method and consequently multiple values of IRR essentially hinder the application of the IRR criterion. In fact, in the case with multiple IRR values, use of the IRR criterion is normally not recommended.

12.10 EFFECT OF DEPRECIATION

Initial cost (C_i): Also referred as **first cost** or **initial value** or **single amount**. This is the installed cost of the system. The cost includes the purchase price, delivery and installation fee and other depreciable direct cost (defined later) incurred to ready the asset for use.
Salvage value (C_{sal}): This is the expected market value at the end of useful life of the asset. It is negative if dismantling cost or carrying away cost is anticipated. It can be zero also. For example, the window glass has zero salvage value.
Depreciation (C_d): An expenditure that decreases in value with time. This must be apportioned over its lifetime. The term used to describe this loss in value is known as depreciation.

$$C_d = C_i - C_{sal}$$

Book value (B): This represents the remaining undepreciated investment on corporate books. It can be obtained after the total amount of annual depreciation charges to date has been subtracted from the first cost (present value/initial cost). The book value is usually determined at the end of each year.

Book value = initial cost (first cost) − accumulated cost

or,

Book value = salvage value + future depreciation

Depreciation rate (D_t): This is the fraction of first cost removed through depreciation from the corporate book. This rate may be the same, *i.e.* straight-line (SL) rate or different for each year of the recovery period.

Mathematically, it can be written for straight-line (SL) depreciation as follows:

$$D_t = \frac{C_i - C_{sal}}{n} \tag{12.36}$$

The book value at nth year can be expressed as

$$B_n = C_i - nD_t \tag{12.37}$$

Recovery period (n): This is the life of the asset (in years) for depreciation and tax purposes. It is also referred to as the **expected life** of the asset in years.
Market value: This is the actual amount that could be obtained after selling the asset in the open market. For example, (i) the market value of a commercial building tends to increase with the

period in the open market but the book value will decrease as depreciation charges are taken into account and (ii) an electronic piece of equipment (*e.g.* a computer system) may have a market value much lower than book value due to the rapid change of technology.

Present value of R_e. 1 of depreciation ($C_d = R_e$. 1) is,

$$F_{SLP,i,n} = \frac{1}{n}\left[\frac{(1+i)^n - 1}{i(1+i)^n}\right] = \frac{1}{n}F_{RP,i,n} \tag{12.38}$$

An expression for the conversion factor from straight-line depreciation to the present value with tax is given by

$$F_{SLP,r,n} = \frac{1}{n}\left[\frac{(1+r)^n - 1}{r(1+r)^n}\right] \tag{12.39}$$

12.11 COST COMPARISONS OF SOLAR DRYERS WITH DURATION

Hossain *et al.*[1] reported that the payback period of a solar tunnel drier is 4 years for a basic mode drier and that for optimum mode driers are 4 years and about 3 years. On the basis of sensitivity analysis, they showed that the design geometry was not very sensitive to minor material costs, fixed cost and operating cost. It is sensitive to costs of major construction materials of the collector, solar radiation and the air velocity in the drier.

For a solar grain dryer incorporating photovoltaic powered air circulation, the variation of the payback period with respect to PV area to air-heater area ratio is shown in Figure 12.3.[2] It is clear from the figure that the optimum PV area to air-heater area ratio is 0.22 for a payback period of 0.5 year, which is less than a year, if used to dry surplus grain for selling at the markets.

Kumar and Kandpal[3] have estimated the potential for solar drying of selected cash crops, namely tobacco, tea, coffee, grapes raisin, small cardamom, chilli, coriander seeds, ginger, turmeric, black pepper, and onion flakes, *etc.* for Indian conditions. They also estimated the potential of net fossil CO_2 emissions mitigation due to the amounts of different fuels that would be saved, along with the unit cost of CO_2 emissions mitigation.

Table 12.2 gives the cost of different types of solar dryers constructed at the Indian Institute of Technology, New Delhi, India. Table 12.3 gives the cost of plastic and conventional solar collectors, having 560 m^2 gross collector area, to deliver useful energy of 203 MWh/yr.[4]

Figures 12.2(a) and (b) show the comparison of annual cost (USD \$) and useful energy cost (USD \$/kWh), respectively, for plastic and conventional solar collectors with respect to the life of the system.[4]

Table 12.2 Cost of different types of solar dryers (in India).[6]

Type of Solar Dryer	*Initial Investment (Rs.)*	*Salvage Value (Rs.)*	*O & M[a] Cost/year (Rs.)*	*Life (yr)*
Cabinet dryer	5000	—	200	5
Greenhouse crop dryer (natural)	2000	—	—	5
Reverse absorber cabinet dryer	8000	—	200	5
Conventional active solar dryer	15 000	2000	200	5
Hybrid PVT integrated greenhouse dryer	43 000	10 000	1000[b]	35
Hybrid PVT solar dryer	39 000	5000	200[c]	30

[a]O & M represents operation and maintenance;
[b]per five year for UV polyethylene sheet replacement and
[c]per three year for glass replacement.

Table 12.3 Cost of plastic and conventional solar collectors (area, 560 m^2) (in India).

Dryer	*Initial Investment (Rs.)*	*Salvage Value (Rs.)*	*O & M Cost/year (Rs.)*	*Life (yr)*
Plastic solar collectors	560 000	56 000	14 772.8	5
	560 000	56 000	9116.8	10
	560 000	56 000	7362.3	15
Conventional solar collectors	1 120 000	112 000	18 233.6	10
	1 120 000	112 000	13 160	20
	1 120 000	112 000	11 883.2	30

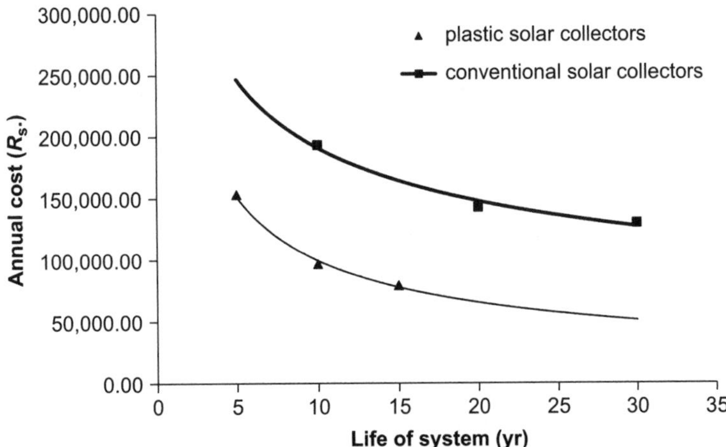

Figure 12.2a Comparison of annual cost (R_s.) with respect to life of the system (years).

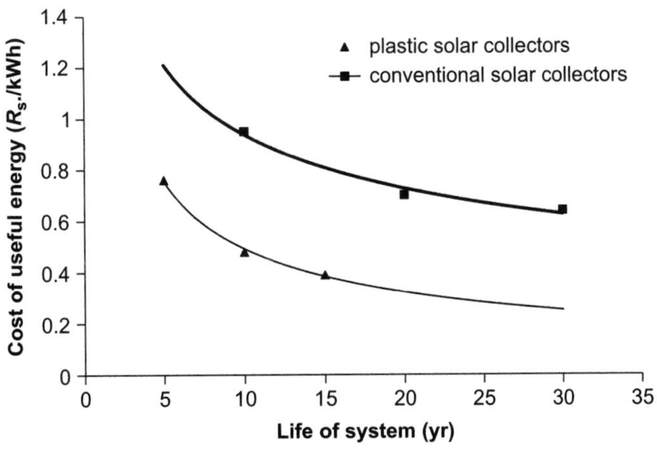

Figure 12.2b Comparison of useful energy cost (R_s./kWh) with respect to life of the system (years).

ADDITIONAL EXERCISES

Exercise 12.1

Calculate future (F_{PS}) and present (F_{SP}) value factors for a given number of years for a 10% rate of interest and show that F_{PS}. $F_{SP}=1$ for each case.

Solution

We have the value expression for future value factor as $F_{PS,i,n}=(1+i)^n$ where i is the rate of interest and n is the number of years.

$$\text{For } n=1, \quad F_{\text{PS, 10\%, 1 yr.}} = (1+0.1)^1 = 1.1$$
$$\text{For } n=2, \quad F_{\text{PS, 10\%, 2 yr.}} = (1+0.1)^2 = 1.21$$
$$\text{For } n=3, \quad F_{\text{PS, 10\%, 3 yr.}} = (1+0.1)^3 = 1.331$$
$$\text{For } n=4, \quad F_{\text{PS, 10\%, 4 yr.}} = (1+0.1)^4 = 1.464$$
$$\text{For } n=5, \quad F_{\text{PS, 10\%, 5 yr.}} = (1+0.1)^5 = 1.6105$$

Similarly, F_{PS} can be calculated for any value of n. We have the expression for the present value factor as

$$F_{SP,i,n} = \frac{1}{(1+i)^n} = (1+i)^{-n}$$

$$\text{For } n=1, \quad F_{\text{SP, 10\%, 1 yr.}} = (1+0.1)^{-1} = 0.900$$
$$\text{For } n=2, \quad F_{\text{SP, 10\%, 2 yr.}} = (1+0.1)^{-2} = 0.8264$$
$$\text{For } n=3, \quad F_{\text{SP, 10\%, 3 yr.}} = (1+0.1)^3 = 0.7513$$
$$\text{For } n=4, \quad F_{\text{SP, 10\%, 4 yr.}} = (1+0.1)^4 = 0.683$$
$$\text{For } n=5, \quad F_{\text{SP, 10\%, 5 yr.}} = (1+0.1)^5 = 0.621$$

Similarly, F_{SP} can be calculated for any value of n.
Now let us show that $F_{PS} \cdot F_{SP}=1$

$$\text{For } n=0, \quad F_{\text{PS}} \cdot F_{SP} = (1+0.1)^0(1+0.1)^0 = 1 \times 1 = 1$$
$$\text{For } n=2, \quad F_{\text{PS}} \cdot F_{SP} = (1+0.1)^2(1+0.1)^{-2} = 1$$
$$\text{For } n=4, \quad F_{\text{PS}} \cdot F_{SP} = (1+0.1)^4(1+0.1)^4 = 1$$
$$\text{For only } n, \quad F_{\text{PS}} \cdot F_{SP} = 1$$

Exercise 12.2

Calculate the effective rate of return for different value of p for 10% rate of interest.

Solution

We have the expression for an effective rate of return as

$$\text{Effective rate of return} = \left(1+\frac{i}{p}\right)^n - 1$$

$$\text{For } p = 1, \text{ Effective rate of return} = \left(1 + \frac{0.1}{1}\right)^1 - 1 = 0.1 = 10.0\%$$

$$\text{For } p = 2, \text{ Effective rate of return} = \left(1 + \frac{0.1}{2}\right)^2 - 1 = 0.1025 = 10.25\%$$

$$\text{For } p = 3, \text{ Effective rate of return} = \left(1 + \frac{0.1}{3}\right)^3 - 1 = 0.10337 = 10.337\%$$

$$\text{For } p = 4, \text{ Effective rate of return} = \left(1 + \frac{0.1}{4}\right)^4 - 1 = 0.1038 = 10.38\%$$

$$\text{For } p = 6, \text{ Effective rate of return} = \left(1 + \frac{0.1}{6}\right)^6 - 1 = 0.10426 = 10.426\%$$

Here, it is important to mention that for $p=2$ means that one year is divided into 2 parts. Each part is for six months.

Exercise 12.3

Calculate capital recovery (F_{PR}) and sinking fund (F_{SR}) factors for different number of years ($n=1$, 5, 10, 15 and 20) for a given rate of interest ($i=0.05$, 0.10, 0.15, and 0.20).

Solution

(a) We have the expression for capital recovery factor as

$$F_{PR,i,n} = \frac{i(1+i)^n}{(1+i)^n - 1}$$

for, $i=0.05$, the capital recovery factor is

$$F_{PR,0.05,1} = \frac{0.05(1+.05)^1}{(1+.05)^1 - 1} = 1.05$$

similarly for other n

$$F_{PR,\ 0.05,\ 5\ \text{yr.}} = 0.2310$$
$$F_{PR,\ 0.05,\ 10\ \text{yr.}} = 0.1295$$
$$F_{PR,\ 0.05,\ 15\ \text{yr.}} = 0.0963$$
$$F_{PR,\ 0.05,\ 20\ \text{yr.}} = 0.080$$

similarly for $i=0.10$

$$F_{PR,\ 0.1,\ 1\ \text{yr.}} = 1.1$$
$$F_{PR,\ 0.1,\ 5\ \text{yr.}} = 0.264$$
$$F_{PR,\ 0.1,\ 10\ \text{yr.}} = 0.163$$
$$F_{PR,\ 0.1,\ 15\ \text{yr.}} = 0.131$$
$$F_{PR,\ 0.1,\ 20\ \text{yr.}} = 0.117$$

For $i = 0.15$

$$F_{PR,\,0.15,\,1\ \text{yr.}} = 1.15$$
$$F_{PR,\,0.15,\,5\ \text{yr.}} = 0.298$$
$$F_{PR,\,0.15,\,10\ \text{yr.}} = 0.1992$$
$$F_{PR,\,0.15,\,15\ \text{yr.}} = 0.1710$$
$$F_{PR,\,0.15,\,20\ \text{yr.}} = 0.1597$$

For $i = 0.2$

$$F_{PR,\,0.2,\,1\ \text{yr.}} = 1.2$$
$$F_{PR,\,0.2,\,5\ \text{yr.}} = 0.334$$
$$F_{PR,\,0.2,\,10\ \text{yr.}} = 0.239$$
$$F_{PR,\,0.2,\,15\ \text{yr.}} = 0.214$$
$$F_{PR,\,0.2,\,20\ \text{yr.}} = 0.205$$

(b) The expression for the sinking-fund factor can be given as

$$\text{Sinking factor } F_{SR,\,i,\,n} = \frac{i}{(1+i)^n - 1}$$

The sinking-fund factor for $i = 0.05$, the value of FSR for different "n" is calculated as

$$F_{SR,\,0.05,\,1\ \text{yr.}} = \frac{0.05}{(1+.05)^1 - 1} = \frac{0.05}{0.05} = 1$$

$$F_{SR,\,0.05,\,5\ \text{yr.}} = 0.181$$
$$F_{SR,\,0.05,\,10\ \text{yr.}} = 0.0796$$
$$F_{SR,\,0.05,\,15\ \text{yr.}} = 0.0464$$
$$F_{SR,\,0.05,\,20\ \text{yr.}} = 0.0303$$

Similarly, the sinking-fund factor for $i = 0.1$ and different values of n is given by

$$F_{SR,\,0.1,\,1\ \text{yr.}} = 1$$
$$F_{SR,\,0.1,\,5\ \text{yr.}} = 0.163$$
$$F_{SR,\,0.1,\,10\ \text{yr.}} = 0.0623$$
$$F_{SR,\,0.1,\,15\ \text{yr.}} = 0.0315$$
$$F_{SR,\,0.1,\,20\ \text{yr.}} = 0.0175$$

The sinking-fund factor for $i = 0.15$ and different values of n, is given as

$$F_{SR,\,0.15,\,1\ \text{yr.}} = 1.0$$
$$F_{SR,\,0.15,\,5\ \text{yr.}} = 0.148$$
$$F_{SR,\,0.15,\,10\ \text{yr.}} = 0.0492$$
$$F_{SR,\,0.15,\,15\ \text{yr.}} = 0.02102$$
$$F_{SR,\,0.15,\,20\ \text{yr.}} = 0.00976$$

The sinking-fund factor for $i = 0.20$ and different values of n is given by

$$F_{SR,\,0.2,\,1\text{ yr.}} = 1.0$$
$$F_{SR,\,0.2,\,10\text{ yr.}} = 0.0385$$
$$F_{SR,\,0.1,\,15\text{ yr.}} = 0.0139$$
$$F_{SR,\,0.2,\,20\text{ yr.}} = 0.00538$$

Exercise 12.4

Prove that $F_{SR} \cdot F_{RS} = 1$.

Solution

We have

$$F_{SR,\,i,\,n} = \frac{i}{[(1+i)^n - 1]}$$

and

$$F_{RS,\,i,\,n} = \frac{[(1+i)^n - 1]}{i}$$

from the above equations

$$F_{SR\,i,\,n} \times F_{RS\,i,\,n} = \frac{i}{(1+i)^n - 1} \times \frac{(1+i)^n - 1}{i} = 1$$

Exercise 12.5

Two swimming pools have been heated by solar water heating systems that have the following cost comparison. Determine which system is more economical if the money is worth 12% per year.

Draw a cash-flow diagram for each system.

Solution

The cash-flow diagram of system I is given below

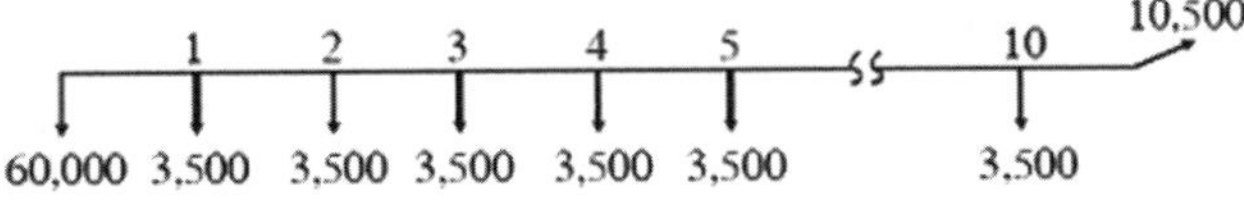

The present net value of system I is given by

$$\text{NPV} = p + 3500 \times F_{RP,\,0.12,\,10} - 10\,500 \times F_{SP,\,0.12,\,10}$$

After substituting the value, one gets

$$\text{NPV} = 60\,000 + 3500\frac{[(1.12)^{10} - 1]}{0.12 \times (1.12)^{10}} + 30\,000 \times 0.5674 - 10\,500 \times 0.322 = \text{Rs. } 79\,083.20$$

Similarly, the cash-flow diagram of system II is given below

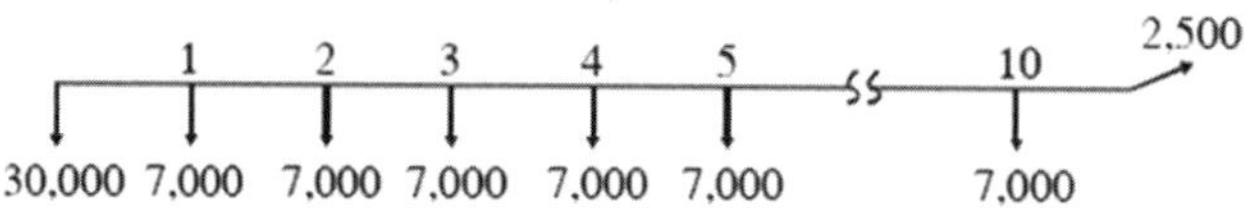

The net present value of system II is

$$\text{NPV} = 30\ 000 + 7000\frac{[(1.12)^{10} - 1]}{0.12\,(1.12)^{10}} + 10\ 000 \times 0.5674 - 2\ 500 \times 0.322 = \text{Rs. } 76\ 393.00$$

The net present value of system II is less; hence system II is more economical.

OBJECTIVE QUESTIONS

12.1 The capital recovery factor (F_{PR}) is always
(a) Less than one
(b) More than one
(c) Equal to one
(d) None

12.2 The compound interest factor (F_{PS}) is always
(a) Less than one
(b) More than one
(c) Equal to one
(d) None

12.3 The present-value factor (F_{SP}) is always
(a) Less than one
(b) More than one
(c) Equal to one
(d) None

12.4 The compound interest operator is always
(a) Less than one
(b) More than one
(c) Equal to one
(d) None

12.5 The sinking-fund factor (SFF), F_{SR} is always
(a) Less than one
(b) More than one
(c) Equal to one
(d) None

12.6 The sinking-fund factor (SFF), F_{SR} is
(a) Proportional to F_{RS}
(b) Inversely proportional to F_{RS}
(c) Equal to F_{RS}
(d) None

12.7 The payback time period depends on
(a) Cash flow
(b) Initial investment

(c) Use of these system
(d) Cases (a) and (b)

12.8 The project is economically accepted if B/C is always
(a) Less than one
(b) Greater than one
(c) Equal to one
(d) None

12.9 The rate of return after tax
(a) Increases with increase of tax
(b) Decreases with increase of tax
(c) Unaffected
(d) None

12.10 The conversion factor from straight-line depreciation to the present value with tax is
(a) Lower without tax
(b) Higher without tax
(c) Equal without tax
(d) None

12.11 The straight line depreciation is
(a) Directly proportional to the life of the system (n)
(b) Inversely proportional to the life of the system (n)
(c) Equal to the life of the system (n)
(d) None

12.12 The expression for total depreciation (C_d) is given by
(a) $C_i - C_{sal}$
(b) $C_i \cdot C_{sal}$
(c) $C_i + C_{sal}$
(d) None

12.13 The expression for future depreciation at end of the mth year (FD_m) is
(a) $C_d\left[1 - \frac{m}{n}\right]$
(b) $C_d \frac{m}{n}$
(c) $C_d\left[1 + \frac{m}{n}\right]$
(d) None

12.14 The expression for the book value (B_m) at the end of mth year is
(a) $C_{sal} + FD_m$
(b) $C_{sal} - FD_m$
(c) $C_{sal} \cdot FD_m$
(d) None
here, $FD_m = C_d\left[1 - \frac{m}{n}\right]$

12.15 The value of unacost after tax is
(a) Reduced
(b) Increased
(c) Unaffected
(d) None

12.16 The value of F_{PR}, r, n is always
(a) More than $F_{PR,\ i,\ n}$
(b) Less than $F_{PR,\ i,\ n}$

(c) Equal to $F_{PR,\,i,\,n}$
(d) None

12.17 The expression for $F_{PR,r,n}$ is
(a) $\frac{r(1+r)^n}{(1+r)^n-1}$
(b) $\frac{(1+r)^n-1}{r(1+r)^n}$
(c) $\frac{i(1+i)^n}{(1+i)^n-1}$
(d) $\frac{(1+i)^n-1}{i(1+i)^n}$

12.18 The expression for $F_{SR,r,n}$ is
(a) $\frac{r}{(1+r)^n-1}$
(b) $\frac{(1+r)^n-1}{r}$
(c) $\frac{i}{(1+i)^n-1}$
(d) $\frac{(1+i)^n-1}{i}$

12.19 The expression for the payback period (n) without interest rate and uniform cash flow is
(a) $\frac{P}{\text{CF}}$
(b) $\frac{\text{CF}}{P}$
(c) P.CF
(d) None

12.20 The tax rate is always
(a) Less than one
(b) More than one
(c) Equal to one
(d) None

ANSWERS

12.1 **(a)**; 12.2 **(b)**; 12.3 **(a)**; 12.4 **(b)**; 12.5 **(a)**; 12.6 **(b)**; 12.7 **(d)**; 12.8 **(b)**; 12.9 **(b)**; 12.10 **(a)**; 12.11 **(b)**; 12.12 **(a)**; 12.13 **(a)**; 12.14 **(a)**; 12.15 **(b)**; 12.16 **(a)**; 12.17 **(a)**; 12.18 **(a)**; 12.19 **(a)**; 12.20 **(a)**.

REFERENCES

1. Hossain, M. A., Woods, J. L., and Bala, B. K. 2005. *Optim. Renew. Energy*, **30**, 729–742.
2. Mumba, J. 1996. *Energy Conver. Manag.*, **37**(5), 615–621.
3. Kumar, A. and Kandpal, T. C. 2005. *Solar Energy*, **78**(2), 321–329.
4. Sodha, M. S., Chandra, R., Pathak, K., Singh, N. P. and Bansal, N. K. 1991. *Energy Conver. Manag.*, **31**(6), 509–513.
5. Tiwari, G. N. 2004. *Solar Energy: Fundamentals, Design, Modeling and Applications*, Narosa Publishing House, New Delhi, India.
6. Barnwal, P. and Tiwari, A. 2008. *Int. J. Agricult. Res.*, **3**(2), 110–120.

APPENDIX – I

Conversion of Units

i) Length, m

1 yd (yard) = 3 ft = 36 in (inches) = 0.9144 m

1 m = 39.3701 in = 3.280839 ft = 1.093613 yd = 1650763.73 wavelength

1 ft = 12 in = 0.3048 m

1 in = 2.54 cm = 25.4 mm

1 mil = 2.54×10^{-3} cm

1 μm = 10^{-6} m

1 nm = 10^{-9} m = 10^{-3} μm

ii) Area, m^2

1 ft^2 = 0.0929 m^2

1 in^2 = 6.452 cm^2 = 0.00064516 m^2

1 cm^2 = 10^{-4} m^2 = 10.764×10^{-4} ft^2 = 0.1550 in^2

1 ha = 10 000 m^2

iii) Volume, m^3

1 ft^3 = 0.02832 m^3 = 28.3168 l (litre)

1 in^3 = 16.39 cm^3 = 1.639×10^2 l

1 yd^3 = 0.764555 m^3 = 7.646×10^2 l

1 UK gallon = 4.54609 l

1 US gallon = 3.785 l = 0.1337 ft^3

1 m^3 = 1.000×10^6 cm^3 = $2.642 \times 10^{1\ 2}$ US gallons = 109 l

1 l = 10^{-3} m^3

1 fluid ounce = 28.41 cm^3

iv) Mass, kg

1 kg = 2.20462 lb = 0.068522 slug

1 ton (short) = 2000 lb (pounds) = 907.184 kg

1 ton (long) = 1016.05 kg

1 lb = 16 oz (ounces) = 0.4536 kg

1 oz = 28.3495 g

1 quintal = 100 kg

1 kg = 1000 g = 10 000 mg

Advanced Renewable Energy Sources
G. N. Tiwari and R. K. Mishra

Published by the Royal Society of Chemistry, www.rsc.org

(*Continued*).

1 μg = 10^{-6} g
1 ng = 10^{-9} g

v) Density and specific volumes, kg/m^3, m^3/kg

1 lb/ft^3 = 16.0185 kg/m^3 = 5.787 × 10^{-4} lb/in^3
1 g/cm^3 = 10^3 kg/m^3 = 62.43 lb/ft^3
1 lb/ft^3 = 0.016 g/cm^3 = 16 kg/m^3
1 ft^3 (air) = 0.08009 lb = 36.5 g at N.T.P.
1 gallon/lb = 0.010 cm^3/kg
1 μg/m^3 = 10^{-6} g/m^3

vi) Pressure, Pa (Pascal)

1 lb/ft^2 = 4.88 kg/m^2 = 47.88 Pa
1 lb/in^2 = 702.7 kg/m^2 = 51.71 mm Hg = 6.894757 × 10^3 Pa = 6.894757 × 10^3 N/m^2
1 atm = 1.013 × 10^5 N/m^2 = 760 mm Hg = 101.325 kPa
1 in H_2O = 2.491 × 10^2 N/m^2 = 248.8 Pa = 0.036 lb/in^2
1 bar = 0.987 atm = 1.000 × 10^6 dynes/cm^2 = 1.020 kgf/cm^2 = 14.50 lbf/in^2 = 10^5 N/m^2 = 100 kPa
1 torr (mm Hg 0°C) = 133 Pa
1 Pascal (Pa) = 1 N/m^2 = 1.89476 kg
1 inch of Hg = 3.377 kPa = 0.489 lb/in^2

vii) Velocity, m/s

1 ft/s = 0.3041 m/s
1 mile/h = 0.447 m/s = 1.4667 ft/s = 0.8690 knots
1 km/h = 0.2778 m/s
1 ft/min = 0.00508 m/s

viii) Force, N

1 N (Newton) = 10^5 dynes = 0.22481 lb wt = 0.224 lb f
1 pdl (poundal) = 0.138255 N (Newton) = 13.83 dynes = 14.10 gf
1 lbf (*i.e.* wt of 1 lb mass) = 4.448222 N = 444.8222 dynes
1 ton = 9.964 × 10^3 N
1 bar = 10^5 Pa (Pascal)
1 ft of H_2O = 2.950 × 10^{-2} atm = 9.807 × 10^3 N/m^2
1 in H_2O = 249.089 Pa
1 mm H_2O = 9.80665 Pa
1 dyne = 1.020 × 10^{-6} kg f = 2.2481 × 10^{-6} lb f = 7.2330 × 10^{-5} pdl = 10^{-5} N
1 mm of Hg = 133.3 Pa
1 atm = 1 kg f/cm^2 = 98.0665 k Pa
1 Pa (Pascal) = 1 N/m^2

ix) Mass flow rate and discharge, kg/s, m^3/s

1 lb/s = 0.4536 kg/s
1 ft^3/min = 0.4720 1/s = 4.179 × 10^{-4} m^3/s
1 m^3/s = 3.6 × 10^6 l/h
1 g/cm^3 = 10^3 kg/m^3
1 lb/h ft^2 = 0.001356 kg/s m^2
1 lb/ft^3 = 16.2 kg/m^2
1 litre/s (l/s) = 10^{-3} m^3/s

x) Energy, J

1 cal = 4.187 J (Joules)
1 kcal = 3.97 Btu = 12 × 10^{-4} kWh = 4.187 × 10^3 J
1 Watt = 1.0 J/s

(*Continued*).

1 Btu = 0.252 kcal = 2.93×10^{-4} kWh = 1.022×10^{3} J
1 hp = 632.34 kcal = 0.736 kWh
1 kWh = 3.6×10^{6} J = 1 unit
1 J = 2.390×10^{-4} kcal = 2.778×10^{-4} Wh
1 kWh = 860 kcal = 3413 Btu
1 erg = 1.0×10^{-7} J = 1.0×10^{-7} Nm = 1.0 dyne cm
1 J = 1 W s = 1 N m
1 eV = 1.602×10^{-19} J
1 GJ = 10^{9} J
1 MJ = 10^{6} J
1 TJ (Terajoules) = 10^{12} J
1 EJ (Exajoules) = 10^{18} J

xi) Power, Watt (J/s)

1 Btu/h = 0.293071 W = 0.252 kcal/h
1 Btu/h = 1.163 W = 3.97 Btu/h
1 W = 1.0 J/s = 1.341×10^{-3} hp = 0.0569 Btu/min = 0.01433 kcal/min
1 hp (F.P.S.) = 550 ft lb f/s = 746 W = 596 kcal/h = 1.015 hp (M.K.S.)
1 hp (M.K.S.) = 75 mm kg f/s = 0.17569 kcal/s = 735.3 W
1 W/ft^2 = 10.76 W/m^2
1 ton (Refrigeration) = 3.5 kW
1 kW = 1000 W
1 GW = 10^{9} W
1 W/m^2 = 100 lux

xii) Specific Heat, J/kg °C

1 Btu/lb°F = 1.0 kcal/kg °C = 4.187×10^{3} J/kg °C
1 Btu/lb = 2.326 kJ/kg

xiii) Temperature, °C and K used in SI

$T_{(\text{Celcius},°\text{C})} = (5/9)\,[T_{(\text{Fahrenheit},°\text{F})} + 40] - 40$
$T_{(°\text{F})} = (9/5)\,[T_{(°\text{C})} + 40] - 40$
$T_{(\text{Rankine},\ °\text{R})} = 460 + T_{(°\text{F})}$
$T_{(\text{Kelvin},\ \text{K})} = (5/9)\ T_{(°\text{R})}$
$T_{(\text{Kelvin},\ \text{K})} = 273.15 + T_{(°\text{C})}$
$T_{(°\text{C})} = T_{(°\text{F})}/1.8 = (5/9)\ T_{(°\text{F})}$

xiv) Rate of heat flow per unit area or heat flux, W/m^2

1 Btu/ft^2 h = 2.713 kcal/m^2 h = 3.1552 W/m^2
1 kcal/m^2 h = 0.3690 Btu/ft^2 h = 1.163 W/m^2 = 27.78×10^{-6} cal/s cm^2
1 cal/cm^2 min = 221.4 Btu/ft^2 h
1 W/ft^2 = 10.76 W/m^2
1 W/m^2 = 0.86 kcal/hm^2 = 0.23901×10^{-4} cal/s cm^2 = 0.137 Btu/h ft^2
1 Btu/h ft = 0.96128 W/m

xv) Heat transfer coefficient, W/m^2 °C

1 Btu/ft^2 h °F = 4.882 kcal/m^2 h °C = 1.3571×10^{-4} cal/cm^2 s °C
1 Btu/ft^2 h °F = 5.678 W/m^2 °C
1 kcal/m^2 h °C = 0.2048 Btu/ft^2 h °F = 1.163 W/m^2 °C
1 W/m^2 K = 2.3901×10^{-5} cal/cm^2 s K = 1.7611×10^{-1} Btu/ft^2 °F = 0.86 kcal/m^2 h °C

xvi) Thermal conductivity, W/m °C

1 Btu/ft h °F = 1.488 kcal/m h °C = 1.73073 W/m °C
1 kcal/m h °C = 0.6720 Btu/ft h °F = 1.1631 W/m °C

(*Continued*).

1 Btu in/ft^2 h °F = 0.124 kcal/mh °C = 0.144228 W/m °C
1 Btu/in h °F = 17.88 kcal/mh °C
1 cal/cm s °F = 4.187 × 10^2 W/m °C = 242 Btu/h ft °F
1 W/cm °C = 57.79 Btu/h ft °F

xvii) Angle, rad

2π rad (radian) = 360° (deg)
1° (degree) = 0.0174533 rad = 60′ (min)
1′ = 0.290888 × 10^{-3} rad = 60″ (s)
1″ = 4.84814 × 10^{-6} rad
1° (h angle) = 4 min (time)

xviii) Illumination

1 lx (lux) = 1.0 lm (lumen)/m^2
1 lm/ft^2 = 1.0 foot candle
1 foot candle = 10.7639 lx
100 lux = 1 W/m^2

xix) Time, h

1 week = 7 days = 168 h = 10 080 minutes = 604 800 s
1 mean solar day = 1440 min = 86 400 s
1 calendar year = 365 days = 8760 h = 5.256 × 10^5 min
1 tropical mean solar year = 365.2422 days
1 sidereal year = 365.2564 days (mean solar)
1 s (s) = 9.192631770 × 10^9 Hertz (Hz)
1 day = 24 h = 360° (hour angle)

xx) Concentration, kg/m^3 and g/m^3

1 g/l = 1 kg/m^3
1 lb/ft^3 = 6.236 kg/m^3

xxi) Diffusivity, m^2/s

1 ft^2/h = 25.81 × 10^{-6} m^2/s

APPENDIX – II

Parameters on Horizontal Surface for Sunshine Hours = 10 for All Four Weather Type of Days for Different Indian Climates

(a): New Delhi

Type of day	*Month▶ Parameters▼*	*January*	*February*	*March*	*April*	*May*	*June*	*July*	*August*	*September*	*October*	*November*	*December*
a	T_R	2.25	2.79	2.85	2.72	3.54	2.47	2.73	2.58	2.53	1.38	0.62	0.72
	α	0.07	0.10	0.17	0.23	0.16	0.28	0.37	0.41	0.29	0.47	0.59	0.54
	K_1	0.47	0.39	0.33	0.28	0.20	0.27	0.41	0.40	0.23	0.21	0.21	0.28
	K_2	−13.17	−6.25	5.61	38.32	65.04	31.86	−40.57	−55.08	39.92	32.77	30.62	9.73
b	T_R	2.28	2.78	2.89	3.15	5.44	4.72	5.58	5.43	3.23	4.56	0.19	1.83
	α	0.15	0.13	0.14	0.17	0.16	0.20	0.24	0.18	0.31	0.22	1.14	0.42
	K_1	0.51	0.54	0.49	0.46	0.45	0.45	0.53	0.39	0.37	0.42	0.35	0.40
	K_2	−21.77	−28.26	−9.22	−11.55	1.54	23.99	−51.61	9.46	14.07	−9.50	17.47	−0.07
c	T_R	5.88	6.36	6.11	7.77	9.20	10.54	7.13	7.97	5.51	5.01	4.93	3.23
	α	0.27	0.37	0.37	0.31	0.07	0.06	0.41	0.51	0.49	1.26	1.06	0.64
	K_1	0.39	0.36	0.33	0.35	0.56	0.48	0.47	0.35	0.39	0.36	0.31	0.43
	K_2	−14.73	−7.97	10.87	20.45	−56.00	−0.37	−52.27	47.70	35.64	−0.68	13.06	−7.04
d	T_R	7.47	8.97	10.77	11.18	13.69	12.47	8.21	8.58	9.40	7.24	4.30	4.02
	α	0.96	1.04	0.24	0.07	0.07	0.61	1.26	1.10	0.84	1.29	1.43	1.70
	K_1	0.35	0.30	0.43	0.49	0.48	0.46	0.43	0.43	0.41	0.36	0.31	0.38
	K_2	−25.89	−6.48	−36.46	−44.07	−42.58	−62.66	−56.75	−61.08	−27.09	3.90	20.10	−11.78

Advanced Renewable Energy Sources
G. N. Tiwari and R. K. Mishra

Published by the Royal Society of Chemistry, www.rsc.org

(b): Bangalore

Type of day	Month▶ Para-meters▼	January	February	March	April	May	June	July	August	September	October	November	December
a	T_R	3.36	3.27	3.63	5.05	4.24	4.32	5.18	4.75	4.10	2.28	1.66	1.65
	α	0.07	0.13	0.06	–0.06	0.10	0.19	0.10	0.18	0.13	0.33	0.35	0.36
	K_1	0.33	0.35	0.33	0.29	0.21	0.25	0.32	0.23	0.20	0.05	0.03	0.12
	K_2	–18.05	–22.11	–5.44	14.54	47.81	22.40	–26.04	10.14	38.54	107.04	103.64	47.70
b	T_R	3.24	5.25	6.21	5.72	5.90	7.35	4.12	5.27	4.83	2.43	1.89	3.68
	α	0.31	0.24	0.21	0.19	0.25	0.17	0.51	0.44	0.62	0.56	0.78	0.39
	K_1	0.50	0.45	0.48	0.50	0.41	0.50	0.46	0.50	0.33	0.26	0.37	0.41
	K_2	–60.12	–60.50	–80.04	–75.59	–28.55	–103.35	–90.54	–115.27	13.80	69.14	9.08	–33.76
c	T_R	3.70	4.51	7.74	5.83	4.95	4.39	5.68	2.67	6.64	4.71	5.68	2.02
	α	0.96	0.94	0.63	0.98	0.96	1.12	1.07	1.35	0.78	1.03	0.93	1.44
	K_1	0.46	0.57	0.36	0.50	0.53	0.58	0.50	0.55	0.48	0.43	0.36	0.43
	K_2	–63.02	–129.68	–20.76	–61.13	–103.14	–156.14	–108.34	–161.61	–52.93	–26.53	–15.95	–47.21
d	T_R	6.13	7.49	7.35	6.86	6.33	4.84	4.45	6.68	3.94	3.91	3.84	2.80
	α	1.61	1.31	1.41	1.48	1.59	2.00	2.32	1.69	2.16	2.00	2.04	2.58
	K_1	0.29	0.30	0.40	0.45	0.53	0.61	0.41	0.50	0.38	0.42	0.55	0.27
	K_2	36.80	83.73	–39.85	–72.22	–99.52	–213.29	–79.79	–146.94	–88.62	–125.35	–177.28	–12.29

(c): Jodhpur

Type of day	Month▶ Para-meters▼	January	February	March	April	May	June	July	August	September	October	November	December
a	T_R	1.26	1.33	1.59	2.82	3.72	3.87	3.25	3.39	3.20	2.26	1.56	1.54
	α	0.37	0.38	0.37	0.27	0.21	0.21	0.27	0.28	0.27	0.33	0.39	0.31
	K_1	0.22	0.14	0.18	0.21	0.20	0.13	0.10	0.17	0.26	0.24	0.23	0.26
	K_2	30.67	63.90	56.40	47.66	50.84	87.88	105.23	59.41	14.42	27.40	22.71	9.48
b	T_R	2.34	2.03	3.00	4.07	5.21	5.50	5.07	4.73	3.81	2.90	2.28	3.43
	α	0.46	0.55	0.42	0.31	0.23	0.28	0.37	0.40	0.35	0.38	0.46	0.24
	K_1	0.33	0.29	0.31	0.34	0.33	0.33	0.34	0.33	0.34	0.30	0.33	0.40
	K_2	12.89	43.13	42.22	23.50	31.22	33.40	35.81	29.57	8.71	24.12	12.35	–11.64
c	T_R	3.81	4.78	4.04	4.97	6.87	5.58	4.90	5.10	3.40	3.71	3.28	4.23
	α	0.93	1.32	0.98	0.64	0.61	0.67	1.02	0.88	0.97	2.05	1.31	1.06
	K_1	0.43	0.40	0.42	0.47	0.47	0.46	0.41	0.50	0.48	0.53	0.44	0.44
	K_2	–33.72	12.44	–19.11	–26.93	–44.76	–35.15	2.06	–60.42	–26.96	–62.06	–35.85	–32.84
d	T_R	2.25	5.20	7.09	9.33	8.01	3.52	9.62	3.17	1.63	7.67	1.71	1.94
	α	1.89	1.64	2.03	1.59	1.66	2.37	2.37	2.77	3.24	0.86	2.89	2.03
	K_1	0.44	0.46	0.42	0.44	0.43	0.28	0.52	0.44	0.44	0.52	0.36	0.39
	K_2	–19.31	–45.44	–89.92	–149.27	–117.01	60.69	–221.29	–87.34	–77.55	–26.47	–15.46	–14.88

(d): Mumbai

Type of day	Month▶ Para-meters▼	January	February	March	April	May	June	July	August	September	October	November	December
a	T_R	1.95	1.80	2.88	3.95	5.40	3.20	3.31	4.25	4.22	3.16	2.97	3.27
	α	0.34	0.37	0.23	0.14	–0.02	0.16	0.61	0.33	0.15	0.30	0.23	0.18
	K_1	0.26	0.19	0.28	0.34	0.28	0.25	0.09	0.12	0.24	0.24	0.26	0.30
	K_2	19.77	53.96	27.13	–0.75	30.06	4.55	27.28	47.27	30.02	15.87	9.11	–4.81
b	T_R	2.96	2.68	3.57	4.98	6.25	6.08	7.74	6.70	4.78	3.93	3.40	4.21
	α	0.43	0.49	0.37	0.25	0.15	0.19	0.20	0.37	0.47	0.47	0.45	0.24
	K_1	0.35	0.31	0.35	0.40	0.42	0.44	0.31	0.39	0.41	0.36	0.34	0.37
	K_2	–0.14	24.17	11.73	–13.57	–13.69	–19.52	61.35	22.16	–14.71	5.99	0.60	–14.17
c	T_R	3.06	2.26	3.24	4.39	5.91	5.97	8.17	4.24	5.36	3.16	2.97	3.75
	α	1.14	1.18	1.10	1.00	0.79	0.86	0.62	1.26	0.98	1.13	1.10	0.91
	K_1	0.59	0.58	0.52	0.54	0.60	0.52	0.54	0.43	0.44	0.47	0.57	0.54
	K_2	–59.86	–47.12	–58.09	–78.37	–111.97	–81.79	–95.21	–34.40	–39.31	–28.02	–48.41	–52.45
d	T_R	3.38	7.42	4.45	2.30	4.71	4.71	6.41	7.40	7.46	3.22	5.13	3.05
	α	1.71	1.73	2.29	2.08	2.95	2.66	2.68	1.81	2.14	2.15	1.53	1.51
	K_1	0.52	0.56	0.50	0.35	0.41	0.38	0.32	0.47	0.34	0.42	0.57	0.53
	K_2	–59.78	–26.16	–82.34	63.52	–101.81	–87.19	–61.50	–108.37	–38.68	–25.89	–78.03	–40.51

(e): Srinagar

Type of day	Month▶ Para-meters▼	January	February	March	April	May	June	July	August	September	October	November	December
a	T_R	1.45	5.37	3.31	4.25	5.41	3.63	5.77	6.45	4.06	2.61	4.03	0.72
	α	0.33	–0.36	–0.03	–0.03	–0.12	0.08	–0.09	–0.23	0.03	0.20	–0.37	0.53
	K_1	0.37	0.63	0.69	0.37	0.51	0.33	0.17	0.37	0.46	0.43	0.66	0.33
	K_2	–6.14	–82.86	–94.01	–10.95	–79.57	–13.73	68.06	–42.79	–60.27	–47.83	–37.00	–6.60
b	T_R	3.09	6.98	4.65	6.92	5.86	6.82	7.40	7.58	6.41	4.04	0.04	0.35
	α	0.38	–0.48	0.23	0.06	0.29	0.11	0.00	–0.13	–0.04	0.19	1.16	1.00
	K_1	0.39	0.83	0.59	0.42	0.32	0.63	0.48	0.38	0.48	0.52	0.37	0.41
	K_2	–23.08	–110.23	–107.74	–49.61	0.26	–167.86	–80.06	–13.91	–66.64	–62.52	–14.63	–12.20
c	T_R	2.35	6.59	6.31	7.57	8.69	8.00	9.72	8.23	7.36	5.02	1.86	0.76
	α	1.64	0.86	1.35	0.57	0.61	0.81	0.69	0.90	0.99	1.49	1.47	1.98
	K_1	0.41	0.42	0.48	0.54	0.50	0.39	0.56	0.49	0.44	0.52	0.41	0.31
	K_2	–37.87	–85.68	–180.45	–120.38	–146.97	–87.44	–228.91	–147.96	–62.10	–93.64	–40.07	–12.15
d	T_R	1.69	1.36	7.52	9.09	9.48	10.79	10.93	8.54	8.16	7.75	3.78	2.44
	α	2.63	2.97	1.87	1.35	1.13	1.56	3.08	1.71	3.15	1.70	1.74	2.04
	K_1	0.43	0.36	0.35	0.62	0.92	0.80	0.45	0.75	0.67	0.55	0.48	0.63
	K_2	–41.27	–44.68	–65.17	–254.24	–467.30	–421.63	–129.49	–356.92	–261.85	–119.53	–49.16	–64.02

APPENDIX – III

Specifications of Solar-cell Material (at Solar Intensity 1000 W/m^2 and Cell Temperature 25 °C) and Cost

Cell technology	*Efficiency (%)*	*Fill factor (FF)*	*Aperture area ($10^{-4} \times m^2$)*	*Life time* (years)*	*Manufacturing cost (\$/kWp in 2007)*	*Selling price (\$/kWp in 2007)*
Monocrystalline silicon	24.7 ± 0.5	0.828	4.0	30	2.5	3.7
Multicrystalline silicon	19.8 ± 0.5	0.795	1.09	30	2.4	3.5
Copper indium diselenide (CIS/CIGS)	18.4 ± 0.5	0.77	1.04	5	1.5	2.5
Thin silicon cell	16.6 ± 0.4	0.782	4.02	25	2.0	3.3
Cadmium telluride (CdTe)	16.5 ± 0.5	0.755	1.03	15	1.5	2.5
Amorphous silicon (a-si)	10.1 ± 0.2	0.766	1.2	20	1.5	2.5

*Based on experience.
Source: B. Agarwal, G.N. Tiwari, Development in environmental durability for photovoltaics, Pira International Ltd., UK, 2008.

Advanced Renewable Energy Sources
G. N. Tiwari and R. K. Mishra

Published by the Royal Society of Chemistry, www.rsc.org

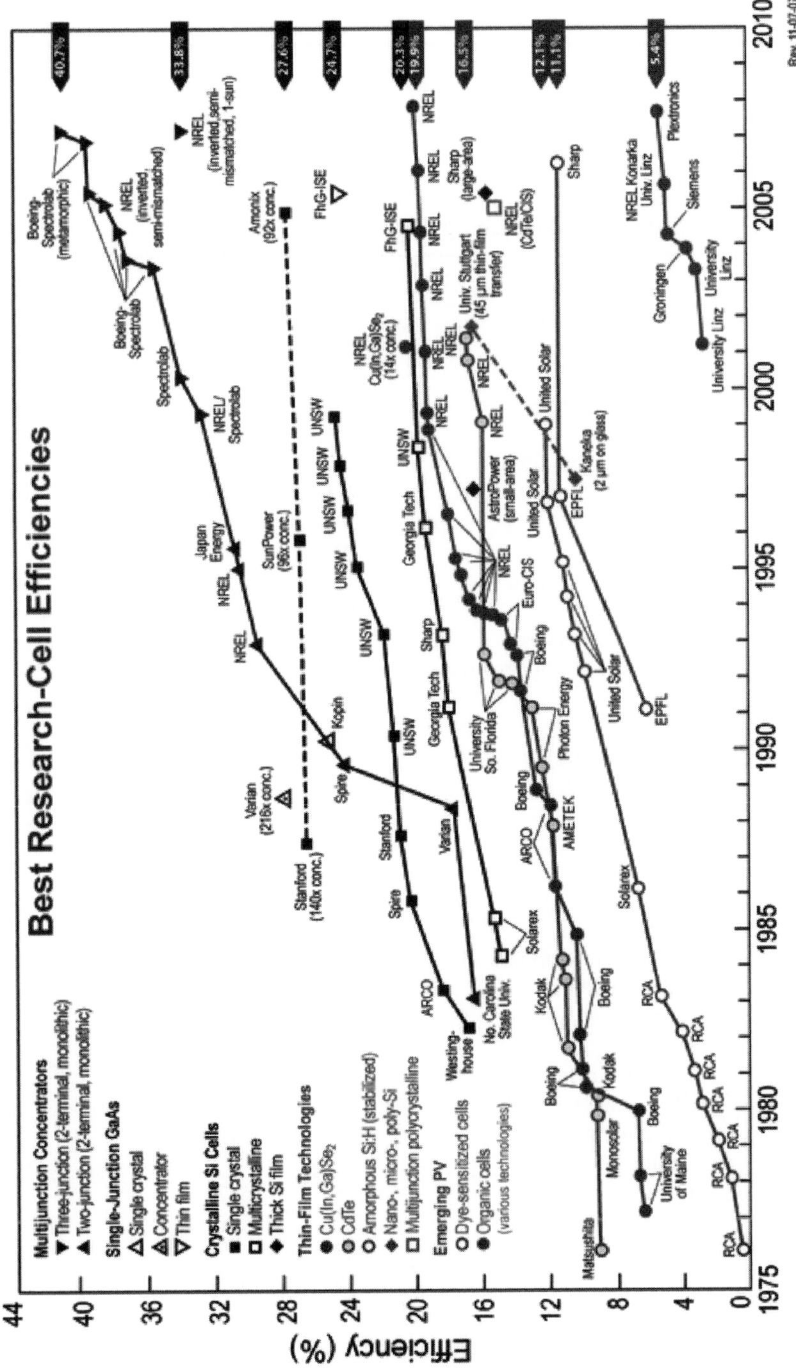

Curtsey: L.L. Kazmerski, NREL, http://en.wikipedia.org/wiki/File:PVeff(rev110707)d.png

APPENDIX – IV

List of Embodied Energy Coefficients

MATERIAL	*MJ/kg*	*MJ/m^3*
Aggregate, general	0.10	150
virgin rock	0.04	63
river	0.02	36
Aluminium, virgin	191	515 700
extruded	201	542 700
extruded, anodised	227	612 900
extruded, factory painted	218	588 600
foil	204	550 800
sheet	199	537 300
Aluminium, recycled	8.1	21 870
extruded	17.3	46 710
extruded, anodised	42.9	115 830
extruded, factory painted	34.3	92 610
foil	20.1	54 270
sheet	14.8	39 960
Asphalt (paving)	3.4	7140
Bitumen	44.1	45 420
Brass	62.0	519 560
Carpet	72.4	–
felt underlay	18.6	–
nylon	148	–
polyester	53.7	–
polyethylterepthalate (PET)	107	–
polypropylene	95.4	–
wool	106	–
Cement	7.8	15 210
cement mortar	2.0	3200
fibre cement board	9.5	13 550
soil-cement	0.42	819
Ceramic		–
brick	2.5	5170
brick, glazed	7.2	14 760
pipe	6.3	–
tile	2.5	5250

Advanced Renewable Energy Sources
G. N. Tiwari and R. K. Mishra

Published by the Royal Society of Chemistry, www.rsc.org

(*Continued*).

MATERIAL	*MJ/kg*	*MJ/m*^{3}
Concrete		–
block	0.94	–
brick	0.97	–
GRC	7.6	14 820
paver	1.2	–
precast	2.0	–
ready mix, 17.5 MPa	1.0	2350
30 MPa	1.3	3180
40 MPa	1.6	3890
roofing tile	0.81	–
Copper	70.6	631 160
Earth, raw		–
adobe block, straw stabilised	0.47	750
adobe, bitumen stabilised	0.29	–
adobe, cement stabilised	0.42	–
rammed soil cement	0.80	–
pressed block	0.42	–
Fabric		–
cotton	143	–
polyester	53.7	–
Glass		–
float	15.9	40 060
toughened	26.2	66 020
laminated	16.3	41 080
tinted	14.9	375 450
Insulation		–
cellulose	3.3	112
fibreglass	30.3	970
polyester	53.7	430
polystyrene	117	2340
wool (recycled)	14.6	139
Lead	35.1	398 030
Linoleum	116	150 930
Paint	90.4	118 per litre
solvent based	98.1	128 per litre
water based	88.5	115 per litre
Paper	36.4	33 670
building	25.5	–
kraft	12.6	–
recycled	23.4	–
wall	36.4	–
Plaster, gypsum	4.5	6460
Plaster board	6.1	5890
Plastics		–
ABS	111	–
high-density polyethelene (HDPE)	103	97 340
low density polyethelene (LDPE)	103	91 800
polyester	53.7	7710

(*Continued*).

MATERIAL	*MJ/kg*	*MJ/m³*
polypropylene	64.0	57 600
polystyrene, expanded	117	2340
polyurethane	74.0	44 400
PVC	70.0	93 620
Rubber		–
natural latex	67.5	62 100
synthetic	110	–
Sand	0.10	232
Sealants and adhesives		–
phenol formaldehyde	87.0	–
urea formaldehyde	78.2	–
Steel, recycled	10.1	37 210
reinforcing, sections	8.9	–
wire rod	12.5	–
Steel, virgin, general	32.0	251 200
galvanised	34.8	273 180
imported, structural	35.0	274 570
Stone, dimension		–
local	0.79	1890
imported	6.8	1890
Straw, baled	0.24	30.5
Timber, softwood		–
air dried, roughsawn	0.3	165
kiln dried, roughsawn	1.6	880
air dried, dressed	1.16	638
kiln dried, dressed	2.5	1380
mouldings, *etc.*	3.1	1710
hardboard	24.2	13 310
MDF	11.9	8330
glulam	4.6	2530
particle bd	8.0	–
plywood	10.4	–
shingles	9.0	–
Timber, hardwood		–
air dried, roughsawn	0.50	388
kiln dried, roughsawn	2.0	1550
Vinyl flooring	79.1	105 990
Zinc	51.0	364 140
galvanising, per kg steel	2.8	–

APPENDIX – V

Table V(a) Properties of air at atmospheric pressure.

T *K*	ρ *(kg/m³)*	C_p *(kJ/Kg K)*	μ *(kg/m s)* × 10^{-5}	ν *(m²/s)* × 10^{-6}	K *(W/m² K)* × 10^{-3}	α *(m²/s)* × 10^{-5}	*Pr*
100	3.6010	1.0259	0.6924	1.923	9.239	0.2501	0.770
150	2.3675	1.0092	1.0283	4.343	13.726	0.5745	0.753
200	1.7684	1.0054	1.3289	7.490	18.074	1.017	0.739
250	1.4128	1.0046	1.488	9.49	22.26	1.3161	0.722
300	1.1774	1.0050	1.983	15.68	26.22	2.216	0.708
350	0.9980	1.0083	2.075	20.76	30.00	2.983	0.697
400	0.8826	1.0134	2.286	25.90	33.62	3.760	0.689

The value of μ, *K*, C_p and Pr are not strongly pressure dependent and may be used over a fairly wide range of pressures.

Table V(b) Properties of water (saturated liquid).

°*F*	°*C*	C_p *(kJ/kg K)*	ρ *(kg/m³)*	μ_k *(kg/m s)*	K *(W/m K)*	Pr	$\frac{g\beta P^2 C_p}{\mu_k}$ *(1/m³ K)*
32	0.00	4.225	999.8	1.79×10^3	0.566	13.25	1.91×10^9
40	4.44	4.208	999.8	1.55	0.575	11.35	6.34×10^9
50	10.00	4.195	999.2	1.31	0.585	9.40	1.08×10^{10}
60	15.56	4.186	998.6	1.12	0.595	7.88	1.46×10^{10}
70	21.11	4.179	997.4	9.8×10^4	0.604	6.78	1.46×10^{10}
80	26.67	4.179	995.8	8.6	0.614	5.85	1.91×10^{10}
90	32.22	4.174	994.9	7.65	0.623	5.12	2.48×10^{10}
100	37.78	4.174	993.0	6.82	0.630	4.53	3.3×10^{10}
110	43.33	4.174	990.6	6.16	0.637	4.04	4.19×10^{10}
120	48.89	4.174	988.8	5.62	0.644	3.64	4.89×10^{10}
130	54.44	4.179	985.7	5.13	0.649	3.30	5.66×10^{10}
140	60.00	4.179	983.3	4.71	0.654	3.01	6.48×10^{10}

Advanced Renewable Energy Sources
G. N. Tiwari and R. K. Mishra

Published by the Royal Society of Chemistry, www.rsc.org

(*Continued*).

°F	°C	C_p (kJ/kg K)	ρ (kg/m³)	μ_k (kg/m s)	K (W/m K)	Pr	$\frac{g\beta P^2 C_p}{\mu_k}$ (1/m³ K)
150	65.55	4.183	980.3	4.3	0.659	2.73	7.62×10^{10}
160	71.11	4.186	977.3	4.01	0.665	2.53	8.84×10^{10}
170	76.67	4.191	973.7	3.72	0.668	2.33	9.85×10^{10}
180	82.22	4.195	970.2	3.47	0.673	2.16	1.09×10^{10}
190	87.78	4.199	966.7	3.27	0.675	2.03	
200	93.33	4.204	963.2	3.06	0.678	1.90	
210	104.40	4.216	955.1	2.67	0.684	1.66	

Table V(c) Properties of metals.

Metal	ρ (kg/m³)	*Properties at 20 °C* C_p (kJ/kg K)	K (W/m K)	α (m²/s) × 10^{-5}
Aluminium:				
Pure	2707	0.896	204	8.418
Al-Si (Silumin, copper bearing) 86% Al, 1% Cu	2659	0.867	137	5.933
Lead	11 400	0.1298	34.87	7.311
Iron:				
Pure	7897	0.452	73	2.034
Steel	7753	0.486	63	0.970
C ≈ 1.5% (C-Carbon steel)				
Copper:				
Pure	8954	0.3831	386	11.234
Aluminum bronze	8666	0.410	383	2.330
95% Cu, 5% Al	8666	0.343	326	0.859
Bronze				
75% Cu, 25% Sn	8714	0.385	61	1.804
Red Brass				
85% Cu, 9% Sn 6% Zn	8600	0.877	85	3.412
Brass				
70% Cu, 30% Zn	8618	0.394	24.9	0.733
German Silver				
62% Cu, 15% Ni, 22% Zn	8922	0.410	22.7	0.612
Constantan				
60% Cu, 40% Ni	1746	1.013	171	9.708
Magnesium Pure				
Nickel, Pure	8906	0.4459	90	2.266
Silver				
Purest	10 524	0.2340	419	17.004
Pure (99.9%)	10 524	0.2340	407	16.563
Tin, Pure	7304	0.2265	64	3.884
Tungsten	19 350	0.1344	163	6.271
Zinc, Pure	7144	0.3843	112.2	4.106

Table V(d) Properties of nonmetals.

	Temperature (°C)	K *(W/m K)*	ρ *(kg/m³)*	C *(kJ/kg K)*	α *(m²/s) × 10⁻⁷*
Asbestos	50	0.08	470	–	–
Building brick	20	0.69	1600	0.84	5.2
Common face		1.32	2000	–	–
Concrete, cinder	23	0.76	–	–	–
Stone 1–2–4 mix	20	1.37	1900–2300	0.88	8.2–6.8
Glass, window	20	0.78 (avg)	2700	0.84	3.4
borosilicate	30–75	1.09	2200	–	–
Plaster, gypsum	20	0.48	1440	0. 84	4.0
Stone:					
Granite	–	1.73–3.98	2640	0.82	8–18
Limestone	100–300	1.26–1.33	2500	0.90	5.6–5.9
Marble	–	2.07–2.94	2500–2700	0.80	10–13.6
Sandstone	40	1.83	2160–2300	0.71	11.2–11.9
Wood (across the grain):					
Fir	23	0.11	420	2.72	0.96
Maple or Oak	30	0.166	540	2.4	1.28
Yellow pine	23	0.147	640	2.8	0.82
Cork board	30	0.043	160	1.88	2–5.3
Cork, regranulated	32	0.045	45–120	1.88	2–5.3
Ground	32	0.043	150	–	–
Glass wool	23	0.38	24	0.7	22.6
Sawdust	23	0.059	–	–	–
Wood shavings	23	0.059	–	–	–

Table V(e) Physical properties of some other materials.

S.No.	*Material*	*Density (ρ) (kg/m³)*	*Thermal conductivity (W/m K)*	*Specific heat (J/kg K)*
1.	Air	11.177	40.026	1006
2.	Alumina	3800	29.0	800
3.	Aluminium	41–45	211	0.946
4.	Asphalt	1700	0.50	1000
5.	Brick	1700	0.84	800
6.	Carbon dioxide	1.979	0.145	871
7.	Cement	1700	0.80	670
8.	Clay	1458	11.28	879
9.	Concrete	2400	1.279	1130
10.	Copper	8795	385	–
11.	Cork	240	0.04	2050
12.	Cotton Wool	1522	–	1335
13.	Fibre board	300	0.057	1000
14.	Glass-Crown	2600	1.0	670
	Window	2350	0.816	712
	Wool	50	0.042	670
15.	Ice	920	2.21	1930
16.	Iron	7870	80	106
17.	Lime stone	2180	1.5	–

(*Continued*).

S.No.	Material	Density (ρ) (kg/m^3)	Thermal conductivity ($W/m\ K$)	Specific heat ($J/kg\ K$)
18.	Mudphuska	–	–	–
19.	Oxygen	1.301	0.027	920
20.	Plaster-board	950	0.16	840
21.	Polystyrene-expanded	25	0.033	1380
22.	P.V.C.-rigid foam	25–80	0.035–0.041	–
	-rigid sheet	1350	0.16	–
23.	Saw dust	188	0.57	–
24.	Thermocole	22	0.03	–
25.	Timber	600	0.14	1210
26.	Turpentine	870	0.136	1760
27.	Water (H_2O)	998	0.591	4190
	(Sea)	1025	–	3900
	(Vapour)	0.586	0.025	2060
28.	Wood wool	500	0.10	1000

Table V(f) Absorptivity of various surfaces for Sun's rays.

Surfaces	*Absorptivity*
White paint	0.12–0.26
Whitewash/glossy white	0.21
Bright aluminium	0.30
Flat white	0.25
Yellow	0.48
Bronze	0.50
Silver	0.52
Dark aluminium	0.63
Bright red	0.65
Brown	0.70
Light green	0.73
Medium red	0.74
Medium green	0.85
Dark green	0.95
Blue/black	0.97
Roofs	
Asphalt	0.89
White asbestos cement	0.59
Cooper sheeting	0.64
Uncoloured roofing tile	0.67
Red roofing tiles	0.72
Galvanised iron, clean	0.77
Brown roofing tile	0.87
Galvanised iron, dirty	0.89
Black roofing tile	0.92
Walls	
White/yellow brick tiles	0.30
White stone	0.40
Cream brick tile	0.50

(*Continued*).

Surfaces	*Absorptivity*
Burl brick tile	0.60
Concrete/red brick tile	0.70
Red sand line brick	0.72
White sand stone	0.76
Stone rubble	0.80
Blue brick tile	0.88
Surroundings	
Sea/lake water	0.29
Snow	0.30
Grass	0.80
Light-coloured grass	0.55
Light green shiny leaves	0.75
Sand Gray	0.82
Rock	0.84
Green leaf	0.85
Earth (black ploughed field)	0.92
White leaves	0.20
Yellow leaves	0.58
Aluminium foil	0.39
Unpainted wood	0.60
Metals	
Polished aluminium/copper	0.26
New galvanised iron	0.66
Old galvanised iron	0.89
Polished iron	0.45
Oxidised rusty iron	0.38

APPENDIX – VI

Heating Values of Various Combustibles and their Conversion Efficiencies

Fuel	*Heating value (kJ/kg)*	*Efficiency of device*
Coal coke	29 000	70
Wood	15 000	60
Straw	14 000–16 000	60
Gasoline	43 000	80
Kerosene	42 000	80
Methane (Natural gas)	50 000	80
Biogas (60% methane)	20 000	80
Electricity	–	95

Advanced Renewable Energy Sources
G. N. Tiwari and R. K. Mishra

Published by the Royal Society of Chemistry, www.rsc.org

APPENDIX – VII

Ideal Gas Specific Heats of Various Common Gases (at 300 K)

Gas	*Formula*	*Gas constant, R (kJ/kg K mol)*	C_p *(kJ/kg K)*	C_v *(kJ/kg K)*	k
Air		0.2870	1.0050	0.7180	1.400
Argon	Ar	0.2081	0.5203	0.3122	1.667
Butane	C_4H_{10}	0.1433	1.7164	1.5734	1.091
Carbon dioxide	CO_2	0.1889	0.8460	0.6570	1.289
Carbon monoxide	CO_2	0.2968	1.0400	0.7440	1.400
Ethane	C_2H_6	0.2765	1.7662	1.4897	1.186
Ethylene	C_2H_4	0.2964	1.5482	1.2518	1.237
Helium	He	2.0769	5.1926	3.1156	1.667
Hydrogen	H_2	4.1240	14.3070	10.1830	1.405
Methane	CH_4	0.5182	2.2537	1.7354	1.299
Neon	Ne	0.4119	1.0299	0.6179	1.667
Nitrogen	N_2	0.2968	1.0390	0.7430	1.400
Octane	C_8H_{18}	0.0729	1.7113	1.6385	1.044
Oxygen	O_2	0.2598	0.9180	0.6580	1.395
Propane	C_3H_8	0.1885	1.6794	1.4909	1.126
Steam	H_2O	0.4615	1.8723	1.4108	1.327

Advanced Renewable Energy Sources
G. N. Tiwari and R. K. Mishra

Published by the Royal Society of Chemistry, www.rsc.org

APPENDIX – VIII

Table VIII(a) Saturated water – temperature table.

		Specific volume m^3/kg		*Internal energy kJ/kg*		*Enthalpy kJ/kg*		*Entropy kJ/kg K*	
Temp. °*C*	*Sat. Pres.* P_{sat} *kPa*	*Sat. liquid* V_f	*Sat. vapor* V_g	*Sat. liquid* U_f	*Sat. vapor* U_g	*Sat. liquid* h_f	*Sat. vapor* h_g	*Sat. liquid* S_f	*Sat. vapor* S_g
0.01	0.6113	0.001000	206.14	0.0	2375.3	0.01	2501.4000	0.0000	9.1562
5	0.8721	0.001000	147.12	20.97	2382.3	20.98	2510.6000	0.0761	9.0257
10	1.2276	0.001000	106.38	42.00	2389.2	42.01	2519.8000	0.1510	8.9008
15	1.7051	0.001001	77.93	62.99	2396.1	62.99	2528.9000	0.2245	8.7814
20	2.3390	0.001002	57.79	83.95	2402.9	83.96	2538.1000	0.2966	8.6672
25	3.1690	0.001003	43.36	104.88	2409.8	104.89	2547.2000	0.3674	8.5580
30	4.2460	0.001004	32.89	125.78	2416.6	125.79	2556.3000	0.4369	8.4533
35	5.6280	0.001006	25.22	146.67	2423.4	146.68	2565.3000	0.5053	8.3531
40	7.3840	0.001008	19.52	167.56	2430.1	167.57	2574.3000	0.5725	8.2570
45	9.5930	0.001010	15.26	188.44	2436.8	188.45	2583.2000	0.6387	8.1648
50	12.3490	0.001012	12.03	209.32	2443.5	209.33	2592.1000	0.7038	8.0763
55	15.7580	0.001015	9.57	230.21	2450.1	230.23	2600.9000	0.7679	7.9913
60	19.9400	0.001017	7.67	251.11	2456.6	251.13	2609.6000	0.8312	7.9096
65	25.0300	0.001020	6.20	272.02	2463.1	272.06	2618.3000	0.8935	7.8310
70	31.1900	0.001023	5.04	292.95	2469.6	292.98	2626.8000	0.9549	7.7553
75	38.5800	0.001026	4.13	313.90	2475.9	313.93	2643.7000	1.0155	7.6824
80	47.3900	0.001029	3.41	334.86	2482.2	334.91	2635.3000	1.0753	7.6122
85	57.8300	0.001033	2.83	355.84	2488.4	355.90	2651.9000	1.1343	7.5445
90	70.1400	0.001036	2.36	376.85	2494.5	376.92	2660.1000	1.1925	7.4791
95	84.5500	0.001040	1.98	397.88	2500.6	397.96	2668.1000	1.2500	7.4159

Advanced Renewable Energy Sources
G. N. Tiwari and R. K. Mishra

Published by the Royal Society of Chemistry, www.rsc.org

Table VIII (b) Saturated water – pressure table.

		Specific volume m^3/kg		*Internal energy kJ/kg*		*Enthalpy kJ/kg*		*Entropy kJ/kg K*	
Press. P_{sat} *kPa*	*Sat. Temp.* T_{sat} *°C*	*Sat. liquid* V_f	*Sat. vapor* V_g	*Sat. liquid* U_f	*Sat. vapor* U_g	*Sat. liquid* h_f	*Sat. vapor* U_g	*Sat. liquid* V_f	*Sat. vapor* V_g
0.6113	0.01	0.001	206.14	0	2375.3	0	2501.4	0	9.1562
1	6.98	0.001	129.21	29.3	2385	29.3	2514.2	0.1059	8.9756
1.5	13.03	0.001001	87.98	54.71	2393.3	54.71	2525.3	0.1957	8.8279
2	17.5	0.001001	67	73.48	2399.5	73.48	2533.5	0.2607	8.7237
2.5	21.08	0.001002	54.25	88.48	2404.4	88.49	2540	0.312	8.6432
3	24.08	0.001003	45.67	101.04	2408.5	101.05	2545.5	0.3545	8.5776
4	28.96	0.001004	34.8	121.45	2415.2	121.46	2554.4	0.4226	8.4746
5	32.88	0.001005	28.19	137.81	2420.5	137.82	2561.5	0.4764	8.3951
7.5	40.29	0.001008	19.24	168.78	2430.5	168.79	2574.8	0.5764	8.2515
10	45.81	0.00101	14.67	191.82	2437.9	191.83	2584.7	0.6493	8.1502
15	53.97	0.001014	10.02	225.92	2448.7	225.94	2599.1	0.7549	8.0085
20	60.06	0.001017	7.649	251.38	2456.7	251.4	2609.7	0.832	7.9085
25	64.97	0.00102	6.204	271.9	2463.1	271.93	2618.2	0.8931	7.8314
30	69.1	0.001022	5.229	289.2	2468.4	289.23	2625.3	0.9439	7.7686
40	75.87	0.001027	3.993	317.53	2477	317.58	2636.8	1.0259	7.67
50	81.33	0.00103	3.24	340.44	2483.9	340.49	2645.9	1.091	7.5939
75	91.78	0.001037	2.217	384.31	2496.7	384.39	2663	1.213	7.4564

Table VIII (c) Ideal-gas properties of air.

T *(K)*	h *(kJ/kg)*	p_r	u *(kJ/kg)*	v_r	s^o *(kJ/kg*K)*
200	199.97	0.3363	142.56	1707.0	1.29559
210	209.97	0.3987	149.69	1512.0	1.34444
220	219.97	0.4690	156.82	1346.0	1.39105
230	230.02	0.5477	164.00	1205.0	1.43557
240	240.02	0.6355	171.13	1084.0	1.47824
250	250.05	0.7329	178.28	979.0	1.51917
260	260.09	0.8405	185.45	887.8	1.55848
270	270.11	0.9590	192.60	808.0	1.59634
280	280.13	1.0889	199.75	738.0	1.63279
285	285.14	1.1584	203.33	706.1	1.65055
290	290.16	1.2311	206.91	676.1	1.66802
295	295.17	1.3068	210.49	647.9	1.68515
300	300.19	1.3860	214.07	621.2	1.70203
305	305.22	1.4686	217.67	596.0	1.71865
310	310.24	1.5546	221.25	572.3	1.73498
315	315.27	1.6442	224.85	549.8	1.75106
320	320.29	1.7375	228.42	528.6	1.76690
325	325.31	1.8345	232.02	508.4	1.78249
330	330.34	1.9352	235.61	489.4	1.79783
340	340.42	2.149	242.82	454.1	1.82790
350	350.49	2.379	250.02	422.2	1.85708
360	360.58	2.626	257.24	393.4	1.88543
370	370.67	2.892	264.46	367.2	1.91313
380	380.77	3.176	271.69	343.4	1.94001

Table VIII (c) (*Continued*).

T (K)	h (kJ/kg)	p_r	u (kJ/kg)	v_r	s^o (kJ/kg*K)
390	390.88	3.481	278.93	321.5	1.96633
400	400.98	3.806	286.16	301.6	1.99194
410	411.12	4.153	293.43	283.3	2.01699
420	421.26	4.522	300.69	266.6	2.04142
430	431.43	4.915	307.99	251.1	2.06533
440	441.61	5.332	315.30	236.8	2.08870
450	451.80	5.775	322.62	223.6	2.11161
460	462.02	6.245	329.97	211.4	2.13407
470	472.24	6.742	337.32	200.1	2.15604
480	482.49	7.268	344.70	189.5	2.17760
490	492.74	7.824	352.08	179.7	2.19876
500	503.02	8.411	359.49	170.6	2.21952
510	513.32	9.031	366.92	162.1	2.23993
520	523.63	9.684	374.36	154.1	2.25997
530	533.98	10.37	381.84	146.7	2.27967
540	544.35	11.10	389.34	139.7	2.29906
550	555.74	11.86	396.86	133.1	2.31809
560	565.17	12.66	404.42	127.0	2.33685
570	575.59	13.50	411.97	121.2	2.35531
580	586.04	14.38	419.55	115.7	2.37348
590	596.52	15.31	427.15	110.6	2.39140
600	607.02	16.28	434.78	105.8	2.40902
610	617.53	17.30	442.42	101.2	2.42644
620	628.07	18.36	450.09	96.92	2.44356
630	638.63	19.84	457.78	92.84	2.46048
640	649.22	20.64	465.50	88.99	2.47716
650	659.84	21.86	473.25	85.34	2.49364
660	670.47	23.13	481.01	81.89	2.50985
670	681.14	24.46	488.81	78.61	2.52589
680	691.82	25.85	496.62	75.50	2.54175
690	702.52	27.29	504.45	72.56	2.55731
700	713.27	28.80	512.33	69.76	2.57277
710	724.04	30.38	520.23	67.07	2.58810
720	734.82	32.02	528.14	64.53	2.60319
730	745.62	33.72	536.07	62.13	2.61803
740	756.44	35.50	544.02	59.82	2.63280
750	767.29	37.35	551.99	57.63	2.64737
760	778.18	39.27	560.01	55.54	2.66176
780	800.03	43.35	576.12	51.64	2.69013
800	821.95	47.75	592.30	48.08	2.71787
820	843.98	52.59	608.59	44.84	2.74504
840	866.08	57.60	624.95	41.85	2.77170
860	888.27	63.09	641.40	39.12	2.79783
880	910.56	68.98	657.95	36.61	2.82344
900	932.93	75.29	674.58	34.31	2.84856
920	955.38	82.05	691.28	32.18	2.87324
940	977.92	89.28	708.08	30.22	2.89748
960	1000.55	97.00	725.02	28.40	2.92128
980	1023.25	105.2	741.98	26.73	2.94468
1000	1046.04	114.0	758.94	25.17	2.96770
1020	1068.89	123.4	776.10	23.72	2.99034

Table VIII (c) (*Continued*).

T (K)	h (kJ/kg)	p_r	u (kJ/kg)	v_r	s^o (kJ/kg*K)
1040	1091.85	133.3	793.36	23.29	3.01260
1060	1114.86	143.9	810.62	21.14	3.03449
1080	1137.89	155.2	827.88	19.98	3.05608
1100	1161.07	167.1	845.33	18.896	3.07732
1120	1184.28	179.7	862.79	17.886	3.09825
1140	1207.57	193.1	880.35	16.946	3.11883
1160	1230.92	207.2	897.91	16.064	3.13916
1180	1254.34	222.2	915.57	15.241	3.15916
1200	1277.79	238.0	933.33	14.470	3.17888
1220	1301.31	254.7	951.09	13.747	3.19834
1240	1324.93	272.3	968.95	13.069	3.21751
1260	1348.55	290.8	986.90	12.435	3.23638
1280	1372.24	310.4	1004.76	11.835	3.25510
1300	1395.97	330.9	1022.82	11.275	3.27345
1320	1419.76	352.5	1040.88	10.747	3.29160
1340	1443.60	375.3	1058.94	10.247	3.30959
1360	1467.49	399.1	1077.10	9.780	3.32724
1380	1491.44	424.2	1095.26	9.337	3.34474
1400	1515.42	450.5	1113.52	8.919	3.36200
1420	1539.44	478.0	1131.77	8.526	3.37901
1440	1563.51	506.9	1150.13	8.153	3.39586
1460	1587.63	537.1	1168.49	7.801	3.41247
1480	1611.79	568.8	1186.95	7.468	3.42892
1500	1635.97	601.9	1205.41	7.152	3.44516
1520	1660.23	636.5	1223.87	6.854	3.46120
1540	1684.51	672.8	1242.43	6.569	3.47712
1560	1708.82	710.5	1260.99	6.301	3.49276
1580	1733.17	750.0	1279.65	6.046	3.50829
1600	1757.57	791.2	1298.30	5.804	3.52364
1620	1782.00	834.1	1316.96	5.574	3.53879
1640	1806.46	878.9	1335.72	5.355	3.55381
1660	1830.96	925.6	1354.48	5.147	3.56867
1680	1855.50	974.2	1373.24	4.949	3.58335
1700	1880.1	1025	1392.7	4.761	3.5979
1750	1941.6	1161	1439.8	4.328	3.6336
1800	2003.3	1310	1487.2	3.994	3.6684
1850	2065.3	1475	1534.9	3.601	3.7023
1900	2127.4	1655	1582.6	3.295	3.7354
1950	2189.7	1852	1630.6	3.022	3.7677
2000	2252.1	2068	1678.7	2.776	3.7994
2050	2314.6	2303	1726.8	2.555	3.8303
2100	2377.7	2559	1775.3	2.356	3.8605
2150	2440.3	2837	1823.8	2.175	3.8901
2200	2503.2	3138	1872.4	2.012	3.9191
2250	2566.4	3464	1921.3	1.864	3.9474

The properties P_r (relative pressure) and v_r (relative pressure) are dimensionless quantities used in the analysis of isentropic process, and should not be confused with the properties pressure and specific volume.
Source: Kenneth Wark, Thermodynamics. 4th edn (New York: McGraw-Hill, 1983), pp. 785–86, Table A-5. Originally published in J.H. Keenan and J. Kaye, Gas Tables (New York: John Wiley & Sons, 1948).

Glossary

Absorber Plate: A component of the solar flat plate collector that absorbs solar radiation and converts it into heat.

Absorptance: It is the ratio between the radiations absorbed by a surface (absorber) and the total amount of solar radiation striking the surface.

Actinometer: This is the instrument used to measure direct radiation from the sun, also known as pyrheliometer.

Active Solar Heating: This refers to heating by solar energy using an additional energy source (usually electricity) for pimping water or blowing air.

Air Heating System: This refers to air heating by solar energy.

Air Mass: It is atmospheric attenuation.

Albedo: The ratio of the amount of light reflected by a surface to the light falling onto it.

Alternating Current (AC): An electric current that alternates direction between positive and negative cycles, usually 50 or 60 times per second. Alternating current is the current typically available from power outlets in a household.

Altitude: The sun's angle above the horizon, as measured in a vertical plane.

Amorphous Silicon: A thin-film PV silicon cell having no crystalline structure. It is manufactured by depositing layers of doped silicon on a substrate.

Ampere: Electrical current; a measure of flowing electrons.

Ampere-hour (Amp-hr): Measure of flowing electron for a period of time.

Anaerobic digestion: It is a process by which organic matter is decomposed by bacteria in absence of oxygen to produce methane and other by product.

Anemometer: This is the instrument used for measuring the wind speed.

Annual Mean daily Insolation: The average solar energy per m^2 available per day over the whole year.

Annual Solar saving: The annual energy saving due to solar energy devices.

Antifreeze: This refers to the substance added to water to lower its freezing point. Solar water heaters usually use a mixture of water and propylene glycol instead of water to prevent freezing.

Antireflection Coating: A thin coating of a material, which reduces the light reflection and increases light transmission, applied to a photovoltaic cell surface.

Array: Any number of Photovoltaic modules connected together electrically to provide a single electrical output. An array is a mechanically integrated assembly of modules or panels together with support structure (including foundation and other components, as required) to form a free-standing field installed unit that produces DC power.

Advanced Renewable Energy Sources
G. N. Tiwari and R. K. Mishra

Published by the Royal Society of Chemistry, www.rsc.org

Audit: An energy audit seeks energy inefficiencies and prescribes improvement.

Automatic tracking: A device that permits solar collector to track or to follow the sun during the day without manual adjustment usually for concentrating collectors.

Azimuth: The horizontal angle between the sun and due south in the northern hemisphere, or between the sun and due north in the southern hemisphere.

Balance of System (BOS): Term used in photovoltaic, which represents all components and costs other than the PV modules.

Band GAP: It is the difference in energy between the state of the highest valance band and conduction band.

Battery: A collection of cells that store electrical energy; each cell converts chemical energy into electricity or vice versa, and is interconnected with other cells to form a battery for storing useful quantities of electricity.

Battery Capacity: The maximum total electrical charge, expressed in ampere-hours (AH), that a battery can deliver to a load under a specific set of conditions.

Battery Cell: The simplest operating unit in a storage battery. It consists of one or more positive electrodes or plates, an electrolyte that permits ionic conduction, one or more negative electrodes or plates, separators between plates of opposite polarity, and a container for all the above.

Battery Available Capacity: The total maximum charge, expressed in ampere-hours, that can be withdrawn from a cell or battery under a specific set of operating conditions including discharge rate, temperature, initial state of charge, age, and cutoff voltage.

Battery Energy Capacity: The total energy available, expressed in watt-hours (kilowatt-hours), that can be withdrawn from a fully-charged cell or battery. The energy capacity of a given cell varies with temperature, rate, age, and cutoff voltage.

Battery Cycle Life: The number of cycles, to a specified depth of discharge, that a cell or battery can undergo before failing to meet its specified capacity or efficiency performance criteria.

Beam Radiation: It is the radiation propagating along the line joining the receiving surface and sun. It is also known as direct radiation.

Biofuel: It is a product from biomass.

Biogas: It is a mixture of methane (CH_4), carbon dioxide (CO_2) and some trace gases.

Biomass: Organic material of a non-fossil organic (living or recently dead plant and animal tissue) including aquatic, herbaceous and woody plants, animal waste and portion of municipal waste.

Biopower: It is electricity produced from biomass fuel.

Black body: A perfect absorber and emitter of radiation. A cavity is a perfect black body. Lampblack is close to a black body, while aluminum (polished) is a poor absorber and emitter of radiation.

Brightness: The subjective human perception of luminance.

BTU: British thermal unit, the amount of heat required raising the temperature of one pound of water one degree Fahrenheit; 3411 BTU equals one kWh.

Cadmium Telluride (CdTe): A polycrystalline thin-film photovoltaic material.

Calorie: The amount of heat required to raise the temperature of one gram of water one degree Celsius.

Calorific Value: The energy content per unit mass (or volume) of a fuel, which will be released in combustion. (kWh/kg, MJ/kg, kWh/m^3, MJ/m^3)

Candela (cd): an SI unit of luminous intensity. An ordinary candle has a luminous intensity of one candela.

Carbon Dioxide (CO_2): The colorless, odorless gas that is formed during normal human breathing. It is also emitted by combustion activities used to produce electricity. CO_2 is a major cause of the greenhouse effect that traps radiant energy near the earth's surface.

Carbonization: It is a process whereby wood is heated with restricted air flow to from high carbon product by removing volatile materials from it.

Cell: A device that generates electricity, traditionally consisting of two plates or conducting surfaces placed in an electrolytic fluid.

Celsius: The international temperature scale in which water freezes at 0 [degrees] and boils at 100 [degrees] and named after Anders Celsius.

Central Power Tower: A configuration of independently tracking solar collectors focusing all the reflected solar radiation onto a receiver placed on a top of tower.

Charge Rate: The current applied to a cell or battery to restore its available capacity. This rate is commonly normalized by a charge control device with respect to the rated capacity of the cell or battery.

Charge Controller: A component of photovoltaic system that controls the flow of current to and from the battery to protect the batteries from over-charge and over-discharge. The charge controller may also indicate the system operational status.

Circuit: A system of conductors (*i.e.* wires and appliances) capable of providing a closed path for electric current.

Clear sky: A sky condition with few or no clouds, usually taken as 0-2 tenths covered with clouds. Clear skies have high luminance and high radiation, and create strong shadows relative to more cloudy conditions. The sky is brightest nearest the sun, whereas away from the sun, it is about three times brighter at the horizon then at the zenith.

Closed Cycle: In this case the working fluid is returned to the initial stage at end of cycle and is recirculated.

Clerestory: A wall with windows that is between two different (roof) levels. The windows are used to provide natural light into building.

Coefficient of Performance (COP): An efficiency term to compare the performance of refrigerators and heat pump.

Cogeneration: Joint production of heat and work, most often electricity and heat.

Collector: The name given to the device, which converts the incoming solar radiation to heat.

Collector Efficiency: This is obtained as the ratio of the useful (heat) energy converted by the solar collector to the radiation incident on the device.

Collector Plate: A component of the solar flat plate collector that absorbs solar radiation and converts it into heat.

Collector Tilt Angle: An angle between the horizontal plan and the surface of solar collector.

Compact Fluorescent Light (CFL): A modern light bulb with integral ballast using a fraction of the electricity used by a regular incandescent light bulb.

Concentrating Collector: A solar collector which reflects the solar radiation (direct radiation) to an absorber plate to produce high temperature.

Condensation: The process of vapor changing into the liquid state. Heat is released in the process.

Conductance (C): A measure of the ease with which heat flows though a specified thickness of a material by conduction. Units are W/m^2 °C.

Conduction: The process by which heat energy is transferred through materials (solids, liquids or gases) by molecular excitation of adjacent molecules.

Conduction Band: This is the energy band in which electron moves freely.

Conductivity: The quantity of heat that will flow through one square meter of material, one meter thick, in one second, when there is a temperature difference of 1°C between its surfaces.

Conductor: A substance or a body capable of transmitting the electricity, heat and sound.

Convection: The transfer of heat between a moving fluid medium (liquid or gas) and a surface, or the transfer of heat within a fluid by movements within the fluid.

Conservation of Energy: The total amount of energy in any closed system remains constant.

Core: The central region of earth, having a radius of 3,470 km., The radius of earth is 6,370 km. outside of which lies the mantle and crust.

Crystalline Silicon: A type of PV cell made from a single crystal or polycrystalline slices of silicon.

Current: The flow of electron through the conductor.

Declination: The angle of the sun north or south of the equatorial plane.

Deep Discharge Battery: A type of battery that is not damaged when a large portion of its energy capacity is repeatedly removed (*i.e.* motive batteries).

Depth of Discharge (DOD): The ampere-hours removed from a fully charged cell or battery, expressed as a percentage of rated capacity. For example, the removal of 25 ampere- hours from a fully charged 100 ampere-hours rated cell results in a 25% depth of discharge.

Design Heat Load: The total heat loss from a building during most severe winter condition the building is likely to experience.

Design Month: The month has the lowest mean daily insolation value, around which many stand alone systems are planed.

Diffuse Radiation: The solar radiation reaching the surface due to reflection and scattering effect.

Diffusion Length: The mean distance through which a free electron or hole moved before recombining with another holes or electron.

Direct Combustion: Burning of biomass in the presence of oxygen.

Direct Current (DC): The complement of AC, or alternating current, presents one unvarying voltage to a load. This is standard in automobiles.

Direct Radiation: Radiation coming in a beam from the Sun, which can be focused.

Direct Solar Gain: The thermal energy gain in building through glazed window.

Discharge: The removal of electric energy from battery.

Diurnal: Recurring every day or having a daily cycle.

Dopants: A chemical impurity added usually in minute's amount to a pure semiconductor in order to alter its electrical properties.

Doping: The addition of dopants to a semiconductor.

Dry bulb temperature: The temperature of a gas of mixture or gases indicated by an accurate thermometer after correction for radiation.

Duct: A pipe, tube, channel that conveys a substance usually warm or cold air.

Efficacy: Special term which refers to the efficiency by which lamps converts electricity to a visible radiation. It is expressed in lumens per watt (lux).

Efficiency: The ratio of output power (or energy) to input power (or energy) expressed as a percentage.

Electromagnetic Spectrum: The entire range of wavelengths or frequencies of electromagnetic radiation extending from gamma rays to the longest radio waves including visible light.

Electronic Ballasts: An improvement over core/coil ballasts used to drive fluorescent lamps.

Embodied Energy: Literally the amount of energy required producing an object in its present form; an inflated balloon's embodied energy includes the energy required to manufacture it and inflate it.

Emissivity: The ratio of the radiant energy emitted by a body to that emitted by a perfect black body.

Emittance: A measure of the ability of a material to give off heat as radiant energy.

Energy Density: Energy per unit area.

Energy Intensity: The ratio of energy use in a sector to activity in that sector, for example, the ration of energy use to constant dollar production in manufacturing.

Energy: The ability to do work.

Equinox: The times of the year when the sun passes over the celestial equator and when the day of length and night are almost equal. It happens twice a year.

Evacuated Tube Collector: A solar collector that uses a vacuum between the absorber to glass to reduce the top loss coefficient (insulate the absorber plate).

EVA: Ethylene-Vinile-Acetate Foil, it will be used by module production for covering the cells.

Fill Factor (FF): For an I-V curve, the ratio of the maximum power to the product of the open-circuit voltage and the short-circuit current. Fill factor is a measure of the "squareness" of the I-V curve.

Flat Plate Collector: A solar collection device for gathering the Sun's heat, consisting of a shallow metal container covered with one or more layers of transparent glass or plastic; either air or a liquid is circulated through the cavity of the container, whose interior is painted "black" and exterior is well insulated.

Focusing Collector: See concentrating collector.

Fresnel Collector: A type of concentrating solar collector consisting of a concentric series of rings with reflecting surface.

Fuel Cell: A device combining a fuel with oxygen in an electrochemical reaction to generate electricity directly without combustion.

Gallium Arsenide (GaAs): A crystalline, high-efficiency semiconductor/photovoltaic material.

Generator: A device that produces electricity.

Geothermal Energy: It is energy contained in the earth's interior.

Glare: The perception caused by a very bright light or a high contrast of light, making it uncomfortable or difficult to see.

Glazing: Transparent or translucent materials, usually glass or plastic, used to cover an opening without impeding (relative to opaque materials) the admission of solar radiation and light.

Global Radiation: The sum of direct, diffuse and reflected radiation.

Greenhouse Effect: The global warming resulting from the absorption of infrared solar radiation by carbon dioxide and other traces of gases present in the atmosphere. (The term is a misnomer in that in actual greenhouses the warming comes primarily from restriction on air flow.)

Greenhouse Gases: Gases which contribute to the greenhouse effect by absorbing infrared radiation in the atmosphere. These gases include carbon dioxide, nitrous oxide, methane, water vapor and variety of chlorofluorocarbons (CFCs).

Grid: A utility term for the network of wires that distribute electricity from a variety of sources across a large area.

Heat Capacity: The quantity of heat required to raise one kilogram of a substance by one degree Centigrade.

Heat Exchanger: Device that passes heat from one substance to another; in a solar hot water heater, for example, the heat exchanger takes heat harvested by a fluid circulating through the solar panel and transfers it to domestic hot water.

Heat Loss: It is a thermal energy loss to atmosphere due to temperature difference.

Hydropower: Energy in falling water is converted into mechanical energy and then electrical energy.

I-V Curve: The plot of current versus voltage characteristics of a solar cell, module or array. I-V curves are used to compare various solar cell modules, and to determine their performance at various levels of insolation and temperatures.

Iluminance: It is defined as the amount of lumens per unit area.

Incandescent Bulb: A light source that produces light by heating a filament until it emits photons.

Incident Radiation: The quantity of radiant energy striking a surface per unit time and unit area.

Infrared Radiation: The part of the electromagnetic radiation (waves) whose wavelength lies between 0.75 to 1000 micrometer.

Insolation (or Incident solar radiation)**:** The amount of sunlight falling on a place.

Insulation: A material that keeps energy from crossing from one place to another: on electrical wire, it is the plastic or rubber that covers the conductor; in a building, insulation makes the walls, floor and roof more resistant to the outside (ambient) temperature.

Insulator: A material that is a poor conductor of electricity or heat.

Inverter: The electrical device that changes direct current (DC) into alternating current (AC).

Irradiance: The solar radiation incident per unit time (W/m^2).

Joule: The unit of energy or work. One joule is equal to one Watt second.

Kilowatt (kW): 1000 Watts, energy consumption at a rate of 1000 Joules per second.

Kilowatt-hour (kWh): One kilowatt of power used for one hour. A typical house uses 750 kW/h per month.

Kinetic Energy: Energy of motion is known as kinetic energy.

Latitude: The angular position of a location north or south of the equator.

Life Cycle Costing: A method for estimating the comparative costs of alternative energy or other systems. Life cycle costing takes into consideration such long term costs as energy consumption, maintenance and repair.

Life Cycle Costs: The entire cost of an energy device, including the capital cost in present dollars and the cost and the benefits discounted to the present.

Light Emitting Diode: An efficient source of electrical lighting, typically lasting 50,000 to 100, 00 hours.

Load: The set of equipments or appliances that use the electrical power from the generating source, battery or PV module.

Longitude: The angular position east or west of Greenwich.

Low-E Window: Window that reflects the infrared (IR) back into the room, instead of absorbing and transmitting the IR heat to outside.

Maximum power point (MPP): The voltage at which a PV array is producing maximum power.

Maximum power point tracker (MPPT): A power conditioning unit that increases the power of a PV system by ensuring operation of the PV generator at its Maximum Power Point (MPP). The ability to do so can depend on climate and the battery's state of charge.

Medium Temperature Solar Collector: A solar thermal collector designed to operate in the temperature range of 80–100 °C.

Megawatt: 1,000,000 W.

Module: The smallest self contained, environmentally protected structure housing interconnected photovoltaic cells and providing a single direct current (DC) electrical output.

Monthly Mean Daily Insolation: The average solar energy per square meter available per day of a given month.

N-Type Semiconductor: A semiconductor produced by doping an intrinsic semiconductor with electron donor impurity (e.g. phosphorous in silicon).

Natural Convection: The natural convection of heat through the fluid in a body that occurs when warm, less dense fluid rises and cold, dense fluid sinks under the influence of gravity.

Newton: It is the basic unit of force.

Night sky radiation: A reversal of the day time insolation principle. Just as the sun radiates energy during the day through the void of space, so also heat energy can travel unhindered at night from the earth's surface back into space. On a clear night, any warm object can cool itself by radiating long wave heat energy to the cooler sky. On a cloudy night, the cloud cover acts as an insulator and prevents the heat from travelling to the cooler sky.

NOCT: Nominal Operating Cell temperature; The estimated temperature of a PV module when operating under 800 W/m^2 irradiance, 20 °C ambient temperature and wind speed of 1 meter per second. NOCT is used to estimate the nominal operating temperature of a module in its working environment.

Nonrenewable Energy Sources: It is energy derived from finite and static stocks of energy.

One-Axis Tracking: A solar system capable of rotating about one axis and track the sun from east to west.

Open Circuit Voltage (V_{oc}): The maximum possible voltage across a solar module or array. Open circuit voltage occurs in sunlight when no current is flowing.

Open Cycle: In this case the working fluid is renewed at the end of each cycle.

Orientation: The arrangement of solar device along a given axis to face in a direction best suited to absorb solar radiation.

OTEC (Ocean thermal energy conversion): A process which exploits the natural temperature difference (gradient) between the shallow and deep ocean water as the driving potential for a simple thermodynamic cycle that can extract work out of temperature gradient.

Off-the-grid: Not connected to power grid.

Overcharging: Leaving batteries on charge after they have reached their full state of charge (100%).

Parabola: The geometrically curved shape to focus sun light on a single point.

Parabolic Concentrating Cooker: A solar cooker that uses a parabolic disk to focus sun light.

Parabolic Mirror: A device with a large, shiny and curved surface that focuses solar radiation on a specific point.

Passive Solar Design: A building design that makes use of structural materials, using no moving part to heat or cool the space in a building.

Passive Solar Heater: A solar water or space heating system that moves heated water or air without using fan/motor/pump.

Passive Solar Water Heater: Solar water heating system with natural/thermosyphon circulation is known as passive solar water heater.

Peak Sun Shine Hours: The number of hours per day during which solar radiation averages 1000 W/m^2.

Peak Watt (W_p): Power out put of PV module under standard test condition *i.e.* 1000 W/m^2 and 25 °C.

Pelletization: It is a process in which wood is compressed and extracted in the form of rods and cubes.

Periodic Motion: Any motion that repeats itself in equal interval of times.

Photon: The elementary particle of electromagnetic energy; light. (Greek photos, light).

Photovoltaic Array: A number of PV modules that are electrically connected in series or parallel to provide required rated power.

Photovoltaic Conversion Efficiency: The ratio of the electric power produced by a photovoltaic device to the power of the sunlight incident on the device.

Photovoltaic Device: A device that converts light directly into DC electricity.

Photovoltaic Module: The basic building block of a photovoltaic device, which consists of a number of interconnected solar cells.

Photovoltaics (PV): A technology for using semiconductors to directly convert light into electricity.

Polycrystalline Silicon: A material used to make PV cells which consist of many crystals as contrasted with single crystal silicon.

Potential Energy: It is the energy that an object possesses as a result of its elevation in a gravitational field.

Power: The rate at which energy is consumed or produced. The unit is Watt.

Power Density: It is power per unit area (W/m^2).

Power Factor: It is the cosine of the phase angle between the current and voltage of a circuit.

ppb: It is parts per billion.

ppm: It is parts per million.

Producer gas: It is a mixture of combustible and non- combustible gas

PV: See photovoltaic.

Pyranometer: A device to measure total (global) radiation.

Pyrheliometer: A device to measure direct radiation.

Pyrolysis: It is canonization at high process temperature.

R-value: It is thermal resistance of material which is inverse of heat transfer coefficient.

Radiant Energy: Energy in the form of electromagnetic waves that travels outward.

Radiation: Electromagnetic waves that directly transport energy through space. Sunlight is a form of radiation.

Reflectivity: The ratio of radiant energy reflected by a body to that falling upon it.

Reflector: A device that can be used to reflect solar radiation.

Renewable Energy Sources: An energy source that renews itself without, effort; fossil fuels, once consumed, are gone forever, while solar energy is renewable in that the Sun we harvest today has no effect on the Sun we can harvest tomorrow.

Resistor: Any electronic component that restricts the flow of electrical current in circuits.

Selective Surface: A special surface which have high absorptions and low emissive power.

Semiconductor: A material such as silicon, which has a crystalline structure that will allow current to flow under certain conditions. Semiconductors are usually less conductive than metals but not an insulator like rubber.

Short Circuit Current (I_{sc}): Current across the terminals when a solar cell or module in strong sunlight is not connected to a load (measured with ammeter).

Silicon: A semiconductor material commonly used make PV cells.

Single-Crystal Structure: A material having a crystalline structure such that a repeatable or periodic molecular pattern exists in all three dimensions.

Solar Altitude: The Sun's angle above the horizon, as measured in vertical plane.

Solar Azimuth: The horizontal angle between the sun and due south in the northern hemisphere, or between the sun and due north in the southern hemisphere.

Solar Cabinet Dryer: A device that uses solar radiation for crop drying.

Solar Cell: It is a device that converts light energy or solar radiation (photons) directly into DC electricity.

Solar Cell Module: Groups of encapsulated solar cells framed in a glass or plastic units, usually the smallest unit of solar electric equipment available to the consumer.

Solar Air Collector: A device that gathers and accumulates solar radiation to produce heat.

Solar Concentrator: A device which uses reflective surfaces in a planar, parabolic trough or parabolic bowl configuration to concentrate solar radiation onto a smaller surface.

Solar Constant: An amount referring to radiation arriving from the Sun at the edge of the Earth's atmosphere. The accepted value is about 1367 Watts per square meter.

Solar Cooker: A device that uses solar radiation for food cooking.

Solar Declination: The angle of the Sun north or south of the equatorial plane.

Solar Distillation: A process in which solar energy is trapped and used for to evaporate impure or salty water.

Solar Electricity: Electricity that is obtained by using solar energy.

Solar Energy: The electromagnetic radiation generated by the Sun.

Solar Incident Angle: The angle at which the incoming solar beam strikes a surface.

Solar Pond: A shallow body of salt water with a black or dark bottom.

Solar Radiation: The radiant energy received from the Sun, from both direct and diffuse or reflected sunlight.

Solar Spectrum: The total distribution of electromagnetic radiation emitted from the Sun.

Solar Still: A device consisting of one or several stages in which brackish water is converted to portable water by successive evaporation and condensation with the aid of solar heat.

Solar Water Heater: A water heater that depends on solar radiation as its source of power.

Standard Test Condition: It is condition having 1000 W/m^2 solar radiation and 25 °C ambient air temperature with air mass of 1.5.

Sustainable: Material or energy sources that, if managed carefully, will provide at current level indefinitely.

Temperature: Degree of hotness or coldness measured on one of several arbitrary scales based on some observable phenomenon (such as the expansion).

Thermal Conductivity: The ability of material to conduct heat.

Thermal Mass: A material used to store heat, thereby slowing the temperature variation within a space.

Thermal Storage Wall: A south facing glazed wall, generally known as Trombe wall.

Thermosyphon: A close loop system in which water automatically circulates between a solar collector and a water storage tank above it due to the natural difference in density between the warmer and cooler portions of a liquid.

Thermal Storage: Any of several techniques to store heat energy by utilizing either heat capacity of material, the latent heat of phase change or heat of chemical dissociation.

Thermosyphon: A closed loop system in which water automatically circulates between the flat plate collector and storage tank placed above flat plate collector due to pressure difference.

Thin-Film Silicon: Most often this is amorphous (non-crystalline) material used to make photovoltaic (PV) cells.

Tidal Power: Energy obtained by using the motion of tides to run water turbines that drives electricity generation.

Tilt Angle: The angle at which solar collector is tilted upward from the horizontal surface to receive the maximum solar radiation.

Tracking: The adjustment made to solar concentrating collector to track or follow the sun's path across the sky.

Transfer Medium: A substance (air, water or antifreeze solution) that carries heat from a solar collector to a storage area or from a storage area to in a collector.

Transmission: Transporting bulk power over long distance.

Transmittance: The ratio of the solar radiation transmitted through a glazing to the total radiant energy falling on its surface.

Trombe Wall: A solar radiation facing glazed thermal storage wall.

Trough: This is a type of concentrating collector with one axis tracking.

Turbine Generator: A device that uses steam, heated gas, water flow or wind to cause spinning motion that activates electromagnetic forces to generate electricity.

U-Value: The amount of heat that flows in or out of a system at steady state, in one hour, when there is a one degree difference in temperature between fluid inside and outside.

Ultra-Violet Radiation: A portion of the electromagnetic radiation in the wavelength range of 4 to 400 nanometers.

Valence Band: The highest energy band in a semiconductor that can be filled with electron.

Ventilation: The exchange of room air between rooms to outside ambient air as per requirement. Generally it measured in the terms of number of air change per hour.

Volt: It is the measure of potential difference between two electrodes.

Wafer: Raw material for a solar cell, a thin sheet of crystalline semiconductor material is made by mechanically sawing it from a single-crystal boule or by casting it.

Water Heating: The process of generating domestic hot water by employing a flat plate collector and utilizing solar radiation.

Watt: Measure of power (or work) equivalent to 1/746 of a horse power.

Watt Hour (Wh): A common energy measure arrived at by multiplying the power times the hours of use. Grid power is ordinarily sold and measured in kilowatt hours.

Wavelength: The distance between two similar points of a given wave.

Wind Turbine: It is a device to convert kinetic energy associated with wind into mechanical energy.

Zenith: The top of the sky dome. It is a point directly overhead, 90° in altitude angle above the horizon.

Subject Index